Methods in Enzymology

Volume 210
NUMERICAL COMPUTER METHODS

METHODS IN ENZYMOLOGY

EDITORS-IN-CHIEF

John N. Abelson   Melvin I. Simon

DIVISION OF BIOLOGY
CALIFORNIA INSTITUTE OF TECHNOLOGY
PASADENA, CALIFORNIA

FOUNDING EDITORS

Sidney P. Colowick and Nathan O. Kaplan

Methods in Enzymology

Volume 210

Numerical Computer Methods

EDITED BY

Ludwig Brand

DEPARTMENT OF BIOLOGY
MCCOLLUM-PRATT INSTITUTE
THE JOHNS HOPKINS UNIVERSITY
BALTIMORE, MARYLAND

Michael L. Johnson

DEPARTMENT OF PHARMACOLOGY
UNIVERSITY OF VIRGINIA
CHARLOTTESVILLE, VIRGINIA

ACADEMIC PRESS, INC.

Harcourt Brace Jovanovich, Publishers

San Diego   New York   Boston
London   Sydney   Tokyo   Toronto

Copyright © 1992 by ACADEMIC PRESS, INC.

All Rights Reserved.

No part of this publication may be reproduced or transmitted in any form or by any means, electronic or mechanical, including photocopy, recording, or any information storage and retrieval system, without permission in writing from the publisher.

Academic Press, Inc.
1250 Sixth Avenue, San Diego, California 92101

United Kingdom Edition published by
Academic Press Limited
24–28 Oval Road, London NW1 7DX

Library of Congress Catalog Number: 54-9110

International Standard Book Number: 0-12-182111-0

PRINTED IN THE UNITED STATES OF AMERICA
92 93 94 95 96 97   EB   9 8 7 6 5 4 3 2 1

Table of Contents

Contributors to Volume 210

Article numbers are in parentheses following the names of contributors.
Affiliations listed are current.

GARY K. ACKERS (19), *Department of Biochemistry and Molecular Biophysics, Washington University School of Medicine, St. Louis, Missouri 63110*

MARCEL AMELOOT (12, 14), *Research Group of Physiology, Limburgs Universitair Centrum, B-3590 Diepenbeek, Belgium*

RONN ANDRIESSEN (14), *Department of Chemistry, Katholieke Universiteit Leuven, B-3001 Heverlee, Belgium*

DAVID L. BAIN (19), *Department of Biochemistry and Molecular Biophysics, Washington University School of Medicine, St. Louis, Missouri 63110*

ŽELJKO BAJZER* (10), *Department of Biochemistry and Molecular Biology, Mayo Clinic/Foundation, Rochester, Minnesota 55905*

DOROTHY BECKETT (19), *Department of Chemistry and Biochemistry, University of Maryland, Baltimore, Maryland 21228*

JOSEPH M. BEECHEM (2, 23), *Department of Molecular Physiology and Biophysics, Vanderbilt University School of Medicine, Nashville, Tennessee 37232*

NOËL BOENS (14), *Department of Chemistry, Katholieke Universiteit Leuven, B-3001 Heverlee, Belgium*

DAVID WAYNE BOLEN (22), *Department of Chemistry and Biochemistry, Southern Illinois University at Carbondale, Carbondale, Illinois 62901*

E. E. BRUMBAUGH (25), *Department of Chemistry, Bridgewater College, Bridgewater, Virginia 22812*

PAUL BRIAN CONTINO (21), *Department of Medicine/Thrombosis, Mount Sinai School of Medicine, New York, New York 10029*

JULIEN S. DAVIS (17), *Department of Biology and McCollum-Pratt Institute, The Johns Hopkins University, Baltimore, Maryland 21218*

FRANS C. DE SCHRYVER (14), *Department of Chemistry, Katholieke Universiteit Leuven, B-3001 Heverlee, Belgium*

ENRICO DI CERA (4), *Department of Biochemistry and Molecular Biophysics, Washington University School of Medicine, St. Louis, Missouri 63110*

LINDSAY M. FAUNT (1, 15), *Biodynamics Institute and Diabetes Center, University of Virginia, Charlottesville, Virginia 22908*

HERBERT R. HALVORSON (3, 28), *Department of Pathology, Henry Ford Hospital, Detroit, Michigan 48202*

E. R. HENRY (8), *Laboratory of Chemical Physics, National Institute of Diabetes and Digestive and Kidney Diseases, National Institutes of Health, Bethesda, Maryland 20892*

PRESTON HENSLEY (18), *Macromolecular Sciences Department, SmithKline Beecham Pharmaceuticals, King of Prussia, Pennsylvania 19406*

J. HOFRICHTER (8), *Laboratory of Chemical Physics, National Institute of Diabetes and Digestive and Kidney Diseases, National Institutes of Health, Bethesda, Maryland 20892*

C. HUANG (25), *Department of Biochemistry, University of Virginia, Charlottesville, Virginia 22908*

MICHAEL L. JOHNSON (1, 5, 6, 7, 15, 26), *Department of Pharmacology, University of Virginia, Charlottesville, Virginia 22908*

* On leave of absence from Rugjer Bošković Institute, Zagreb, Croatia.

W. CURTIS JOHNSON, JR. (20), *Department of Biochemistry and Biophysics, Oregon State University, Corvallis, Oregon 97331*

W. T. KATZ (29), *Department of Biomedical Engineering, University of Virginia, Charlottesville, Virginia 22908*

JAY R. KNUTSON (16), *Laboratory of Cell Biology, National Heart, Lung and Blood Institute, National Institutes of Health, Bethesda, Maryland 20892*

KENNETH S. KOBLAN (19), *Department of Biochemistry and Molecular Biophysics, Washington University School of Medicine, St. Louis, Missouri 63110*

WILLIAM R. LAWS (21), *Department of Biochemistry, Mount Sinai School of Medicine, New York, New York 10029*

HENRY H. MANTSCH (9), *Steacie Institute for Molecular Sciences, National Research Council of Canada, Ottawa, Ontario K1A 0R6, Canada*

M. B. MERICKEL (29), *Department of Biomedical Engineering, University of Virginia, Charlottesville, Virginia 22908*

WEBB MILLER (27), *Department of Computer Science and Institute of Molecular Evolutionary Genetics, Pennsylvania State University, University Park, Pennsylvania 16802*

DOUGLAS J. MOFFATT (9), *Steacie Institute for Molecular Sciences, National Research Council of Canada, Ottawa, Ontario K1A 0R6, Canada*

GLENN NARDONE (18), *Edvotek, Inc., West Bethesda, Maryland 20827*

WILLIAM R. PEARSON (27), *Department of Biochemistry, University of Virginia, Charlottesville, Virginia 22908*

FRANKLYN G. PRENDERGAST (10), *Department of Biochemistry and Molecular Biology, Mayo Clinic/Foundation, Rochester, Minnesota 55905*

CATHERINE A. ROYER (23), *School of Pharmacy, University of Wisconsin–Madison, Madison, Wisconsin 53706*

DONALD F. SENEAR (22), *Department of Biochemistry and Molecular Biology, University of California, Irvine, Irvine, California 92717*

MADELINE A. SHEA (19), *Department of Biochemistry, University of Iowa College of Medicine, Iowa City, Iowa 52242*

ENOCH W. SMALL (11), *Department of Chemistry and Biochemistry, Eastern Washington University, Cheney, Washington 99004*

JEANNE RUDZKI SMALL (24), *Department of Chemistry and Biochemistry, Eastern Washington University, Cheney, Washington 99004*

J. W. SNELL (29), *Department of Biomedical Engineering, University of Virginia, Charlottesville, Virginia 22908*

CAROLE J. SPANGLER (13), *Center For Molecular Genetics, University of California, San Diego, La Jolla, California 92093*

MARTIN STRAUME (5, 7), *Biocalorimetry Center, Department of Biology, The Johns Hopkins University, Baltimore, Maryland 21218*

VIVIANE VAN DEN BERGH (14), *Department of Chemistry, Katholieke Universiteit Leuven, B-3001 Heverlee, Belgium*

JOHANNES D. VELDHUIS (26), *Departments of Internal Medicine and Pharmacology, University of Virginia Health Sciences Center, and National Science Foundation Center for Biological Timing, Charlottesville, Virginia 22908*

MERYL E. WASTNEY (18), *Department of Pediatrics, Georgetown University Medical Center, Washington, D.C. 20007*

F. EUGENE YATES (30), *Medical Monitoring Unit, Department of Medicine, University of California, Los Angeles, Los Angeles, California 90025*

Preface

In the past decade microcomputers have revolutionized biomedical research. Almost every new scientific instrument is "computer controlled." Almost every researcher has a personal computer easily available or is readily linked to a mainframe. The improvements in computer software development and advances in methods of analysis have paralleled the computer hardware improvements.

It is clear that new ways of evaluating experimental data have enhanced the type of conclusions that may be drawn and have changed the way in which experiments are conceived and done. The biochemical community must be aware of new developments in data analysis and computer usage. The primary aim of this volume is to inform biomedical researchers of the modern data analysis methods that have developed concomitantly with computer hardware.

The process of collecting experimental data, analyzing the data, and then publishing the data and results is not a one-way street. All methods of data analysis make assumptions about the nature of the data. Specifically, they make assumptions about the types and magnitudes of the experimental uncertainties contained within the data. A biomedical researcher should carefully design the experimental data collection procedures such that they are compatible with the desired method of data analysis.

A common procedure used in the past for the analysis of nonlinear systems was to rearrange the equation describing the process into a linear form and then to use linear least-squares to determine slopes and intercepts related to the parameters of interest. Typical examples include the Lineweaver–Burk plot for analysis of Michaelis–Menton kinetic data and the Scatchard plot for analysis of equilibrium binding data.

Consider a Scatchard plot as an example. The objective of this, and many other types of plots, is to transform a set of experimental data into a straight line form; in this case, a plot of the amount of bound ligand divided by the free ligand (Y axis) as a function of bound ligand (X axis). For a ligand binding problem with a single class of noninteracting binding sites this transformation will provide a straight line. The next step is to "fit" a least-squares straight line to the transformed data points. The slope of this line is related to the ligand binding affinity and the X-axis intercept is the binding capacity. However, this approach makes an invalid assumption about the nature of the uncertainties contained in the experimental data. Fitting a least-squares straight line to the transformed data assumes that the experimental uncertainties follow a random distribution and are parallel to the Y axis. However, in a Scatchard plot the uncertainties are nearly parallel to the Y axis at low fractional saturations and nearly parallel to the X axis at high fractional saturations. Consequently, the use of a least-squares method is not valid for the analysis of Scatchard plots. Note that this does not preclude the use of a Scatchard plot to help a researcher visualize an experiment if the actual data analysis is performed by another method.

So how can the data be analyzed? The best approach is to fit the original data, without any transformations, by nonlinear least-squares. For a more complete discussion of Scatchard plots refer to Klotz,[1,2] Munson and Rodbard,[3,4] and Johnson and Frasier.[5]

So why was the Scatchard plot developed? The Scatchard plot was developed in the 1940s before the availability of digital computers. Some nonlinear least-squares techniques were available at the time, i.e., the Gauss–Newton Method.[6] However, nonlinear least-squares techniques require too many operations to be performed without a computer in a reasonable length of time. At the time the Scatchard plot was the only "show in town." *Now that high-speed computers are available there is no reason to attempt to analyze transformed data.*

Almost every type of transformation "plot" to analyze experimental data was developed because high-speed digital computers were not available to perform the correct calculation. Incidently, this includes a semilog plot for the analysis of exponential decays.[5] These plots fail to meet the statistical requirements of linear least-squares methods. This failure is due to the required transformation of the data. The reason that these plots are still used for the analysis of data is primarily due to a lack of information about the available methods of data analysis. One purpose of this volume is to provide this information to biomedical researchers.

"On the other side of the coin," many biomedical researchers have learned to revere computers as oracles. They assume that if a computer analyzed their data then the results must be correct. *Computers are not oracles!* The results of any computer analysis are no better than the computer programs used for the analysis. Data analysis computer programs are created by people who make assumptions about the nature of

[1] I. M. Klotz, *Science* **217,** 1247 (1982).
[2] I. M. Klotz, *Science* **220,** 981 (1983).
[3] P. J. Munson and D. Rodbard, *Anal. Biochem.* **107,** 220 (1980).
[4] P. J. Munson and D. Rodbard, *Science* **220,** 979 (1983).
[5] M. L. Johnson and S. G. Frasier, this series, Vol. 117, p. 301.
[6] M. L. Johnson and L. M. Faunt, "Parameter estimation by least-squares methods," this volume [1].

the experimental data. They subsequently make assumptions about the best method of analysis based on their assumptions about the experimental data. These assumptions may not be acceptable for your experimental data. They also make compromises to save computer time and space in the memory of the computer. Computer programmers can also make mistakes. Thus, *computer programs sometimes include unwarranted assumptions and can make mistakes!*

Consequently, biomedical researchers cannot simply insert data into a computer and accept the results as gospel. Researchers must be aware of the assumptions used by their data analysis programs. They must be certain that they are using methods that are appropriate for their particular type of data. They need to validate their computer programs with real and synthetic data to ascertain that the computer programs are producing the results they expect. They should always question the results of a computer analysis, i.e., do they have physical meaning? The purpose of this volume is to help biomedical researchers meet these needs.

The chapters in this book are written for biomedical researchers by biomedical researchers. The volume is divided into three basic categories. First, basic methods such as nonlinear least-squares and maximum likelihood are described. Second, specific examples of the use of some of these methods are presented. The volume ends with introductory discussions about methods that are currently being developed, such as neural networks and fractals.

We are grateful to Nathan O. Kaplan who recognized the importance of data analysis in enzymology and conceived the idea for this volume.

LUDWIG BRAND
MICHAEL L. JOHNSON

METHODS IN ENZYMOLOGY

[1] Parameter Estimation by Least-Squares Methods

By MICHAEL L. JOHNSON and LINDSAY M. FAUNT

Introduction

This chapter presents an overview of least-squares methods for the estimation of parameters by fitting experimental data. We demonstrate that least-squares methods produce the estimated parameters with the highest probability (maximum likelihood) of being correct if several critical assumptions are warranted. We then discuss several least-squares parameter-estimation procedures, as well as methods for the evaluation of confidence intervals for the determined parameters. We conclude with a discussion of the practical aspects of applying least-squares techniques to experimental data.

Biologists are often called upon to evaluate "constants" from tabulated experimental observations. For example, relaxation rate constants are evaluated from stopped-flow and temperature-jump experiments, and binding constants are evaluated from ligand binding experiments. This chapter provides an overview of several least-squares methods that can be applied to the evaluation of constants from experimental data.

In the past, biological researchers usually evaluated parameters by graphical methods such as the Scatchard plot, Hill plot, log plot, and double-reciprocal plot. The common feature of these methods is that they transform the experimental data into a representation that yields a straight line in certain limiting cases. For the more realistic cases usually found in biological research their usefulness is questionable, even for the limiting cases that yield straight lines. So why were these procedures developed, and why are they used? For the most part, they were developed before the proliferation of computer resources in the 1980s, when they were the only commonly available methods. Today, when virtually everyone has access to significant computer resources, there is no longer the need to make the assumptions and approximations inherent in these graphical methods.

In the present research environment, many researchers will use software packages with blind faith. They assume that since the experimental data are analyzed by a computer, the results must be meaningful and correct. This approach is quite often worse than using a graphical method. At least with the graphical methods, the researcher must actually look at a graph of the experimental data. This may raise questions if the graph is not a straight line, or otherwise, as expected.

Computers are not oracles! What a researcher gets out of a computer depends critically on what he or she and the computer programmer put into the program. All computer programs used to analyze experimental data make assumptions about the nature of the experimental data and the process being studied. The researcher must understand the methods and underlying assumptions implicit in a computer program and how they relate to the particular analysis problem being studied by the researcher. It is possible that the methods and assumptions used by a particular computer program are not compatible with the experimental data; consequently the results obtained may be meaningless.

Computer programs quite often contain mistakes. The more complex the program, the more mistakes it is likely to contain. The distinguishing feature of a good computer program is that it contains, and makes, few mistakes. With increasing skill and investment of time on the part of the programmer, a computer program can be made to contain fewer mistakes. It is not possible for a programmer to test a program with all possible combinations of experimental data; therefore, it is the responsibility of the researcher using the program to be certain that the computer program is producing the answers appropriate to the experimental application. To validate a computer program a researcher must (1) understand the numerical methods and underlying assumptions used by the program, (2) understand the nature of the experimental data and associated uncertainties, (3) understand how items (1) and (2) are interrelated, and (4) test the computer program with realistic data.

Understanding the nature of the experimental data and the associated uncertainty is the key to evaluating the parameters (rate constants, binding constants, etc.) with the highest probability of being correct. Virtually all analysis methods make assumptions about the nature of the experimental uncertainties. The experimental uncertainties in the data dictate the method of analysis to be used to obtain statistically correct results. This is not a one-way street, however. The numerical methods available for the data analysis also influence the choice of experimental protocol.

One commonly overlooked aspect of the analysis of experimental data is the fact that the researcher is actually interested in obtaining two things from the analysis of experimental data. One goal of the analysis is the determination of the values of the experimental parameters having the highest probability (maximum likelihood) of being correct. It is of equal or greater importance to obtain a realistic measure of the statistical confidence of the determined parameters. This allows comparison of results from different experiments and enables conclusions to be drawn about the significance of the results.

Parameter-Estimation Methods

This section outlines some of the least-squares methods available to evaluate the set of parameters with the highest probability of being correct, given a set of experimental data. For purposes of discussion, a data set will consist of a group of points x_i, $y(x_i)$, and σ_i, where x_i is the independent variable, $y(x_i)$ is the dependent variable, and σ_i is the standard error of the mean (SEM) of $y(x_i)$. The subscript i denotes a particular one of N data points. This data set is to be "fit" to an equation $f(x_i,\mathbf{a})$ which is a function of the independent variable and the vector $\mathbf{a}$ of parameters to be evaluated. The "fit" is to be performed such that the dependent variable can be approximated by the function evaluated at the corresponding independent variable and the parameter values having the maximum likelihood (highest probability) of being correct: that is, $y(x_i) \cong f(x_i,\mathbf{a})$.

As an example, consider a temperature-jump kinetic experiment. The data comprise a series of measurements of optical density as a function of time. Time is the independent variable, x_i, and the optical density is the dependent variable, $y(x_i)$. There is an infinite number of possible fitting equations, but one of the simplest is a single exponential with an additive constant:

$$y(x_i) \cong f(x_i,\mathbf{a}) = amplitude \cdot e^{-kx_i} + baseline \qquad (1)$$

In this case, the vector of fitting parameters $\mathbf{a}$ contains three elements: *amplitude, k,* and *baseline.* This simple equation will be used as an example throughout this chapter.

What Is Least-Squares?

Nonlinear least-squares analysis actually comprises a group of numerical procedures that can be used to evaluate the "optimal values" of the parameters in vector $\mathbf{a}$ for the experimental data. Several assumptions are implicit. In general, the nonlinear least-squares procedures consist of an algorithm that uses an initial approximation vector $\mathbf{g}$ of the parameters to generate a "better" approximation. These better answers are then used as initial approximations in the next iteration to yield an even better approximation. This process is continued until the approximations converge to a stable set of answers, $\mathbf{a}$.

For least-squares the definition of "better" approximation is one for which the weighted sum of the squares of the differences between the fitted function and the experimental data decreases. Including the weighting factors for the relative precision of each data point, this norm of the data is given by

$$\chi^2 = \sum_{i=1}^{N} \left(\frac{y(x_i) - f(x_i, \mathbf{a})}{\sigma_i} \right)^2 = \sum_{i=1}^{N} r_i^2 \tag{2}$$

As the iterative process continues, the weighted sum of the squared differences, χ^2, approaches a minimum. As discussed later, obtaining a set of parameters by minimizing the least-squares norm χ^2 does not always provide the set of parameters with the maximum likelihood of being correct.

Linear least-squares is a special case of the more general nonlinear least-squares. In this context linear does not imply a straight line. A function is linear when its second, and higher, order derivatives with respect to the fitting parameters are all zero:

$$\frac{\partial^2 f(x_i, \mathbf{a})}{\partial a_j \partial a_k} = 0 \tag{3}$$

where the subscripts j and k denote individual parameters. For example, a quadratic equation of the form

$$y = a_1 + a_2 x + a_3 x^2 + \cdots \tag{4}$$

is linear because the second- and higher-order derivatives with respect to the parameters are zero. In this work we will consider the more general case of nonlinear least-squares and note where differences from linear least-squares exist.

What Are Inherent Assumptions of Least-Squares Methods?

In the derivations to follow, the data $[x_i, y(x_i),$ and $\sigma_i]$ will be considered for simplicity as scalar quantities. However, each of these quantities could well be a vector of observations. For example, for two-dimensional data like that from NMR or electrophoresis studies, each of the x_i values is a vector of length two containing the two independent variables. For phase-modulation fluorescence lifetime measurements, each of the dependent variables $y(x_i)$ (and its associated SEM, σ_i) is a vector of length two containing the two dependent variables, phase and modulation, at each frequency x_i. In this case the fitting equation $f(x_i, \mathbf{a})$ would be two different equations, one for the phase, and another for the modulation. The derivations to follow can be readily expanded with no loss of generality to include multiple independent and dependent variables by including the appropriate summations. However, in the remainder of this work we restrict our discussion to scalar independent and dependent variables.

To demonstrate that the least-squares method is appropriate and will yield the parameters $\mathbf{a}$ having the maximum likelihood of being correct,

Least Squares Minimizes $\sum r_i{}^2$

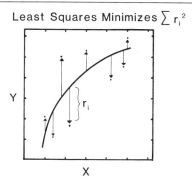

Fig. 1. Graphical representation of the residuals r_i of a least-squares parameter estimation. (Redrawn from Johnson and Frasier[1] with permission.)

several interrelated assumptions must be made.[1] We must assume (1) that all of the experimental uncertainty can be attributed to the dependent variables (ordinate or Y axis), (2) that the experimental uncertainties of the data can be described by a Gaussian (bell-shaped) distribution, (3) that no systematic error exists in the data, (4) that the functional form $f(x_i, \mathbf{a})$ is correct, (5) that there are enough data points to provide a good sampling of the experimental uncertainties, and (6) that the observations (data points) are independent of each other. Each of these assumptions implies important requirements for the data.

The first assumption is presented graphically in Fig. 1. The least-squares method finds values of the parameters such that the sum of the squares of the difference between the fitting function and the experimental data is minimized. These differences, or "residuals," are shown as r_i in Fig. 1 and are defined in Eq. (2). For the least-squares method to be valid, the uncertainties in the independent variables x_i must be significantly smaller than the uncertainties in the dependent variables y_i. Furthermore, if small uncertainties do exist in the independent variables, they must be independent of (i.e., not correlated with) the uncertainties in the dependent variables. In general, there is no method of circumventing this requirement with the least-squares method. It cannot be corrected for by "appropriate weighting factors."[2] Other non-least-squares methods have been developed that do not require this assumption.[2,3]

The second assumption means that in the limit of an infinite number of independent measurements, the amplitudes of the experimental uncertain-

[1] M. L. Johnson and S. G. Frasier, this series, Vol. 117, p. 301.
[2] M. L. Johnson, *Anal. Biochem.* **148**, 471 (1985).
[3] Y. Bard, "Nonlinear Parameter Estimation," p. 67. Academic Press, New York, 1974.

ties follow a normal (Gaussian or bell-shaped) distribution. This is a reasonable assumption for many, but not all, experimental protocols. For example, in hormone binding studies with radioactively labeled ligands, the measured amount bound will follow a Poisson distribution. In the limit of a large number of observed radioactive decays, a Poisson distribution can be approximated by a Gaussian distribution, and so least-squares can be used. However, when the number of observed counts is small, the Poisson distribution cannot be approximated by a Gaussian; therefore, it is not valid to use a least-squares method to analyze such data sets. For a more complete discussion of the properties of the Poisson distribution, the reader is referred to Bevington.[4]

The third assumption is that no systematic uncertainties exist in the data. The only means of circumventing this assumption is to include terms in the fitting function $f(x_i, \mathbf{a})$ describing the systematic errors in the data. In hormone binding experiments the "nonspecific binding" represents a systematic uncertainty in the experimental data. A discussion of how to correctly include a term in the fitting equation to account explicitly for the nonspecific binding has been presented elsewhere.[1,5,6]

The fourth assumption is that the functional form is correct. This requires more than simply finding a functional form that seems to describe the experimental data. The reason for this assumption can be described by an example. A simulated Scatchard plot for the binding of rhodamine 6G to glucagon[6,7] is shown in Fig. 2. This Scatchard plot exhibits the characteristic positive (upward) curvature indicative of multiple classes of binding sites. Analysis of the data using a molecular model with two classes of binding sites results in equilibrium constants with values of 1.79×10^3 and $40.7 \ M^{-1}$, and corresponding binding capacities of 1.8 and 1.2 μM.[1,6] This provides a reasonable fit of the data. However, the equilibrium constants are different from those presented in the legend to Fig. 2. The source of the error is the use of the wrong molecular mechanism and consequently of an inappropriate fitting function. The net result of using the wrong fitting function is that the resulting parameters, the equilibrium constants, have no physical meaning.

The binding of oxygen to human hemoglobin A_0 presents another example of the use of the wrong fitting equation leading to incorrect answers. It is commonly believed that the oxygen binding constants for

[4] P. R. Bevington, "Data Reduction and Error Analysis for the Physical Sciences," p. 36. McGraw-Hill, New York, 1969.

[5] P. J. Munson and D. Rodbard, *Anal. Biochem.* **107**, 220 (1980).

[6] M. L. Johnson and S. G. Frasier, *in* "Methods in Diabetes Research" (J. Larner and S. Pohl, eds.), Vol. 1, Part A, p. 45. Wiley, New York, 1984.

[7] M. L. Johnson, S. Formisano, and H. Edelhoch, *J. Biol. Chem.* **253**, 1353 (1978).

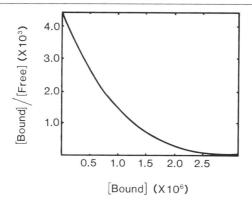

FIG. 2. Simulated Scatchard plot for the binding of rhodamine 6G to 10 μM glucagon at pH 10.6 in 0.6 M phosphate. The graph is based on two equilibrium reactions,[7] namely, a monomer to trimer association reaction of the glucagon with an equilibrium constant of $4.8 \times 10^7 \ M^{-1}$ and the binding of a single rhodamine 6G to the trimeric species with an equilibrium constant of $9.4 \times 10^4 \ M^{-1}$. (Redrawn from Johnson and Frasier[6] with permission of John Wiley and Sons, Inc. Copyright © 1984 John Wiley and Sons, Inc.)

tetrameric hemoglobin can be measured at "high" hemoglobin concentrations. The problem is complicated, however, by the fact that hemoglobin tetramers exist in a reversible equilibrium with hemoglobin dimers. It has been shown elsewhere[8,9] that neglecting the existence of these dimers can lead to significant errors in the values obtained for the tetrameric equilibrium constants even at hemoglobin concentrations as high as 3 mM. The problem of the wrong fitting equation has also been discussed in reference to the binding of insulin to its cell surface receptor.[10]

The fifth assumption is that the number of data points is sufficiently high to provide a good sampling of the random experimental noise superimposed on the data. The minimum number of independent (noncorrelated) data points is, by definition, equal to the number of parameters being estimated. However, because the experimental data contain experimental uncertainties, the number of data points required for a quality analysis is significantly greater than the minimum. There is no *a priori* way to predict the value of N such that N data points will be sufficient while $N - 1$ data points will not.

The number of data points required to determine "goodness-of-fit" is another factor that should be considered. Once a fit to a set of data is

[8] M. L. Johnson and G. K. Ackers, *Biophys. Chem.* **7**, 77 (1977).

[9] M. L. Johnson and A. E. Lassiter, *Biophys. Chem.* **37**, 231 (1990).

[10] C. DeLisi, *in* "Physical Chemical Aspects of Cell Surface Events in Cellular Regulation" (C. Delisi and R. Blumenthal, eds.), p. 261. Elsevier/North-Holland, New York, 1979.

accomplished, one of the next steps should be to establish whether the fitted function actually describes the experimental data. Most goodness-of-fit criteria are based on analyses of the randomness of the weighted residuals, r_i, in Eq. (2). Clearly, it is best to have as many data points as possible. In practice, however, it is not always possible to collect a large number of data points. Consequently, an intermediate number of data points must often be used. For a more complete discussion of goodness-of-fit criteria, the reader is referred to another chapter of this volume.[11]

Why Least-Squares?

One objective of data analysis is the determination of the set of parameters with the highest probability (maximum likelihood) of being correct. Why does that coincide with the set with the minimum least-squares norm? The answer is that least-squares methods do not always provide the set of parameters with the highest probability of being correct. To understand when and how least-squares methods can correctly be applied, we must consider the relationship between least-squares and maximum-likelihood methods. This relationship is based on the six assumptions outlined in the previous section.

It is commonly assumed that the assumptions discussed in the previous section are not required for least-squares methods to be valid because the least-squares methods will yield parameter values minimizing the variance of the fit even when the assumptions do not apply. While it is true that the least-squares methods will minimize the variance of fit, a minimum variance of fit is not necessarily what is desired. The set of parameter values with the highest probability of being correct is what we desire, and this may or may not correspond to the set which yields the minimum variance of fit.

If we assume (1) that the experimental uncertainties of the data are all in the dependent variables, (2) that the experimental uncertainties of the data follow a Gaussian distribution, (3) that no systematic errors exist, and (4) that the fitting function is correct, we can write the probability of a particular set of parameters **a** based on the value of the ith data point as

$$P_i[\mathbf{a}, x_i, y(x_i), \sigma_i] = \frac{1}{\sigma_i(2\pi)^{1/2}} e^{-1/2[y(x_i)-f(x_i,\mathbf{a})/\sigma_i]^2} = \frac{1}{\sigma_i(2\pi)^{1/2}} e^{-1/2(r_i^2)} \quad (5)$$

where r_i is the residual defined in Eq. (2), and σ_i corresponds to the standard error of the observed $y(x_i)$. This defines the appropriate weighting factor for each data point as the standard error (SEM) of the individual

[11] M. Straume and M. L. Johnson, this volume [5].

observation. For a more complete discussion of weighting functions, the reader is referred to another chapter of this volume.[12]

If we further assume that the N data points are independent observations, we can write the probability of observing a set of parameters **a**, based on the entire set of data, as the product of the individual probabilities:

$$P(\mathbf{a}) = \prod_{i=1}^{N} P_i[\mathbf{a}, x_i, y(x_i), \sigma_i] \propto e^{(\Sigma_{i=1}^{N} r_i^2)} \tag{6}$$

To obtain the parameters **a** with the highest probability of being correct, we maximize the probability $P(\mathbf{a})$ in Eq. (6). This is accomplished when the summation in Eq. (6) is minimized since it occurs as a negative exponent. The summation in Eq. (6) is identical to the least-squares norm given in Eq. (2). Consequently, the method of least-squares will yield parameter values having the highest probability (i.e., maximum likelihood) of being correct if the assumptions are valid. Conversely, if the assumptions are not met, the least-squares methods will yield parameter values that do not have the highest probability of being correct.

How Is Least-Squares Parameter Estimation Performed?

There are numerous numerical algorithms that can be used to find a set of parameters satisfying the least-squares criterion. We will not attempt to describe them all. We will, however, review several of the more common algorithms: the Gauss–Newton method and derivatives, and the Nelder–Mead simplex method. Each of these methods has advantages and disadvantages. The most commonly used algorithm for nonlinear problems is the Marquardt method, a derivative of the Gauss–Newton method. In our experience, an alternative derivative of the Gauss–Newton method works better for nonlinear cases. For linear problems, the classic Gauss–Newton method is clearly the best choice.

For nonlinear least-squares problems, all the general methods listed above are iterative methods requiring an initial approximation to the parameters and providing successively better approximations. The iterative process is repeated until the parameters do not change to within specified limits. For linear least-squares problems the Gauss–Newton method requires just a single iteration, and the initial estimates for the values of the parameters can all be zero.

For nonlinear problems the possibility of multiple minima in the least-squares norm exists. This means there may be more than one set of parameters, each of which yields a relative minimum in the sum of the

---

[12] E. Di Cera, this volume [4].

squares of the residuals. The values of the least-squares norm at these multiple minima may or may not have the same value. Once any of the least-squares methods has converged to a minimum, there is no guaranteed means of determining whether it is a unique or even the global (lowest) minimum.

One way to be more certain that a minimum is actually the global minimum desired is to start the nonlinear least-squares procedure at several different starting estimates for the parameters. If the final results are independent of the initial values, there is more confidence that the final parameter values describe a global minimum.

Parabolic Extrapolation of χ^2 Method

The most straightforward method of determining the minimum or maximum of a function is to find where the first derivatives of the function are zero. This method also can be used to find the set of parameters that yields a minimum least-squares norm. This is accomplished by setting each of the derivatives of the least-squares norm, χ^2 in Eq. (2), to zero

$$\frac{\partial \chi^2}{\partial a_k} = 0 \tag{7}$$

and solving the resulting system of equations for the values of the parameters, the vector $\mathbf{a}$. The system consists of M equations, one for each fitting parameter. This approach has the advantage that it can be used for virtually any norm of the data, not just the least-squares norm. Its disadvantage is the general computational complexity of solving the system of equations.

Differentiating Eqs. (2) and (7), we get the system of equations given by

$$-2 \sum_{i=1}^{N} \frac{y(x_i) - f(x_i, \mathbf{a})}{\sigma_i} \frac{1}{\sigma_i} \frac{\partial f(x_i, \mathbf{a})}{\partial a_k} \equiv G_k(\mathbf{a}) = 0 \tag{8}$$

This system of equations can be solved explicitly for some specific forms of the fitting function f; however, a fast general solution is more difficult. The most common general solution of a system of equations like that of Eq. (8) is by Newton's iteration. Newton's iteration involves the expansion of each of the $G_k(\mathbf{a})$ in a Taylor series about the function $G_k(\mathbf{g})$

$$G_k(\mathbf{a}) = G_k(\mathbf{g}) + \sum_{j=1}^{M} \frac{\partial G_k(\mathbf{g})}{\partial g_j} (a_j - g_j) + \cdots \tag{9}$$

where $\mathbf{a}$ and $\mathbf{g}$ refer to the vectors of desired parameters and the approximations to those parameters, respectively. The sum is over the parameters being estimated. For nonlinear equations, this expansion is truncated at first order, and for linear equations, these higher terms are identically zero. These first-order equations are then solved in the same iterative fashion as described in the later section on the Gauss–Newton method. The computationally intensive problem is the calculation of the derivatives of the G function. The G function itself contains derivatives of the fitting function f; therefore, this method involves evaluating the fitting function, all M first derivatives with respect to the fitting parameters, and all M^2 second derivatives with respect to combinations of the fitting parameters. Furthermore, as discussed later, Newton's method may not always converge for nonlinear problems. Several methods for improving the convergence properties of Newton's iteration are outlined later in this chapter. For a more complete discussion of this method, the reader is referred to Bevington.[13]

Gauss–Newton Method

The Gauss–Newton method is a less general but computationally simpler method of performing least-squares parameter estimations. It is less general because it can only minimize the least-squares norm, whereas the parabolic extrapolation of χ^2 method can be used to minimize any norm of the data and fitting function. The Gauss–Newton method exhibits a "quadratic convergence" which, simply put, means that the uncertainty in the parameters after $p + 1$ iterations is proportional to the square of the uncertainty after p iterations. Once these uncertainties begin to get small they decreased quite rapidly. An additional advantage of the Gauss–Newton method is that only first-order derivatives of the fitting function are required, as opposed to the parabolic extrapolation of χ^2 method which requires second-order derivatives. The major problem with the Gauss–Newton method is that it sometimes diverges instead of converging. Methods of modifying the Gauss–Newton method to solve this problem are presented in a later section.

The Gauss–Newton least-squares method is formulated as a system of Taylor series expansions of the fitting function. The dependent variable for each data point is approximated by the fitting function evaluated at the maximum-likelihood parameter values, **a**. This in turn is approximated by a Taylor series expansion of the fitting function evaluated at an estimate of the parameters values, **g**

[13] P. R. Bevington, "Data Reduction and Error Analysis for the Physical Sciences," p. 222. McGraw-Hill, New York, 1969.

$$y(x_i) \cong f(x_i,\mathbf{a}) = f(x_i,\mathbf{g}) + \sum_{j=1}^{M} \frac{\partial f(x_i,\mathbf{g})}{\partial g_j} (a_j - g_j) + \cdots \qquad (10)$$

It is generally assumed that this Taylor series can be truncated after the first-order derivatives. For linear fitting equations this is clearly correct. However, this truncation sometimes causes a divergence when the Gauss–Newton method is used to solve nonlinear equations. Truncating the series and rearranging the terms, we can express Eq. (10) as

$$\sum_{j=1}^{M} \frac{1}{\sigma_i} \frac{\partial f(x_i,\mathbf{g})}{\partial g_j} (a_j - g_j) = \frac{y(x_i) - f(x_i,\mathbf{g})}{\sigma_i} \qquad (11)$$

The σ_i in this expression represents weighting factors for the individual data points; they are introduced to account for the possibility of variable precision in the measurements of the dependent variable.

Given an estimate $\mathbf{g}$ of the fitting parameters, we can utilize Eq. (11) to obtain a better estimate $\mathbf{a}$ of the fitting parameters. This procedure is applied iteratively until the values of $\mathbf{a}$ do not change to within specified limits. Equation (11) can be written in matrix notation as

$$\mathbf{Ae} = \mathbf{D} \qquad (12)$$

The elements of the matrix $\mathbf{A}$ are the weighted partial derivatives evaluated at the current estimate of the parameters:

$$\mathbf{A} = \begin{pmatrix} \dfrac{1}{\sigma_1} \dfrac{\partial f(x_1,\mathbf{g})}{\partial g_1} & \cdots & \dfrac{1}{\sigma_1} \dfrac{\partial f(x_1,\mathbf{g})}{\partial g_M} \\ \vdots & \ddots & \vdots \\ \dfrac{1}{\sigma_N} \dfrac{\partial f(x_N,\mathbf{g})}{\partial g_1} & \cdots & \dfrac{1}{\sigma_N} \dfrac{\partial f(x_N,\mathbf{g})}{\partial g_M} \end{pmatrix} \qquad (13)$$

The elements of the vector $\mathbf{e}$ are the weighted differences between the better approximation of the fitting parameters and the current approximation of the fitting parameters:

$$\mathbf{e} = \begin{pmatrix} a_1 - g_1 \\ \vdots \\ a_M - g_M \end{pmatrix} \qquad (14)$$

The elements of vector $\mathbf{D}$ are the weighted differences between the data points and the fitting function evaluated at the current estimate of the parameters, $\mathbf{g}$:

$$D \; = \; \begin{pmatrix} \dfrac{y(x_1) - f(x_1, \mathbf{g})}{\sigma_1} \\ \vdots \\ \dfrac{y(x_N) - f(x_N, \mathbf{g})}{\sigma_N} \end{pmatrix} \tag{15}$$

When convergence is reached, the elements of **D** are the residuals as shown in Fig. 1.

One method of solving Eq. (12) for the better estimate of **a** is singular value decomposition (SVD).[14] However, if this method is to be programmed for a computer with a limited amount of memory, such as an IBM PC running DOS, SVD may not be the method of choice. The SVD method requires several large arrays of size equal to the number of fitting parameters times the number of data points. These arrays can easily exceed the memory capacity of an IBM PC running DOS.

Our preferred method of solving the problem is to convert Eq. (12) to a form solvable by matrix-inversion techniques.[15] This method requires much smaller arrays, and it provides results which are identical to those provided by SVD. The matrix **A** cannot be inverted directly because it is not a square matrix (a square matrix has the same number of rows and columns). By multiplying both sides of Eq. (12) by $\mathbf{A}^T$, the transpose of **A** (the transpose of a matrix is the matrix with the rows and columns interchanged), we obtain

$$(\mathbf{A}^T \mathbf{A}) \, \mathbf{e} \; = \; \mathbf{A}^T \mathbf{D} \tag{16}$$

where $\mathbf{A}^T \mathbf{A}$ is a square matrix that can usually be inverted. This matrix is sometimes referred to as the information matrix, and at other times it is referred to as the Hessian matrix. This matrix can be evaluated directly without the need of creating the **A** and $\mathbf{A}^T$ matrices, both of which are much larger. The number of elements in the $\mathbf{A}^T \mathbf{A}$ matrix is equal to the number of parameters squared. In general, if there are enough independent data points, and the parameters being estimated are not perfectly correlated, then Eq. (16) can be solved for **e**.

Equation (16) is the classic linear algebra form for the problem of M linear equations in M unknowns. Consequently, there are numerous methods to solve it. Careful consideration should be given to the method of solution, however, as $\mathbf{A}^T \mathbf{A}$ will usually be a nearly singular matrix. Equation (16) could be solved for **e** by inverting $\mathbf{A}^T \mathbf{A}$

[14] G. E. Forsythe, M. A. Malcolm, and C. B. Moler, "Computer Methods for Mathematical Computations," p. 192. Prentice-Hall, Englewood Cliffs, New Jersey, 1977.
[15] V. N. Faddeeva, "Computational Methods of Linear Algebra," p. 83. Dover, New York, 1959.

$$\mathbf{e} = (\mathbf{A}^T\mathbf{A})^{-1}(\mathbf{A}^T\mathbf{D}) \tag{17}$$

but matrix inversion is intrinsically less efficient, and more sensitive to truncation and round-off errors, than solving for $\mathbf{e}$ directly. The choice of method for the direct solution of Eq. (16) is based on the properties of the matrix $\mathbf{A}^T\mathbf{A}$. It is a positive-definite symmetric matrix and is usually nearly singular. We recommend the square root method[15] for the solution of Eq. (16).

Once $\mathbf{e}$ has been evaluated, a better approximation of the parameters $\mathbf{a}$ can be evaluated as

$$\mathbf{a} = \mathbf{g} + \mathbf{e} \tag{18}$$

This better approximation of the fitting parameters is then used as the initial approximation $\mathbf{g}$ for the next iteration. The entire process is repeated until the parameter values do not undergo significant changes between iterations.

Some researchers assume that convergence is reached when the least-squares norm, χ^2, does not change by more than 0.1% between iterations. This criterion can in some cases indicate that convergence has been reached when in fact it has not. We prefer a more stringent convergence criterion involving two tests. First, the least-squares norm should not change by more than 0.01% from one iteration to the next. Second, the fitting parameter values should not change by more than 0.01% from one iteration to the next; this is equivalent to requiring that $\mathbf{e}$ approach zero at convergence. Convergence is accepted when both criteria are satisfied over several successive iterations.

At no point in the above derivation has anything been assumed about minimizing the least-squares norm or maximizing the probability of the parameter values. Why then is the Gauss–Newton method a least-squares method? The iterative nature of the Gauss–Newton method requires that $\mathbf{e}$ be equal to zero at convergence. Thus from Eq. (16) we know that either $\mathbf{A}^T\mathbf{A}$ is infinite or $\mathbf{A}^T\mathbf{D}$ is equal to zero. Because the $\mathbf{A}^T\mathbf{A}$ matrix will in general not be infinite, $\mathbf{A}^T\mathbf{D}$ must approach zero at convergence. The form of the jth element of the vector $\mathbf{A}^T\mathbf{D}$ at convergence is

$$(\mathbf{A}^T\mathbf{D})_j = \sum_{i=1}^{N} \frac{y(x_i) - f(x_i,\mathbf{a})}{\sigma_i} \frac{1}{\sigma_i} \frac{\partial f(x_i,\mathbf{a})}{\partial a_j} \tag{19}$$

The right-hand side of this equation is proportional to the derivative of the least-squares norm with respect to the jth fitting parameter, following Eqs. (2) and (8). Because $\mathbf{A}^T\mathbf{D}$ approaches zero at convergence, the derivatives of the least-squares norm, χ^2, with respect to the fitting parameters must also be equal to zero at convergence. Therefore, the Gauss–Newton method produces a least-squares parameter estimation, as in Eq. (7).

$$(\mathbf{A}^{\mathrm{T}}\mathbf{D})_j \propto \frac{\partial \chi^2}{\partial a_j} \tag{20}$$

The reader is reminded that the unmodified Gauss–Newton method will sometimes diverge. This is a problem when the initial "guess" $\mathbf{g}$ of the parameter-value vector is far from the maximum-likelihood value $\mathbf{a}$. Because of this, many modifications of the basic Gauss–Newton method have been developed. Two of these are presented in later sections.

Steepest Descent Method

Also known as the gradient search method,[16] the steepest descent technique searches along the gradient (or direction of maximum variation) of χ^2 to obtain a better estimate of the parameters. This method is especially useful for large problems as it requires the least computer memory. However, the steepest descent method does require many more iterations than the Gauss–Newton method. It converges linearly (i.e., the error after $p + 1$ iterations is proportional to the error after p iterations).

The gradient $\nabla \chi^2$ is a vector pointing in the direction of maximum slope (increase) in the least-squares norm. The elements of this vector are

$$(\nabla \chi^2)_j = \frac{\partial \chi^2}{\partial a_j} \tag{21}$$

The gradient is normalized to unit length to obtain the amount by which to change the parameters in the next iteration. The vector $\mathbf{e}$ thus calculated is analogous to the vector $\mathbf{e}$ in the derivation of the Gauss–Newton method:

$$\mathbf{e}_j = \frac{-\dfrac{\partial \chi^2}{\partial a_j}}{\left[\displaystyle\sum_{k=1}^{M} \left(\dfrac{\partial \chi^2}{\partial a_k}\right)^2\right]^{1/2}} \tag{22}$$

The minus sign indicates that the vector is in the direction of decreasing χ^2. Once the vector $\mathbf{e}$ is obtained, it is used in an iterative fashion analogous to its use in Eq. (18) until convergence is reached.

[16] P. R. Bevington, "Data Reduction and Error Analysis for the Physical Sciences," p. 215. McGraw-Hill, New York, 1969.

Marquardt Method

The Marquardt method[17,18] is the most commonly used procedure of improving the convergence properties of the Gauss–Newton method. It is essentially a linear combination of the steepest descent and Gauss–Newton methods. The Marquardt method retains the robust, but linear, convergence properties of the steepest descent method when the parameter values are far from their final values, and it still has the rapid quadratic convergence properties of the Gauss–Newton method when the parameter values are close to the final converged values.

The Marquardt method may be expressed as a modification of the Gauss–Newton method [Eq. (16)] by

$$(\mathbf{A}^T\mathbf{A})'\mathbf{e} = (\mathbf{A}^T\mathbf{D}) \tag{23}$$

where the elements of the $(\mathbf{A}^T\mathbf{A})'$ matrix are defined as

$$
\begin{aligned}
(\mathbf{A}^T\mathbf{A})'_{j,k} &= (\mathbf{A}^T\mathbf{A})_{j,j}(1 + \lambda) \qquad \text{for } j = k \\
&= (\mathbf{A}^T\mathbf{A})_{j,k} \qquad\qquad \text{for } j \neq k
\end{aligned}
\tag{24}
$$

The parameter λ is initially a large number like 10^7 and is adjusted such that χ^2 decreases with each iteration. For each iteration, λ is divided by three if χ^2 decreases and multiplied by three if χ^2 increases. The large initial value of λ makes the diagonal elements of $\mathbf{A}^T\mathbf{A}$ dominate at first, giving the convergence properties of the steepest descent method. As the parameters converge, λ approaches zero and Eq. (23) reduces to Eq. (16), giving the rapid convergence characteristic of the Gauss–Newton method.

The Marquardt method is relatively easy to implement, usually converges rapidly, and does not require large storage arrays. However, if the fitting parameters are highly correlated (which implies that the $\mathbf{A}^T\mathbf{A}$ matrix is nearly singular), the Marquardt method can require a tremendous number of iterations. This is because the large initial value of λ forces the counterassumption, according to Eq. (24), that the fitting parameters are *not* correlated (i.e., that the parameters are orthogonal). A large value of λ in Eqs. (23) and (24) is equivalent to setting all the off-diagonal elements of $(\mathbf{A}^T\mathbf{A})'$ to zero, in other words, assuming that $(\mathbf{A}^T\mathbf{A})'$ is a diagonal matrix. The inverse of a diagonal matrix is also a diagonal matrix; therefore, all the off-diagonal elements of the inverse matrix will also be zero. Thus, from Eq. (24), for large values of λ, the cross-correlation between fitting parameters a_j and a_k ($j \neq k$) will be zero. The cross-correlation coefficient

[17] D. W. Marquardt, *SIAM J. Appl. Math* **14**, 1176 (1963).
[18] P. R. Bevington, "Data Reduction and Error Analysis for the Physical Sciences," p. 246. McGraw-Hill, New York, 1969.

(CC) between two fitting parameters is defined in terms of the inverse of the $\mathbf{A}^T\mathbf{A}$ matrix

$$(\mathbf{CC})_{j,k} = \frac{(\mathbf{A}^T\mathbf{A})_{j,k}^{-1}}{[(\mathbf{A}^T\mathbf{A})_{j,j}^{-1}(\mathbf{A}^T\mathbf{A})_{k,k}^{-1}]^{1/2}} \tag{25}$$

The values of the cross-correlation coefficients range from minus to plus one. If the parameters are orthogonal (i.e., can be independently determined), their cross-correlation coefficients are zero. Conversely, the larger the magnitude of the cross-correlation coefficients, the more difficult will be the parameter-estimation process.

The cross-correlation coefficients, as defined in Eq. (25), cannot be used to infer anything about possible relationships in the chemistry of the problem. These cross-correlations between the estimated parameters are a consequence of the process of measuring the dependent variables at a finite number of data points, over a limited range of the independent variable, and the subsequent fitting of the data to estimate the parameters. As an example, consider an experiment that measures a "spectrum" sensitive to both intracellular Mn and Ca concentrations. By an analysis of such spectra it is possible to estimate the values of intracellular Mn and Ca concentrations. The analysis will also provide an apparent cross-correlation between the Mn and Ca concentrations. This cross-correlation is actually a measure of the difficulty of the fitting procedure and cannot be used to infer any relationship in the mechanisms of the cell for the regulation of Mn and Ca concentrations.

Preferred Method

We prefer a least-squares method that does not make the assumption of orthogonality at the initial stages of convergence. Under conditions where the parameters are highly correlated, this method retains the quadratic convergence properties of the Gauss–Newton method, but does not diverge.

The Gauss–Newton method provides the vector $\mathbf{e}$ that is the best direction in which to look for a smaller χ^2. However, the truncation of the higher-order derivatives in Eq. (10) may yield a vector $\mathbf{e}$ having too large a magnitude. When this occurs we decrease the magnitude of $\mathbf{e}$ but do not change its direction.

Our method introduces λ into Eq. (18) as a scaling parameter according to

$$\mathbf{a} = \mathbf{g} + \lambda\mathbf{e} \tag{26}$$

For each iteration, λ is set initially to unity. The approximation $\mathbf{a}$ for the next iteration is calculated according to Eq. (26). If the value of χ^2 evaluated at $\mathbf{a}$ is less than that at $\mathbf{g}$, we proceed with the next iteration as in the Gauss–Newton method. If, on the other hand, the χ^2 value at $\mathbf{a}$ is greater, we divide λ by two and reevaluate $\mathbf{a}$ with the same initial guesses $\mathbf{g}$. We continue decreasing λ by factors of two until the new fitting parameters $\mathbf{a}$ yield a lower χ^2 than is obtained with $\mathbf{g}$. Once a lower χ^2 is found we proceed with the next iteration as in the Gauss–Newton method.

Nelder–Mead Simplex Method

The Nelder–Mead algorithm[19] is a geometric rather than a numeric method. The advantage of this method is that the derivative of the fitting function (dependent variable) need not be calculated. The simplex method is suited to the minimization or maximization of a variety of norms, Ψ, including the χ^2 least-squares norm as defined in Eq. (2). Because it exhibits a linear convergence, it is somewhat slower than some of the other, more restricted algorithms.

As mentioned, the Nelder–Mead algorithm is a geometric construct. This method depends on the comparison of the norm values at $M + 1$ vertices of a general simplex. In other words, there are $M + 1$ points, P_1, $P_2, \ldots, P_{M+1}$, in an M-dimensional parameter space; each vertex point P_i represents a unique set of choices for the M fitting parameters, and hence a specific value Ψ_i of the norm. Furthermore, no three vertex points in this space should be collinear. If this is the case, the dimensionality of the parameter space is in effect reduced. Although not explicitly mentioned in the derivation, the quantity of data points, N, must be of sufficient number.[20]

The initialization of the parameter values is usually accomplished by generating random numbers for the parameter space coordinates; each vertex point corresponds to a unique set of parameter values. The norm is then evaluated at every vertex point in the M-dimensional space. Those points where it takes on the minimum and maximum values are denoted P_1 and P_h. In the case that we wish to minimize the norm, we set aside the vertex P_h at which the norm takes on its maximum value, and define the centroid in terms of the M remaining vertices. This is represented symbolically as

[19] J. A. Nelder and R. Mead, *Comp. J.* **7,** 308 (1965).
[20] The Nelder–Mead algorithm is independent of the number N of data points. However, for statistical significance of the results, there must be a sufficient number of data points for the number M of parameters and the particular fitting function chosen.

$$\bar{P} = \frac{1}{M} \sum_{i=1}^{M}{}' P_i \tag{27}$$

where the prime on the summation indicates the specific exclusion of the vertex P_h.

In general, the procedure proceeds iteratively in three sequential steps: reflection, expansion, and contraction. First, the excluded vertex, P_h, is reflected with respect to the centroid $\bar{P}$ to the new position P^* on the line connecting P_h and $\bar{P}$:

$$P^* = (1 + \alpha)\bar{P} - \alpha P_h \tag{28}$$

where α is the reflection coefficient, a positive constant generally taken to be one. At this point, the procedure branches to one of three distinct paths, depending on the magnitude of Ψ^* (Ψ evaluated at the reflection point) relative to the other vertices:

$\Psi_1 < \Psi^* < \Psi_h$. If Ψ^* falls within the range of the other $M + 1$ vertices, P_h is replaced by P^* and the process is repeated with the new simplex.

$\Psi^* < \Psi_1$. If Ψ^* is a new minimum, then P^* is expanded to P^{**} according to

$$P^{**} = \gamma P^* + (1 - \gamma)\bar{P} \tag{29}$$

where the expansion coefficient γ must be greater than one and is usually assigned a value of two. At this point, if $\Psi^{**} < \Psi_1$, then P_h is replaced by P^{**}, and the process is restarted, with the new simplex. On the other hand, the expansion fails if $\Psi^{**} > \Psi_1$. In this case, P_h is replaced instead by P^*, and again the process is restarted.

$\Psi^* > \Psi_i$. If Ψ^* is a new maximum, namely, it is larger than the function values at the other M vertices from which the centroid is calculated (the vertex P_h is specifically excluded from the comparison), then it is compared with Ψ_h. If it is *smaller*, $\Psi^* < \Psi_h$, then P_h is replaced by P^*. In either case (whether it is smaller or not), a new vertex is calculated by contracting P_h:

$$P^{**} = \beta P_h + (1 - \beta)\bar{P} \tag{30}$$

where the contraction coefficient β has value between 0 and 1 and is usually 0.5. At this point, Ψ^{**} is compared with the smaller of Ψ_h and Ψ^*. It is hoped that the vertex P^{**} will be closer to a minimum than the starting point of the current iteration. If in fact Ψ^{**} is smaller than the two comparison values, then P_h is replaced by P^{**}, and the process is restarted. Otherwise, the contraction has failed, and all $M + 1$ vertices P_i are replaced by quantities $(P_i + P_l)/2$. Again, the process is restarted.

The iterative nature of the simplex method is evident from the above discussion. The means proposed by Nelder and Mead for halting the procedure is comparison of the "standard error" (S.E.), or "scatter," of the simplex vertices

$$\text{S.E.} = \left(\sum_{i=1}^{M+1} \frac{(Q_i - \bar{Q})^2}{M} \right)^{1/2} \tag{31}$$

with a preset value.

The Nelder–Mead minimization procedure, as outlined above, does not provide an estimate of the $\mathbf{A}^T\mathbf{A}$ information matrix. As shown in Eq. (25), the information matrix is needed to evaluate the cross-correlation between fitting parameters and, as shown later, is also needed to evaluate the confidence intervals of the fitted parameters. A method of approximating the $\mathbf{A}^T\mathbf{A}$ information matrix is presented in an appendix to the original Nelder–Mead method.[19]

Although it sometimes converges quite slowly, the Nelder–Mead algorithm always converges. Furthermore, although it may not be as fast as some of the other minimization procedures discussed here, this method does have the advantage of not requiring a specific form for the norm, nor even that it be differentiable.

Confidence Interval Estimation

Of critical importance is the determination of the precision to which the maximum likelihood parameters are determined. We might determine that a protein molecular weight is 100,000. If, however, the statistical confidence in this molecular weight is ±90,000 we know relatively little. On the other hand, if the statistical confidence is ±1000 we know a great deal. It is the confidence intervals on the determined parameters that allow us to compare experiments for statistical significance and to test hypotheses about mechanism.

Before discussing the methodology of finding estimates for the confidence intervals of estimated parameters, there are two important points to note. First, the researcher should always be aware of the distinction between precision and accuracy. The determined confidence interval is a measure of the precision (reproducibility) of the estimated parameter, based on a single set of experimental data. However, if the basic assumptions of the methodology are followed, the confidence intervals serve as measures not only of the precision but also of the accuracy of the determined parameters. The absolute accuracy of the fitted parameters is after all the object of the analysis. The second point is that most of the methods for the evaluation of the confidence intervals utilize an F-statistic as a measure of significant change in the least-squares norm. This use of the F-statistic is based on linear theory. Therefore, the approximation is valid only insofar as the fitting equation is linear.

There are numerous methods of evaluating the confidence intervals of determined parameters.[21] The most commonly used method, which uses the variance–covariance matrix, will usually significantly underestimate the actual confidence intervals, but it is fast and easy to program. This method may provide wrong conclusions about the statistical significance of the determined parameters and so should be avoided. Some other methods, like Monte Carlo, provide reliable estimates of the actual confidence intervals but require large amounts of computer time. In this section we compare some of these methods.

The most serious complication in the evaluation of confidence intervals is the cross-correlation of the maximum likelihood parameters, as in Eq. (25). When the parameters are cross-correlated, or not orthogonal, an uncertainty in one of the determined parameters will induce a systematic uncertainty in the other parameters. Thus, the confidence interval for a given parameter will usually include a contribution from the uncertainties in the other parameters.

Monte Carlo Method

The Monte Carlo method is one of the most precise methods for the evaluation of confidence intervals of determined parameters. A separate chapter of this volume is dedicated to this method[22] so we describe it only briefly here. Once a set of maximum likelihood parameters has been determined, by one of the methods previously described, a set of synthetic data based on those parameters can be generated. This set of data should be as close an approximation to the actual experimental data set as possible using the same independent variables, x_i. Realistic pseudorandom experimental uncertainties are then added to the synthetic data set, and the maximum-likelihood parameter-estimation procedure is repeated to obtain a new set of parameters. The process of generating random noise and parameter estimation is repeated many times to obtain a distribution of parameter values. Confidence intervals can then be generated from the parameter probability distribution obtained for each of the parameters.

Two aspects of the Monte Carlo method should be noted. First, validity of the method is dependent on how realistically the sets of pseudorandom

[21] M. Straume, S. G. Frasier-Cadoret, and M. L. Johnson, in "Fluorescence Spectroscopy, Volume 2: Principles" (J. Lakowicz, ed.). Plenum, New York, 1992.
[22] M. Straume and M. L. Johnson, this volume [7].

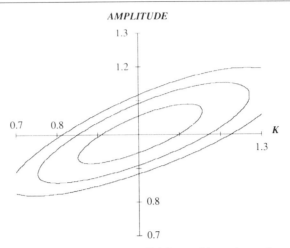

FIG. 3. Example of the 68%, 95%, and 99% joint confidence intervals evaluated by the grid search method.

uncertainties model the uncertainties in the actual experimental data. This point is critical, so care must be taken to model accurately the actual experimental uncertainties. The second point is that many different sets of pseudorandom noise must be generated. The number required depends on several factors, such as the number of parameters being estimated. For this reason, it is difficult to define precisely; a typical number might be 500 or more. Because of the need to analyze so many synthetic data sets, this method of confidence interval determination requires proportionally more computer time than is needed to compute the parameter values themselves.

Grid Search Method

A second method for the evaluation of confidence intervals involves creating a grid of parameter values and evaluating the criterion function at each of the grid vertices.[23] The confidence intervals are the regions surrounding the minimum over which the criterion function does not increase significantly with regard to the minimum value. The question of what is significant will be discussed later. This method, too, is quite precise but also requires substantial computer time.

An example of the grid search method for the evaluation of confidence intervals is shown in Fig. 3. Thirty synthetic data points were generated

---

[23] J. Beechem, this volume [2].

to represent a single exponential decay of the type defined by Eq. (1). The independent variables were equally spaced between 0.0 and 1.5. The values of both *amplitude* and k were 1.0, and the value of *baseline* was 0.0. Gaussian distributed pseudorandom noise with a 0.1 standard deviation was added to these synthetic data. Next we calculated the apparent variance of the fit, $s^2 = \chi^2/N$, for a grid of *amplitude* ranging from 0.7 to 1.3 and k ranging from 0.7 to 1.3. The *baseline* was held constant so the graph would contain only two dimensions. The general method will in principle work with any number of dimensions. Figure 3 presents a series of contours for which the least-squares norm is constant; each contour represents an increase in the value of s^2 with respect to the optimal value. This increase is 8.5% for the inner curve, 24.5% for the middle curve, and 38.9% for the outer curve. Calculation of this contour map involved evaluating s^2 at 230,400 pairs of *amplitude* and k. The contours define the joint confidence intervals of *amplitude* and k.

The most striking characteristic of these contours is the large cross-correlation between the values of *amplitude* and k. For this set of synthetic data, the cross-correlation, evaluated from Eq. (25), is 0.78. If the value of k is changed by a small amount away from the optimal value (the origin of Fig. 3), s^2 will increase. However, the *amplitude* can be changed from the optimal value to partially compensate for the increase in s^2 caused by the variation in k.

The variation of the least-squares norm is used to define the probability of a significant variation in the parameters. The value of s^2 at any point in the parameter space of Fig. 3 is the sum of the value at the minimum and a contribution due to the variation of the parameters:

$$s^2 = s^2_{\text{minimum}} + s^2_{\text{parameters}} \tag{32}$$

An F-statistic is used to determine the probability of a particular ratio of the value of s^2 due to the parameters, and the value at the minimum

$$\frac{s^2_{\text{parameters}}}{s^2_{\text{minimum}}} = F(M, N - M, 1 - P) \tag{33}$$

where M is the number of parameters, N is the number of data points, and P is the probability density that the two measures of s^2 are different. Furthermore, it should be noted that $N - M$ is the number of degrees of freedom for s^2_{minimum}, and M is the number of degrees of freedom for $s^2_{\text{parameters}}$. By combining Eqs. (32) and (33), we can express the fractional increase in s^2 in terms of the probability:

$$\frac{s^2}{s^2_{\text{minimum}}} = 1 + \frac{M}{N - M} F(M, N - M, 1 - P) \tag{34}$$

For a 0.32 probability of difference (one standard deviation), the fractional variance increase in this example is 1.085. For a 0.05 probability of difference (two standard deviations), the fractional increase is 1.245. For a 0.01 probability of difference (2.6 standard deviations), the fractional increase is 1.389. The inner contours in Fig. 3 define the one- and two-standard deviation joint confidence intervals for the parameters.

Although the grid search method provides accurate values for the confidence intervals of determined parameters, it is extremely slow. The above example took 3 min on a 25-MHz, 486-based PC, whereas the evaluation of the minimum point by our nonlinear least-squares method required just 1 sec. The amount of computer time required for this method is proportional to the linear number of grid elements to the power of the number of parameters being estimated. Consequently, if we had also varied *baseline*, about 24 hr of computer time would have been required. Clearly, a more efficient method is needed to evaluate confidence intervals in the case that multiple parameters are to be estimated.

Asymptotic Standard Errors

The most common method of evaluating the confidence intervals of estimated parameters is to use the diagonal elements of $\mathbf{VC}$, the variance–covariance matrix:

$$\mathbf{VC} = s^2_{\text{minimum}}(\mathbf{A}^{\text{T}}\mathbf{A})^{-1} \tag{35}$$

where s^2_{minimum} is the value of the least-squares norm at the minimum (optimal parameter values). The square roots of the diagonal elements of the $\mathbf{VC}$ matrix are the asymptotic standard errors of the estimated parameters. The off-diagonal elements are the covariances between the corresponding parameters.

The validity of this formulation of the $\mathbf{VC}$ matrix is based on the assumption that the variance space is quadratic—an assumption valid only for linear fitting equations. An additional assumption made by many programs is that of zero covariance between the estimated parameters. The covariance between two parameters is given by an off-diagonal element of the $\mathbf{VC}$ matrix. The variances of the estimated parameters are given directly by the diagonal elements of the $\mathbf{VC}$ matrix. When the off-diagonal elements of $\mathbf{VC}$ can be neglected and the fitting equation is linear, these variances define the joint confidence intervals of the parameters. This is equivalent to assuming orthogonality of the fitting parameters. For some applications like Fourier series analysis,[24] these may be reasonable as-

---

[24] L. M. Faunt and M. L. Johnson, this volume [15].

sumptions, but in general the fitting equations are nonlinear and the parameters are far from orthogonal. Consequently, with very few exceptions, the diagonal elements of the **VC** matrix should not be used to evaluate confidence intervals for estimated parameters.

For the grid search example presented above, the asymptotic standard errors of the estimated parameters are 0.094 for k and 0.0535 for *amplitude*. Comparison of these values with Fig. 3 clearly indicates that these values underestimate the actual acceptable range by about a factor of two. The covariance between k and *amplitude* is 0.00393, and the cross-correlation coefficient is 0.78. The underestimate of the confidence intervals comes from neglecting the covariance, or cross-correlation, between the fitting parameters. Even for this simple example the method of asymptotic standard errors does not provide quality estimates of the confidence intervals. This method, however, is easy to program and requires very little computer time; it is as a consequence commonly used.

Linear Joint Confidence Intervals

The linear joint confidence interval method is a rather well-known, linear procedure for including the covariance of the parameters in the evaluation of the confidence intervals.[25,26] It is, unfortunately, not commonly used in biological research. If the fitting equation is linear, the confidence profile can be approximated by an "elliptically shaped" joint confidence region. This elliptical joint confidence region, which is not necessarily aligned with the parameter axis (see Fig. 3), includes all vectors Ω of parameter values that satisfy the inequality

$$(\mathbf{a} - \Omega)^{\mathrm{T}}(\mathbf{A}^{\mathrm{T}}\mathbf{A})(\mathbf{a} - \Omega) \leq M s_{\mathrm{minimum}}^2 F(M, N - M, 1 - P) \qquad (36)$$

where $\mathbf{a}$, $\mathbf{A}$, N, M, s_{minimum}^2, and P are as previously defined. The solution of this inequality is straightforward but not simple. Often, the solutions of Eq. (36) provide good values for the precision with which a fitting parameter may be estimated. However, in many cases the confidence regions cannot be described by an elliptical joint confidence region; the confidence regions may be asymmetrical and not elliptically shaped. An example from the literature[26] of such an asymmetrical, nonelliptical confidence region is presented in Fig. 4. Furthermore, careful inspection of Fig. 3 demonstrates that even for this simple example the confidence region is not symmetrical. This formulation of elliptically shaped confidence regions is based on the assumption—not valid in many cases—that the fitting function is linear.

[25] G. E. P. Box, *Ann. N.Y. Acad. Sci.* **86,** 792 (1960).
[26] M. L. Johnson, *Biophys. J.* **44,** 101 (1983).

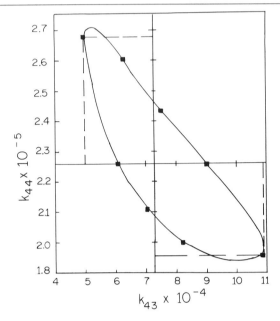

FIG. 4. Example of an asymmetrical, nonlinear confidence region taken from the litera-ture.[26] This example is for the evaluation of the third and fourth Adair oxygen binding constants (k_{43} and k_{44}) of human hemoglobin A_0. (Redrawn from Johnson[26] with permission of Rockefeller University Press. Copyright © 1983 Rockefeller University Press.)

Preferred Method

Our preferred method allows for a variance space that is asymmetrical and nonlinear.[26] This method provides only an approximation to the joint confidence region. The basic idea is to search the variance space for values of the parameters that yield a significant increase, defined by Eq. (34), in the apparent variance of fit s^2. The confidence region for a given parameter is defined by the extreme values found for that parameter, after searching in "all" directions, which correspond to an s^2 less than the critical value defined by the F-statistic. In practice we search the variance space in carefully selected specific directions from the point of minimum s^2 instead of in all possible directions. Because only specific directions are searched, the amount of computer time required is proportional to the number of directions searched. In the grid search method, the search is over a volume in parameter space, and so the time required is proportional to the number of linear grid elements to the power of the number of fitting parameters.

The search directions are defined by the axes of two different coordinate systems. One of the coordinate systems is the parameter space itself.

We search each of the parameters, in both directions, without varying any of the other parameters. Referring to the example of Fig. 3, we would increase the value of k from the optimal value until a variance ratio (F-statistic) of 1.085 is found. We then decrease the value of k until the same variance ratio is found. For both searches, we hold the value of *amplitude* constant at its optimal value. Next, we perform two more searches holding k constant and varying the *amplitude* in two directions.

The other coordinate system to be searched is the set of axes of the "elliptically shaped" confidence region defined by Eq. (36). The eigenvectors[27,28] of the correlation matrix, **CC** defined in Eq. (25), are vectors in the direction of these axes in the correlation space. To convert a correlation space eigenvector to the parameter space, each element of the particular eigenvector is multiplied by the square root of the corresponding diagonal element of the variance–covariance matrix, **VC** defined in Eq. (35). The relative lengths of each of the eigenvectors is given by the corresponding eigenvalue of the correlation matrix. Each axis is searched independently in both directions.

The use of eigenvectors to define a new coordinate system is equivalent to defining a new coordinate system corresponding to linear combinations of the desired fitting parameters. These linear combinations of parameters are orthogonal. Furthermore, because the parameters of the new coordinate system are orthogonal, each of the new parameters can be evaluated independently of the others.

Searching each axis of both coordinate systems in each direction involves $4M$ searches. Each of these searches will require about 10 or 20 evaluations of s^2 or χ^2. Thus, for the example in Fig. 3, a total of about 120 evaluations of s^2 was required. This is three orders of magnitude fewer than was required for the basic grid search method. Furthermore, the amount of computer time required is proportional to the number M of fitting parameters, whereas the time for the grid search method is proportional to the grid size raised to the power M.

For the example in Fig. 3 this search method predicts that the one-standard deviation confidence regions are $0.858 < k < 1.146$ and $0.919 < amplitude < 1.084$. These intervals are more realistic than those predicted by the asymptotic standard error method and agree with those from the grid search method presented in Fig. 3.

[27] The evaluation of eigenvalues is a standard linear algebra operation for which there are many standard software packages available. We use the routines *TRED2* and *TQL2* from the public domain EISPACK routines.

[28] B. T. Smith, J. M. Boyle, J. J. Dongarra, B. S. Garbow, Y. Ikebe, V. C. Klema, and C. B. Moler, "Matrix Eigensystem Routines—EISPACK Guide," 2nd Ed. Springer-Verlag, New York, 1976.

We have assumed that the linear fitting equation approximation provides a reasonable estimate of the directions, but not the magnitudes, of the eigenvector axes. Consequently, this method provides results which are intermediate between the grid search method and the solution of Eq. (36).

Propagation of Confidence Intervals

Once a set of parameters and their confidence intervals has been evaluated by one of the parameter-estimation procedures, it is common to use those values to evaluate other quantities. For example, given the values of *amplitude* and k from the example shown in Fig. 3, we might wish to evaluate the area under the decay curve. For fluorescence lifetimes, the area under the curve corresponds to the steady-state intensity of the fluorescence. The *area* under the exponential part of Eq. (1) is simply

$$area = \frac{amplitude}{k} \tag{37}$$

The question to be addressed is how to properly evaluate the confidence interval of a derived quantity such as the *area*.

If the confidence intervals of the *amplitude* and k are expressed as variances, the corresponding *area* is[29]

$$\frac{\sigma_{area}^2}{area^2} = \frac{\sigma_{amplitude}^2}{amplitude^2} + \frac{\sigma_k^2}{k^2} - 2\frac{\sigma_{covariance}^2}{k \cdot amplitude} \tag{38}$$

Equation (38) includes a term with the covariance $\sigma_{covariance}^2$ between k and the *amplitude*. Inclusion of this term is often neglected by programmers coding least-squares algorithms. Furthermore, this formulation assumes that the fitting function is linear.

A preferable method for the propagation of uncertainties is to map the confidence profile, shown in Fig. 3, into a different coordinate system that contains the parameters of interest. In our example, each pair of k and *amplitude* values from the grid search is used to calculate a corresponding *area*. The confidence region of the *area* is thus bounded by the extreme values corresponding to a reasonable variance ratio. This still requires a large amount of computer time.

If, on the other hand, the confidence intervals are evaluated by our preferred method, the only points which need be mapped into the new coordinate system are the $4M$ contour points determined by the search as

---

[29] P. R. Bevington, "Data Reduction and Error Analysis for the Physical Sciences," p. 62. McGraw-Hill, New York, 1969.

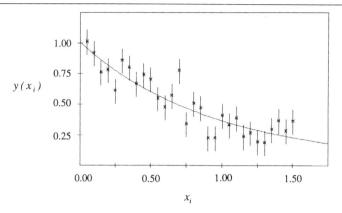

FIG. 5. Simulated experimental data used for the example analysis of a single exponential decay. See text for discussion.

defining the confidence intervals of the original fitting parameters. The example confidence interval calculated in this manner is $0.910 < area < 1.100$. This procedure is quite fast and accounts for the nonlinear behavior of the fitting equation.

Sample Implementation of Least-Squares Procedure

In the previous section we presented an example of the analysis of synthetic data in terms of a single exponential decay. In presenting that example we postponed discussing several important aspects of the least-squares methodology. In this section we discuss these in more detail.

We will continue to use the synthetic data set used in the section on confidence interval evaluations. These data are presented in Fig. 5. We have until now assumed that the value of the *baseline* was zero; this made it simpler to represent graphically the contour plot in two dimensions: k and *amplitude*. This proved to be a convenient, if not particularly realistic, simplification. In this section we discuss the more complex problem where the *baseline* value is unknown.

The first step in formulating a least-squares analysis is the choice of numerical method. The advantages and disadvantages of the various methods have been discussed in previous sections and so will not be repeated. For the particular example under study, any of the methods that we have described is satisfactory. In addition to the choice of particular method, the researcher also must decide on the numerical method for evaluating the confidence intervals for the fitted parameters. This is an important choice because the normally used method, using the variance–covariance matrix, does not always provide accurate values for the confidence intervals.

Next, the researcher must decide which of the parameters to estimate. The form of Eq. (1) suggests the obvious choices *amplitude, k,* and *base-line.* Equation (1) could, however, just as well be written in terms of a half-life instead of a rate constant. The choice of parameters is somewhat arbitrary, but in general it is best to express the fitting equation directly in terms of the parameters of interest, rather than in terms of some other intermediate parameters. For example, the *k* in our example represents an average rate constant for a chemical reaction. Alternatively, we could formulate the fitting equation in terms of the microscopic forward and reverse reaction rates of a specific molecular mechanism. Various good-ness-of-fit criteria[11] could then be used to test the validity of the particular molecular mechanism.

An additional point about the choice of fitting parameters is that care must be taken to ensure that the parameters are not totally correlated. Consider a fitting equation of the form

$$f(x_i, \mathbf{a}) = amplitude \cdot e^{-(cd)x_i} + baseline \tag{39}$$

where *c, d, amplitude,* and *baseline* are fitting parameters. In this case *c* and *d* are completely correlated. Any change in the value of *c* can be completely compensated by a change in *d*, leaving χ^2 unchanged. In this case a value for either *c* or *d* must be assumed in order for the other three parameters to be estimated.

For more complex fitting equations, the problem may not be so obvious as with *c* and *d* in Eq. (39). If the researcher were to attempt simultaneous estimation of two completely correlated fitting parameters, the matrix $\mathbf{A}^T\mathbf{A}$ would contain one row and one column, each of which is identically zero. Such a singular matrix cannot be inverted; consequently, the next estimate of the fitting parameters would be infinite. Thus, the parameter-estimation procedure can fail in dramatic fashion.

A similar problem can occur when fitting to the sum of two exponentials. In this case four fitting parameters are estimated simultaneously: two amplitudes and two rate constants. If in the iterative process one of the amplitudes becomes negligible, then its associated rate constant cannot be determined and the estimation process will fail with symptoms similar to those encountered with Eq. (39).

The choice of fitting parameters also may be influenced by a desire to impose physical constraints on the problem. If we are determining an equilibrium constant, we might wish to constrain the equilibrium constant to positive values. In determining the value of a fraction of α helix, we might wish to constrain the fraction to lie between zero and one. There are numerous methods of performing such constrained estimations of the parameters.[30] Transformation of the fitting parameters is the simplest, most easily programmed, and perhaps the best of these methods. In the transformation, only allowed values of the parameters can exist in the new coordinate system. Consider a one site ligand-binding isotherm of the form

$$B = \frac{nKX}{1 + KX} \qquad (40)$$

where B is the amount bound, X is the free ligand concentration, n is the number of binding sites, and K is the binding constant. To be physically meaningful K must be positive, but if the data were fit to Eq. (40), the analysis might yield a negative value for K. To force K to be positive we can instead fit to the variable Z that is the logarithm of K. This would, of course, require that the fitting equation be altered:

$$B = \frac{n e^Z X}{1 + e^Z X} \qquad (41)$$

The new variable Z can have any real value from minus to plus infinity, and the resulting equilibrium constant, $K = e^Z$, will always be positive. Similarly, a variable Q could be constrained to lie between A and B by a transformation to a new fitting variable Z where

$$Q = A + (B - A) \frac{e^Z}{1 + e^Z} \qquad (42)$$

The fitting parameters should be normalized so that they have similar magnitudes, both about unity. This is not an intrinsic requirement of the least-squares methods that we have outlined. Rather, it is a consequence of computers' truncation of numbers to limited precision and their having a limited dynamic range for numbers. The desired normalization can readily be accomplished by prudent choice of the parameter units. For example, when fitting to a ligand binding problem, as in Eq. (40), the choice of scales for bound and free ligand concentrations is theoretically arbitrary. However, in practice, the concentration scales should be chosen such that the values of n and K are within a few orders magnitude of each other and preferably near unity. After the parameter estimation has been performed, the units can be changed as desired.

Another important decision is the choice of independent (x axis) and dependent (y axis) variables. In general, the fitting equation can be written in several forms, each of which corresponds to a different transformation of the independent and dependent variables. An example of a transforma-

[30] Y. Bard, "Nonlinear Parameter Estimation," p. 141. Academic Press, New York, 1974.

tion that is inappropriate for our synthetic data is a log plot, that is, a plot of the logarithm of $y(x_i)$ − *baseline* as a function of x_i.

The choice of independent variable is almost arbitrary. Because we have assumed that the values of the independent variables are known to great precision, we can make virtually any transformation of the independent variables, and a corresponding transformation of the fitting equation, without diminishing the effectiveness of the least-squares methods. The only proviso is that the independent variable must not contain any appreciable experimental uncertainty. For this reason, the data must be collected in a manner which is noise-free for the independent variables, and any transformation used must not introduce experimental uncertainty into the independent variables. The Scatchard plot is a commonly used transformation that violates the requirement of no experimental uncertainty in the independent variable. In general, it is best to avoid transformations of the experimental data.

The choice of dependent variable is not arbitrary. All least-squares methods assume that the experimental uncertainties in the dependent variables are Gaussian-distributed. The dependent variables must be chosen so this assumption is valid. Almost any transformation of a dependent variable with Gaussian uncertainties will generate a new dependent variable with non-Gaussian uncertainties. For example, our synthetic data contains the appropriate Gaussian-distributed random uncertainties, but a log plot of the data does not. The log transformation of a Gaussian distribution is not a Gaussian distribution,[1] so it is not an appropriate transformation for our example. If possible, experimental data with Gaussian uncertainties should be analyzed without transformations of the dependent variables. On the other hand, it is possible that if the experimental uncertainties of the original data are not Gaussian-distributed there exists a transformation that will convert the uncertainty distribution to a form readily approximated by the Gaussian form.

There is a second reason that a log plot is not appropriate for our example. The log plot requires that we know the value of *baseline* with great precision. It is evident from a cursory examination of the data shown in Fig. 5 that the baseline is approximately zero, but with some considerable uncertainty. If the value of *baseline* is taken to be 0.2, a log plot would indicate that the rate constant k is 1.6; if the value is taken to be −0.2, k would be 0.7. It is clear from this example that the value of *baseline* must not be assumed *a priori*.

The solid line in Fig. 5 was generated from a least-squares estimation of all three parameters. The resulting optimal values based on the synthetic data set are given in Table I. The difficulty in this least-squares analysis is borne out by the large confidence intervals shown in Table I and the

TABLE I

OPTIMAL VALUES AND CONFIDENCE INTERVALS
FOR SINGLE EXPONENTIAL EXAMPLE WITH 30
DATA POINTS AND $0.0 < x_i < 1.5$

Parameter	Fitted value	Confidence interval	
Amplitude	0.969	0.832	1.093
k	1.083	0.718	1.487
Baseline	0.039	−0.120	0.215
Area	0.895	0.560	1.522

large cross-correlation coefficients in Table II. This difficulty is due to the limited range of the independent variable and the large amount of experimental uncertainty.

The asymmetry of the confidence intervals presented in Table I is of interest. This asymmetry is a consequence of the nonlinear nature of the fitting equation.

An important point must be made concerning the range of values of the independent variable. In the present example, the 30 values of x_i are equally spaced between 0.0 and 1.5. The range of x_i values was specifically limited to increase the cross-correlation between the parameters. As the range of independent variables increases, the cross-correlation between the estimated parameters decreases. The size of the joint confidence intervals also will decrease as the range of independent variables increases. Consequently, the collection of real data should span the largest range of independent variables that time and money permit.

Tables III and IV show the optimal values, confidence intervals, and cross-correlation coefficients for a repeat analysis of the synthetic data of our previous example, except that the range of the independent variable is from 0.0 to 3.0. It is evident from a comparison of Tables III and IV

TABLE II

CROSS-CORRELATION MATRIX FOR SINGLE
EXPONENTIAL EXAMPLE WITH 30 DATA POINTS
AND $0.0 < x_i < 1.5$

	Amplitude	k	Baseline
Amplitude	1.000	−0.858	−0.940
k	−0.858	1.000	0.975
Baseline	−0.940	0.975	1.000

TABLE III

OPTIMAL VALUES AND CONFIDENCE INTERVALS
FOR SINGLE EXPONENTIAL EXAMPLE WITH 30
DATA POINTS AND $0.0 < x_i < 3.0$

Parameter	Fitted value	Confidence interval	
Amplitude	1.003	0.910	1.097
k	1.120	0.858	1.378
Baseline	0.053	−0.008	0.113
Area	0.895	0.732	1.161

with Tables I and II that the confidence interval ranges and the cross-correlation between the parameters are substantially improved by increasing the range of the independent variable.

The number of data points is another important consideration. In general, the precision of the estimated parameters increases in proportion to the square root of the number of data points. Furthermore, many data points are required to apply the goodness-of-fit criterion.[11] Of course, the amount of computer time required to estimate the parameters and evaluate the associated confidence intervals is directly proportional to the number of data points. The researcher should nevertheless attempt to collect as many data points as time and money permit.

Tables V and VI show the optimal values, confidence intervals, and cross-correlation coefficients for a repeat analysis of our example with the number of data points increased from 30 to 300. The range of the independent variable is from 0.0 to 1.5. It is evident from a comparison of Tables V and VI with Tables I and II that the confidence interval range is improved by increasing the number of data points. However, simply increasing the number of data points does not significantly improve the cross-correlation coefficients.

TABLE IV

CROSS-CORRELATION MATRIX FOR SINGLE
EXPONENTIAL EXAMPLE WITH 30 DATA POINTS
AND $0.0 < x_i < 3.0$

	Amplitude	k	Baseline
Amplitude	1.000	0.215	−0.155
k	0.215	1.000	0.881
Baseline	−0.155	0.881	1.000

TABLE V

OPTIMAL VALUES AND CONFIDENCE INTERVALS
FOR SINGLE EXPONENTIAL EXAMPLE WITH 300
DATA POINTS AND $0.0 < x_i < 1.5$

Parameter	Fitted value	Confidence interval	
Amplitude	0.952	0.898	1.008
k	1.185	0.997	1.365
Baseline	0.070	−0.002	0.138
Area	0.803	0.658	1.011

Obviously, a compromise must be reached between time, money, number of data points, and range of the independent variable. For the simulations presented here it is obvious that increasing the range of the independent variable by a factor of two accomplished as much as increasing the number of data points 10-fold. In general, it is important to collect data over as wide a range as is possible and significant. One compromise is to allow unequally spaced sampling of the independent variables. There is, unfortunately, no easy method to predict the optimal sampling distribution of independent variables. Furthermore, the optimal distribution of values depends on the values and precisions of the fitted parameters. Consequently, it is impossible to predict the optimal distribution of x_i values before the experiment has actually been performed and analyzed.

For our example, if *baseline* were known and we were interested only in the value of *amplitude*, the optimal distribution of x_i values would be with all x_i values equal to zero. This degenerate distribution of independent variables would prove useless in determining the rate constant k. If, however, we wish to evaluate all three of the parameters, it is important to collect data over a wide range of values. For an exponential decay it is important to collect a reasonable number of points in the "baseline"

TABLE VI

CROSS-CORRELATION MATRIX FOR SINGLE
EXPONENTIAL EXAMPLE WITH 300 DATA POINTS
AND $0.0 < x_i < 1.5$

	Amplitude	k	Baseline
Amplitude	1.000	−0.824	−0.923
k	−0.824	1.000	0.970
Baseline	−0.923	0.970	1.000

region. A more complete discussion of the spacing of independent variables is presented in another chapter in this volume.[11]

Note that derivatives of the fitting functions with respect to the fitting parameters are required by several of the least-squares methods. In general, any method that provides the correct numerical value is acceptable. The values can be obtained by analytically differentiating the functions and then evaluating the resulting equation, or they can be obtained by numerically differentiating the original function. Both analytical and numerical derivatives are functionally equivalent for these applications. In general the analytical derivatives require less computer time to evaluate, but they require more time on the part of the programmer (and scientist) to perform the calculus, as well as to write and test the derivative-calculating code. If the purpose of the least-squares analysis is to allow the user to test many different mechanistic models (i.e., fitting functions), then it is more reasonable to evaluate the derivatives numerically. We generally use a five-point Lagrange differentiation[31]

$$\frac{\partial f}{\partial a_j} \cong \frac{f(a_j - 2\Delta) - 8f(a_j - \Delta) + 8f(a_j + \Delta) - f(a_j + 2\Delta)}{12\Delta} \quad (43)$$

where Δ is a small increment of a_j, typically chosen such that

$$|{\sim}0.001f(a_j)| < |f(a_j + \Delta) - f(a_j)| < |{\sim}0.1f(a_j)| \quad (44)$$

The error in evaluating the derivative by this method is proportional to the fourth power of Δ and to the fifth derivative of the function evaluated at ξ, where ξ is in the range $(a_j - \Delta < \xi < a_j + \Delta)$.

Conclusion

In this chapter we have considered some of the methods available for least-squares estimation of parameters from experimental data. The techniques presented were chosen because of their usefulness in analyzing various types of experimental data. In addition, there are many other methods available in the literature. For further reading, we refer the novice to books by Bevington[4] and Acton,[32] as well as to some of our earlier publications.[1,2,26] For a more complete and rigorous treatment of the subjects discussed, we recommend books by Bard[3] and Hildebrand,[31] and several of the original articles.[17,19,25] Software is available in LINPACK,[33]

[31] F. B. Hildebrand, "Introduction to Numerical Analysis," p. 82. McGraw-Hill, New York, 1956.
[32] F. S. Acton, "Analysis of Straight Line Data." Wiley, New York, 1959.
[33] J. J. Dongarra, C. B. Moler, J. R. Bunch, and G. W. Stewart, "LINPACK User's Guide." Society for Industrial and Applied Mathematics, Philadelphia, Pennsylvania, 1979.

EISPACK,[28] and in books by Bevington[4] and Forsythe *et al.*[14] Software is also available on written request from the authors.

Acknowledgments

This work was supported in part by National Institutes of Health Grants GM-28928 and DK-38942, National Science Foundation Grant DIR-8920162, National Science Foundation Science and Technology Center for Biological Timing, the Diabetes Endocrinology Research Center of the University of Virginia, and the Biodynamics Institute of the University of Virginia.

[2] Global Analysis of Biochemical and Biophysical Data

By Joseph M. Beechem

Introduction

The goal of experimental science is to relate laboratory observations (i.e., data) into biological information (e.g., rate constants, binding constants, and diffusion coefficients). Regretfully, the laboratory techniques utilized in biochemistry and biophysics very rarely *directly* yield useful biological information. Instead, the investigator must "analyze" the experimental observables in such a manner as to extract biological information from the data.

When first approaching a data analysis problem, one has to determine what general class of computer algorithms should be applied to the data. The relationship between experimental observables and biological information in almost all of biochemistry and biophysics is nonlinear. Therefore, one needs to determine what is the best methodology to fit a nonlinear model to experimental data. While there is still considerable debate over the best type of fitting methodology, for all practical purposes this is a nonquestion. For greater than 90% of all of the data collected in the fields of biochemistry and biophysics, the proper technique to analyze the data is nonlinear least-squares. No attempt will be made to justify the preceding statement; it has simply been arrived at from personal experience with a variety of data analysis projects and by communications with other biophysical data analysis groups. A very thorough description of the assumptions inherent in applying nonlinear least-squares techniques is given by Johnson and Frasier along with a general review.[1]

---

[1] M. L. Johnson and S. G. Frasier, this series, Vol. 117, p. 301.

Certainly the wrong place to start data analysis is by attempting to transform data in a manner such that a nonlinear model is transformed into a linear model. There are many transforms of this type which have been utilized within the field of biochemistry, including Scatchard-type or Eadie–Hofstee transformations, Lineweaver–Burke double-reciprocal plots, and logarithmic transforms. Although it is still common practice in many biochemical studies to determine binding constants, kinetic rates, etc., by using graphical transforms, these methodologies should be avoided if at all possible. If there is one general rule concerning the analysis of data which should always be remembered, it is that data should not be transformed in any manner (besides trivial linear operations; multiply/ divide or add a constant value) prior to performing a data analysis. The graphical transformation analysis methodologies in common use in many biochemistry laboratories were developed prior to the existence of general purpose laboratory computers and good nonlinear analysis software. Within the current laboratory environment, these conditions no longer exist, and these methodologies should definitely *not* be considered viable analysis techniques. The role of the classic biochemical data transformations should be exclusively for graphical representation of the data. All of the data analysis should be performed in a nonlinear fashion on the actual (raw) data.

The nonlinear data analysis requirements of different laboratories in the field of biochemistry and biophysics vary greatly. For the laboratory which occasionally needs to fit small numbers of data (say 10–500 data points) to a model which can be easily described analytically (e.g., sums of exponentials, simple binding isotherms), prepackaged nonlinear data analysis programs will probably be sufficient. An example of a general purpose software of this type would be Sigmaplot 4.0 (Jandel Scientific, Corte Madera, CA). This data analysis and graphics package is very useful for "on-the-fly" fitting. Within this software, one can specify models which contain up to 25 fitting parameters with as many as 10 independent variables. These types of data analysis packages are very useful for quick initial data analyses. However, if your laboratory routinely analyzes larger numbers of data points and/or multiple experiments with more complicated sets of fitting functions, it will be extremely advantageous to obtain a nonlinear data analysis program which can be specifically modified and optimized for your particular experimental configuration.

Global Analysis Approach

To the average biochemist and biophysicist, the field of data analysis, and its associated language and methodologies, can appear so complex that one simply relegates data analysis to a "specialist," or utilizes prepackaged software to analyze the data. The main problem with this approach is the fact that data analysis specialists and prepackaged software companies often have no real concept of the science behind the experiments. The major emphasis of this chapter is to reveal that one can often effect a major transformation of the amount of biological information which can be obtained from a set of experiments by direct incorporation of "scientific constraints" into the analysis of the data.

The term scientific constraints is meant to denote direct application (into the analysis program) of the mathematical constraints imposed on the data through experimental design and accessory information. For instance, for many classes of data, one is often interested in recovering closely spaced lifetimes (τ) from a sums-of-exponentials model:

$$F(t) = \sum_{j=1}^{n} \alpha_i e^{-t/\tau_i} \tag{1}$$

If one consults data analysts (or mathematicians) and describes that one needs to recover multiple lifetimes that differ by as little as 5–10% from realistic data with noise, they will simply say that it is impossible. However, what they do not realize is that, experimentally, one may be able to manipulate the system under investigation such that very closely spaced exponentials *can* be resolved.

As an example of multiple experiments constraining the data analysis, in the field of time-resolved fluorescence spectroscopy, multiple data sets can be obtained which all have a common set of lifetimes but differing preexponential terms. These data are obtained simply by collecting individual kinetics measurements at multiple emission wavelengths. If one performs these multiple experiments, and analyzes all of the various data sets as if they were independent of one another, then the multiple closely spaced relaxation times cannot be recovered. However, if one performs a data analysis of all of the experiments simultaneously, applying the scientific constraint that all of the data sets have internally consistent sets of relaxation times (with varying amplitude terms), then the correct relaxation times can be recovered.

It is very important to understand why the scientific constraint needs to be applied directly within the data analysis program and not *a posteriori*. First and foremost is the realization that all of the information content of an experiment is contained in the "error surface" of a data set. The error surface of an experiment is represented as a plot of the χ^2 statistic along the z axis versus all of the fitting parameters along the other axes. The χ^2 statistic [Eq. (2)] is a single-valued function which approaches unity for properly weighted data and good fits and grows larger as the data fit worsens:

$$\chi^2 = \sum_{k=1}^{nexps} \sum_{i=1}^{ndata(k)} \frac{[data(i,k) - model(i,k)]^2}{\sigma^2(i,k)(N - m - 1)} \qquad (2)$$

The experimental data and the proposed fit to the data are denoted by *data* and *model*, respectively. N is the number of data points in the data surface, $\sigma(i,k)$ is the standard deviation in the ith data point in the kth experiment, m is the number of total fitting parameters, *nexps* is the total number of experiments in the data surface, and *ndata* is the number of data points in the kth experiment.

If one examines the error surface for a particular experiment and discovers that it is very "flat" (i.e., the χ^2 statistic does not significantly change as the values of the fitting parameters are altered), then the experiment will be unable to resolve the fitting parameters of the model. In other words, the link between experimental data and biological information is very weak. In Fig. 1A,B is a plot of a typical error surface of a double-exponential model, $0.25 \exp(-t/4) + 0.75 \exp(-t/5)$, applied to a high signal-to-noise ratio data set (~140 : 1) consisting of 1000 time points over a time range of 5 times the mean lifetime. Because a double-exponential decay is specified with four parameters, the actual error surface is five dimensional (four fitting parameters and the χ^2 statistic). Because it is very difficult to visualize a five-dimensional surface, what is plotted in this graph are the two lifetime parameters versus the χ^2 statistic. The other two dimensions, however, are not ignored. At each point on this graph an entire nonlinear data analysis was performed with the amplitude terms allowed to adjust to yield the minimum possible χ^2 value. Therefore, this surface represents the "best fit" error surface for the two lifetime parameters.

An important feature to notice in Fig. 1A,B is the apparent discontinuity running diagonally across the error surface. This discontinuity is actually the $\tau_1 = \tau_2$ plane, and therefore represents the best possible monoexponential fit to the double-exponential data. Figure 1A is scaled in units appropriate for the monoexponential decay analysis. One can see immediately how well defined the single-exponential data surface is compared to the double-exponential case. Figure 1A illustrates how dramatically error surfaces can change as one adds additional fitting parameters into the analysis. There is a mirror plane of symmetry in this error surface, because the model function is symmetric in the lifetimes.

The double-exponential error surface (see Fig. 1B) is very ill-defined, as evidenced by the large "covariance valleys" where both lifetimes can vary over quite a large region with very little χ^2 penalty. The horseshoe-shaped contour along the lower surface represents an approximate 67% confidence level for resolving the two lifetimes. Note that along this sur-

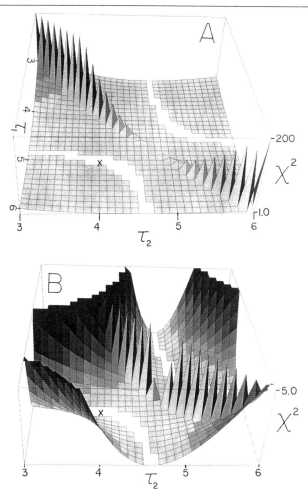

FIG. 1. χ^2 error surfaces generated from the analysis a biexponential decay of $0.25 \exp(-t/4) + 0.75 \exp(-t/5)$. The white horseshoe-shaped region represents the 67% confidence region for the recovery of the two lifetime parameters. X marks the actual solution space desired. The surface has a 2-fold mirror symmetry reflecting the model fitting function. (A) Single-experiment analysis of biexponential decay scaled to reveal the dramatic change in error surface which occurs on going from single-exponential analysis to a double-exponential fit. The ridge along the diagonal represents the intersection of the single-exponential error surface with the double-exponential error surface. The single-exponential analysis would involve moving from spike to spike along an approximate parabola to the minimum at the weighted average lifetime of 4.85. (B) Same as (A), except with a scaling appropriate for double-exponential analysis. (C) Error surface for the global simultaneous analysis of the experiment used in (A) and (B) with an additional experiment of $0.75 \exp(-t/4) + 0.25 \exp(-t/5)$. Note the drastic shrinking of the 67% confidence level around the actual solution.

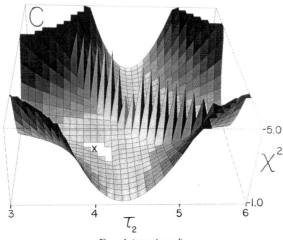

FIG. 1 (*continued*)

face the relaxation times can vary from less than 3 to greater than 6 without a statistically significant increase in χ^2 [traveling along a single horseshoe at the minimum, one proceeds from the pair ($\tau_1 = 4.8$, $\tau_2 = 3.0$) to ($\tau_1 = 6.0$, $\tau_2 = 4.7$)]. The error surface is independent of the type of nonlinear least-squares methodology utilized to analyze the data. Hence, there is absolutely *nothing substantial* to gain by applying different types of data analysis routines on this data set. The 67% confidence region spans the entire surface, and analysis routines cannot change that! Only through additional experimentation can one expect to better resolve these two relaxation times.

Granted, this is not exactly what the experimentalist wants to hear. However, this is no reason to give up on the problem. Instead, this is just where the problem begins to get interesting. The next question to ask, therefore, is, What can be done to improve the data? The answer to this question takes two distinct pathways: (1) increase the signal-to-noise ratio of the experiment currently examined; (2) combine together multiple experiments which will better resolve the model. Although both pathways are extremely important, the basic emphasis of global analysis techniques is placed on combining together multiple experiments to better determine model fitting parameters.

A direct comparison of the increase in information content of a given set of experiments on negotiating pathways 1 and 2 (above) will now be described. Figure 1C reveals how the error surface of Fig. 1A,B can be transformed by combining two experiments together where the weighting factors for the two relaxation times are inverted [data set #1 = 0.25

$\exp(-t/4) + 0.75 \exp(-t/5)$; data set #2 = 0.75 $\exp(-t/4) + 0.25$ $\exp(-t/5)$]. The single lifetime plane is now clearly separated from the double-exponential error surface, and, in addition, one can see how dramatically the 67% confidence region has shrunk around the actual solution.

These high dimensional data surfaces can be reduced to only two dimensions and examined in more detail in the following manner. To examine the errors associated with any given fitting parameter (say the ith one), one can perform a series of nonlinear analyses, systematically altering the ith parameter over any range. *All* other fitting parameters are allowed to adjust so as to obtain the minimum possible χ^2 value. At the end of the minimization, the value of the fixed parameter and the value of the χ^2 obtained for this fit is stored. By plotting the fixed parameter value versus the minimum χ^2, one obtains a rigorous confidence interval on the ith parameter. Plots of this type for the double-exponential analysis of the single- and the global double-exponential analysis of the two-data set experiment described above are shown in Figs. 2A and 3. Note that these graphs are actually replots of the original error surfaces shown in Fig. 1. By forcing the program to proceed as described above, one is actually traveling along one of the horseshoe paths and recording the χ^2 values on proceeding. There is a χ^2 penalty effected on "traveling over" the single-exponential ridge in these surfaces, and this shows up as a spike in the rigorous error analysis graphs. These rigorous error analysis graphs represent the "most truthful" representation of errors in the recovered parameters that are possible.

Figure 2A shows how ill-resolved the two-exponential components are from single-experiment analysis. The horizontal line represents a statistically significant increase in the χ^2 at the 67% confidence level. Compare this ill-defined minima with the well-defined minima from the global analysis resolution of the two exponentials (Fig. 3). Note that now these two relaxation times are rigorously recovered at both the 67% and 98% confidence levels. Also included in Fig. 2B,C are the results of simply increasing the signal-to-noise ratio of the single experiment. The increased signal-to-noise experiments are simulated for photon (or particle) counting data with 4.5, 18, and 45 million events detected, respectively. Note that even collecting counting data for a factor of 10 longer fails to resolve the shorter lifetime, and barely resolves the longer component at the 67% level. The 98% confidence levels are still not even close to being obtained. Just as important, comparing the three parts of Fig. 2 reveals that the increased signal-to-noise experiments are not really changing the overall character of the error surfaces, compared to the dramatic change on performing a global analysis (Fig. 3).

The above results often hold in general. Increasing the signal-to-noise

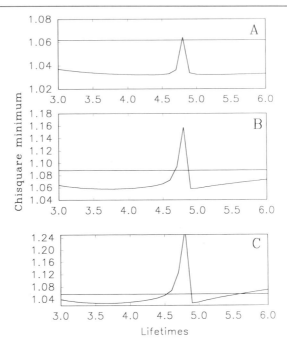

FIG. 2. Rigorous confidence intervals around the solution of the biexponential with lifetimes of 4 and 5 as described in the text. The horizontal line represents a statistically significant increase in χ^2 at an approximate 67% confidence level using the F-statistic. (A) Confidence region for data set containing 4.5×10^6 events. (B) 18×10^6 events. (C) 45×10^6 events.

ratio of a single experiment only results in very gradual changes in the error surface, once the signal-to-noise ratio reaches a particular value. However, by combining multiple experiments together in a single analysis, much more dramatic changes in the error surface can be obtained. It is a very useful exercise for a laboratory to perform the above type of error-surface mapping and rigorous error analysis on a typical data set to establish the overall nature and characteristics of the error surface. Examination of how this error surface is altered on performing a hypothetical set of multiple experiments can greatly assist in deciding what combination of experiments yield better defined error surfaces. In this respect, the global analysis tools can be utilized for experimental design as well as data analysis.

Building Global Nonlinear Least-Squares Analysis Packages

There are certainly a wide variety of very good, basic nonlinear least-squares analysis routines in the literature. Probably the single most im-

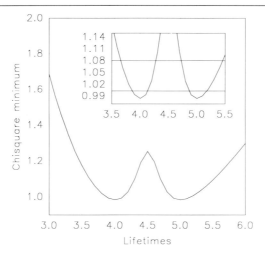

FIG. 3. Rigorous confidence intervals around the solution of the global analysis solution of the double exponential of Fig. 1C. Horizontal lines represent a statistically significant increase in χ^2 at approximate 67% (lower horizontal line) and 98% (higher horizontal line) levels.

portant introductory text concerning data analysis in the physical sciences is the classic work of P. R. Bevington.[2] This text both is a very good introduction to data analysis and contains a wide variety of very useful FORTRAN programs for nonlinear least-squares curve fitting. The subroutine CURFIT provides a very good Marquardt–Levenberg type nonlinear least-squares routine whose FORTRAN code can be entered, compiled, and made running in a single afternoon (only about 100 lines of code). As it stands, however, this code is not really designed for the simultaneous analysis of multiple experiments. To convert this code to performing global-type analyses a few minor modifications are needed.

All Marquardt–Levenberg type nonlinear least-squares procedures determine "how-to-move-downhill" by the following method. Given an initial set of guesses for the fitting parameters in a vector $\mathbf{P}^n$ (the superscript denotes the current iteration number), one obtains $\mathbf{P}^{n+1}$ by forming

$$\mathbf{P}^{n+1} = \mathbf{P}^n + \delta \tag{3}$$

The parameter improvement vector (δ) is determined from the shape of the χ^2 hypersurface around the current fitting parameters by solving the linearized system of equations:

[2] P. R. Bevington, "Data Reduction and Error Analysis for the Physical Sciences." McGraw-Hill, New York, 1969.

$$\alpha\delta = \beta \tag{4}$$

where

$$\alpha_{jk} = \sum_{i=1}^{ndata} \frac{1}{\sigma(i)^2} \frac{\partial model(i)}{\partial \mathbf{P}_j} \frac{\partial model(i)}{\partial \mathbf{P}_k} + \lambda I \tag{5}$$

$$\beta_k = \sum_{i=1}^{ndata} \frac{[data(i) - model(i)]^2}{\sigma(i)^2} \frac{\partial model(i)}{\partial \mathbf{P}_k} \tag{6}$$

and δ_k is the proposed change for parameter k, λ is the Marquardt scaling factor, and I is the identity matrix.

Although these equations look rather formidable, they are certainly very easy to program and calculate from the data. A couple of items are worth noting from these equations. Note that the only place that the actual data enter these equations is through the vector β. In this vector, the differences between the data and fit are multiplied by the derivatives of the model with respect to the various fitting parameters. This operation can be visualized as the inner product of the residual vector with the vector of the sensitivity of the model to each particular fitting parameter. This means that the kth element of β is large only when there is both a mismatch in the data and when moving the kth parameter actually effects the fit in this region. This process is exactly what one would intuitively perform, if data analysis were being performed manually. The α matrix provides a fine-tuning of the projected parameter increments by taking into consideration the correlation between the individual parameters.

The Marquardt scaling factor (λ) is simply a numerical trick to allow the program to decide which is most important, β or the cross-correlation terms. λ is often initially given the value of 0.01 at the start of a fit and is then decreased (or increased) during the course of analysis as dictated by the success (or failure) to find a decrease in χ^2. When λ is large, the "brute force" β dominates the solution (the so-called gradient search mode), whereas closer to the solution the cross-correlation terms become more important, λ is decreased, and α is utilized to a greater extent. This process can be visualized by referring back to any of the error surface plots (e.g., Fig. 1B). Imagine that a blind person was instructed to rush to the bottom of this error surface. Initial guesses of $\tau_1 = 3.3$, $\tau_2 = 5.7$ will place this person most of the way up the hill in the far right of Fig. 1B. Initially, the surface is very steep, and it is very easy to determine how to proceed down. The person can just extend the arms in two directions (say along the parameter axes) and utilize this (rather limited) information to immediately proceed downhill. This search mode corresponds to the β-dominated solution with relatively substantial λ. However, as the person begins to approach the valley, the surface becomes more flat, and simply extending the arms in two directions does not yield enough information to proceed downhill very swiftly. To compensate for this, the person can extend the arms in a circular region centered around the current location, to better determine the direction downhill. This mode of operation corresponds to the small λ case of Marquardt fitting, and it becomes very useful for the program as it approaches the minimum. One of the main reasons that the Marquardt–Levenberg algorithm has proved so successful is because of the seamless transition that it is able to make on navigation of the complex topology associated with error surfaces by simply scaling up and down λ.

For global types of analyses, one wants to analyze simultaneously multiple experiments in terms of internally consistent sets of fitting parameters. The most frequently asked question concerning this type of analysis is, "How do I link the fitting parameters between the two data sets?" Luckily, the underlying structure of the nonlinear least-squares routines are almost infinitely flexible in this regard: any set of linkages (either by directly linking parameters or by linking parameters through complicated functions) can be accomplished in a very simple manner. Consider the above-described global analysis of the two sets of experiments defined by the following biexponential decay:

$$\text{Data set \#1} = amp_{11} \exp(-t/\tau_{11}) + amp_{21} \exp(-t/\tau_{21}) \qquad (7)$$
$$\text{Data set \#2} = amp_{12} \exp(-t/\tau_{12}) + amp_{22} \exp(-t/\tau_{22}) \qquad (8)$$

where an extra index has been given for the amplitudes (amp) and lifetime terms (read the subscripts as lifetime number, experiment number). Now, consider that one wants to link the two relaxation times between the two experiments. One can proceed as follows. The fitting parameter vector $\mathbf{P}$ [see Eq. (3)] will have the following form:

$$\mathbf{P}^{\text{T}} = [amp_{11}, \tau_{11}, amp_{21}, \tau_{21}, amp_{12}, amp_{22}] \qquad (9)$$

where the superscript T denotes transpose. Note that the τ_{12} and τ_{22} terms do not appear in the fitting vector because the lifetime terms from the first experiment will be used for both data sets. Therefore, the fitting parameter space vector for the two experiments if analyzed independently would consist of eight total terms, and the global analysis fitting vector for the two experiments will only contain six total elements.

In performing the global analysis, one proceeds to loop over the two experiments, filling up the elements of the nonlinear least-squares equations [Eqs. (5) and (6)]. For the first experiment, the α matrix elements can be symbolically represented as follows (α is symmetric, and only the lower triangular is shown):

$$\alpha(\text{data set \#1}) = \begin{bmatrix} \dfrac{\partial}{\partial amp_{11}}\dfrac{\partial}{\partial amp_{11}} & \cdot & & \cdot & & \cdot \\[2ex] \dfrac{\partial}{\partial amp_{11}}\dfrac{\partial}{\partial \tau_{11}} & \dfrac{\partial}{\partial \tau_{11}}\dfrac{\partial}{\partial \tau_{11}} & & \cdot & & \cdot \\[2ex] \dfrac{\partial}{\partial amp_{11}}\dfrac{\partial}{\partial amp_{21}} & \dfrac{\partial}{\partial \tau_{11}}\dfrac{\partial}{\partial amp_{21}} & \dfrac{\partial}{\partial amp_{21}}\dfrac{\partial}{\partial amp_{21}} & & \cdot \\[2ex] \dfrac{\partial}{\partial amp_{11}}\dfrac{\partial}{\partial \tau_{21}} & \dfrac{\partial}{\partial \tau_{11}}\dfrac{\partial}{\partial \tau_{21}} & \dfrac{\partial}{\partial amp_{21}}\dfrac{\partial}{\partial \tau_{21}} & \dfrac{\partial}{\partial \tau_{21}}\dfrac{\partial}{\partial \tau_{21}} \end{bmatrix} \quad (10)$$

A symbolic notation will be used in the following matrix descriptions, whereby all partial derivatives are with respect to the model function and the sum over the i channels of data and weighting terms are omitted for clarity. This notation is intended to emphasize the parameter relationships inherent in these matrices, of which the global analysis methodology will ultimately utilize.

Therefore, for the first experiment, no changes to a standard single curve nonlinear analysis have to be made. All matrix elements as specified in Eqs. (5) and (6) are accumulated normally. However, unlike single experiment analysis, one does not solve these equations to obtain parameter increments. Instead, one saves these matrix elements and begins another program loop over experimental data set #2, assembling another α matrix and β vector, specific for this experiment. The α matrix elements for experiment 2 can be symbolically represented as follows:

$$\alpha(\text{data set \#2}) = \begin{bmatrix} \dfrac{\partial}{\partial amp_{12}}\dfrac{\partial}{\partial amp_{12}} & \cdot & & \cdot & & \cdot \\[2ex] \dfrac{\partial}{\partial amp_{12}}\dfrac{\partial}{\partial \tau_{12}} & \dfrac{\partial}{\partial \tau_{12}}\dfrac{\partial}{\partial \tau_{12}} & & \cdot & & \cdot \\[2ex] \dfrac{\partial}{\partial amp_{11}}\dfrac{\partial}{\partial amp_{22}} & \dfrac{\partial}{\partial \tau_{12}}\dfrac{\partial}{\partial amp_{22}} & \dfrac{\partial}{\partial amp_{22}}\dfrac{\partial}{\partial amp_{22}} & & \cdot \\[2ex] \dfrac{\partial}{\partial amp_{12}}\dfrac{\partial}{\partial \tau_{22}} & \dfrac{\partial}{\partial \tau_{12}}\dfrac{\partial}{\partial \tau_{22}} & \dfrac{\partial}{\partial amp_{22}}\dfrac{\partial}{\partial \tau_{22}} & \dfrac{\partial}{\partial \tau_{22}}\dfrac{\partial}{\partial \tau_{22}} \end{bmatrix} \quad (11)$$

Now, important to note are the elements that are common among the matrices shown in Eqs. (10) and (11). Because one is linking τ_{11} with τ_{12} and also τ_{21} with τ_{22}, the corresponding matrix elements associated with these two parameters represent derivatives with respect to just two fitting parameters. To effect the parameter linkage, one wants to combine (i.e., add together) those matrix elements which are common to both experiments before solving Eq. (4). Therefore, after forming these "local" individual experiment matrix elements (just as in regular nonlinear data analysis), one combines all of the local matrices together, adding together those elements that contain information linked between the individual experiments. The final "global" α matrix (with implemented parameter linkages) would thus appear as follows:

$$\alpha = \begin{bmatrix} \dfrac{\partial}{\partial amp_{11}}\dfrac{\partial}{\partial amp_{11}} & \cdot & \cdot & \cdot & \cdot & \cdot \\[1em] \dfrac{\partial}{\partial amp_{11}}\dfrac{\partial}{\partial \tau_{11}} & \dfrac{\partial}{\partial \tau_{11}}\dfrac{\partial}{\partial \tau_{11}}+\dfrac{\partial}{\partial \tau_{12}}\dfrac{\partial}{\partial \tau_{12}} & \cdot & \cdot & \cdot & \cdot \\[1em] \dfrac{\partial}{\partial amp_{11}}\dfrac{\partial}{\partial amp_{21}} & \dfrac{\partial}{\partial \tau_{11}}\dfrac{\partial}{\partial amp_{21}} & \dfrac{\partial}{\partial amp_{21}}\dfrac{\partial}{\partial amp_{21}} & \cdot & \cdot & \cdot \\[1em] \dfrac{\partial}{\partial amp_{11}}\dfrac{\partial}{\partial \tau_{21}} & \dfrac{\partial}{\partial \tau_{11}}\dfrac{\partial}{\partial \tau_{21}}+\dfrac{\partial}{\partial \tau_{12}}\dfrac{\partial}{\partial \tau_{22}} & \dfrac{\partial}{\partial amp_{21}}\dfrac{\partial}{\partial \tau_{21}} & \dfrac{\partial}{\partial \tau_{21}}\dfrac{\partial}{\partial \tau_{21}}+\dfrac{\partial}{\partial \tau_{22}}\dfrac{\partial}{\partial \tau_{22}} & \cdot & \cdot \\[1em] 0 & \dfrac{\partial}{\partial \tau_{12}}\dfrac{\partial}{\partial amp_{12}} & 0 & \dfrac{\partial}{\partial \tau_{22}}\dfrac{\partial}{\partial amp_{12}} & \dfrac{\partial}{\partial amp_{12}}\dfrac{\partial}{\partial amp_{12}} & \cdot \\[1em] 0 & \dfrac{\partial}{\partial \tau_{21}}\dfrac{\partial}{\partial amp_{22}} & 0 & \dfrac{\partial}{\partial \tau_{12}}\dfrac{\partial}{\partial amp_{22}} & \dfrac{\partial}{\partial amp_{22}}\dfrac{\partial}{\partial amp_{22}} & \dfrac{\partial}{\partial amp_{22}}\dfrac{\partial}{\partial amp_{22}} \end{bmatrix}$$

(12)

$$\beta = \begin{bmatrix} [data(i,1)-model(i,1)]^2 \dfrac{\partial}{\partial amp_{11}} \\[1.5em] [data(i,1)-model(i,1)]^2 \dfrac{\partial}{\partial \tau_{11}} + [data(i,2)-model(i,2)]^2 \dfrac{\partial}{\partial \tau_{12}} \\[1.5em] [data(i,1)-model(i,1)]^2 \dfrac{\partial}{\partial amp_{21}} \\[1.5em] [data(i,1)-model(i,1)]^2 \dfrac{\partial}{\partial \tau_{21}} + [data(i,2)-model(i,2)]^2 \dfrac{\partial}{\partial \tau_{22}} \\[1.5em] [data(i,2)-model(i,2)]^2 \dfrac{\partial}{\partial amp_{12}} \\[1.5em] [data(i,2)-model(i,2)]^2 \dfrac{\partial}{\partial amp_{22}} \end{bmatrix}$$

(13)

Note the terms shown in matrices (12) and (13) which have contributions from both data sets. One can see that within β, the residuals from both experiments "feed into" elements 2 and 4 weighted by the derivatives with respect to the linked lifetimes. In the α matrix, elements α_{22}, α_{44}, and α_{42} all contain summations over both experiments. This is the way that individual experiments talk to each other! Keeping in mind that the linking together of multiple experiments requires a *model-dependent* summation, the matrices defined in Eqs. (5) and (6) can be rewritten:

$$\alpha_{jk} = \sum_{k=1}^{nexps} \sum_{i=1}^{ndata(k)} \frac{1}{\sigma(i,k)^2} \frac{\partial model(i,k)}{\partial \mathbf{P}_j} \frac{\partial model(i,k)}{\partial \mathbf{P}_k} + \lambda I \qquad (14)$$

$$\beta_k = \sum_{k=1}^{nexps} \sum_{i=1}^{ndata(k)} \frac{[data(i,k)-model(i,k)]^2}{\sigma(i,k)^2} \frac{\partial model(i,k)}{\partial \mathbf{P}_k} \qquad (15)$$

always recognizing that the outer summation is dependent on the model applied to link the data sets together.

Some nonlinear data analysis routines have been designed to take advantage of specific linkage structures between multiple experiments.

For instance, in the case presented above, where all of the relaxation times are linked, this problem can be separated into linear and nonlinear subproblems, and experiments can very efficiently be combined using a variable projection algorithm,[3,4] or through singular value decomposition (SVD) type analysis (see E. R. Henry and J. Hofrichter [8], this volume). However, inherent in these specialized algorithms are linkages between experiments which follow a highly specific pattern. The methodology described above allows the linkage of parameters (or functions of parameters) for *all* possible cases.

More complicated linkage patterns, where parameters are linked through functions, can be implemented as above with no changes. For instance, say that the relaxation time $\tau_{22} = f(\tau_{21}, [Q])$, where f is any possible function which maps the relaxation time of experiment 1 to a new relaxation time in experiment 2. One proceeds in exactly the same manner as above. However, the derivative terms must be handled slightly differently. If one is using analytical derivatives, to obtain the elements of the matrices (14) and (15), the derivative terms with respect to τ_{21} will have to be operated on using the chain rule for partial derivatives, before these elements are added together. The application of chain-rule derivatives can be an easy source of error and will result is very slow χ^2 minimization. Much easier is simply to utilize numerical derivatives. In this case, the derivative of τ_{21} in experiment 2 is obtained as:

$$\frac{\partial model(i,2)}{\partial \tau_2} \cong \frac{model[f(\tau_{21} + \Delta\tau_{21})] - model[f(\tau_{21} - \Delta\tau_{21})]}{2\Delta\tau_{21}} \quad (16)$$

In other words, one alters τ_{21} by Δ, operates on this value with f to generate the predicted value of τ_{22}. Use this τ_{22} to generate the observable for experiment 2. Subtract Δ and repeat the process, then divide through by the parameter increment. By mapping the derivatives through the function linking the experiments together, all of the proper chain rules are automatically applied.

To perform this model-dependent summation of matrix elements three major types of algorithms can be written, dependent on the type of global analysis program one wishes to create. If the number and type of experiments which are going to be examined are relatively limited, one can directly hardwire specific linkages into the software. For instance, given the case above, one could automatically sum the matrix elements shown in Eqs. (12) and (13) and have a program which would always link double-exponential lifetimes with varying preexponential components. If this is

[3] G. H. Golub and V. Pereyra, *SIAM J. Numer. Anal.* **10**, 413 (1973).
[4] L. Kaufman, *BIT* **15**, 49 (1975).

the only type of analysis that would ever need to be performed, then there is no reason to expend the effort required for a more general purpose global analysis program. The advantage of this type of hardwiring is that the program size and speed will be optimized. The major disadvantage is the lack of flexibility.

For added flexibility, one can design the algorithm such that each particular model is entered as a specific subroutine. Within this subroutine, the logic for each specific linkage type is programmed. For instance, subroutines could be written to perform simple lifetime linking, as well as linking through specific functions of the lifetimes. Prior to performing an analysis, the user would decide on which specific model was going to be applied to the data, and the program would be recompiled with the specific model-dependent subroutines inserted. The advantage of this approach is that one can maintain essentially a single global analysis KERNEL set of routines, which can be loaded with many different "specific case" type subroutines for different models. The major disadvantage is that the user is required to be capable of programming and to have a reasonably good understanding of the logic required to link the experiments together.

For the largest degree of flexibility, a general purpose indirect addressing linking methodology can be employed. With this technology, each access to any fitting parameter within the program occurs through a series of indirect pointers which are "decoded" during the data analysis. By decoding it is meant that each fitting parameter has associated with it two indices, i1 and i2. i1 provides an index into a table of functions, which operate on the fitting parameter in order to "map it" from one experiment to the next. i2 represents an index which indicates if this fitting parameter is utilized in this particular experiment. The logic required to link the experiments together can therefore be specified by the user "on the fly" while performing the data analysis by altering the index values i1 and i2. In this manner, alternative modeling at both the fitting function and parameter linkage levels can be performed without the need to have multiple programs or any recompilation steps. As a program of this type "evolves," each new fitting function utilized extends the capabilities of the program until, eventually, the program becomes an "expert system" type of data analysis tool. A detailed description of a global analysis program of this type can be found in Beechem et al.[5]

For global types of analyses, often large numbers of data sets are simultaneously analyzed. The total number of fitting parameters over

[5] J. M. Beechem, E. Gratton, M. Ameloot, J. Knutson, and L. Brand, in "Fluorescence Spectroscopy: Principles" (J. R. Lakowicz, ed.), Vol. 2, Chap. 5. Plenum, New York, 1991.

the entire data surface may become quite large (often in the hundreds). Therefore, in the solution of the nonlinear least-squares normal equations [Eq. (4)], matrix dimensions greater than 10 are often encountered. The matrix inversion technique utilized in Bevington (MATINV)[2] becomes extremely unstable for these types of matrices and should be replaced by a more robust methodology. Utilization of the square-root method[6] can greatly decrease numerical instabilities, as pointed out by Johnson and Frasier,[1] and is absolutely essential when fitting for more than approximately 5–7 parameters. Singular value decomposition (SVD) routines (e.g., SVDRS[7]) also work very well for large matrices. The SVD routines also have the advantage that when the fitting matrix becomes singular, the program can still solve the system of equations by automatically fixing the least determined set of fitting parameters (all terms with singular values below a user-definable value are not altered). If these singular values increase again during the analysis, the fixed terms will immediately become free again to assume new values.

Conclusion

Global analysis of data represents the simultaneous analysis of multiple experiments in terms of internally consistent sets of fitting parameters. This type of analysis has been previously applied in many different fields and by many different groups, and no attempt has been made here to review all of the historical developments. Instead, what has been presented is a motivation for performing global types of analyses and the importance of rigorous error analysis.

So often, in reviews of nonlinear data analysis, all that is described are the various methodologies for performing the nonlinear minimization (i.e., "how to go downhill") and how to approximate the errors associated with the recovered fitting parameters. In my opinion, with the advent of fast laboratory computers and very good general purpose nonlinear analysis software these have both become nonquestions. Choosing any of the current Marquardt–Levenberg or modified Gauss–Newton type minimization procedures, "how to go downhill" is almost invariably *not* the problem. Instead, the predominant analysis problems in biochemistry and biophysics are ill-defined error surfaces (multiminima error surface analysis techniques are beyond the scope of this review). Emphasis has been

[6] D. K. Faddeev and V. N. Faddeeva, "Computational Methods of Linear Algebra." Freeman, San Francisco, California, 1963.
[7] C. L. Lawson and R. J. Hanson, "Solving Least Squares Problems." Prentice-Hall, Englewood Cliffs, New Jersey, 1974.

placed on directly combining multiple experiments in a single analysis. In many examples, this type of analysis has been shown to alter drastically the overall shape characteristics of error surfaces.

Implementation of global analysis routines involves rather simple modifications of existing nonlinear least-squares packages. The main change in the algorithm is providing an additional step whereby model-dependent summation of the normal nonlinear least-squares equations can be performed. The methodology utilized to solve the linearized equations should be updated from numerical inverses to either system solvers (e.g., the square-root method) or generalized inverses (e.g., SVD).

Error analysis methodologies can be broadly classified in terms of the approximations that are required. These approximations range from the extreme (no correlation between the parameters) to allowing all correlations between parameters. Historically, when laboratory computers were not very fast, some type of approximations had to be performed to obtain error estimates. However, it is not apparent that this is any longer the case. Very seldom does the researcher require rigorous error estimates on all of the fitting parameters. Instead, some subset of the total number of fitting parameters actually become published data (in tables, etc.). It is with these "publishable" parameters that rigorous error analysis should be performed. By following the error analysis procedure described in this chapter, *no* approximations to the error in the recovered parameters need to be made. The only approximation involved is deciding what level is a statistically significant increase in the χ^2 term (i.e., where to put the horizontal line across in Figs. 2 and 3). Utilization of the F-statistic, as first proposed by Box[8] and Beale[9] appears to be a satisfactory and internally consistent methodology. Publishing the actual confidence curvature (as in Figs. 2 and 3) would allow the reader to have a very accurate picture of the uncertainties in the published values. Also, if one utilizes a relatively flexible global analysis program, there will never be any need to perform approximate error propagation methods, which attempt to map the recovered uncertainty in a fitting parameter to some final derived result. One simply *always* performs data analysis using as fitting parameters the final terms that are of interest in the study.

Postscript

This chapter has provided a "cookbook" approach to developing (and using) nonlinear least-squares analysis programs. As a cookbook (and not a review article), no attempt has been made to reference any of the incredibly

[8] G. E. P. Box, *Ann. N.Y. Acad. Sci.* **86,** 792 (1960).
[9] E. M. L. Beale, *J. R. Stat. Soc. Ser. B* **22,** 41 (1960).

diverse literature on this subject. There are literally an infinite number of ways in which multiple experiments can be linked together into a single analysis. What has been presented is an attempt at providing a methodology which is as intuitive as possible.

A version of a global nonlinear data analysis software, based on modification of Bevington's CURFIT program as described in this chapter, is available in FORTRAN 77 from the author on request. A self-addressed stamped envelope with formatted blank media (either IBM DOS or MAC type floppies) should be sent. This software will also be available through electronic mail; address requests to BEECHEM@VULHMRBA.

Acknowledgments

The author gratefully acknowledges enlightening discussions with the following: Drs. Ludwig Brand, Enrico Gratton, Marcel Ameloot, Jay R. Knutson, R. P. DeToma, Michael Johnson, Herbert Halvorson, Benjamin W. Turner, Elisha Haas, Catherine Royer, J. R. Lakowicz, Zeljko Bajzer, R. Dale, C. W. Gilbert, Eric Henry, Doug Smith, Jean Claude Brochon, Zeljko Jericevic, J. B. A. Ross, W. Laws, M. Zuker, and Tom Ross. The author is a Lucille P. Markey Scholar, and this work was supported in part by a grant from the Lucille P. Markey Charitable Trust. Other funding includes National Institutes of Health Grant GM 45990.

[3] Padé–Laplace Algorithm for Sums of Exponentials: Selecting Appropriate Exponential Model and Initial Estimates for Exponential Fitting

By HERBERT R. HALVORSON

Introduction

Most chapters in this volume address the issue of parameter estimation, reflecting the importance attached to that topic. The companion issue of model selection is often accorded less attention. One purpose of this chapter is to present a method that allows these two statistical issues to be dealt with separately. If the problem is amenable to the techniques of linear statistics (e.g., determining the appropriate degree of a graduating polynomial), there are well-established methods[1] for reaching a decision, methods that are sharpened by the use of orthogonal functions. Attempts

---

[1] P. R. Bevington, "Data Reduction and Error Analysis for the Physical Sciences." McGraw-Hill, New York, 1969.

to extend this decision-making process into the domain of problems that are nonlinear in the parameters force one to deal with two fundamental problems. First, standard statistical tests are not guaranteed to be appropriate, ultimately because the concept of degrees of freedom is a linear one. Second, the fitting functions are intrinsically nonorthogonal generally, and profoundly so for sums of exponentials.

The second purpose is to emphasize the importance of a procedure for generating the initial estimate of the parameter set. Ultimately some iterative procedure (nonlinear least-squares[2] or maximum likelihood) is required to refine the initial values to some best set. (Opinions on this issue differ, but I hold that the primary reason for doing least-squares analysis is to get an estimate of the precision of the parameters.) Unfortunately, convergence in the large is not guaranteed, and the algorithms that have the best probability of finding a global extremum from an arbitrary starting point proceed very slowly in the vicinity of the solution. Additionally, the preconceived notions of an investigator about the solution may trap the parameter estimating routine in a local optimum. (Such an outcome is worse than an outright failure.) Finally, parameter refinement is greatly accelerated by beginning in an appropriate region of parameter space.

The benefits of grappling with this issue as a preliminary to nonlinear least-squares parameter estimation are that it can save much time otherwise wasted trying to fit to noise or to an inadequate model. Moreover, it can provide a suitable (albeit generally biased) initial estimate for the fitting routine.

The approaches described here arose from efforts to deal with data of a particular kind, namely, chemical relaxation kinetics employing repetitive small pressure perturbations. The small perturbations assure that even the weakly populated intermediate species will undergo small relative changes in concentration, validating the essential simplification of chemical relaxation kinetics: the response is comprised of a sum of exponential decays. Small perturbations also produce a small signal, so signal averaging must be used. For technical reasons that are outside the scope of this chapter, the final record can display low-frequency distortions. These make it difficult to estimate the number of decays that are present, and they obscure the values of the parameters (amplitudes and decay rates) that would constitute good initial guesses to be supplied to the iterative fitting routine.

Brief Survey of Methods

Without embarking on a comprehensive review of noniterative methods that have been applied to the analysis of sums of exponentials, it

[2] M. L. Johnson and T. M. Schuster, *Biophys. Chem.* **2**, 32 (1974). See also [1], this volume.

is useful to list the general categories of techniques that have been tried.

Graphical "Curve Peeling"

Historically, the familiar graphical technique of curve peeling[3] was the first approach to the problem of analysis of sums of exponentials. This entails no more than plotting $\ln[y(t)]$ against t and taking the terminal slope as the slowest rate. The linear portion is extrapolated back to zero time to get the amplitude. The extrapolated curve is subtracted from the data, and the logarithms of the residuals are then plotted against time. This is repeated as necessary. The selection of segments of the data for linear extrapolation is subjective to the point of being arbitrary. The method is thus difficult to implement on the computer, besides having certain intrinsic shortcomings, and it is not discussed further.

Inverse Laplace Transform

A sum of exponentials is equivalent to the Laplace transform of the distribution function for the decay rates. Accordingly, an inverse Laplace transform of the data should regenerate the distribution of decay constants, either discrete or continuous. Unfortunately, the inverse Laplace transform is technically an ill-posed problem (a small perturbation of the start can produce large effects at the end), and the procedure of Provencher[4] requires elaborate efforts to avoid these difficulties.

Iterative Integration

A variant of the Cornell procedure (discussed below), the method of successive integration,[5] forms running sums and sums of sums, etc. The problem is readily cast in the form of a sequence of matrix equations of increasing degree. Although the method is more robust than that of Cornell and the parameter estimates are generally closer to those derived from nonlinear least-squares analysis, numerical instabilities in the matrix inversion require care in the coding, and the complexity of the expressions increases rapidly with the number of exponentials. A commercial version of this software is available (OLIS Inc., Jefferson, GA).

[3] R. E. Smith and M. F. Morales, *Bull. Math. Biophys.* **6,** 133 (1944).
[4] S. W. Provencher, *Biophys. J.* **16,** 27 (1976).
[5] I. B. C. Matheson, *Anal. Instrum.* (*N.Y.*) **16,** 345 (1987).

Other Integral Transforms

Most of the techniques for coping with sums of exponentials involve the use of an integral transform to convert the transcendental problem to an algebraic problem that is more amenable to analysis. These transforms are closely related. In the following brief summary e^{-kt} is freely substituted for $y(t)$ to show a simple result when it exists. These transforms are all linear (the transform of a constant times a function is the product of the constant and the transform of the function, the transform of a sum of functions is the sum of the transforms of the functions), so the generalization to a sum of exponential decays is transparent.

Forward Laplace Transform. The Laplace transform is probably the most familiar of these transforms to the general scientific reader. The definition of the Laplace transform of $y(t)$ is

$$\mathscr{L}[y](s) = \int_0^\infty e^{-st} y(t) \, dt = 1/(s + k) \tag{1}$$

in the case of an exponential decay, where s must be taken large enough that the integral converges. Most of this chapter is devoted to the use of this transform, developed by Claverie and co-workers.[6,7] An alternative method[8] has been described earlier.

Mellin or "Moments" Transform. The Mellin transform with integer argument is the basis for the method of moments.[9] The Mellin transform is defined as

$$M[y](s) = \int_0^\infty t^{s-1} y(t) \, dt \tag{2}$$

When s takes on integer values $M[y](n + 1)$ is the nth moment of $y(t)$ $(n!k^{n+1})$, but s may have any value that leads to convergence of the integral. Noninteger Mellin transforms of the exponential function do not have a closed form expression.

z Transform. The z transform is widely familiar to electrical engineers working in control theory, filter design, or closely related problems in sampled data. It is related to the Laplace transform through a replacement of the transform variable s by $z = e^{sT}$, where T is the sampling interval. The subsequent development is complicated by the explicit consideration of the discreteness of the sampled function. The consequence is that

[6] E. Yeramian and P. Claverie, *Nature (London)* **326**, 169 (1987).
[7] J. Aubard, P. Levoir, A. Denis, and P. Claverie, *Comput. Chem.* **11**, 163 (1987).
[8] P. Colson and J. P. Gaspard, *in* "Dynamic Aspects of Conformation Changes in Biological Macromolecules" (C. Sadron, ed.), p. 117. Reidel, Dordrecht, The Netherlands, 1973.
[9] I. Isenberg and R. D. Dyson, *Biophys. J.* **9**, 1337 (1969). See also [11], this volume.

$$Z[y](z) = \Sigma_j \, y(jT)(1/z)^j = z/(z - e^{-kT}) \tag{3}$$

for the z transform of an exponential decay. Most commonly the z transform is implicit, and one is concerned with finding the poles of $Z[y]$ (the roots of the denominator), as in Prony's method.[10]

The same can be said for Cornell's method of successive sums,[11,12] which exploits the representation of the discrete sampling of an exponential function as a geometric series. The data are first partitioned into groups of equal size ($2n$ groups for n exponentials, $2n + 1$ with an unknown offset from zero), and the data within each group are summed. These sums are then manipulated to form a polynomial of degree n (the nature of the manipulation depending on the presence or absence of an offset), and the roots of the polynomial provide estimates of the decay constants for the exponentials. Successful extraction of n roots provides assurance of the presence of at least n exponentials, and the parameter values are good (if biased) starting values for the nonlinear least-squares iteration. Failure can occur simply from an untoward error distribution. The numerical problems associated with root finding increase rapidly with increasing degree, also limiting the applicability of the method.

Padé–Laplace Algorithm of Claverie

The primary purpose of this chapter is to describe an approach based on the use of the (forward) Laplace transform and Padé approximants. This method has been applied to the analysis (parameter estimation) of exponentials in different contexts.[6,7] A relatively simple modification of the procedure makes it quite useful for model determination. Among the attractive features of this approach are that the coding is independent of the number of exponentials sought and that the algorithm readily converges to complex[13] decay constants (damped sinusoids) in the event that too many components have been requested. The final decision (model selection) is not completely rigorous statistically, but this defect is common to all aspects of nonlinear problems.

The approach devised by Claverie and co-workers[6,7] is sophisticated, so an overview is presented before considering the modifications and details of practical implementation. The Padé–Laplace algorithm proceeds

[10] M. E. Magar, "Data Analysis in Biochemistry and Biophysics." Academic Press, New York, 1972.

[11] R. G. Cornell, *Biometrics* **18**, 104 (1962).

[12] M. H. Klapper, *Anal. Biochem.* **72**, 648 (1976).

[13] The word "complex" is used in the mathematical sense of describing a quantity with both real and imaginary parts: $z = x + iy$, where x and y are real and i is $(-1)^{1/2}$.

by first forming the forward Laplace transform of the signal (a sum of exponentials), mathematically developing the result in two different directions, and then equating the results of the two developments.

Laplace Transform of Sum of Exponentials

More concretely,

$$y(t) = \Sigma_i\, a_i \exp(-r_i t) \tag{4}$$

$$\mathscr{L}(y) = \int_0^\infty e^{-st} y(t)\, dt \tag{5}$$

$$= \Sigma_i\, a_i/(s + r_i) \tag{6}$$

The latter sum can be expressed as the ratio of two polynomials simply by combining over a common denominator of degree n (the numerator is of degree $n - 1$):

$$\mathscr{L}(y) = A_{n-1}(s)/B_n(s) \tag{7}$$

Because the denominator is merely the product of the denominators of the sum [Eq. (6)], the roots of the denominator constitute the set of decay constants.

Taylor Series Expansion of the Laplace Transform

The second mathematical path is an expansion of Eq. (5) in a Taylor (or Maclaurin) series. For reasons that become much clearer later on, s is replaced by $s' + s_0$, and e^{-st} is expanded as a power series in s' [$e^{-s_0 t} \Sigma_j(-t)^j/j!\, s'^j$]. This series is substituted into Eq. (5) and the order of integration and summation are interchanged by integrating term by term. Thus $\mathscr{L}(y)$ can be expressed as a Taylor series in s about some s_0

$$\mathscr{L}(y) = \Sigma_j(s - s_0)^j \int \frac{(-t)^j}{j!} e^{-s_0 t} y(t)\, dt \tag{8}$$

The expressions for $\mathscr{L}(y)$ [Eqs. (6)–(8)] are equivalent. The use of a new transform variable ($s' = s - s_0$) is immaterial until the solution for the decay rates is reported, at which time the new decay rate is corrected ($r_j' = r_j + s_0$), as can be seen from Eqs. (5) and (6). The variable s in the Laplace transform must be chosen so that all integrals converge. For Eq. (8) to be meaningful, it is necessary both that the series for e^{-st} converge ($|s'| < 1$) and that all the integrals in Eq. (8) exist. This affects the choice of the parameter s_0 (see below).

The first step of the Claverie algorithm entails numerical evaluation of the coefficients of the Taylor expansion of the Laplace transform of the data [the sequence of integrals in Eq. (8)]. (This limits the kind of problems that can be addressed to those which have suitable Laplace transforms.) These coefficients must then be related to the coefficients of the two polynomials in Eq. (7).

Padé Approximants

The Padé approximants $\mathcal{P}[n,m]$ are rational functions (ratios of polynomials) that, for specified degrees n in the numerator and m in the denominator, have the property of best approximating a given power series in a least-squares sense. The first $n + m + 1$ coefficients of the series are used to make the approximant, but these functions characteristically agree with the parent function of the series much better than the series truncated after $n + m + 1$ terms. The key feature here is that for the proper n the agreement would be exact (in the absence of noise).

Padé approximants occupy a somewhat obscure and specialized niche in numerical analysis, so some simple examples may be helpful. Consider the following power series expansion

$$\ln(1 + x) = x - \tfrac{1}{2}x^2 + \tfrac{1}{3}x^3 - \tfrac{1}{4}x^4 + \tfrac{1}{5}x^5 - \cdots \qquad (9)$$

which converges for $|x| < 1$, but poorly. The $\mathcal{P}[2,2]$ approximant to $\ln(1 + x)$ is constructed by solving the following equation

$$\frac{a_0 + a_1 x + a_2 x^2}{1 + b_1 x + b_2 x^2} = c_0 + c_1 x + c_2 x^2 + c_3 x^3 + c_4 x^4 \qquad (10)$$

for the coefficients a_j and b_j (b_0 is set to 1 by convention to make the problem deterministic). The brute force attack is to multiply through by the denominator, equate coefficients of powers of x (e.g., $a_0 = c_0 = 0$), and solve the resulting set of equations. (The labor involved is a major factor in the obscurity of Padé approximants.) For this example,

$$\mathcal{P}[2,2] \ln(1 + x) = \frac{x + \tfrac{1}{2}x^2}{1 + x + \tfrac{1}{6}x^2} \qquad (11)$$

which has a power series expansion

$$\mathcal{P}[2,2] = x - \tfrac{1}{2}x^2 + \tfrac{1}{3}x^3 - \tfrac{1}{4}x^4 + \frac{1}{5\tfrac{1}{7}}x^5 - \frac{1}{6\tfrac{6}{11}}x^6 + \cdots \qquad (12)$$

The approximant has an error of $\tfrac{1}{180}x^5 - \tfrac{1}{72}x^6 + \cdots$, whereas the polynomial from which it was derived (the truncated series) has an error of $\tfrac{1}{5}x^5 - \tfrac{1}{6}x^6 + \cdots$.

A second example, closer to the interests of this chapter, is provided by

$$\frac{1}{1 + x} + \frac{1}{1 + 2x} = 2 - 3x + 5x^2 - 9x^3 + 17x^4 - 33x^5 + \cdots \quad (13)$$

The $\mathcal{P}[0,1]$ approximant to this is

$$\frac{2}{1 + \frac{3}{2}x} = 2 - 3x + 4\tfrac{1}{2}x^2 - 6\tfrac{3}{4}x^3 + 10\tfrac{1}{8}x^4 - \cdots \quad (14)$$

which deviates from the true value by $\tfrac{1}{2}x^2 - 2\tfrac{1}{4}x^3 + \cdots$.

The $\mathcal{P}[1,2]$ approximant, constructed from just the first four terms $(2 - 3x + 5x^2 - 9x^2)$, is

$$\mathcal{P}[1,2] = \frac{2 + 3x}{1 + 3x + 2x^2} \quad (15)$$

which is identical to the parent function. (Exact agreement is possible only when the starting function is itself a rational function.)

The second major computational step in the Claverie algorithm is thus to construct the $\mathcal{P}[n - 1, n]$ approximant from the first $2n$ terms of the Taylor series as a representation of the Laplace transform of a sum of n exponentials. (Henceforth $\mathcal{P}[n - 1, n]$ is abbreviated as $\mathcal{P}[n]$.) That is, within experimental error, the Padé approximant to the Taylor series expansion applied to the data [Eq. (8)] is equivalent to the Laplace transform of the correct model [Eq. (7)].

Solving for Rates and Amplitudes

The third computational task of the Claverie algorithm is then to determine the roots of the denominator of the Padé approximant (the poles of the transform) and thereby to determine the decay constants for the n exponentials. One benefit of this approach is that the decay rates being sought never appear explicitly as arguments to the exponential function, providing much greater latitude as to acceptable or "legal" solutions. Knowing the decay constants then permits the respective amplitudes to be determined from the coefficients of the polynomial in the numerator of the approximant. A key contribution of Claverie and co-workers is the recognition that all the steps in their algorithm are valid in the half of the complex plane to the right of the critical value of s. That is, complex decay rates (sinusoidal oscillations, possibly damped) are acceptable solutions. Accordingly, the method for solution must be done with complex arithmetic.

Model Selection via Padé Approximants

In the preceding section we saw that if the data originated from the sum of n exponentials, then the Taylor series constructed from a sequence of integrals of the data should agree exactly with the $\mathcal{P}[n]$ Padé approximant and hence its power series expansion. This observation suggests the feasibility of testing the quality of a given Padé approximant, independently of explicit knowledge of the roots (decay constants). Reference to the second example of Padé approximation [Eqs. (13)–(15)] will help to clarify the idea and the procedure. One begins by evaluating the Taylor series to as many terms as possible. Start with some small number of exponentials, say $n = 1$. The $2n$ coefficients C_j of the series [the respective integrals of Eq. (8)] are used to make $\mathcal{P}[n]$. Now reverse the procedure and expand $\mathcal{P}[n]$ in a power series. Next evaluate the mean square deviation between the coefficients of the original Taylor series (C_j) and those of the Padé expansion (D_j) (the divisor is the number of Taylor coefficients not used in forming the Padé approximant). This function decreases as significant exponential terms are incorporated, leveling off or even increasing when only noise is being considered. If the mean square deviation appears to be decreasing, then increment n and try again.

It is necessary to make two refinements to this simple idea. First, the way in which the C_j coefficients are evaluated from the data introduces a nonuniformity of variance for C_j as j changes and, worse, a correlation between C_j and C_k. The data can be described as

$$y_i = \Sigma_k \, a_k \exp(-r_k t_i) + \varepsilon_i \tag{16}$$

where the summation over k (to n) represents the model ("signal") and ε_i is a random variable ("noise"). It is assumed that the distribution of ε is stationary, unbiased, and white. These conditions are fulfilled by a Gaussian noise distribution, $N(0,\sigma^2)$. However, the coefficients C_j are formed by integrating $t^j y$ over time, the independent variable. As a consequence, $\mathrm{var}(C_j) \neq \mathrm{var}(C_k)$ for $j \neq k$, and the coefficients are serially correlated. The elements of the covariance matrix are given by

$$V_{jk} = \frac{\sigma^2}{2s_0} \left(\frac{-1}{2s_0}\right)^{j+k} {}_{j+k}C_j \tag{17}$$

where ${}_{j+k}C_j$ is the binomial coefficient $[(j + k)!/j! \, k!]$.

Bard[14] describes a procedure for decorrelation when the underlying relations are known, as in this case. There is a simpler alternative. The sequence of coefficients represents integrals over successive powers of

14 Y. Bard, "Nonlinear Parameter Estimation." Academic Press, New York, 1974.

the time. The integrations could be performed equally well in a sequence of orthogonal polynomials of increasing degree. The polynomials appropriate to this problem are the Laguerre polynomials:

$$L_n(x) = e^x/n! \; d^n(x^n \, e^{-x})/dx^n \tag{18}$$

which have the important property that

$$\int_0^\infty e^{-x} L_m(x) L_n(x) \, dx = \delta_{mn} \tag{19}$$

The coefficients of this hypothetical expansion, C_j^*, are related to those actually determined, C_k, by

$$C_j^* = \Sigma_k \, _jC_k(2x_0)^k C_k \tag{20}$$

Equation (20) also describes the relation between the errors of the hypothetical expansion, E_j^*, and those measured, E_k, after correcting for multiple comparisons. Referring to the second example given previously, the sequence of E_k is $\{0, 0, \frac{1}{2}, -2\frac{1}{4}, 6\frac{7}{8}, -17\frac{13}{16}, 42\frac{7}{32}, \ldots\}$ and the sequence of E_j^* is $\{0, 0, \frac{1}{2}, -\frac{3}{4}, \frac{7}{8}, -\frac{15}{16}, \frac{31}{32}, \ldots\}$. It is the sum of E_j^{*2} that is used in statistical tests.

Discrepancies between C_j and D_j are of three kinds: (1) model bias [n(assumed) $\neq \nu$(true)], (2) method bias (inadequacy of the numerical integration), and (3) unbiased or random (arising from the noise ε_i). The null hypothesis under test is that $n = \nu$. Method bias is eliminated by collecting data to sufficient time and using an appropriate integration routine. The remaining question concerns assessing the likelihood that the observed discrepancies arise from chance.

Three different procedures can be adopted. If the variance of the raw data is known, then it is possible to test the sum of the squares of E_j^* as χ^2. The variance of the data may not be known well enough, particularly if the data have been smoothed to expedite the integrations. Second, the improvement in the sum of squares can be tested with the familiar incremental F-test. One assumes n exponentials, evaluates M coefficients, and determines SSR_n (the sum of the squares of the E_j^* values with ($M - 2n$) degrees of freedom. This is repeated for $n + 1$ exponentials, with a new set of M coefficients giving SSR_{n+1} with ($M - 2n - 2$) degrees of freedom. The significance of the improvement with respect to the residual variance is then given by

$$F = \frac{(SSR_n - SSR_{n+1})/2}{SSR_{n+1}/(M - 2n - 2)} \tag{21}$$

which is compared with tabulated values of $F(2, M - 2n - 2)_\alpha$ at the desired confidence level. These two tests depend on the assumed error distribution and use a linear treatment of degrees of freedom.

The second refinement is the use of the Akaike information theory criterion (AIC)[15]

$$AIC = M \ln(SSR_m) + 2m \qquad (22)$$

where M is the (total) number of coefficients and SSR_m is the sum of the squares of the residuals evaluated using m parameters. Akaike developed this test from a maximum likelihood treatment of an information theory perspective, and it does not depend on the existence of a particular error distribution. To use this test, one merely looks for the appearance of a first minimum as m is increased (no tables are necessary). Because the rates and amplitudes are allowed to be complex in this application, there are four parameters for each exponential term.

Twofold Importance of Parameter s_0

The mathematically inclined reader will find much to enjoy in a careful study of the Claverie algorithm. It provides a novel perspective to the properties of exponential decays in the complex frequency domain (s plane). In this context the primary role of s_0 is to shift the circle of convergence of the series expansion in order to avoid the poles [points where the representation of Eq. (6) becomes infinite]. Because it is presumed that $y(t)$ is a sum of exponential decays, the validity of Eq. (8) is assured simply if s_0 is not negative.

When the problem is viewed in the time domain, s_0 is seen to permit the evaluation of high-order coefficients in the expansion because it forces the integrand toward zero at large (but finite) time. In other words, to perform the calculation it is not sufficient that the integrals in Eq. (8) merely exist in the mathematical sense; they must be representable within the computer. This practical computational benefit was introduced into the method of moments by Isenberg as "exponential depression."[16] Comparing the Mellin transform with the Taylor expansion of the Laplace transform shows the relation between the two algorithms.

Tests on Synthetic and Experimental Data

The ability of the algorithm to discriminate between two decay rates depends on their ratio, the relative amplitudes, the completeness of the decay, and the overall noise level. I have explored a limited region of this space

[15] H. Akaike, *IEEE Trans. Autom. Control* **AC-19**, 716 (1974).
[16] I. Isenberg and E. W. Small, *J. Chem. Phys.* **77**, 2799 (1982).

to resolve questions pertinent to my experimental interests. The simulated experimental data all contained one exponential decay (5/sec) sampled at 1-msec intervals to 0.95 sec (950 points), at which time the decay had gone to 1% of the starting value. One group of simulations contained a second faster decay of the same initial amplitude but varying rates. All simulations had superimposed Gaussian noise of varying standard deviation (expressed in terms of the total initial amplitude). The purpose of the exercise was 2-fold: under what conditions would the Padé–Laplace procedure report extraneous decays, and what were the limiting combinations of noise and ratio of rates that permitted resolution of two decays?

The Padé–Laplace–AIC algorithm never reported the existence of multiple decays in noisy single-exponential data. With the standard deviation of the noise at 5% of the total amplitude (10% of the amplitude of each component), two decays could be resolved if the faster decay was at least 11/sec (a ratio of 2.2). Decays of 5/sec and 10/sec could be resolved by AIC only if the noise was 1.35% of total amplitude or less. One can anticipate an equally sharp cutoff for unequal amplitudes, but the location would depend on the ratio of the amplitudes as well.

Solving the Padé approximant for rates and amplitudes yields parameters that are similar to the results of nonlinear least-squares analysis or to the parameters used to synthesize the data. In general, they lie within the confidence region estimated by the nonlinear least-squares routine. Even when they lie outside this region, their location in parameter space is simply connected with the optimum, and convergence is rapid. Attempts to analyze excessively noisy simulations (i.e., where the *AIC* test underestimated the number of decays) for the true number of decays often could be made to converge, but with ambiguous precision and marginal improvement in the goodness of fit. In some instances the nonlinear least-squares routine could converge on an extraneous solution (two decays found in noisy single-exponential data), but the quality of the estimates was poor.

This procedure has also been applied to experimental data of the type mentioned in the Introduction. This does not constitute a test, in the sense that it does with simulated data, but the overall performance can be assessed. The primary benefit was the automated generation of a set of initial guesses that led to rapid convergence. Secondarily, the presence of low-frequency sinusoids did not "hang up" the routine, although in some instances it was necessary to enter the AIC selection manually without the sinusoid.

Notes on Implementation

The practical implementation of this approach involves some details that are now considered. First, one should reduce the offset of the data by subtracting off the mean of the last 5–100 points. This is not because the algorithm has any difficulty with roots of zero, but rather keeps the coefficients of the Taylor series well bounded at long times. Calculation of these coefficients is further aided by introducing a dummy time scale such that the total time for N points is $N^{1/2}$ (time increment is $1/N^{1/2}$). This helps to keep the calculation of the higher order coefficients within bounds at both short and long times.

The coefficients of the Taylor series expansion were evaluated by simple trapezoidal integration of the data, as speed and simplicity were regarded as being more important than accuracy (these are, after all, just initial estimates). Additionally, all coefficients can be evaluated on a single pass through the data, evaluating the integrand for each coefficient recursively from that for the previous coefficient. That is, at fixed time, integrand$_j$ = integrand$_{j-1}[(-t)/j]$. This inversion of the normal loop structure circumvents overflow/underflow problems associated with trying to form the relation $t^n/n!$ directly.

A suitable value for s_0 is chosen by displaying the first 20 Taylor coefficients and interactively adjusting s_0 until the coefficients are of the same order of magnitude. This value must be found empirically, as it depends on the length of the record in decay times and on the number of data points (because of the dummy time scale). Values between 0.6 and 0.9 were generally suitable in the tests of simulations.

Once the set of Taylor coefficients has been evaluated, the Padé approximants can be constructed from published FORTRAN code.[17] For small computers it is more convenient to use a variant of Euclid's factorization algorithm applied to polynomials.[18] Either of these recursive approaches is preferable to solving the set of simultaneous equations directly. Back-expansion of the Padé approximant is by synthetic division.[18,19]

Obtaining the complex roots of a polynomial with possibly complex coefficients is a problem that was solved by Laguerre. The algorithm is ingenious, but it is not easy to find. Published routines (ZROOT and LAGUER)[19] are available.

Concluding Assessment

The Padé–Laplace algorithm has several attractive features. It is robust with respect to pecularities of the error distribution that lead to the infer-

[17] I. M. Longman, *Int. J. Comput. Math.* **B3**, 53 (1971).

[18] D. E. Knuth, "The Art of Computer Programming: Seminumerical Algorithms," Vol. 2, 2nd Ed. Addison-Wesley, Reading, Massachusetts, 1981.

[19] W. H. Press, B. P. Flannery, S. A. Teukolsky, and W. T. Vetterling, "Numerical Recipes." Cambridge Univ. Press, London and New York, 1986.

ence of negative or complex decay rates. It can be made applicable to the study of complex data (phase-sensitive detection) simply by defining the data array and the polynomial coefficients as complex quantities. The coding is reasonably straightforward and, once implemented, applies to any number of exponential terms. It is ideally suited for use as the "initial guess" routine of an iterative program that analyzes exponential data, although smaller computers may require that it be overlaid. Such a routine presents a menu of parameters for differing numbers of exponential terms, with the *AIC* determining the default choice.

Two final caveats are appropriate. First, the Claverie algorithm, like the other noniterative algorithms, entails finding the roots of a function of the experimental data in order to extract the desired parameters. Such an operation amplifies the effect of experimental uncertainties ("noise") in an unpredictable way. Although the impact can be reduced by presmoothing[20] the data, this author strongly prefers to regard the numbers obtained as no more than preliminary estimates to be subsequently refined by an iterative least-squares procedure. Second, the procedure outlined in this chapter can do no more than return the minimum number of exponential terms necessary to explain the data. It has been pointed out before[21,22] that this numerical result may grossly underestimate the complexity of the physical situation. The method described here was developed for a narrowly defined function: to assist in the planning of chemical relaxation kinetics experiments at a stage when the overall mechanism remains elusive. It performs that function well. Final analysis of the data is best done by fitting the total reaction scheme to all of the experimental records simultaneously.[23]

Acknowledgments

This work has been supported by a grant from the U.S. Office of Naval Research.

[20] A. Savitzky and M. J. E. Golay, *Anal. Chem.* **36,** 1627 (1964).
[21] S. L. Laiken and M. P. Printz, *Biochemistry* **9,** 1547 (1970).
[22] H. R. Halvorson, *Biopolymers* **20,** 241 (1981).
[23] J. R. Knutson, J. M. Beechem, and L. Brand, *Chem. Phys. Lett.* **102,** 501 (1983). See also [2], this volume.

[4] Use of Weighting Functions in Data Fitting

By ENRICO DI CERA

Introduction

Data analysis plays a central role in the description of a number of physical, chemical, and biochemical phenomena. Our interpretation of the experimental facts, phenomenological or mechanistic as it may be, is very often based on a set of physicochemical parameters involved in the particular model being proposed or tested. Best-fit parameter estimates are derived from minimization of a suitable functional, or sum of squares, that can be written as[1]

$$\Phi = \sum_{j=1}^{n} [y_j - F(x_j, \{\psi\})]^2 \tag{1}$$

Here n is the number of experimental points, y_j is the jth experimental observation, while F is the deterministic model, or fitting function, used to interpolate the experimental measurements through the independent variable x and a set of s parameters $\{\psi\} = \psi_1, \psi_2, \ldots, \psi_s$.

When dealing with repeated experimental observations Eq. (1) is extended to the number of different experiments $i = 1, 2, \ldots, m$ as follows:

$$\Phi = \sum_{i=1}^{m} \sum_{j=1}^{n} [y_{ij} - F(x_{ij}, \{\psi\})]^2 \tag{2}$$

where y_{ij} is the jth experimental observation in the ith experiment, while x_{ij} is the value of the independent variable x coupled to y_{ij}. The significance of Eq. (2) and its simplest form given by Eq. (1) is independent of the particular form of the fitting function F and hinges solely on its computational aspects related to minimization of the difference between experimental data and theoretical predictions. In so doing we make no distinction among different experimental points and implicitly assume that they are perfectly equivalent in the minimization procedure and contribute equally to parameter estimation. This assumption is certainly appealing when considering the computational aspects of least-squares minimization.

In many cases of practical interest experimental data points cannot be treated as equivalent. In some cases experimental points carry different

[1] Y. Bard, "Nonlinear Parameter Estimation." Academic Press, New York, 1974.

METHODS IN ENZYMOLOGY, VOL. 210

experimental error. In other cases minimization of Eq. (2) involves experimental determinations of different scales. Application of Eqs. (1) and (2) in these cases may lead to unsatisfactory estimates of the parameters. The solution of the problem is one and the same for all cases and hinges on the definition of a weighting factor, w, for each experimental data point. In so doing one takes explicitly into account the exact probability distribution of experimental measurements, and the relevant functionals to be minimized become

$$\Phi = \sum_{j=1}^{n} w_j [y_j - F(x_j, \{\psi\})]^2 \tag{3}$$

in the case of single determinations, and

$$\Phi = \sum_{i=1}^{m} \sum_{j=1}^{n} w_{ij} [y_{ij} - F(x_{ij}, \{\psi\})]^2 \tag{4}$$

for multiple experimental determinations. The weight w_j of point j in Eq. (3) expresses the relative contribution of that point to the sum of squares. When all weights w are equal in Eq. (3), then the weighting factor becomes a scaling factor that has no effect on the minimization procedure. In this case one deals with a functional such as Eq. (1) that implies uniform weighting of all experimental data points. A differential weighting is introduced as soon as the experimental data points have different weights. In the case of multiple determinations each data point is weighted according to its contribution within the data set to which it belongs and the relative contribution of that data set to the total sum of squares.

It is clear from Eqs. (3) and (4) that the definition of weighting factors for each experimental data point is critical for correct data analysis. However, very little attention is often paid to aspects of data analysis that deal with the formulation of a correct weighting scheme for the data set under consideration. The issue of correct data weighting is commonly dismissed with generic assumptions on the distribution of experimental errors, and data analysis is usually concentrated on parameter resolution and tests of different models, interested as we are in the interpretation of the experimental facts. The definition of weighting factors in the functionals given in Eqs. (3) and (4) is *conditio sine qua non* for correct data analysis and parameter resolution in practical applications. In this chapter we shall discuss the importance of properly casting the fitting problem in terms of weighting functions. We shall also discuss the consequences of improper or incorrect weighting schemes in parameter resolution and show a specific application to the analysis of steady-state enzyme kinetics, which provides an example of general interest.

Weighting Factors and Difference between Errors and Residuals

Under a set of assumptions the weighting factor of point j can uniquely be defined in terms of its actual experimental error. In principle, if we knew the exact model to be used as fitting equation and the true values $\{\psi^*\}$ of the parameter set $\{\psi\}$, then the difference

$$\varepsilon_j = y_j - F(x_j, \{\psi^*\}) \tag{5}$$

would actually give us the experimental error of point j. In practice, even if we knew the exact fitting function we could only access the best-fit estimates of $\{\psi\}$ and not the true values $\{\psi^*\}$. The difference

$$\rho_j = y_j - F(x_j, \{\psi\}) \tag{6}$$

is equal to the residual of point j. The fundamental difference between errors and residuals is that the standard deviation of ε_j is independent of F, while that of ρ_j is not.

The only way we can practically obtain a value for the experimental error of point j is by carrying out repeated measurements $y_{1j}, y_{2j}, \ldots,$ y_{mj} of y_j at x_j. Under the assumptions that m is large and the y_j values are randomly scattered around the mean $\langle y_j \rangle = (y_{1j} + y_{2j} + \ldots + y_{mj})/m$, the standard deviation

$$\sigma_j = \left(\sum_{i=1}^{m} [y_{ij} - \langle y_j \rangle]^2/m \right)^{1/2} \tag{7}$$

can be taken as the experimental error of point j. The important definition of weighting factor as

$$w_j = \sigma_j^{-2} \tag{8}$$

leads to least-variance estimates of the model parameters $\{\psi\}$ in Eq. (3). The reader should be able to demonstrate that the same expression in Eq. (7) can be arrived at by computing the standard deviation of ε_j according to Eq. (5). In fact, the value of $F(x_j, \{\psi^*\})$ cannot change in different determinations of y_j, since $\{\psi^*\}$ represents the true parameter values. On the other hand, the best-fit value of $F(x_j, \{\psi\})$ is a function of y_{ij}, and therefore the standard deviation of ρ_j is a function of the particular model we have chosen to fit the data. For this reason residuals are *not* experimental errors and must never be used to assess the distribution of weighting factors for uniform or differential weighting purposes.

Experimental errors reflect the property of a distribution of experimental measurements and have nothing to do with the fitting function that is going to be used to interpolate the data. Residuals reflect the difference between experimental determinations and theoretical predictions of a particular model that we have selected for our analyses. The weighting factor is defined as the inverse of the experimental error squared and likewise reflects a property of the experimental data alone. It can be proved mathematically that such a definition of weighting factors guarantees least-variance parameter estimates when F is a linear function of $\{\psi\}$. Although in the general case of nonlinear least-squares, where F is a nonlinear function of $\{\psi\}$, a similar proof does not exist, nevertheless there is reason to believe that Eq. (8) still provides an optimal weighting scheme.[1,2] In computing the standard deviation of the distribution of experimental measurements we have implicitly assumed that each experimental point is independent of any other. In general, such a restriction can be dropped and the weighting factor can be calculated as the element of the covariance matrix associated with the measurements.[1-3]

Weighting Factors and Data Transformations

Once experimental data have been collected and information on the distribution of experimental errors is available, the minimization procedure can be started using the functional

$$\Phi = \sum_{j=1}^{n} w_j[y_j - F(x_j,\{\psi\})]^2 = \sum_{j=1}^{n} \sigma_j^{-2}[y_j - F(x_j,\{\psi\})]^2 \qquad (9)$$

The nature of the fitting function F is critical for correct parameter estimation. It is often found more convenient to transform the stretch of original data points y into a new set of values y' to simplify the form of the fitting function. In many cases such a transformation is aimed at linearizing the fitting problem, that is, to yield a new $F(y',\{\psi'\})$ that is linear in the parameters $\{\psi'\}$. If the experimental error of y_j is σ_j, what is the value of σ_j' for y_j'? It is clear that the distribution of the errors in y' values is *not* the same as that of the errors in data points y, and therefore if a given weighting scheme is applied in the case of Eq. (9), a different one should be used when minimizing:

$$\Phi = \sum_{j=1}^{n} w_j'[y_j' - F(x_j,\{\psi'\})]^2 = \sum_{j=1}^{n} \sigma_j'^{-2}[y_j' - F(x_j,\{\psi'\})]^2 \qquad (10)$$

The calculation of w_j' from w_j on the transformation $y_j \rightarrow y_j'$ hinges on the derivation of the error σ_j' from the error σ_j. It is important to stress that such a problem has *not* a general and exact solution and can only be

[2] D. A. Ratkowski, "Nonlinear Regression Modeling." Dekker, New York, 1983.
[3] M. E. Magar, "Data Analysis in Biochemistry and Biophysics." Academic Press, New York, 1972.

solved to a first approximation.[1] If the errors are small and the transformation $y_j \to y_j'$ is continuous and bounded and so are all its derivatives, then to a first approximation, the Jacobian transformation

$$\sigma_j'^2 = (\partial y_j'/\partial y_j)^2 \sigma_j^2 \tag{11}$$

along with the definition of weighting factor given in Eq. (8) allow calculation of the new set of weights. Equation (11) is the familiar formula for error propagation.[3]

Data transformations are widely used in the analysis of many systems of biochemical interest. Typical examples are given in the following sections.

Relaxation Kinetics

In the process $A \to B$, the concentration of A, c, at time t is given by

$$c = c_0 \exp(-t/\tau) \tag{12}$$

where c_0 is the initial concentration of A and τ is the relaxation time for the irreversible conversion of A to B. If data are collected as c values as a function of time using spectroscopic methods, as in the case of stopped-flow kinetics, the exact form to be minimized is

$$\Phi = \sum_{j=1}^{n} w_j [c_j - c_0 \exp(-t_j/\tau)]^2 \tag{13}$$

and the two parameters c_0 and τ need to be resolved by nonlinear least-squares. If the independent variable t is assumed to be errorless and the accuracy of the spectrophometer is the same over the concentration range $0-c_0$ of A, then one can reasonably conclude that all w values are equal in Eq. (13) and therefore apply uniform weighting in the minimization procedure. Taking the logarithm of both sides of Eq. (12) yields

$$\ln c = \ln c_0 - t/\tau \tag{14}$$

so that

$$\Phi = \sum_{j=1}^{n} w_j' [\ln c_j - \ln c_0 + t_j/\tau]^2 \tag{15}$$

which is a linear function in the parameters $\ln c_0$ and $1/\tau$. In view of Eq. (11) it is clear that the distribution of the errors in $\ln c$ is not the same as that of the errors in c, and therefore if uniform weighting is applied in the case of Eq. (13), differential weighting should be used in Eq. (15). Hence, in the transformation $c \to \ln c$ we should minimize

$$\Phi = \sum_{j=1}^{n} w_j c_j^2 [\ln c_j - \ln c_0 + t_j/\tau]^2 \tag{16}$$

since $w_j' = 1/\sigma_j'^2 = 1/[(\partial y_j'/\partial y_j)^2\sigma_j^2] = w_j/[(\partial \ln c_j/\partial c_j)^2] = w_j c_j^2$. Whatever the value of w_j, the value of w_j' will depend explicitly on c_j. In the case of uniform weighting of the c measurements ($w_j = 1$ for all j), the $\ln c$ values should be weighted according to c^2. Therefore, $c_1 = 10$ and $c_n = 1$ receive the same weight in Eq. (13), but $\ln c_1$ has 100 times more weight than $\ln c_n$ in Eq. (16).

Steady-State Kinetics

In the case of steady-state kinetics the velocity, v, of product formation is measured as a function of the substrate concentration, x. For a system obeying a simple Michaelis–Menten equation

$$v = e_T k_{cat} x/(K_m + x) \tag{17}$$

where e_T is the total enzyme concentration, and the values of the catalytic constant k_{cat} and Michaelis–Menten constant K_m are to be resolved from analysis of the experimental data. If the values of x are again assumed to be errorless and the distribution of the errors on the v values is uniform, then

$$\Phi = \sum_{j=1}^{n} [v_j - e_T k_{cat} x_j/(K_m + x_j)]^2 \tag{18}$$

is the functional to be minimized. There are several possible linearizations of the Michaelis–Menten equation. The familiar double-reciprocal transformation of Lineweaver–Burk, namely,

$$1/v = 1/(e_T k_{cat}) + [K_m/(e_T k_{cat})](1/x) \tag{19}$$

is the most widely used procedure to analyze steady-state kinetics. It is important to recognize that this transformation is no longer accompanied by a uniform distribution of errors. The correct weighting factor of $1/v_j$ is obtained from Eq. (11) and is equal to v_j^4. The correct form to be minimized is therefore

$$\Phi = \sum_{j=1}^{n} v_j^4 \{1/v_j - 1/(e_T k_{cat}) - [K_m/(e_T k_{cat})](1/x_j)\}^2 \tag{20}$$

Interestingly, the weighting factor for the $v \rightarrow 1/v$ transformation is similar to that for the $c \rightarrow \ln c$ transformation discussed in the case of relaxation kinetics. Consequently, a velocity measurement around K_m that yields a value of $v = e_T k_{cat}/2$ is equivalent to a measurement taken under saturating conditions, that is, $v \cong e_T k_{cat}$ when using the Michaelis–Menten equation, as opposed to the Lineweaver–Burk plot where the measurement around K_m should be weighted 16 times less.

Ligand Binding Equilibria

In the case of ligand binding studies the fractional saturation, Y, of the macromolecule is measured as a function of the ligand activity, x. Experimental data are fitted to the equation

$$Y = \sum_{j=0}^{t} jA_j x^j / t \sum_{j=0}^{t} A_j x^j = F(x,\{A\}) \tag{21}$$

where t is the number of sites, and the A's are the overall association constants to be resolved by nonlinear least-squares. If the distribution of experimental errors on Y is uniform, what is the correct weighting scheme for the transformation $Y \rightarrow \ln[Y/(1 - Y)]$ that leads to the familiar Hill plot? Application of Eq. (11) yields $w' = Y^2(1 - Y)^2$. Therefore, when fitting data in the form of the Hill plot, a point collected at half-saturation should be weighted about 8 times more than a point at 90% saturation, and 638 times more than a point at 99% saturation.

Global Data Fitting

In a number of cases it is necessary to interpolate globally observations of different scale. For example, such a situation arises when k_{cat} and K_m values for an enzyme are collected over a wide pH range and need to be globally interpolated with a thermodynamic scheme to pull out the pK values of the ionizable groups involved in the control of binding and catalytic events, as recently shown in the case of thrombin.[4] In this case $F(x,\{\psi\})$ is the fitting function for k_{cat}, $G(x,\{\psi\})$ is the fitting function for K_m, and the set of parameters $\{\psi\}$ shared in both functions is resolved by global data fitting. The form to be minimized is

$$\Phi = \sum_{j=1}^{n} w_{1j}[y_{1j} - F(x_j,\{\psi\})]^2 + \sum_{j=1}^{n} w_{2j}[y_{2j} - G(x_j,\{\psi\})]^2 \tag{22}$$

where the y_1's and y_2's refer to k_{cat} and K_m values, respectively. Typically, k_{cat} values are 5–8 orders of magnitude bigger than K_m values, and so are their respective errors. In this case it would make no sense to minimize a functional such as Eq. (22) using the same weight for both k_{cat} and K_m values, because the information relative to the k_{cat} values would over-

[4] R. De Cristofaro and E. Di Cera, *J. Mol. Biol.* **216,** 1077 (1990).

whelm that relative to the K_m values in the sum of squares. Use of correct weighting factors according to the actual experimental errors carried by the two sets of observations guarantees meaningful parameter estimation in the global data fitting.

Consequences of Incorrect Data Weighting

In science asking the right question is often more important than giving the right answer. Indeed, a key question arises quite naturally at this point as to the consequences of incorrect data weighting. Why should one worry about computing a correct weighting scheme for each experimental data set and any tranformations thereof? Why should the budding enzymologist be concerned with a correct weighting scheme for his/her beloved double-reciprocal transformation? With perfect data and the correct fitting function the difference $y_j - F(x_j,\{\psi\})$ in Eq. (3) is identically zero for all points, and the true values of the parameters are recovered in the minimization procedure independently of any weighting scheme. If the fitting function is incorrect there is very little need to worry about the quality of data or correct weighting schemes, since the best-fit parameter estimates are meaningless anyway. Therefore, the situation where one should worry about correct weighting arises when the fitting function is correct and the data carry a finite experimental error, which is exactly the most common situation encountered in practice. It is clear that there is no magic formula to answer the question of the consequences of incorrect weighting in the general case. As already pointed out, we do not even have a mathematical proof that an optimal weighting scheme should exist when dealing with nonlinear least-squares. Therefore, we shall address the problem by means of Monte Carlo simulations of an *ad hoc* chosen example of practical interest, namely, the analysis of steady-state kinetics. The case of ligand binding equilibria has been dealt with elsewhere.[5]

Consider the set of steady-state measurements reported in Fig. 1, and assume that nothing is known about the distribution of experimental errors. A list of the experimental data is given in Table I to allow the interested reader to reproduce the results. How should we proceed to analyze this data set? The simplest choice would be to assume that the substrate concentration is errorless and the experimental error is uniformly distributed among all velocity measurements. This implies that the exact form to be minimized needs to be cast in terms of the Michaelis–Menten (MM) equation [Eq. (17)] with uniform weighting. The best-fit values of the parameters are in this case $k_{cat} = 104.7 \pm 0.6 \text{ sec}^{-1}$ and $K_m = 2.89 \pm 0.07$

[5] E. Di Cera, F. Andreasi Bassi, and S. J. Gill, *Biophys. Chem.* **34**, 19 (1989).

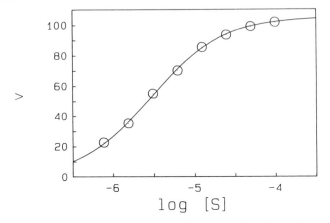

FIG. 1. Steady-state velocity measurements of human α-thrombin amidase activity as a function of the logarithm of substrate concentration (data obtained by Dr. Raimondo De Cristofaro). Velocity values are expressed per unit thrombin concentration. Experimental conditions are as follows: 1.2 nM thrombin, 50 mM Tris-HCl, 0.1 M NaCl, 0.1% polyethylene glycol (PEG) 6000, pH 8.0, 25°C. The substrate used for these determinations is the synthetic tripeptide S-2238.

μM. The standard error of the fit, σ, is given by $[\Phi/(n - s)]^{1/2}$, where Φ is given in Eq. (3), n is the number of data points, and s is the number of independent parameters. The value of σ is 0.9 sec^{-1} for the data reported in Fig. 1, which shows that the data are very accurate and have less than 1% error. To assess the consequences of incorrect weighting the same data can then be converted to a double-reciprocal plot according to the Lineweaver–Burk (LB) transformation [Eq. (19)] and fitted using uniform weighting. The best-fit values of the parameters are in this case k_{cat} = 103.0 ± 2.7 sec^{-1} and K_m = 2.76 ± 0.11 μM.

Under the assumption that the error is uniformly distributed among the velocity values there is a definite, although not significant, difference between the values of K_m and k_{cat} obtained with the MM and LB equations. When the data are correctly weighted in the LB transformation according to Eq. (20) we obtain k_{cat} = 104.7 ± 0.6 sec^{-1} and K_m = 2.88 ± 0.07 μM. These values are practically identical to those obtained using the MM equation with uniform weighting.

Another linearization of Eq. (18) is given by the Eadie–Hofstee (EH) transformation

$$v = e_T k_{cat} - K_m v/x \tag{23}$$

A plot of v versus v/x yields a straight line with slope K_m and intercept

TABLE I
EXPERIMENTAL VALUES OF STEADY-STATE
VELOCITY MEASUREMENTS OF HUMAN
α-THROMBIN AMIDASE ACTIVITY[a]

x (μM)	v (sec^{-1})
98.000	101.67
49.000	98.90
24.500	93.45
12.250	85.40
6.125	70.06
3.062	54.90
1.531	35.16
0.766	22.62

[a] Under the experimental conditions given in the legend to Fig. 1. Data are averages of duplicate determinations. Velocity values have been normalized for the enzyme concentration.

$k_{cat}e_T$. This transformation is the kinetic parallel of the Scatchard plot in ligand binding equilibria and introduces an error in the independent variable. The EH transformation is extremely interesting for the following reason. If we assume that the error in the EH plot is uniformly distributed, then the functional

$$\Phi = \sum_{j=1}^{n} [v_j - (e_T k_{cat} - K_m v_j/x_j)]^2 \tag{24}$$

is the correct form to resolve the independent parameters. This form can be rearranged to yield

$$\Phi = \sum_{j=1}^{n} [(K_m + x_j)/x_j]^2 [v_j - e_T k_{cat} x_j/(K_m + x_j)]^2 \tag{25}$$

and is therefore identical to fitting data with an error proportional to the velocity in the MM plot. Therefore, if the error is not uniform and is proportional to v, then one should either use differential weighting in the MM plot or uniform weighting in the EH Plot. Transformation of the data in the EH plot has in this case a variance-equalizing effect and corrects for the nonuniform error distribution.[6]

When the data in Fig. 1 are analyzed according to Eq. (23) with uniform weighting the best-fit values of the parameters are $k_{cat} = 104.3 \pm 1.2$ sec^{-1}

[6] J. A. Zivin and D. R. Waud, *Life Sci.* **30**, 1407 (1982).

and $K_m = 2.85 \pm 0.08\ \mu M$. Again, very little difference is observed with respect to the MM equation. Therefore, experimental data taken in the substrate concentration range from $K_m/4$ to $32K_m$ and with 1% error on the velocity values show no significant difference in the best-fit values of k_{cat} and K_m obtained from the MM, LB, and EH plots. Because with perfect data we would see no difference at all in the parameter values obtained in the three plots, one may naturally wonder what experimental error or substrate concentration range would reveal a significant difference with incorrect weighting schemes.

A Monte Carlo study has been conducted as follows. Steady-state velocity measurements, normalized by the enzyme concentration e_T, were simulated according to the equation

$$v = k_{cat}x/(K_m + x) + \varepsilon[-2\ln(\mathrm{RND_1})]^{1/2}\cos(2\pi\mathrm{RND_2}) \qquad (26)$$

where $\mathrm{RND_1}$ and $\mathrm{RND_2}$ are two random numbers in the range 0–1. A pseudorandom error distribution[7] $N(0,\varepsilon)$ with 0 mean and a standard deviation equal to ε was assumed for all velocity measurements, independent of x, to simulate a uniform experimental error. For each data set eight points were generated in the substrate concentration range $K_m/4$ to $32K_m$ using a value of x_j as follows

$$x_j = 32K_m/2^{j-1} \qquad (j = 1, 2, \ldots, 8) \qquad (27)$$

to yield points equally spaced in the logarithmic scale, as in the case of the data shown in Fig. 1. The simulated values of k_{cat} and K_m were both equal to 1. This choice was made solely for the sake of simplicity and without loss of generality. Ten values of ε were used in the simulations, from 0.01 to 0.1, to yield an average standard error of the fit from 1% to 10%. For each value of ε, 10,000 data sets were generated according to Eq. (26). Each set was analyzed according to the MM, EH, and LB equations using uniform weighting in all cases. Simulation and fitting of 10,000 data sets took on the average less than 70 sec on our Compaq 486/320 computer. Owing to the (known) error distribution, this weighting scheme is correct for the MM transformation and incorrect for the EH and LB equations.

The mean, m, and standard deviation, σ, of the distribution of k_{cat} and K_m values were computed in the three cases, and the results of the simulation study are shown in Fig. 2. The actual values of the parameter bias and percent error are listed in Table II. The simulation study confirms the result observed in the analysis of real experimental data with 1% error (0.01 in Fig. 2 and Table II), namely, that very little difference exists

---

[7] G. E. P. Box and M. E. Muller, *Ann. Math. Stat.* **29**, 610 (1958).

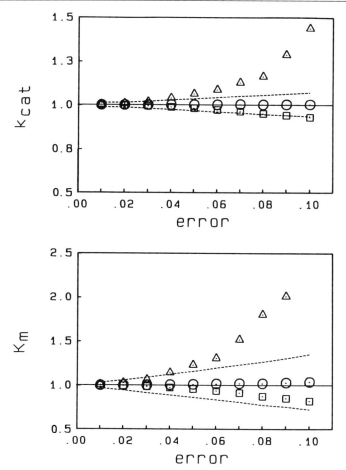

FIG. 2. Results of the Monte Carlo study discussed in the text for 10,000 simulated data sets of 8 points each carrying a uniform error. The mean values of k_{cat} (top) and K_m (bottom) were obtained with the MM (○), EH (□), or LB (△) equation. The one standard deviation (68%) confidence interval for the parameter values obtained by means of the MM equation is depicted by a dashed line.

among the values of K_m and k_{cat} obtained with the MM, EH, and LB equations. However, as soon as the error exceeds the 4% level, then a significant difference is observed with the LB transformation. The bias on the K_m and k_{cat} lies well outside the one standard deviation confidence interval of the values determined with the MM equation, whereas the percent error shows that both parameters are unresolved. Likewise, a small but definite bias is observed with the EH transformation, although

TABLE II
RESULTS OF THE MONTE CARLO STUDY DISCUSSED IN THE TEXT FOR
10,000 SIMULATED DATA SETS OF 8 OR 6 (REPORTED IN PARENTHESES) POINTS
EACH CARRYING A UNIFORM ERROR[a]

	MM		EH		LB	
ε	%B	%E	%B	%E	%B	%E
K_m (simulated value = 1)						
0.01	0 (0)	3 (4)	−0 (−0)	4 (5)	1 (1)	8 (10)
0.02	0 (0)	6 (8)	−1 (−1)	8 (11)	2 (4)	17 (23)
0.03	0 (1)	9 (12)	−1 (−2)	13 (16)	6 (12)	29 (*)
0.04	0 (1)	12 (16)	−3 (−4)	17 (21)	14 (15)	73 (*)
0.05	1 (2)	15 (20)	−4 (−7)	20 (26)	23 (34)	* (*)
0.06	2 (3)	18 (24)	−6 (−10)	24 (30)	30 (*)	* (*)
0.07	2 (3)	21 (28)	−9 (−15)	27 (36)	52 (*)	* (*)
0.08	2 (5)	24 (32)	−13 (−19)	31 (41)	80 (*)	* (*)
0.09	3 (5)	27 (36)	−15 (−23)	35 (47)	* (*)	* (*)
0.10	4 (6)	30 (38)	−18 (−28)	39 (52)	* (*)	* (*)
k_{cat} (simulated value = 1)						
0.01	0 (0)	1 (2)	−0 (−0)	1 (2)	0 (0)	3 (4)
0.02	0 (0)	1 (2)	−0 (−1)	2 (3)	1 (1)	6 (9)
0.03	0 (0)	2 (4)	−1 (−1)	3 (5)	2 (5)	10 (71)
0.04	0 (0)	3 (5)	−1 (−2)	4 (6)	4 (5)	25 (*)
0.05	0 (0)	3 (6)	−2 (−4)	4 (8)	6 (14)	* (*)
0.06	0 (1)	4 (7)	−3 (−5)	6 (10)	9 (28)	* (*)
0.07	0 (1)	5 (9)	−4 (−7)	6 (12)	13 (43)	* (*)
0.08	0 (1)	6 (10)	−5 (−9)	7 (13)	16 (64)	* (*)
0.09	0 (1)	6 (11)	−6 (−11)	8 (15)	29 (97)	* (*)
0.10	0 (1)	7 (12)	−7 (−13)	9 (16)	44 (*)	* (*)

[a] The mean, m, and standard deviation, σ, of each parameter obtained by using the MM, EH, or LB transformation are used to calculate the percent bias = $100(m − \psi^*)/\psi^*$ (%B) and percent error = $100\sigma/m$ (%E). Values of %B and %E exceeding 100 are indicated by an asterisk (*).

the mean values of K_m and k_{cat} are within the confidence interval of the MM determinations.

A similar simulation study has been carried out using only 6 velocity values per data set in the substrate concentration range $K_m/4$ to $8K_m$, in order to investigate the effect of a narrower substrate range on parameter resolution with incorrect weighting schemes. The results are shown in Fig. 3, and the relevant statistics are summarized in Table II (values given in parentheses). As expected, the bias and standard deviation of the distribution of each parameter increase, but qualitatively the results resemble those shown in Fig. 2. The LB transformation yields meaningless results

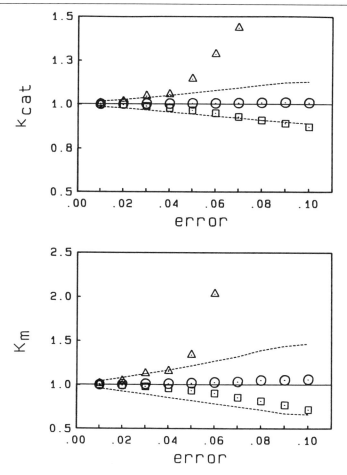

FIG. 3. Results of the Monte Carlo study discussed in the text for 10,000 simulated data sets of 6 points each carrying a uniform error. The mean values of k_{cat} (top) and K_m (bottom) were obtained by using the MM (○), EH (□), or LB (△) equation. The one standard deviation (68%) confidence interval for the parameter values obtained by means of the MM equation is depicted by a dashed line.

as soon as the error exceeds 3%. In both cases the MM equation yields lowest-variance and lowest-bias parameter estimates and performs better than the EH transformation. Therefore, when using incorrect weighting schemes we do observe difficulties in parameter resolution, as shown by large bias and percent error, but this effect strongly depends on the experimental error. In general, however, the LB transformation provides the least reliable parameter estimates for errors less than 3% and yields meaningless results for bigger errors. This conclusion concurs with the results of previous Monte Carlo studies.[8-10]

In many circumstances of practical interest one is not aware of the exact distribution of experimental errors, and therefore a meaningful Monte Carlo study of steady-state kinetics should also consider the consequences of incorrect weighting when the error is not uniformly distributed. A necessary condition for this case to be directly compared to the one already analyzed is that the standard deviation of the fit must be comparable to that observed in the simulation of velocity measurements with uniform error. Simulated values that carry the same average noise, independent of the error distribution, perfectly mimic the real situation where we have a finite noise but we know nothing about the error distribution. Another Monte Carlo study has thus been carried out as follows. Steady-state velocity measurements, normalized by the enzyme concentration e_T, were simulated according to the equation

$$v = k_{cat}x/(K_m + x) + \varepsilon[-2 \ln(RND_1)]^{1/2} \cos(2\pi RND_2)k_{cat}x/(K_m + x) \tag{28}$$

that is, with a pseudorandom error distributed as $N(0,\varepsilon v)$ and proportional to the velocity value. This error simulates a nonuniform distribution that necessarily demands differential weighting in the minimization procedure.

For each data set eight data points were generated in the substrate concentration range $K_m/4$ to $32K_m$, as in the case of uniform error distribution, with K_m and k_{cat} both equal to 1. Likewise, 10 values of ε were used in the simulations from 0.02 to 0.2. These values yielded the same standard deviation of the fit as ε values in the range 0.01–0.1 in Eq. (26). For each value of ε, 10,000 data sets were generated according to Eq. (28), and each set was analyzed according to the MM, EH, and LB equations using uniform weighting in all cases. In this case, the weighting scheme is correct for the EH transformation and incorrect for the MM and LB equations. The results of the simulation study are shown in Fig. 4 and Table III. When velocity measurements in the substrate range $K_m/4$ to $32K_m$ carry an experimental error proportional to v, then there is no significant difference in the parameter values obtained with the MM, EH, and LB equations using uniform weighting. This drastic difference versus what was observed in the case of uniform error has not been pointed out before. The same conclusion is reached for velocity measurements in the substrate range $K_m/4$ to $8K_m$, as shown in Fig. 5 and Table III (values given in parentheses). The parameter bias is approximately the same for all transformations,

[8] J. E. Dowd and D. S. Riggs, *J. Biol. Chem.* **240**, 863 (1965).
[9] G. L. Atkins and I. A. Nimmo, *Biochem. J.* **149**, 775 (1975).
[10] J. G. W. Raaijmakers, *Biometrics* **43**, 793 (1987).

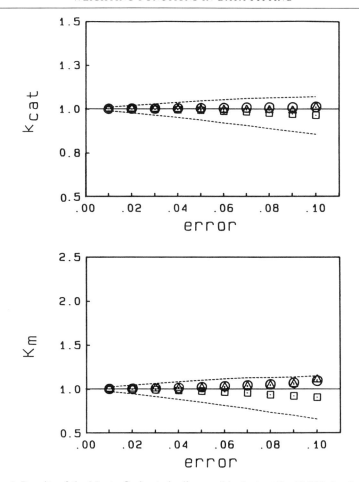

FIG. 4. Results of the Monte Carlo study discussed in the text for 10,000 simulated data sets of 8 points each carrying an error proportional to the velocity value. The mean values of k_{cat} (top) and K_m (bottom) were obtained by using the MM ($\bigcirc$), EH ($\square$), or LB ($\triangle$) equation. The one standard deviation (68%) confidence interval for the parameter values obtained by means of the EH equation is depicted by a dashed line.

whereas the percent error indicates that the EH equation performs best, as implied by the nature of the distribution of errors.

Conclusions

Correct data weighting guarantees least-variance estimates of the parameter values obtained in least-squares minimization. Weighting functions can be derived from the distribution of experimental errors, and data

TABLE III

RESULTS OF THE MONTE CARLO STUDY DISCUSSED IN THE TEXT FOR 10,000 SIMULATED DATA SETS OF 8 OR 6 (REPORTED IN PARENTHESES) POINTS EACH CARRYING AN ERROR PROPORTIONAL TO THE VELOCITY VALUE[a]

	MM		EH		LB	
$\varepsilon/2$	%B	%E	%B	%E	%B	%E
K_m (simulated value = 1)						
0.01	0 (0)	4 (5)	−0 (−0)	3 (3)	0 (0)	4 (4)
0.02	0 (1)	7 (9)	−0 (−1)	5 (7)	0 (0)	7 (9)
0.03	1 (1)	11 (14)	−1 (−1)	8 (10)	0 (1)	11 (14)
0.04	1 (2)	15 (19)	−2 (−2)	10 (13)	1 (1)	15 (19)
0.05	2 (3)	19 (24)	−2 (−3)	13 (17)	1 (3)	19 (26)
0.06	3 (5)	23 (29)	−3 (−5)	16 (20)	2 (5)	24 (32)
0.07	4 (7)	27 (36)	−4 (−7)	18 (24)	3 (7)	28 (61)
0.08	5 (9)	32 (42)	−7 (−9)	21 (28)	4 (9)	60 (93)
0.09	7 (12)	37 (51)	−8 (−11)	24 (32)	6 (15)	92 (*)
0.10	10 (14)	43 (61)	−9 (−13)	27 (35)	11 (22)	* (*)
k_{cat} (simulated value = 1)						
0.01	0 (0)	1 (2)	−0 (−0)	1 (2)	0 (0)	1 (2)
0.02	0 (0)	2 (4)	−0 (−0)	2 (3)	0 (0)	3 (5)
0.03	0 (0)	4 (6)	−0 (−1)	3 (5)	0 (0)	5 (7)
0.04	0 (1)	5 (8)	−1 (−1)	4 (7)	−0 (0)	6 (9)
0.05	0 (1)	6 (10)	−1 (−2)	6 (8)	−0 (0)	8 (13)
0.06	0 (1)	7 (12)	−1 (−2)	7 (10)	1 (1)	10 (16)
0.07	1 (2)	9 (14)	−2 (−3)	8 (12)	1 (2)	12 (27)
0.08	1 (2)	10 (16)	−2 (−4)	9 (13)	1 (2)	24 (37)
0.09	1 (3)	11 (19)	−3 (−5)	10 (15)	2 (4)	30 (58)
0.10	1 (4)	12 (21)	−4 (−6)	11 (16)	4 (6)	56 (89)

[a] The mean, m, and standard deviation, σ, of each parameter obtained by using the MM, EH, or LB transformation are used to calculate the percent bias = $100(m - \psi^*)/\psi^*$ (%B) and percent error = $100\sigma/m$ (%E). Values of %B and %E exceeding 100 are indicated by an asterisk (*).

should be analyzed in the form they are taken experimentally, using a correct weighting scheme. Data transformation leads to changes in the error distribution and consequently demands correct reformulation of the weighting scheme according to Eq. (11). The reader should not be encouraged, however, to apply systematically such a procedure in data fitting because Eq. (11) gives only an approximate estimate of the new weighting factor as a function of the old one and strongly depends on the analytical form of the transformation.

In general, one should not overlook the fact that different weighting schemes have no influence on parameter estimation when the fitting func-

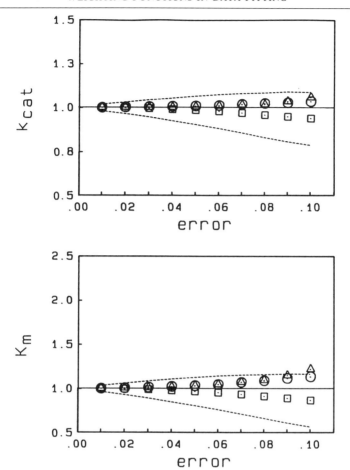

FIG. 5. Results of the Monte Carlo study discussed in the text for 10,000 simulated data sets of 6 points each carrying an error proportional to the velocity value. The mean values of k_{cat} (top) and K_m (bottom) were obtained by using the MM (○), EH (□), or LB (△) equation. The one standard deviation (68%) confidence interval for the parameter values obtained by means of the EH equation is depicted by a dashed line.

tion is correct and data points are errorless, and therefore one should expect very little influence of different weighting schemes on parameter determinations when the fitting function is correct and experimental data are very accurate. When data are not accurate and the distribution of experimental errors is unknown, then it is difficult to predict the effect of incorrect weighting schemes in the general case. Each case should be thoroughly dissected with the help of Monte Carlo simulations in order to assess specifically the influence of weighting functions on parameter determinations for a particular model, under particular experimental conditions. Also, a direct experimental measure of the error distribution is strongly recommended. In some cases, the use of nonparametric procedures of parameter estimation might be useful.[11,12]

The specific case of the analysis of steady-state kinetics dealt with in this chapter has revealed the consequences of incorrect weighting schemes as a function of experimental error. Of particular practical interest is the fact that the Lineweaver–Burk transformation always yields the least reliable parameter estimates, regardless of the nature and extent of experimental error. The budding enzymologist should thus be encouraged to abandon such a procedure in practical applications. The Michaelis–Menten equation performs best with uniformly distributed errors and so does the Eadie–Hofstee transformation when errors are proportional to the velocity value. However, when errors are proportional to the velocity value the Michaelis–Menten equation performs better than the Eadie–Hofstee transformation in the case of uniformly distributed errors. This suggests that the Michaelis–Menten equation should be preferred in the analysis of steady-state kinetics whenever the exact distribution of experimental errors is unknown. Although when dealing with errors of 1% or less the values of K_m and k_{cat} are not significantly affected by the particular transformation, nevertheless the reader should be very cautious in generalizing such a result. The analytical form of the Michaelis–Menten equation is particularly simple and does not comply with intrinsic difficulties in parameter resolution. When the fitting function contains parameters that are difficult to resolve, as in the case of ligand binding equilibria of multimeric proteins such as hemoglobin, then even an experimental error as small as 0.3% may lead to significantly different parameter estimates when using different weighting schemes.[5]

A final remark should be made as to the operational aspects of weighting functions. Throughout this chapter we have linked the concept of weighting factor to the experimental error associated with a given experimental determination. Looking back to Figs. 2–5 one sees that when the fitting problem is correctly cast, then it makes very little difference to determine K_m and k_{cat} in the substrate range $K_m/4$ to $32K_m$ or $K_m/4$ to $8K_m$. This means that we could drop two points at high enzyme saturation from the data set in Fig. 1 without affecting the results, and we in fact obtain in this case $k_{cat} = 104.7 \pm 1.2 \text{ sec}^{-1}$ and $K_m = 2.89 \pm 0.11 \ \mu M$. Dropping experimental data is mathematically equivalent to weighting those data

[11] A. Cornish-Bowden and R. Eisenthal, *Biochem. J.* **139**, 721 (1974).
[12] E. Di Cera and S. J. Gill, *Biophys. Chem.* **29**, 351 (1988).

points zero, which implies that those points carry an infinite experimental error. Because we know that this is not the case, then it is clear that the two points we dropped in Fig. 1 have very little, if any, influence on the determination of K_m and k_{cat}. This is because the fitting function is not sufficiently sensitive to experimental points collected for substrate concentrations over $8K_m$. Indeed, one can assign an informational content to any experimental data point by consideration of the information stored in the fitting equation. The informational content depends solely on the particular form of the fitting function and is independent of experimental errors. Consequently, it must not be used to assess correct weighting factors that are set solely by experimental errors, as we have seen. Rather, the informational content of a data point should be seen as an operationally useful concept characterizing the particular fitting problem under consideration that may indicate the range of the independent variable to be explored experimentally in the resolution of model parameters.

Acknowledgments

I am grateful to Dr. Michael L. Doyle for assistance in preparation of the figures.

[5] Analysis of Residuals: Criteria for Determining Goodness-of-Fit

By MARTIN STRAUME and MICHAEL L. JOHNSON

Introduction

Parameter-estimation procedures provide quantitation of experimental data in terms of model parameters characteristic of some mathematical description of the relationship between an observable (the dependent variable) and experimental variables [the independent variable(s)]. Processes such as least-squares minimization procedures[1,2] will produce the maximum likelihood model parameter values based on minimization of the sum of squared residuals (i.e., the sum of the squares of the differences between the observed values and the corresponding theoretical values calculated by the model employed to analyze the data). There are assumptions regarding the properties of experimental uncertainty distributions contained in

[1] M. L. Johnson and L. M. Faunt, this volume [1].
[2] M. L. Johnson and S. G. Frasier, this series, Vol. 117, p. 301.

the data that are implicit to the validity of the least-squares method of parameter estimation, and the reader is referred to Refs. 1 and 2 for a more detailed discussion. The widespread availability of computer hardware and software (particularly that implementing parameter-estimation algorithms such as least-squares) translates into commonplace implementation of parameter-estimation algorithms and, on occasion, perhaps a not-close-enough look at the appropriateness of particular mathematical models as applied to some experimental data.

Of course, just how critical a determination of the appropriateness of fit of a model is required will vary depending on the significance of the data, the phenomenon, and the interpretation being considered. When looking at simple, routine analytical applications (linear or polynomial empirical fits of protein assay standard curves, for example, or perhaps analysis for single-exponential decay in kinetic enzyme assays for first-order rate constant estimates to use for defining specific activities during steps of purification procedures), it may not be particularly important to examine carefully the quality of fit produced by the model used to analyze the data. An empirical or "lower-order" estimate of the behavior of some system property in these cases is fully sufficient to achieve the goals of the analysis. However, when quantitatively modeling detailed aspects of biomolecular properties, particularly when asking more advanced theoretical models to account for experimental data of ever increasing quality (i.e.,more highly determined data), many sophisticated numerical methods and complex mathematical modeling techniques are often implemented. In these cases, a careful eye must be directed toward consideration of the ability of the model to characterize the available experimentally determined system properties reliably, sometimes to quite exquisite levels of determination.

To perform these types of detailed analyses (and, in principle, for any analysis), data must be generated by experimental protocols that provide data (1) possessing experimental uncertainties that are randomly distributed and (2) free of systematic behavior not fully accounted for by the mathematical model employed in analysis. A mathematical model must be defined to describe the dependence of the observable on the independent variable(s) under experimental control. The definition of an appropriate mathematical model involves considerations of how to transform the experimentally observed system behavior into a mathematical description that permits physical interpretation of the model parameter values. In this way, information about the biomolecular phenomena underlying the system response is quantitatively defined. Such modeling efforts can become quite specific when addressing molecular level interpretations of the functional, structural, and thermodynamic properties of biological systems.

Ongoing biochemical and biophysical studies to elucidate the molecular and thermodynamic foundations of macromolecular structure–function relationships have been producing data from experiments designed to test, to ever finer levels of detail, behavior predicted or theorized to exist as based on modeling efforts. All complementary experimental information available about a particular system must be incorporated into comprehensive mathematical models to account fully for all the known properties of a system. Therefore, data regarding structural properties, functional properties, influences of experimental conditions (e.g., ionic strength, pH, and ligand concentration), and any other specifically relevant system variables must, in principle, all be consistent with a common model descriptive of the system under study to be comprehensively valid. Approximations in data analysis applications such as these are therefore no longer tolerable so as to achieve an accurate and precise characterization of biochemical or biophysical properties. Nor are approximations necessary given the recent increases in computational capacity both in terms of hardware capabilities as well as software availability and theoretical advancements. Analyses of better determined experimental data sometimes indicate deficiencies in current interpretative models, thereby prompting a closer look at the system and how it is best modeled mathematically. The consideration of residuals (the differences between observed and calculated dependent variable values) becomes a very important element in the overall data analysis process in cases where attempts to model detailed molecular system properties mathematically are being pursued.

The significance of subtle behavior in residuals may suggest the presence of a significant system property that is overlooked by the current mathematical model. But a more fundamental role served by examination of residuals is in providing information on which to base a judgment about the appropriateness of a particular mathematical description of system behavior as a function of some independent, experimental variable(s). If an examination of the residuals obtained from a parameter-estimation procedure on some experimental data yields the conclusion that the data are reliably characterized by the mathematical model (i.e., that a good fit to the data is obtained possessing no unaccounted for residual systematic behavior in the data), this is not to say that this represents justification for necessarily accepting the model as correct. Rather, it indicates that the model employed is sufficient to characterize the behavior of the experimental data. This is the same as saying that the data considered provide no reason to reject the current model as unacceptable. The residuals for a case in which a "good fit" is obtained then, in principle, represent the experimental uncertainty distribution for the data set. However, if an examination of the residuals indicates inconsistencies between the data and the behavior predicted by the analysis, then the current model may correctly be rejected and considered unacceptable (unless some other source for the residual systematic behavior is identified).

When considering residuals, a qualitative approach is often the most revealing and informative. For example, generating plots to represent visually trends and correlations provides a direct and often unambiguous basis for a judgment on the validity of a fit. Of course, quantitative methods to test more rigorously particular properties of residuals sometimes must be considered in order to quantitate the statistical significance of conclusions drawn as a result of data analysis. Some of the available methods for considering residuals will be discussed below with the aid of illustrative examples.

Scatter Diagram Residual Plots

Visualizing residuals is commonly performed by generating scatter diagrams.[3] Residuals may be plotted as a function of various experimental variables to permit convenient identification of trends that may not have been accounted for by the analytical model. Residuals are most commonly plotted as a function of either the values of the independent variable(s) (e.g., time in kinetic experiments or ligand concentration in ligand binding experiments) or the calculated values of the dependent variable (i.e., the values of the experimental observable calculated from the model). However, residual plots versus some other functional relationship of the independent variable(s) or some other potentially significant variable that was not explicitly considered in the original model may also provide information about important relationships that have not been previously identified.

In Fig. 1 is presented a simulated scatter diagram to illustrate the type of information potentially provided by visual inspection of residual plots. The circles represent pseudo-Gaussian distributed residuals with a standard deviation of 1.0 and a mean of 0.0. The points denoted by crosses represent similar bandwidth noise as seen in the pseudo-Gaussian distributed points but possessing higher-order structure superimposed on them. Visual inspection of such plots permits ready identification of deficiencies in the ability of an analytical model to describe adequately the behavior

---

[3] P. Armitage, "Statistical Methods in Medical Research," 4th Printing, p. 316. Blackwell, Oxford, 1977.

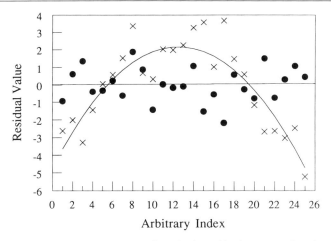

FIG. 1. Scatter diagrams for two sets of synthetic residuals generated to demonstrate a normally distributed set of residuals and another that exhibits a trend in the behavior of the residual values as a function of residual number (presented as an arbitrary index in this example).

of experimental data if nonrandom residuals are obviously present (as in the residuals represented by the crosses in Fig. 1, for example). This type of observation would suggest that there exists some systematic behavior in the data (as a function of the variable against which the residuals were plotted) that was not well accounted for by the model employed in analysis.

An examination of the trend in residuals as a function of some particular variable space may even provide information about the type of quantitative relationship that must be accommodated by the analytical model that currently is not. However, correctly accounting for any newly incorporated relationships into currently existing analytical models requires re-evaluation of the data set(s) originally considered. This is necessary so as to simultaneously estimate values for all of the model parameters characteristic of the model, both previously existing and newly incorporated. Quantifying phenomena originally omitted from consideration by a model must not be attempted by analyzing the resulting residuals. Correlation among parameters must be accommodated during a parameter-estimation procedure so as to produce the true best-fit parameter values that accurately characterize the interdependence between parameters of the model and between these parameters and the properties of the data being analyzed (the dependence of the experimental observable on the independent experimental variables as well as on the distribution of experimental uncertainty in the data).

Cumulative Probability Distributions of Residuals

Another visual method for examining residuals involves generating a cumulative frequency plot.[4] The information provided by this form of consideration of residuals is related to the randomness of the distribution of residual values. The process requires that the residuals be ordered and numbered sequentially such that

$$r_1 < r_2 < r_3 < \cdots < r_n$$

where r_i is the ith residual value. A quantity P_i is then defined such that

$$P_i = (i - 0.5)/n$$

Here, P_i, the cumulative probability, represents a statistical estimate of the theoretical probability of finding the ith residual (out of n total residuals) with a value of r_i if they are distributed randomly (i.e., Gaussian or normally distributed residuals). A graph of the standard normal deviate, or Z-value (which represents the number of standard deviations from the mean), corresponding to the cumulative probability P_i versus the values of the ordered residuals will then produce a straight line of points, all of which will be very near the theoretical cumulative probability line if the residuals are distributed randomly. The Z-values corresponding to particular levels of probability may be obtained from tabulations in statistics books or calculated directly by appropriate integration of the function defining Gaussian distributed probability.

The cumulative probability plots corresponding to the two sets of simulated residuals presented in Fig. 1 are shown in Fig. 2. The points for the pseudorandom residuals (circles) form a linear array with all points in close proximity to the theoretical line. The slope of this line is 1.0 in the manner plotted in this graph, corresponding to a standard deviation of 1.0 for this particular distribution of residuals. The points for the distribution exhibiting residual structure in the scatter diagram of Fig. 1 (crossed points) can be seen to generally follow along their theoretical line (with a slope of 2.5 corresponding to an apparent standard deviation of 2.5); however, they show systematic behavior and occasionally deviate considerably from the line (relative to the near superposition for the pseudorandomly distributed residuals). This level of deviation in a cumulative probability plot suggests that the data are not well characterized by the model used to describe their behavior because the resulting residuals clearly exhibit nonrandom behavior.

[4] Y. Bard, "Nonlinear Parameter Estimation," p. 201. Academic Press, New York, 1974.

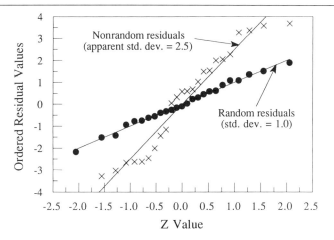

FIG. 2. Cumulative frequency plots for the two sets of residuals presented in Fig. 1. The ordered residual values are plotted relative to the Z-value (corresponding to the number of estimated standard deviations from the mean, in this case zero) characteristic of Gaussian distributed residuals. The estimated standard deviations are 1.0 and 2.5 for the Gaussian and nonrandom residuals, respectively, as reflected in the slopes of the theoretical lines.

χ^2 Statistic: Quantifying Observed versus Expected Frequencies of Residual Values

To assess the properties of distributions of residuals more quantitatively, one may generate a discrete, theoretical residual probability distribution (based on an assumption of randomly distributed residuals) and compare the distribution of observed residual values with these expected frequencies.[5,6] A histogram is in effect created in which the range of residual values is divided into a number of intervals such that at least one residual (out the n total residuals being considered) is expected to exist in each interval. The expected frequencies are then compared to the observed frequencies by the relationship

$$\chi^2 = \Sigma \left[(O_i - E_i)^2/E_i \right]$$

which is summed over each interval considered. Here, O_i represents the observed number of residuals possessing values within the range defined by interval i. Analogously, E_i is the expected number of residuals in this interval if the residuals are randomly distributed. The value of this

[5] P. Armitage, "Statistical Methods in Medical Research," 4th Printing, p. 391. Blackwell, Oxford, 1977.

[6] W. W. Daniel, "Biostatistics: A Foundation for Analysis in the Health Sciences," 2nd Ed. Wiley, New York, 1978.

calculated parameter will be distributed approximately as the χ^2 statistic (for the desired level of confidence and the number of degrees of freedom). The significance of the χ^2 statistic is that it represents a method for quantifying the probability that the distribution of residuals being considered is not random.

When the χ^2 statistic is applied to the residual distributions presented in Fig. 1 (see Table I), we find that the pseudo-Gaussian residual distribution produces a χ^2 value of 1.326, whereas that which possessed residual structure had a χ^2 value of 7.344. In this case, nine intervals were considered, the central seven being finite with a width of one-half the (apparent, in the case of the nonrandom residuals) standard deviation, with the two extreme intervals considering the remaining probability out to $\pm\infty$. For a total of 25 residuals, this choice of interval width and number produced expected frequencies of at least one for each of the intervals considered (1.0025 being the lowest, for the two end intervals). When considering this type of analysis of a residual distribution, small expected frequencies must be dealt with so as to produce intervals with at least one for an expected

TABLE I
RESIDUAL PROBABILITY DISTRIBUTION PER INTERVAL[a]

	Observed		
Z-Interval	Pseudo-Gaussian	Structured	Expected
$-\infty$, -1.75	1	1	1.0025
-1.75, -1.25	2	1	1.6375
-1.25, -0.75	2	6	3.0250
-0.75, -0.25	6	6	4.3675
-0.25, 0.25	4	4	4.9350
0.25, 0.75	4	3	4.3675
0.75, 1.25	3	4	3.0250
1.25, 1.75	2	0	1.6375
1.75, $+\infty$	1	0	1.0025

[a] For $\chi^2 = \Sigma [(O_i - E_i)^2/E_i]$, where O_i is the observed number of residuals with values in interval i and E_i is the expected number of residuals with values in interval i. χ^2 (pseudo-Gaussian) = 1.326; χ^2 (structured) = 7.344. The probabilities associated with these values of χ^2 (for seven degrees of freedom, nine intervals minus the two constraints for estimating the mean and variance of the distributions) verify that the Gaussian distributed residuals are correctly identified as being Gaussian distributed [χ^2 (pseudo-Gaussian) = 1.326], whereas the nonrandom residuals are confirmed to be non-Gaussian [χ^2 (structured) = 7.344].

frequency. With small numbers of residuals, this may become a necessary concern.

A χ^2 value of 1.326 means that there is between a 1 and 2.5% chance that the pseudo-Gaussian residuals in Fig. 1 are not randomly distributed. This is the derived level of confidence indicated by this χ^2 value with 7 degrees of freedom [in this case, the number of degrees of freedom is 9 (the number of intervals) minus 1 (for the requirement that $\Sigma\ O_i = \Sigma\ E_i$) minus 1 (for the estimation of an apparent standard deviation) equals 7]. The considerably larger χ^2 value of 7.344 for the structured residuals of Fig. 1 indicates a significantly higher probability that the residuals are indeed not randomly distributed, supporting the conclusion drawn by inspection of the scatter diagrams in Fig. 1.

Kolmogorov–Smirnov Test: An Alternative to the χ^2 Statistic

As an alternative to the χ^2 method for determining whether the residuals generated from an analyis of data by a mathematical model are randomly distributed, one may apply the Kolmogorov–Smirnov test.[6] The Kolmogorov–Smirnov test has a number of advantages over the χ^2 treatment. Whereas the χ^2 approach requires compartmentalization of residuals into discrete intervals, the Kolmogorov–Smirnov test has no such requirement. This relaxes the constraint of possessing a sufficient number of residuals so as to significantly populate each of the intervals being considered in the χ^2 analysis. And to provide a closer approximation to a continuous distribution, the χ^2 approach requires consideration of a large number of intervals. The Kolmogorov–Smirnov approach requires no discrete approximations but rather provides a quantitative basis for making a statistical comparison between the cumulative distribution of a set of residuals and any theoretical cumulative probability distribution (i.e., not limited to only a Gaussian probability distribution).

The statistic used in the Kolmogorov–Smirnov test, D, is the magnitude of the greatest deviation between the observed residual values at their associated cumulative probabilities and the particular cumulative probability distribution function with which the residuals are being compared. To determine this quantity, one must consider the discrete values of the observed residuals, r_i, at the cumulative probability associated with each particular residual, P_i,

$$P_i = [(i - 0.5)/n]; \qquad 1 \leq i \leq n$$

relative to the continuous theoretical cumulative probability function to which the distribution of residuals is being compared (e.g., that of a Gaussian distribution possessing the calculated standard deviation, as visually represented in the cumulative probability plots of Fig. 2). The continuous nature of the theoretical cumulative probability function requires that both end points of each interval defined by the discrete points corresponding to the residuals be considered explicitly. The parameter D is therefore defined as

$$D = \max_{1 \le i \le n}\{\max[|r_i(P_i) - r_{\text{theory}}(P_i)|, |r_{i-1}(P_{i-1}) - r_{\text{theory}}(P_i)|]\}$$

The deviations between the observed and theoretical values are thus considered for each end of each interval defined by the observed values of $r_i(P_i)$ and $r_{i-1}(P_{i-1})$. The value of this statistic is then compared with tabulations of significance levels for the appropriate number of residuals (i.e., sample size), a too-large value of D justifying rejection of the particular theoretical cumulative probability distribution function as incorrectly describing the distribution of residuals (at some specified level of confidence).

Runs Test: Quantifying Trends in Residuals

The existence of trends in residuals with respect to either the independent (i.e., experimental) or dependent (i.e., the experimental observable) variables suggests that some systematic behavior is present in the data that is not accounted for by the analytical model. Trends in residuals will often manifest themselves as causing too few runs (consecutive residual values of the same sign) or, in cases where negative serial correlation occurs, causing too many runs. A convenient way to assess quantitatively this quality of a distribution of residuals is to perform a runs test.[4] The method involves calculating the expected number of runs given the total number of residuals as well as an estimate of variance in this expected number of runs.

The expected number of runs, R, may be calculated from the total number of positive and negative valued residuals, n_p and n_n, as

$$R = \{[2n_p n_n/(n_p + n_n)] + 1\}$$

The variance in the expected number of runs, σ_R^2, is then calculated as

$$\sigma_R^2 = \{[2n_p n_n(2n_p n_n - n_p - n_n)]/[(n_p + n_n)^2(n_p + n_n - 1)]\}$$

A quantitative comparison is then made between the expected number of runs, R, and the observed number of runs, n_R, by calculating an estimate for the standard normal deviate as

$$Z = |(n_R - R + 0.5)/\sigma_R|$$

When n_p and n_n are both greater than 10, Z will be distributed approximately as a standard normal deviate. In other words, the calculated value of Z is the number of standard deviations that the observed number of runs is from the expected number of runs for a randomly distributed set of residuals of the number being considered. The value of 0.5 is a continuity correction to account for biases introduced by approximating a discrete distribution with a continuous one. This correction is $+0.5$ (as above) when testing for too few runs and is -0.5 when testing for too many runs. The test is therefore estimating the probability that the number of runs observed is different from that expected from randomly distributed residuals. The greater the value of Z, the greater the likelihood that there exists some form of correlation in the residuals relative to the particular variable being considered.

In Table II is presented an application of the runs test to the residuals of Fig. 1. The results clearly indicate that the number of runs expected and observed for the pseudo-Gaussian distributed residuals agree quite well (Z values of 0.83 and 1.24), whereas the agreement between expected and observed numbers of runs with the structured residuals is very different (Z values of 4.05 and 4.48). The probability that the distributions of residuals exhibit the "correct" number of runs is therefore statistically acceptable in the former case (less than 0.83 standard deviations from expected) and statistically unacceptable in the latter (more than 4 standard deviations from expected). A cutoff value for acceptability may be consid-

TABLE II
RUNS TEST APPLIED TO RESIDUALS[a]

| | Residuals | |
Parameter	Pseudo-Gaussian	Structured
n_p	12	15
n_n	13	10
R	13.48	13
σ_R^2	5.97	5.5
n_R	16	3
Z_{tf}	1.24	4.05
Z_{tm}	0.83	4.48

[a] The Z values derived from this analysis suggest that the expected number of runs is encountered in the Gaussian distributed residuals ($Z_{tf} = 1.24$ and $Z_{tm} = 0.83$), a situation that is not the case for the nonrandom residual distribution ($Z_{tf} = 4.05$ and $Z_{tm} = 4.48$). Subscripts refer to testing for too few or too many runs.

ered to be 2.5 to 3 standard deviations from the expected value (corresponding to probabilities of approximately 1 to 0.25%). The values calculated in this illustrative example fall well to either side of this cutoff range of Z values.

Serial Lag$_n$ Plots: Identifying Serial Correlation

A situation in which too few runs are present in the residual distribution indicates positive serial correlation. Too many runs, on the other hand, is a situation characteristic of negative serial correlation in residuals. Serial correlation suggests systematic behavior with time and, in fact, is often considered in this context. However, this sort of serial dependence may also be of significance when considering parameter spaces other than time.

Visualization of this phenomenon is best achieved by lag$_n$ serial correlation plots.[7] The residual values are plotted against each other, each value plotted versus the one occurring n units before the other. As demonstrated in Fig. 3 for the two residual distributions presented in Fig. 1, considerable correlation is suggested in the structured residuals in the lag$_1$, lag$_2$, and lag$_3$ serial plots but is no longer obvious in the lag$_4$ plot, whereas the pseudo-Gaussian residual distribution produces no obvious trend in any of the serial lag$_n$ plots. The presence of positive serial correlation (i.e., adjacent residuals with the same sign) is evidenced by positive sloping point distributions (as demonstrated by the distribution exhibiting residual structure in Fig. 3), whereas negative slopes characterize negative serial correlation (i.e., adjacent residuals with opposite sign). In the absence of correlation, the points will be randomly clustered about the origin of the plot (as demonstrated by the pseudo-Gaussian distributed residuals in Fig. 3).

Durbin–Watson Test: Quantitative Testing for Serial Correlation

The Durbin–Watson test provides a quantitative, statistical basis on which to judge whether serial correlation exists in residuals.[7] The test permits estimation of the probability that serial correlation exists by attempting to account for effects of correlation in the residuals by the following formula

$$r_i = \rho r_{i-1} + Z_i$$

Here, r_i and r_{i-1} correspond to the ith and $(i - 1)$th residuals, and ρ

---

[7] N. R. Draper and R. Smith, "Applied Regression Analysis," 2nd Ed., p. 153. Wiley, New York, 1981.

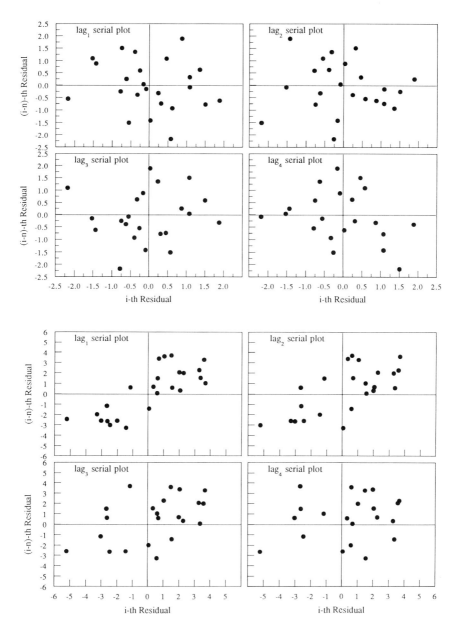

FIG. 3. Lag$_n$ serial plots for the residual distributions presented in Fig. 1 are displayed for lags of 1, 2, 3, and 4. It is apparent that the nonrandom residuals of Fig. 1 (bottom plots) possess some positive serial correlation as demonstrated by positive slopes in the point distributions for the lag$_1$, lag$_2$, and lag$_3$ cases but disappearing by the lag$_4$ plot. The Gaussian distributed residuals (top plots), on the other hand, exhibit no indication of serial correlation in any of the four serial lag$_n$ plots.

and Z_i provide for quantitation of any effects of serial correlation. This approach is based on the null hypothesis that the residuals are distributed in a manner free of any serial correlation and possessing some constant variance. The above expression relating adjacent residuals is a means to quantify deviation from the hypothesized ideal case. The underlying assumptions are that the residual values, the r_i's, as well as the values for the Z_i's in the above equation each possess constant variances, which for the case of the residuals is given by

$$\sigma^2 = \sigma_r^2/(1 - \rho^2)$$

The case in which no serial correlation is indicated is given by $\rho = 0$. The variance then reduces to that estimated from the original distribution (and the null hypothesis is accepted).

A parameter is calculated to provide a statistical characterization of satisfying the null hypothesis and therefore addressing whether any correlation is suggested in the residuals being considered. This parameter is given by

$$d = \sum_{i=2}^{n} (r_i - r_{i-1})^2 \bigg/ \sum_{i=1}^{n} r_i^2$$

In assigning a probability to the likelihood of serial correlation existing in the considered residuals, two critical values of d (a lower and upper value, d_l and d_u) are specified, thus defining a range of values associated with a specified probability and the appropriate number of degrees freedom. Tables of these critical values at various levels of confidence may be consulted[7] and permit testing of the following three conditions: $\rho > 0$, $\rho < 0$, or $\rho \neq 0$. When considering the first case ($\rho > 0$), $d < d_l$ is significant (at the confidence level specified in the table used), and serial correlation with $\rho > 0$ is accepted. A value of $d > d_u$ indicates that the case $\rho > 0$ may be rejected, that is, that one is *not* justified in assigning any positive serial correlation to the distribution of residuals being considered. Intermediate values of d (between d_l and d_u) produce an "inconclusive" test result. The case of $\rho < 0$ is considered in an analogous manner except that the value of $(4 - d)$ is used in comparisons with tabulated values of d_l and d_u. The same process as outlined above for the first case applies here as well. The test for $\rho \neq 0$ is performed by seeing whether $d < d_l$ or $(4 - d) < d_l$. If so, then $\rho \neq 0$ at twice the specified level of confidence (it is now a two-sided test). If $d > d_u$ *and* $(4 - d) > d_u$, then $\rho = 0$ at twice the specified level of confidence. Otherwise, the test is considered "inconclusive." To resolve this "inconclusive" occurrence, one may assume the conservative approach and reject once the more stringent criterion of the two is exceeded.

Autocorrelation: Detecting Serial Correlation in
  Time Series Experiments

Experimental data collected as a time series typically exhibit serial correlations. These serial correlations arise when the random uncertainties superimposed on the experimental data tend to have values related to the uncertainties of other data points that are close temporally. For example, if one is measuring the weight of a test animal once a month and the data are expressed as a weight gain per month, negative serial correlation may be expected. This negative serial correlation is expected because a positive experimental error in an estimated weight gain for one month (i.e., an overestimate) would cause the weight gain for the next month to be under-estimated.

A basic assumption of parameter-estimation procedures is that the experimental data points are independent observations. Therefore, if the weighted differences between experimental data points and the fitted function (the residuals) exhibit such a serial correlation, then either the observations are not independent or the mathematical model did not correctly describe the experimental data. Thus, the serial correlation of the residuals for adjacent and nearby points provides a measure of the quality of the fit.

The autocorrelation function provides a simple method to present this serial correlation for a series of different lags, k.[8] The lag refers to the number of data points between the observations for a particular autocorrelation. For a series of N observations, Y_t, with a mean value of μ, the autocorrelation function is defined as

$$\beta_k = \hat{\sigma}_k / \hat{\sigma}_0$$

for $k = 0, 1, 2, \ldots, K$, where the autocovariance function is

$$\hat{\sigma}_k = \frac{1}{n} \sum_{t=1}^{n-k} (Y_t - \mu)(Y_{t+k} - \mu)$$

For $k = 0, 1, 2, \ldots, K$. In these equations, K is a maximal lag less than n. The autocorrelation function has a value between -1 and $+1$. Note that the autocorrelation function for a zero lag is equal to 1 by definition.

The expected variance[9] of the autocorrelation coefficient of a random process with independent, identically distributed random (normal) errors is

---

[8] G. E. P. Box and G. M. Jenkins, "Time Series Analysis Forecasting and Control," p. 33. Holden-Day, Oakland, California, 1976.
[9] P. A. P. Moran, *Biometrika* **34**, 281 (1947).

$$\mathrm{var}(\beta_k) = \frac{n - k}{n(n + 2)}$$

where μ is assumed to be zero.

Autocorrelations are presented graphically as a function of k. This allows an investigator to compare easily the autocorrelation at a large series of lags k with the corresponding associated standard errors (square root of the variance) to decide if any significant autocorrelations exist.

χ^2 Test: Quantitation of Goodness-of-Fit

After verification that the residuals resulting from a model parameter-estimation process to a set of data are indeed free of any systematic trends relative to any variables of significance (i.e., dependent or independent variables), a quantitative estimate of the adequacy of the particular model in describing the data is possible. Calculation of the χ^2 statistic is a common quantitative test employed to provide a statistical estimate of the quality of fit of a theoretical, mathematical description of the behavior of a system to that measured experimentally.[10] The value of the χ^2 statistic varies approximately as the number of degrees of freedom in situations where the mathematical description is correct and only random fluctuations (i.e., experimental noise) contribute to deviations between calculated and observed dependent values. The χ^2 statistic is defined as

$$\chi^2 = \sum_{i=1}^{n} \left[\frac{Y_{\mathrm{obs},i} - Y_{\mathrm{calc},i}}{\sigma_i} \right]^2$$

that is, as the sum over all n data points of the squared, normalized differences between each observed and calculated value of the dependent variable ($Y_{\mathrm{obs},i} - Y_{\mathrm{calc},i}$), normalized with respect to the error estimate for that particular point (σ_i).

The required knowledge of an accurate estimate for the uncertainty associated with each observed value makes it challenging sometimes to implement this test. It is just these estimated uncertainties that give the χ^2 test its statistical significance, by appropriately normalizing the residuals. By dividing this calculated value of χ^2 by the number of degrees of freedom, the reduced χ^2 value is obtained. The number of degrees of freedom is defined as the number of data points (n) minus the number of parameters estimated during analysis (p) minus 1 (i.e., $NDF = n - p - 1$). The value of the reduced χ^2 value will quite nearly approximate 1 if both (1) the estimated uncertainties, σ_i, are accurate and (2) the mathematical model

[10] P. R. Bevington, "Data Reduction and Error Analysis for the Physical Sciences," p. 187. McGraw-Hill, New York, 1969.

used in analysis accurately describes the data. With accurate knowledge of the experimental uncertainty, it is possible to define statistically the probability that a given model is an accurate description of the observed behavior.

Outliers: Identifying Bad Points

In any experimental measurement, occasionally values may be observed that produce an unusually large residual value after an analysis of the data is performed. The existence of an outlier (or "bad point") suggests that some aberration may have occurred with the measurement of the point. The presence of such a point in the data set being analyzed may influence the derived model parameter values significantly relative to those that would be obtained from an analysis without the apparent outliers. It is therefore important to identify such "bad points" and perhaps reconsider the data set(s) being analyzed without these suspect points.

Visual inspection of residual scatter diagrams often reveals the presence of obvious outliers. Cumulative frequency plots will also indicate the presence of outliers, although perhaps in a less direct manner. Visualization methods may suggest the presence of such points, but what method should be used to decide whether a point is an outlier or just a point with a low probability of being valid? A method to provide a quantitative basis for making this decision derives from estimating the apparent standard deviation of the points after analysis. This is calculated as the square root of the variance of fit obtained from analysis of an unweighted data set. The variance of fit is defined as the sum of the squared residuals divided by the number of degrees of freedom (the number of data points minus the number of parameters being estimated). In the case that the model employed is capable of reliably characterizing the data (i.e., capable of giving a "good fit"), the distribution of residuals will, in principle, represent the distribution of experimental uncertainty. Any residuals possessing values that are more than approximately 2.5 to 3 standard deviations from the mean have only a 1 to 0.25% chance of being valid. When considering relatively large data sets (of the order of hundreds of points or more), the statistical probability of a residual possessing a value 3 standard deviations from the mean suggests that such a point should be expected about once in every 400 data points.

Identifying Influential Observations

The presence of outliers (as discussed in the previous section) may produce derived model parameter values that are biased as a result of the influence of outliers. Methods to test for influential observations may be applied to determine the influence of particular data points or regions of independent variable space on the parameters of the analytical model.[7] The influence a potential bad point may have on the resulting model parameter values will be dependent on whether there exist other data points in the immediate vicinity of the suspect point (i.e., in an area of high data density) or whether the point is relatively isolated from others. And if there are regions of low data density, influential observations may not be made apparent by looking for outliers. That is because the relatively few points defining part of an independent parameter space may be largely responsible for determination of one (or a few) particular model parameters but have very little influence on other model parameters. These points will then represent a particularly influential region of independent parameter space that may strongly affect the outcome of an analysis but may at the same time be difficult to identify as being bad points.

One approach is to omit suspected influential regions of data from consideration during analysis to see if any portion of the complete data set can be identified as being inconsistent with results suggested by consideration of other regions of independent parameter space. A difficulty that may be encountered is that particular regions of independent parameter space may be almost exclusively responsible for determining particular model parameters. Omitting such regions of data from analysis may not permit a complete determination of all the parameters characteristic of the model. If such a situation is encountered, it indicates that a higher level of determination is necessary in this region of independent parameter space and that the experimental protocol during acquisition of data should be modified to permit more data to be accumulated in this "influential window."

The various quantitative methods that have been developed to address influential observations[7] generally involve reconsideration of multiple modified data sets in which some points have been omitted from consideration. The variation in the values of the derived model parameters arising from considering multiple such modified data sets then indicates the degree to which particular regions of data influence various model parameters. If an influential region of independent parameter space is identified, a relatively easy fix to the dilemma is to change the data acquisition protocol to take more experimental measurements over the influential region of independent parameter space.

Conclusions

Qualitative and quantitative examination of residuals resulting from analysis of a set (or sets) of experimental data provides information on which a judgment can be made regarding the validity of particular mathematical formulations for reliably characterizing the considered experimental data. With the advances in biochemical and biophysical instrumentation as well as computer hardware and software seen in recent years (and the anticipated advances from ongoing development), quantitative descriptions of biological system properties are continuously being better determined. Deficiencies in current models characteristic of system behavior are often recognized when more highly determined experimental data become available for analysis. Accommodation of these recognized deficiencies then requires evolution of the particular mathematical description to more advanced levels. In so doing, a more comprehensive understanding of the biochemical or biophysical properties of the system often results.

An interpretation of derived model parameter values implicitly relies on the statistical validity of a particular mathematical model as accurately describing observed experimental system behavior. The concepts and approaches outlined in the present chapter provide a survey of methods available for qualitatively and quantitatively considering residuals generated from data analysis procedures. In those cases where very precise interpretation of experimental observations is required, a thorough, quantitative consideration of residuals may be necessary in order to address the statistical validity of particularly detailed mathematical models designed to account for the biochemical or biophysical properties of any experimental system of interest.

Acknowledgments

This work was supported in part by National Institutes of Health Grants RR-04328, GM-28928, and DK-38942, National Science Foundation Grant DIR-8920162, the National Science Foundation Science and Technology Center for Biological Timing of the University of Virginia, the Diabetes Endocrinology Research Center of the University of Virginia, and the Biodynamics Institute of the University of Virginia.

[6] Analysis of Ligand-Binding Data with Experimental Uncertainties in Independent Variables*

By MICHAEL L. JOHNSON

Introduction

The statistically correct application of any curve-fitting procedure used to obtain estimates of derived parameters requires that a number of assumptions be satisfied. For instance, the parameter-estimation procedure must be tailored to suit the shape and magnitude of the particular distribution of random experimental uncertainties which are inherent in both the dependent variable (the ordinates) and the independent variables (the abscissas). The parameter-estimation procedure must also consider any correlations between the experimental uncertainties of the various dependent and independent variables.

The parameter-estimation procedure commonly known as nonlinear least-squares makes a number of limiting assumptions about the distributions of experimental uncertainties. In particular, the least-squares method assumes (1) that negligible experimental uncertainty exists in the independent variables, (2) that the experimental uncertainties of the dependent variables are Gaussian in their distribution with a mean of zero, and (3) that no correlation exists between the experimental uncertainties of the dependent and independent variables. Any application using least-squares where these assumptions are not met will yield incorrect answers. The magnitude of the errors so introduced is impossible to predict *a priori*, because it is a function of the particular data points and their experimental uncertainties as well as the functional form of the equation being fitted. The assumptions cannot be overcome by an appropriate weighting of the data.

The method of least-squares is often used for applications where these assumptions are not met. For example, the common method for determining a "standard curve" will usually neglect the experimental uncertainties of the independent variables (i.e., the x axis). In column chromatographic studies it is common to generate a standard curve of elution volume versus log molecular weight for a number of standard proteins. The molecular weight estimates of the standard proteins are not known without experimental uncertainties. In electrophoresis experiments log molecular weight

---

* Reprinted from *Analytical Biochemistry* **148**, 471 (1985).

is plotted against electrophoretic mobility. In this case, electrophoretic mobility should not be considered without its relative experimental uncertainty. Other "plots" which are commonly used in biochemical literature that violate one or more of the assumptions of least-squares include Hill plots, Scatchard plots, and double-reciprocal plots.

The primary purpose of this chapter is to describe a method of parameter estimation which allows for experimental uncertainties in the independent variables. The method, as presented, still assumes that the experimental uncertainties follow a Gaussian distribution and are independent of each other. However, the generalization of this method to include cross-correlated non-Gaussian distributions is also discussed.

Numerical Methods

A parameter-estimation procedure takes an equation of an assumed functional form and a set of data and generates a new function, called a NORM, which shows a maximum or minimum when the parameter values, the desired "answers," show the highest probability of being correct. For the standard least-squares technique, the NORM of the data is given by

$$\text{NORM}(\boldsymbol{\alpha}) = \sum_{i=1}^{n} \left[\frac{Y_i - G(\boldsymbol{\alpha}, X_i)}{\sigma_i} \right]^2 \tag{1}$$

where $\boldsymbol{\alpha}$ is any vector of parameters for an arbitrary function G and n data points (X_i, Y_i) with each Y_i having a unique experimental uncertainty (standard error) of σ_i. The maximum likelihood estimate of the parameters, $\boldsymbol{\alpha}$, will correspond to a minimum of the NORM in this case.

A number of assumptions are implicit in the derivation of this least-squares NORM and must also be considered essential to the method derived in this chapter.[1] It must be assumed that there are enough data points to give a random sampling of the experimental uncertainty, and that the function, G, correctly describes the phenomenon occurring. If Gaussian-distributed random experimental uncertainty is assumed on both the ordinate and abscissa, and these experimental uncertainties are assumed to be independent of each other, the statistically correct NORM will be similar to Eq. (1) [see Eq. (5) below]. With the other previously mentioned assumptions it can be shown that the probability, P_i, for observing a particular data point (X_i, Y_i) at any value of the parameters, $\boldsymbol{\alpha}$, is proportional to

---

[1] M. L. Johnson and S. G. Frasier, this series, Vol. 117, p. 301.

$$P_i(\alpha) \approx \frac{1}{2\pi\sigma_{X_i}\sigma_{Y_i}} \exp\left\{ -\frac{1}{2}\left[\frac{Y_i - G(\alpha,\bar{X}_i)}{\sigma_{Y_i}} \right]^2 \right\}$$

$$\times \exp\left\{ -\frac{1}{2}\left[\frac{X_i - \bar{X}_i}{\sigma_{X_i}} \right]^2 \right\} \quad (2)$$

where σ_{X_i} and σ_{Y_i} represent the standard deviations of the Gaussian distributed random experimental uncertainty at the particular data point and $\bar{X}_i$ is the "optimal" value of the independent variable. In order to derive Eq. (2), and as a consequence of Eqs. (3)–(5), it has been assumed that the σ_{X_i} and σ_{Y_i} values are independent of each other. It has not been assumed that there is any relationship, such as constant coefficients of variation, between σ_{X_i} and X_i or σ_{Y_i} and Y_i. The probability of making a series of measurements at n independent data points is then proportional to

$$P(\alpha) \approx \Pi\, P_i(\alpha) \approx \left[\Pi\left(\frac{1}{2\pi\sigma_{X_i}\sigma_{Y_i}} \right) \right] \exp\left[-\frac{1}{2}\Sigma\left(\frac{Y_i - G(\alpha,\bar{X}_i)}{\sigma_{Y_i}} \right)^2 \right]$$

$$\times \exp\left[-\frac{1}{2}\Sigma\left(\frac{X_i - \bar{X}_i}{\sigma_{X_i}} \right)^2 \right] \quad (3)$$

where the product and summation are taken for each of the n data points with subscript i. This equation can be reorganized to the following form:

$$P(\alpha) = \left[\Pi\left(\frac{1}{2\pi\sigma_{X_i}\sigma_{Y_i}} \right) \right]$$

$$\times \exp\left\{ -\frac{1}{2}\Sigma\left[\left(\frac{Y_i - G(\alpha,\bar{X}_i)}{\sigma_{Y_i}} \right)^2 + \left(\frac{X_i - \bar{X}_i}{\sigma_{X_i}} \right)^2 \right] \right\} \quad (4)$$

The maximum likelihood estimates for the parameters, α, with the current assumptions will be those values of α which maximize the probability given by Eq. (4). This can be accomplished by minimizing the summation in the exponential term in Eq. (4). The NORM to minimize is then

$$\text{NORM} = \Sigma\left[\left(\frac{Y_i - G(\alpha,\bar{X}_i)}{\sigma_{Y_i}} \right)^2 + \left(\frac{X_i - \bar{X}_i}{\sigma_{X_i}} \right)^2 \right] \quad (5)$$

This is an extension of the least-squares NORM in Eq. (1), to include the possibilities of experimental uncertainties in the independent variables.

The new statistical NORM [Eq. (5)] can be minimized by a recursive application of a curve-fitting algorithm such as Nelder–Mead.[2,3] Equation (5) can be written as

$$\text{NORM} = \sum_{i=1}^{n} D_i^2 \tag{6}$$

where D_i is the weighted distance between the given data point (X_i, Y_i) and the point of closest approach to the fitted line $[\bar{X}_i, G(\alpha, \bar{X}_i)]$.

The minimization of this norm can be performed by a nested, or recursive, minimization procedure. The parameter-estimation procedure to evaluate the values of α is the standard Nelder–Mead simplex algorithm.[2,3] An initial estimate of α is arbitrarily chosen. This estimate is employed to calculate the D_i^2 at each point. The D_i^2 values are then used to predict new values for the parameters being established, α. This cyclic process is repeated until the values do not change within some specified limit. An inconvenience arises in the evaluation of D_i^2 at each data point and iteration because it requires the value of $\bar{X}_i$. This value $\bar{X}_i$ is the value of the independent variable, $\bar{X}_i$, at the point of closest approach of the function to the particular data point evaluated at the current estimate of the parameters, α. The evaluation of this weighted distance of closest approach, D_i, includes the relative precision of the data point, σ_{X_i} and σ_{Y_i}, as per Eq. (5). This implies that the values of $\bar{X}_i$ will be different for each iteration. Because of this, the values of $\bar{X}_i$ must be reevaluated for each iteration.

If the original parameter-estimation procedure is carefully developed, it can be used recursively to evaluate $\bar{X}_i$. That is, at each iteration and data point of the parameter-estimation procedure the routine calls itself to evaluate D_i^2. For this recursive application of the algorithm, the parameters being estimated are now the $\bar{X}_i$ values, and the function being minimized is D_i^2. The values of $\bar{X}_i$ or the previous values of $\bar{X}_i$ are used as starting values for this iterative process. Function minimization algorithms such as Marquardt–Levenberg, Gauss–Newton, and steepest descent cannot be used for this purpose since they make assumptions about the functional form of the statistical NORM.[1]

It should be noted that this method is not restricted to a two-dimensional problem, Y versus X. As with the least-squares NORM described in Eq. (1), all that is required to expand this method to include multiple dependent and multiple independent variables is to consider each of the

[2] M. S. Caceci and W. P. Cacheris, *Byte Mag.* **9**(5), 340 (1984).
[3] J. A. Nelder and R. Mead, *Comput. J.* **7**, 308 (1965).

Y_i, X_i and $\bar{X}_i$ terms as vectors instead of scalars, and add the appropriate additional terms in the summations.

Furthermore, an equation analogous to Eq. (4) can be generated to include a non-Gaussian distribution of experimental uncertainties and/or any cross-correlation between the various dependent and independent variables. Then this equation is used as the NORM to be maximized to obtain the maximum likelihood parameter estimates. In a similar manner an additional term can be included in Eq. (5) to allow for the possibility of cross-correlation between the dependent and independent variables.

It is worth noting that this maximum likelihood approach allows for an extremely liberal choice of dependent and/or independent variables and the form of the fitting equation. For example, experimental data from the measurement of hormone binding to cell surface receptors usually consist of a series of measurements of amount bound versus total added hormone concentration. The most convenient way to formulate the fitting function, G, is as amount bound versus the free, not the total, hormone concentration. Some authors assume that the total added hormone concentration is equal to the free concentration. Other authors calculate the free hormone concentrations as the total minus the bound hormone concentration. Neither of these approaches is statistically sound.[1,4] A few authors take the more complex approach of fitting bound versus total hormone concentration.[1,4] This approach involves evaluating the free concentration as the numerical root of a conservation of mass equation relating total and free hormone concentrations and the current best estimates of the fitting parameters, namely, binding constants and capacities. This procedure does not allow for experimental uncertainties in the total hormone concentration.

The maximum likelihood method allows a different approach. The function G can be written for two dependent variables, total and bound hormone concentrations, as a function of the free hormone concentration. The free hormone concentration can be assumed to be any reasonable value with its standard error being plus or minus infinity. The algorithm will then find the optimal value of $\bar{X}_i$, in this case the free concentration, which best describes the dependent and independent variables, here the total and bound hormone concentrations, and their experimental uncertainties. This optimal value is then used for the calculation of the standard NORM. The net effect of this approach is that the data can be fit as a function of a quantity which was never measured, yet the procedure is still statistically correct!

If the experimental uncertainties are independent and Gaussian, then the joint confidence intervals of the derived parameters can be determined

[4] P. J. Munson and D. Rodbard, *Anal. Biochem.* **107**, 220 (1980).

by searching for combinations of the parameters which yield values of the NORM [Eq. (5)] which are increased by a multiplicative factor proportional to the desired F-statistic.[1,5] This F-statistic (variance ratio) value is uniquely determined by the desired confidence probability and the number of degrees of freedom. The Hessian matrix which is required for this procedure was evaluated by the method outlined in the appendix to the original Nelder and Mead paper.[3]

The technique presented here is a generalization of a technique developed by Acton[6] for use with straight line data. Acton's method allows for covariance between the independent and dependent variables but does not allow for multiple independent or dependent variables. Furthermore, the method presented by Acton can only be used for a fitting function which is a straight line. The generalization I am presenting allows for multiple independent and dependent variables, nonlinear fitting functions, and a method of evaluating joint confidence intervals of the determined parameters.

The simulated data which I used to test this method included Gaussian distributed pseudorandom noise which was generated by averaging 12 evenly distributed random numbers over a range of ± 0.5. These numbers were obtained from the Digital Equipment Corporation (Maynard, MA) RT-11 FORTRAN 77 library function RANDU.

Each of the tests of the maximum likelihood method includes a comparison of the same simulated data analyzed by a standard weighted nonlinear least-squares technique. The least-squares method which I utilized is a modification of the Gauss–Newton procedure and has been described elsewhere.[1,5]

Test Examples and Results

To demonstrate the functionality of this method, I present two examples of its use. The examples are simulations of ligand-binding problems with either one or two classes of independent binding sites. In both cases, the number of binding sites in each class and the binding affinity of each class is assumed to be unknown.

Gaussian-distributed pseudorandom noise was superimposed on each set of data. To reduce the possibility of inadvertently using a nonrandom set of noise, and thus biasing the results, each calculation was performed 10 times with different sets of random noise.

The first example simulates an experimental system with a single class of binding sites with a binding affinity, K_a, of $10^5\ M^{-1}$ and a maximal

[5] M. L. Johnson, *Biophys. J.* **44**, 101 (1983).
[6] F. S. Acton, "Analysis of Straight Line Data," p. 129. Wiley, New York, 1959.

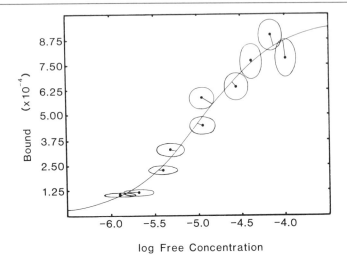

log Free Concentration

FIG. 1. Simulated data representing an experimental ligand-binding system with a single class of binding sites. The binding affinity, K_a, is $10^5 M^{-1}$ and the maximal bound, B_{max}, is 10^{-3} M. The ellipses represent the Gaussian-distributed pseudorandom experimental uncertainty corresponding to 1 SD on both the ordinate and the abscissa. The curve is the calculated "best" fit of the data to the function by the maximum likelihood method of analysis. The short lines connecting the data points to the curve are the D_i's in Eq. (6), whose sum of squares is being minimized by the analysis procedure.

bound, B_{max}, of 10^{-3} M. Ten logarithmically spaced data points were simulated over a concentration range of 10^{-6} to 10^{-4} M. This concentration range and affinity correspond to a fractional saturation ranging between 10 and 90%. The data were then perturbed with multiple sets of pseudorandom noise.

The simulated error for the dependent variable, the amount bound, is 10% of the actual amount bound. The corresponding error in the independent variable, the free concentration, is 7% of the free concentration, that is, a constant coefficient of variation. An example of one such set of data is presented in Fig. 1. The ellipses in Fig. 1 were generated such that their major and minor axes correspond to one standard deviation experimental uncertainty.

For the first example, the desired parameters, the vector α, are the binding affinity, K_a, and the maximal bound, B_{max}. The function G relates the amount bound, Y_i, with the free concentration, X_i, and α as

$$\text{Bound} = G(\alpha, X_i) = B_{max} \frac{K_a X_i}{1 + K_a X_i} \tag{7}$$

The implementation of the maximum likelihood method presented here determines the logs of B_{max} and K_a rather than the actual values. By allowing the logarithms of B_{max} and K_a to assume any real value, I am generating a number system in which the actual values of B_{max} and K_a are restricted to physically meaningful (positive) values.

The data were analyzed 10 times with separate sets of random noise by both a standard nonlinear least-squares technique and the maximum likelihood method which I am presenting. The average values and standard deviations of $\log K_a$ and $\log B_{max}$ as determined by each procedure can be compared to measure the reliability and functionality of the analysis methods. The values determined by the maximum likelihood method were $\log K_a = 5.01 \pm 0.07$ and $\log B_{max} = -3.01 \pm 0.03$. I then analyzed exactly the same data with the superimposed pseudorandom noise by a weighted least-squares method. The values of σ_{Y_i} were used as a weighting factor for the analysis, that is, the experimental uncertainties in the x axis were ignored. The values obtained were $\log K_a = 4.98 \pm 0.12$ and $\log B_{max} = -3.01 \pm 0.06$. The average values as determined by both methods are excellent. However, the reproducibility of the values, measured by an F-test, is decidedly better for the maximum likelihood method ($P \sim 95\%$ for each of $\log K_a$ and $\log B_{max}$).

The calculated "best" fit of the data presented in Fig. 1 to the function by the maximum likelihood method of analysis is shown by the curve in Fig. 1. The short lines connecting each data point, which is the center of its ellipse, with the curve are the corresponding distances D_i in Eq. (6), whose sum of squares is being minimized by this maximum likelihood method. These distances, D_i, are perpendicular neither to the best curve nor to either axis. They are determined by a combination of the slope of the best curve and the relative errors in the dependent and independent variables. In Fig. 1, when the error is predominately in the x axis, these lines are nearly horizontal; when the error is predominately in the y axis, these lines are nearly vertical, and when the error is nearly equal, the line is almost perpendicular to the best curve.

The second example I chose to test is a simulation of a ligand-binding system with two classes of binding sites which differ in affinity by a factor of 5 and have equal binding capacities, namely, K_a values of 10^5 and $5 \times 10^5 \ M^{-1}$ and B_{max} values of $10^{-3} \ M$. Ten data points were simulated with a concentration range of 6×10^{-8} to $3 \times 10^{-4} \ M$ and logarithmic spacing. The magnitude of the simulated experimental uncertainty was assumed to be the same as for the first example. Again, the data were analyzed 10 times with differing sets of superimposed pseudorandom noise. One of these analyses is shown in Fig. 2.

The numerical values obtained by using maximum likelihood and

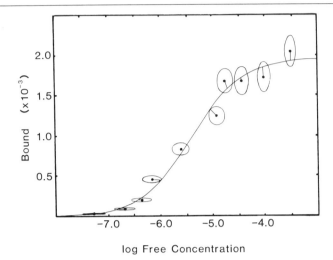

FIG. 2. Simulated data showing a ligand-binding system with two independent classes of binding sites. K_a values are 10^5 and $5 \times 10^5 \, M^{-1}$, and B_{max} is $10^{-3} \, M$. Experimental uncertainty was imposed the same way as in Fig. 1, and the curve was calculated by the maximum likelihood analysis method.

weighted least-squares methods to analyze this second group of simulated data are shown in Table I. The maximum likelihood method gives more consistent results for both the high ($P > 99.9\%$) and low ($P \sim 95\%$) affinity classes of sites. The region of the data which contributes most to the determination of the low-affinity site is the upper half of the saturation curve. It is this upper portion of the curve which comes the closest to satisfying the least-squares criterion of negligible error in the x axis. Therefore, it is expected that the low-affinity class of sites would be evaluated with reasonable precision by the least-squares method. Conversely, the error on the lower portion of the curve is almost totally horizontal, which dramatically violates the error distribution assumption of the least-squares

TABLE I
ANALYSIS OF SIMULATED DATA FOR
TWO CLASSES OF BINDING SITES

	Correct values	Maximum likelihood values	Least-square values
Log K_1	5.00	4.92 ± 0.23	4.90 ± 0.56
Log B_{max1}	−3.00	−2.97 ± 0.09	−2.80 ± 0.17
Log K_2	5.70	5.71 ± 0.08	6.32 ± 1.93
Log B_{max2}	−3.00	−3.01 ± 0.05	−4.15 ± 1.15

analysis method. Therefore, it is expected that, for this example, the high-affinity class of sites cannot be evaluated by least-squares.

The method of determining joint confidence intervals of the parameters was also tested for the two simulated examples. For each set of data, the maximum likelihood algorithm is capable of predicting a 1 standard deviation (SD) confidence interval. In an ideal case, the average span of these predicted confidence intervals should be equal to twice the standard deviation of the determined parameter values averaged over the 10 different sets of pseudorandom noise. As measured by this criterion, the joint confidence intervals of the determined parameters are overestimated by approximately 75%. It has been previously shown that the expected joint confidence intervals of nonlinear parameters will be both asymmetrical and highly correlated.[5] Consequently, the determination of a symmetrical standard deviation for these parameters from the results obtained with multiple sets of random noise is not optimal.

Discussion

The first test case was specifically chosen as a simple two-parameter problem. The cross-correlation[7] between the two unknown parameters was relatively low, approximately 0.87. The second test was chosen because it poses a particularly difficult data analysis problem. The cross-correlation between the parameters for the second test was very high, about 0.99. The second test represents a class of problems commonly referred to as mathematically ill-posed problems.

The convergence properties of the maximum likelihood method appear to be very good. In 2 of the 10 calculations performed by the least-squares technique on the second test example, the analysis would not converge, even when the correct answers were used as initial starting values. The maximum likelihood method converged on all of these in about 30 sec on our microprocessor (DEC LSI-11/73) with no apparent problem.

The mathematical procedure presented here was specifically formulated to treat data analysis problems with experimental uncertainties in both the independent and dependent variables. Experimental uncertainties in the independent variables have previously been treated by a number of methods. The most prevalent method to treat these types of experimental errors is to ignore them. Some investigators minimize the sum of the squares of the perpendicular distance to the fitted curve. This is statisti-

---

[7] The cross-correlation between parameters is a measure of the ability of one or more parameters to compensate for the variation of another parameter in a parameter-estimation process. If the cross-correlation coefficient is $+1$ or -1, then for any reasonable value of one of the parameters, a set of the remaining parameters can be found which will yield the same numerical value for the statistical NORM. Consequently, a value of $+1$ or -1 indicates that unique values of the parameters cannot be estimated.[1,5]

cally correct only if the experimental uncertainties of the independent and dependent variables are equal.

Other investigators attempt to treat uncertainties in an independent variable by reflecting it to a corresponding uncertainty in the dependent variable and then using an "appropriate weighting factor." This weighting factor is generally taken to be the inverse of the standard error of the dependent variable. When the fitted function in the region of a particular data point has a near-zero slope, this procedure is equivalent to ignoring the experimental uncertainties in the independent variable. If, however, the fitted function has a significant slope in the neighborhood of the data point, this procedure will transform small uncertainty in the independent variable into a large corresponding uncertainty in the dependent variable. This large dependent variable uncertainty will translate into such a small "appropriate weighting factor" that the net effect is that the data point will be ignored by the least-squares procedure. Furthermore, if the fitted function has a significant curvature in the neighborhood of a particular data point it will introduce an asymmetrical non-Gaussian uncertainty in the dependent variable. This reflection procedure is incorrect as per Eq. (5).

The maximum likelihood approach, which minimizes the statistical NORM presented in Eq. (5), has been previously employed by other investigators (e.g., see Refs. 6 and 8). The application of the method by Acton is limited to only straight-line data.[6] Bard presents a different mathematical procedure to minimize the same statistical NORM[8]; the procedure determines the $\bar{X}_i$ values and the α values simultaneously in a single iterative approach rather than with the nested approach which was presented here. For each iteration Bard's approach requires the inversion of a sparse matrix of order equal to the number of independent variables times the number of data points plus the number of parameters being determined. As the number of data points increases the order of the matrix will increase, and as a consequence the matrix will become difficult and time consuming to invert. In addition, as the order of the matrix increases its inversion will be more prone to problems caused by computational round-off errors. The numerical method presented here may have some advantages over Bard's method as the number of data points increases even though the final result is mathematically equivalent.

It should be noted that even though the maximum likelihood method is considerably better than the more classic least-squares method for these two test examples, this is not a proof that the maximum likelihood method is always better. It does, however, indicate that when Gaussian-distributed

[8] Y. Bard, "Nonlinear Parameter Estimation," p. 67. Academic Press, New York, 1974.

experimental uncertainties exist in both the independent and dependent variables this method of maximal likelihood estimation of the unknown parameters can prove to be very useful.

Acknowledgments

This work was supported in part by National Institutes of Health Grants GM28929, AM30302, and AM22125.

[7] Monte Carlo Method for Determining Complete Confidence Probability Distributions of Estimated Model Parameters

By Martin Straume and Michael L. Johnson

Introduction

The quantitative analysis of experimental data generally involves some numerical process to provide estimates for values of model parameters (least-squares,[1] method of moments,[2] maximum entropy,[3] Laplace transforms,[4] etc.). The derived parameter values are, in turn, interpreted to provide information about the observed properties of the experimental system being considered. This fundamental process applies for the simplest of analyses (e.g., protein determinations employing standard curves) as well as for the highly sophisticated modeling algorithms in use today for interpretation of a broad spectrum of complex biomolecular phenomena.

The primary objective of a quantitative analysis is derivation of the values corresponding to the best estimates for the parameters of the model employed to characterize the experimental observations. System properties may then be inferred by a physical interpretation of the significance of the model parameter values. However, the level of confidence one can have in the interpretation of derived parameter values depends strongly on the nature and magnitude of the confidence probability distribution of the parameter values about their most probable (or best-fit) values.

Determination of reliable estimates of confidence intervals associated with model parameters may be critical in discerning between alternative

---

[1] M. L. Johnson and L. M. Faunt, this volume [1].

[2] E. W. Small, this volume [11].

[3] W. H. Press, B. P. Flannery, S. A. Teukolsky, and W. T. Vetterling, "Numerical Recipes: The Art of Scientific Computing," p. 430. Cambridge Univ. Press, Cambridge, 1986.

[4] M. Ameloot, this volume [12].

interpretations of some biomolecular phenomena (e.g., the statistical justification for existence of quaternary enhancement in human hemoglobin oxygen-binding behavior[5]). In a case such as this, the most probable derived value is significant, but the shape and breadth of the distribution of expected parameter values, given the experimental uncertainties associated with the data sets being analyzed, are also of critical importance with regard to arriving at a statistically significant conclusion. Knowledge of complete confidence probability distributions as well as the correlation that exists among parameters or between parameters and the experimental independent variable(s) is also of value for identifying influential regions of independent parameter space (e.g., extent of binding saturation in a ligand-binding experiment) as well as for pointing out the relative behavior of parameters between different models used to interpret the same data (e.g., models that explicitly account for ligand-linked cooperative binding versus those allowing nonintegral binding stoichiometries to accommodate effects arising from cooperativity).[6]

The determination of confidence intervals for parameters estimated by numerical techniques can be a challenging endeavor for all but the simplest of models. Methods for estimation of parameter confidence intervals vary in the level of sophistication necessarily employed to obtain reliable estimates.[7] Implementation of parameter spaces that minimize statistical correlation among the parameters being determined may permit extraction of moderately accurate estimates of confidence intervals with relative ease. However, the great majority of parameter estimation procedures employed in interpretation of biophysical data are cast in terms of complex mathematical expressions and processes that require evaluation of nonorthogonal, correlated model parameters.

Accommodation of statistical thermodynamic equations like those describing multiple, linked equilibria (as in the case of oxygen-linked dimer–tetramer association in human hemoglobin as a function of protein concentration, for example) or processes such as iterative interpolation or numerical integration involves solving complex mathematical relationships using nontrivial numerical methods. Additionally, comprehensive modeling of multidimensional dependencies of system properties (e.g., as a function of temperature, pH, ionic strength, and ligand concentration) often requires relatively large numbers of parameters to provide a full description of system properties. Mathematical formulations such as these therefore often involve mathematical relationships and processes suffi-

[5] M. Straume and M. L. Johnson, *Biophys. J.* **56,** 15 (1989).

[6] J. J. Correia, M. Britt, and J. B. Chaires, *Biopolymers* in press (1991).

[7] J. M. Beecham, this volume [2].

ciently complex as to obscure any obvious correlations among model parameters as well as between the parameters and data (through effects of regions of influential observations, for example[6,8]). It therefore becomes difficult to identify conveniently parameter spaces that minimize correlation, creating a potentially more challenging situation with regard to confidence interval determination.

The numerical procedures that have been developed for estimating confidence intervals all involve some approximations, particularly about the shape of the confidence probability distribution for estimated parameters.[7] Sometimes, these approximations may produce grossly incorrect estimates, particularly with more simplistic methods applied to situations exhibiting correlation. Errors in estimates of confidence intervals usually arise from the inability of the estimation procedure to account for high levels of sometimes complex, nonlinear correlation among the parameters being estimated. Improving the accuracy of confidence interval estimates therefore requires implementation of more thorough mathematical procedures designed to eliminate or reduce the influence of approximations regarding the shape of parameter variance space that reduce the reliability of lower-order methods.

Monte Carlo Method

Of course, the ultimate objective is to have available the entire joint confidence probability distributions for each of the parameters being estimated in an analysis. The Monte Carlo approach is unique in the sense that it is capable of determining confidence interval probability distributions, in principle, to any desired level of resolution and is conceptually extremely easy to implement.[9,10] The necessary information for application of a Monte Carlo method for estimating confidence intervals and probability distribution profiles is 2-fold: (1) an accurate estimate of the distribution of experimental uncertainties associated with the data being analyzed and (2) a mathematical model capable of accurately characterizing the experimental observations.

The Monte Carlo method is then applied by (1) analysis of the data for the most probable model parameter values, (2) generation of "perfect" data as calculated by the model, (3) superposition of a few hundred sets of simulated noise on the "perfect" data, (4) analysis of each of the noise-

[8] M. Straume and M. L. Johnson, this volume [5].

[9] W. H. Press, B. P. Flannery, S. A. Teukolsky, and W. T. Vetterling, "Numerical Recipes: The Art of Scientific Computing," p. 529. Cambridge Univ. Press, Cambridge, 1986.

[10] Y. Bard, "Nonlinear Parameter Estimation," p. 46. Academic Press, New York, 1974.

containing, simulated data with subsequent tabulation of each set of most probable parameter values, and finally (5) assimilation of the tabulated sets of most probable parameter values by generating histograms. These histograms represent discrete approximations of the model parameter confidence probability distributions as derived from the original data set and the distribution of experimental uncertainty contained therein.

The level of resolution attainable in determining confidence probability profiles by this method is dependent on the number of Monte Carlo "cycles" performed (i.e., the number of noise-containing, simulated data sets considered). The more cycles carried out, the more accurate will be the resolution of the probability distribution. In practice, this means that the amount of computer time needed to generate a probability distribution will be of the order of 100–1000 times that required for an individual parameter estimation (i.e., after ~100–1000 Monte Carlo cycles). This method must therefore be considered a "brute force" type of approach to the determination of parameter confidence intervals. Although the computational time required by the Monte Carlo method can be substantial, no other method is so easy to implement yet capable of providing information as complete about profiles of confidence probability distributions associated with estimated model parameters.

Generating Confidence Probability Distributions for Estimated Parameters

Implementation of the Monte Carlo confidence probability determination method requires the initial estimation of the set of most probable parameter values that best characterize some set(s) of experimental observations according to a suitable mathematical model (i.e., one capable of reliably describing the data). [At this point, we will proceed under the assumption that the mathematical model being used to analyze the data is "valid." The reader is referred to discussions addressing concepts related to judging the validity of analytical models as descriptors of experimental data in terms of either statistical probability[8] or theoretical prediction[1] (as opposed to simply empirical "fitting").] With this set of best-fit model parameter values in hand, a set of "noise-free" data is next generated to produce a data set made up of simulated "experimental points" calculated at exactly the same independent variable values as those occurring in the original data. For example, suppose that in a ligand-binding experiment measurements of some experimental observable are made as a function of ligand concentration at, say, 0.1, 0.2, 0.25, 0.3, 0.33, 0.37, and 0.4 μM ligand. After the data are analyzed by an applicable model for the most probable parameter values characteristic of this data set, theoretical values of the "expected" observable quantity are calculated from the model at 0.1, 0.2, 0.25, 0.3, 0.33, 0.37, and 0.4 μM ligand using the best-fit parameter values. The calculated dependent variable values (the simulated "experimental points") therefore correspond to those values produced by evaluating the analytical model at the same independent variable values encountered in the original data and employing the derived best-fit parameter values.

In performing an analysis of the experimental data (to obtain the most probable model parameter values), uniform, unit weighting of each experimental data point is usually employed (i.e., each data point possesses a weighting factor of 1). In cases where independent estimates of uncertainties are available for each of the observed experimental values, weighting of the data by their estimated standard deviation is desirable because a more statistically accurate parameter estimation will result.[1] This provides a basis for directly calculating the variance of fit of the analytical model to the experimental data. The square root of this variance of fit represents the estimated standard deviation in the experimental data. In cases where variable weighting is employed, the square root of the variance becomes a relative indicator of the quality of fit (relative to the absolute values of the uncertainties used in weighting the data during the analysis). The assumptions underlying this assignment are (1) that the model employed in analysis is capable of accurately describing the data, (2) that the experimental uncertainty in the data is randomly distributed, and (3) that there is no systematic behavior in the data that is not accounted for by the analytical model. When these three conditions are satisfied, this estimate of the standard deviation of the experimental data permits realistic approximations of the actual experimental uncertainty to be synthesized and superimposed on the noise-free, simulated dependent variable values.

Pseudorandom noise,[11] with a distribution width defined by the estimated standard deviation, is generated to be consistent with the actual experimental uncertainty encountered in the original data. This pseudorandom noise is added to the noise-free data set to produce simulated data possessing a distribution of experimental uncertainty throughout the data. With variably weighted data, the magnitude of the pseudorandom noise that is added for a particular data point is proportional to the estimated uncertainty associated with the data point. A data set such as this corresponds to one possible distribution of noise on the simulated, noise-free data and accounts for both average system properties as well as experimental uncertainties. A few hundred such simulated, noise-containing data sets

[11] G. E. Forsythe, M. A. Malcolm, and C. B. Molter, "Computer Methods for Mathematical Computations," p. 240. Prentice-Hall, Englewood Cliffs, New Jersey, 1977.

are generated and subsequently analyzed in the same manner and by the same analytical model as was the original experimental data. The most probable model parameter values derived from the analysis of these simulated, noise-containing data sets are then recorded as a group for each case considered.

An alternative way to generate synthetic noise sets is to rely on the residuals actually produced as a result of the parameter estimation. With this approach, the residuals obtained from an analysis are "reshuffled" to redistribute them among the independent parameter values encountered in the original data. Again, uniform, unit weighting is straightforward and direct, whereas variably weighted data must take into account the variable relative uncertainties associated with data obtained at different values of independent parameter space. This approach may in some sense be viewed as "more correct" in that the actual noise distribution obtained from analysis of the data is used—it is just redistributed among the available independent variable values. No assumptions about the shape of the actual probability distribution function are involved.

At this point exists a tabulation of a few hundred sets of most probable model parameter values obtained from analysis of a spectrum of simulated data sets. The properties of this group of data sets are meant to represent statistically what would be expected had this many actual experiments been done. This information may be assimilated in terms of probability distributions by generating histograms of relative probability of occurrence as a function of parameter value (as in Figs. 1 and 2). These examples involved determinations of 500 simulated data sets, the results of which were distributed into 51-element histograms to produce readily discernible confidence probability distributions.[5] The resolution of the determined probability distribution is dependent on the number of simulated data sets considered and may be improved by analyzing a greater number. In the example presented herein, 51-element histograms were employed because they were judged as providing sufficient resolution as well as providing intervals sufficiently populated to offer a statistically significant sample size.

Implementation and Interpretation

Knowledge of the full confidence probability distribution of model parameters provides a most rigorous way to address questions regarding resolvability of parameters characteristic of a mathematical model. The distribution of parameter confidence probability is dependent on the scatter or noise present in the experimental data as well as on the correlation between parameters of the model. The mathematical linkage of these coupled properties of the data and the analytical model parameters must be accounted for when estimating parameter confidence intervals and when propagating uncertainties between parameter spaces.

Consider the example of propagating uncertainty for the case of a difference between two derived free energy changes, as in the case for oxygen binding to human hemoglobin.[5] The quaternary enhancement effect in human hemoglobin (as quantified by the quaternary enhancement free energy change, Δg_{QE}) may be defined as the difference between the free energy changes associated with oxygenation of the last available site of hemoglobin tetramers (Δg_{44}) and that for binding oxygen to the last available site in dissociated hemoglobin dimers (Δg_{22}, or Δg_{2i} for the case of noncooperative oxygen binding by hemoglobin dimers). The quaternary enhancement free energy difference is therefore $\Delta g_{QE} = \Delta g_{44} - \Delta g_{2i}$. The significance of this parameter at the molecular level is that it quantifies the cooperative oxygen-binding free energy gained by the macromolecular association of hemoglobin dimers to triply ligated tetramers.

The equilibrium for the molecular association of dimers to tetramers is coupled to the oxygen binding properties of human hemoglobin. Mathematical modeling of the behavior of oxygen-linked dimer–tetramer association involves estimating parameters characteristic of the thermodynamic linkage scheme for this system.[5] Oxygen-binding isotherms obtained over a range of protein concentrations represent the two-dimensional data considered. When analyzed, six model parameters require estimation. The actual parameter spaces employed were those empirically judged to provide the most robust parameter estimation (of those examined).[12,13]

The analysis provides the most probable values for the oxygen-binding free energy changes associated with binding at each step in the thermodynamic linkage scheme. Two of these are Δg_{44} and Δg_{2i}, the parameters by which the quaternary enhancement effect is most obviously defined. Estimates of joint confidence intervals for these derived model parameters are also possible; however, they are difficult to obtain reliably using numerical methods that search the analytical variance space. An estimate of Δg_{QE} now requires subtracting $\Delta g_{44} - \Delta g_{2i}$. But what about the confidence interval associated with this best-estimate value of Δg_{QE}? If confidence intervals for Δg_{44} and Δg_{2i} are determined, a propagation of these uncertainties to that of Δg_{QE} is possible. To account for correlation, however, rigorous methods to map the variance spaces of Δg_{44} and Δg_{2i} to that of Δg_{QE} would have to be performed. This can be a quite challenging task

[12] M. L. Johnson, H. R. Halvorson, and G. K. Ackers, *Biochemistry* **15**, 5363 (1976).
[13] M. Straume and M. L. Johnson, *Biochemistry* **27**, 1302 (1988).

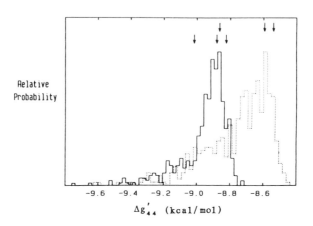

Relative Probability

$\Delta g'_{44}$ (kcal/mol)

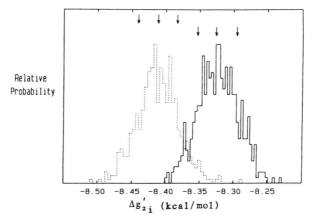

Relative Probability

$\Delta g'_{2i}$ (kcal/mol)

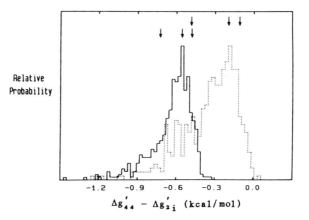

Relative Probability

$\Delta g'_{44} - \Delta g'_{2i}$ (kcal/mol)

with an analytical model as involved as the thermodynamic linkage scheme considered here for human hemoglobin.

In Fig. 1, we see the confidence probability distributions for Δg_{44} and Δg_{2i} as determined by application of the Monte Carlo method to two different sets of oxygen-binding isotherms.[5] The distributions for Δg_{2i} are seemingly symmetric, whereas those for Δg_{44} are noticeably skewed toward negative free energy space. The distributions obtained for Δg_{QE} ($\Delta g_{44} - \Delta g_{2i}$, see Fig. 1) are also (not surprisingly) noticeably skewed toward negative free energy space. The significant point here is that the confidence probability distributions of Δg_{QE} for either data set remain entirely in the negative free energy domain. This result supports the conclusion that association of hemoglobin dimers to form triligated tetramers is accompanied by a cooperative free energy change for oxygen binding. In this case, the molecular structural changes experienced by hemoglobin tetramers (relative to dissociated dimers) are responsible for the enhanced average oxygen affinity of triligated tetramers. This conclusion about thermodynamic properties, in turn, provides information that contributes to elucidating the molecular mechanisms for transduction of information

---

FIG. 1. Derived confidence probability distributions obtained from application of the Monte Carlo method are presented here for three free energy change parameters characteristic of oxygen binding by human hemoglobin tetramers. The parameter Δg_{44} is the intrinsic free energy change for addition of the last (i.e., fourth) oxygen to hemoglobin tetramers, whereas Δg_{2i} is that for oxygenation of dimer binding sites. Because oxygen binding by dimers has been experimentally shown to be noncooperative, both free energy changes Δg_{21} (for binding of the first oxygen) and Δg_{22} (for binding to the second site) are equal and therefore identified as Δg_{2i}. The quaternary enhancement effect (see text for further details) is quantified by the difference $\Delta g_{44} - \Delta g_{2i}$. The quaternary enhancement free energy change is therefore a composite parameter that requires evaluation of the difference between the values of the two constituent parameters by which it is defined, Δg_{44} and Δg_{2i}. The confidence probability distribution for the quaternary enhancement free energy change is demonstrated by these results to reside exclusively in negative free energy space. This leads to the conclusion that, given the experimental data sets considered, quaternary enhancement is indeed indicated to exist under the conditions of the experimental observations.

The two distributions presented in each graph correspond to the results obtained by considering two independent variable protein concentration oxygen-binding data sets [the solid lines are derived from the data of A. H. Chu, B. W. Turner, and G. K. Ackers, *Biochemistry* **23**, 604 (1984) (four binding isotherms at four protein concentrations for a total of 283 data points), and the dotted lines are for the data of F. C. Mills, M. L. Johnson, and G. K. Ackers, *Biochemistry* **15**, 5350 (1976) (five binding isotherms at four protein concentrations for a total of 236 data points)]. The arrows in the upper parts of the graphs correspond to estimates of the most probable and the upper and lower 67% confidence limits for the distributions from the data of Chu *et al.* (lower set of arrows) and Mills *et al.* (upper set of arrows). [Reproduced from M. Straume and M. L. Johnson, *Biophys. J.* **56**, 15 (1989), by copyright permission of the Biophysical Society.]

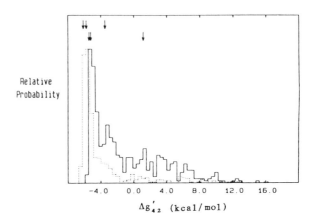

Relative Probability

$\Delta g_{42}'$ (kcal/mol)

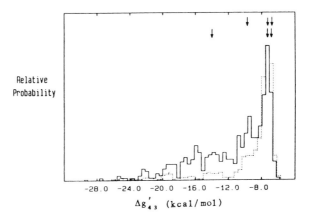

Relative Probability

$\Delta g_{43}'$ (kcal/mol)

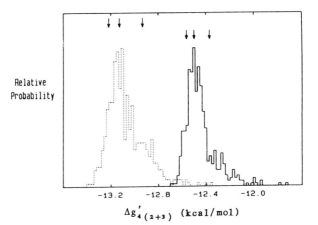

Relative Probability

$\Delta g_{4(2+3)}'$ (kcal/mol)

which ultimately modifies a functionally significant biological property of this system.

Although the confidence probability distributions for Δg_{44}, Δg_{2i}, and Δg_{QE} do not exhibit strong effects of parameter correlation or evidence of highly asymmetric variance spaces, the same is not the case for Δg_{42}, Δg_{43}, and $\Delta g_{4(2+3)}$ (see Fig. 2). Here, Δg_{42} is the average free energy change for adding the second oxygen to hemoglobin tetramers, Δg_{43} is that for adding the third, and $\Delta g_{4(2+3)}$ is that for adding the second and third oxygens (i.e., for proceeding from singly ligated tetramers to triligated ones). As clearly shown, Δg_{42} and Δg_{43} show very broad and highly asymmetric (in opposite directions) confidence probability distributions. However, the probability distributions for $\Delta g_{4(2+3)}$ (the sum of Δg_{42} and Δg_{43}) are symmetric and span much narrower ranges of free energy space than do either Δg_{42} or Δg_{43}. Here is a case where effects of both strong correlation and highly asymmetric parameter variance spaces are demonstrated. The conclusion from the standpoint of a physical interpretation is that it is possible to quantify with considerable confidence the free energy change associated with going from singly to triply ligated tetramers but not how this free energy change is partitioned between adding the second and adding the third oxygens (at least from the particular data being considered in this analysis).

FIG. 2. Derived confidence probability distributions for the intermediate tetramer oxygen-binding free energy changes Δg_{42}, Δg_{43}, and $\Delta g_{4(2+3)}$ are presented. Binding of the second oxygen to singly ligated hemoglobin tetramers is characterized by Δg_{42}, and binding of the third oxygen to doubly ligated tetramers is determined by Δg_{43}. These two free energy change parameters exhibit very broad and highly asymmetric confidence probability distributions. The distributions for the free energy change associated with binding of two oxygens to singly ligated tetramers to produce triply ligated tetramers, $\Delta g_{4(2+3)}$, however, is only moderately asymmetric and spans a much narrower range of free energy space. This property of the parameter confidence probability distributions leads to the conclusion that the free energy change for adding two oxygens to singly ligated hemoglobin tetramers may be confidently determined, whereas the partitioning of this free energy change between the two steps (singly-to-doubly ligated and doubly-to-triply ligated) is very poorly resolvable (from the experimental data considered in this analysis). Propagation of the highly correlated and asymmetric uncertainties of Δg_{42} and Δg_{43} to estimate those of $\Delta g_{4(2+3)}$ would require performing a sophisticated mapping of the three variance spaces relative to each other to provide reliable uncertainty estimates for $\Delta g_{4(2+3)}$. By using the Monte Carlo method, propagation of uncertainties is quite straightforward because the method implicitly accounts for all parameter correlation effects. Possessing the tabulated results from a Monte Carlo confidence probability determination therefore permits generation of complete probability profiles for any other parameter space of interest, as long as it may be obtained from the parameters for which distributions have been determined.

Conclusion

The application of ever more sophisticated analytical protocols to interpretation of experimental data has been made possible largely from ongoing advances in computer technology, both in terms of computational power and speed as well as affordability. Biological scientists thus now have convenient access to analytical capabilities superior in many ways to that available in the past. Continued developments in both computer hardware and software will undoubtedly lead to more widespread use of sophisticated parameter-estimation algorithms that may, in principle, be applied to any analytical situation.

The estimation of most probable (or best-fit) model parameter values is, of course, the primary objective of the great majority of analytical procedures. However, the statistical validity of an interpretation of system properties (based on the most probable derived parameter values) may be critically dependent on the nature of the confidence probability distributions associated with these parameters. In those cases where detailed knowledge of entire confidence probability distributions is needed, the Monte Carlo method is capable of providing the necessary information while minimizing the number of assumptions that are implicit (to varying degrees) in other confidence interval estimation protocols.

The total computer time needed to carry out a Monte Carlo confidence probability determination is directly proportional to the number of Monte Carlo "cycles" needed to produce the desired level of resolution in the probability profile (typically in the range of ~ 500 estimations). Therefore, although other, more approximate methods will produce estimates of parameter confidence intervals using considerably less computer time, the Monte Carlo approach described here circumvents the approximations implicit in these methods and produces the most accurate, experimentally based and numerically derived profiles of entire confidence probability distributions associated with estimated parameters of any analytical model as applied to any particular data set(s).

The Monte Carlo method also implicitly fully accounts for all correlation among model parameters. After the original most probable parameter values obtained from a Monte Carlo analysis are tabulated, it is possible to generate directly complete confidence probability distributions for any composite parameters (e.g., Δg_{QE} or $\Delta g_{4(2+3)}$) from knowledge of the distributions of and correlations between constituent parameters (i.e., Δg_{44} and Δg_{2i} or Δg_{42} and Δg_{43} for Δg_{QE} and $\Delta g_{4(2+3)}$, respectively). Propagating uncertainties in this way requires no assumptions about the correlation among parameters and obviates the need for complex mapping of variance spaces to convert from one parameter space to another.

Acknowledgments

This work was supported in part by National Institutes of Health Grants RR-04328, GM-28928, and DK-38942, National Science Foundation Grant DIR-8920162, the National Science Foundation Science and Technology Center for Biological Timing of the University of Virginia, the Diabetes Endocrinology Research Center of the University of Virginia, and the Biodynamics Institute of the University of Virginia.

[8] Singular Value Decomposition: Application to Analysis of Experimental Data

By E. R. HENRY and J. HOFRICHTER

I. Introduction

The proliferation of one- and two-dimensional array detectors and rapid scanning monochromators during the 1980s has made it relatively straightforward to characterize chemical and biochemical systems by measuring large numbers of spectra (e.g., absorption or emission spectra) as a function of various condition parameters (e.g., time, voltage, or ligand concentration). An example of such a data set is shown in Fig. 1. These data were obtained by measuring absorption difference spectra as a function of time after photodissociation of bound carbon monoxide from a modified hemoglobin. The difference spectra are calculated with respect to the CO-liganded equilibrium state. We will use this data set as an illustrative example at several points in the following discussion. As such experiments have become easier to carry out, two alternative approaches, one based on singular value decomposition (SVD)[1-3] and the other called global analysis,[4-7] have emerged as the most general approaches to the quantitative analysis of the resulting data. Before beginning a detailed

[1] G. Golub and C. VanLoan, "Matrix Computations." John Hopkins Univ. Press, Baltimore, Maryland, 1983.

[2] R. A. Horn and C. R. Johnson, "Matrix Analysis." Cambridge Univ. Press, Cambridge, 1985.

[3] C. L. Lawson and R. J. Hanson, "Solving Least-Squares Problems." Prentice-Hall, Englewood Cliffs, New Jersey, 1974.

[4] G. H. Golub and V. Pereyra, *SIAM J. Numer. Anal.* **10,** 413 (1973).

[5] M. L. Johnson, J. J. Correira, D. A. Yphantis, and H. R. Halvorson, *Biophys. J.* **36,** 575 (1981).

[6] J. F. Nagel, L. A. Parodi, and R. H. Lozier, *Biophys. J.* **38,** 161 (1982).

[7] J. R. Knutson, J. M. Beechem, and L. Brand, *Chem. Phys. Lett.* **102,** 501 (1983).

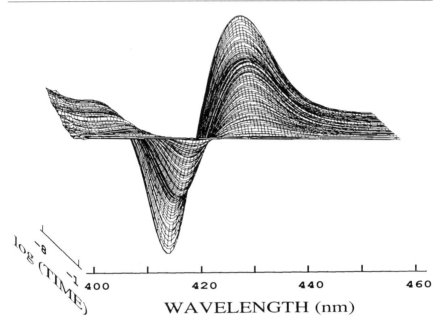

WAVELENGTH (nm)

FIG. 1. Time-resolved absorption difference spectra measured after photodissociation of $\alpha_2(Co)\beta_2(FeCO)$hemoglobin by 10 nsec, 532 nm laser pulses. The original data consisted of 91 sets of intensities measured for both photodissociated and reference (equilibrium sample) portions of the same sample at 480 channels (wavelengths) using an optical multichannel analyzer (OMA) and vidicon detector [J. Hofrichter, J. H. Sommer, E. R. Henry, and W. A. Eaton, *Proc. Natl. Acad. Sci. U.S.A.* **80,** 2235 (1983); J. Hofrichter, E. R. Henry, J. H. Sommer, and W. A. Eaton, *Biochemistry* **24,** 2667 (1985); L. P. Murray, J. Hofrichter, E. R. Henry, M. Ikeda-Saito, K. Kitagishi, T. Yonetani, and W. A. Eaton, *Proc. Natl. Acad. Sci. U.S.A.* **85,** 2151 (1988)]. Background counts from the vidicon measured in the absence of the measuring flash and baseline intensities measured in the absence of the photodissociating flash were also collected. The spectra were calculated by subtraction of the backgrounds from each set of measured intensities and calculation of the absorbance difference spectra as the logarithm of the ratio of the corrected intensities. These spectra were then corrected for the appropriate baseline. The resulting spectra were averaged using a Gaussian filter having the spectral bandwidth of the spectrograph (4 pixels) and then truncated to 101 wavelength points at approximately 0.8 nm intervals to produce the results shown. Positive signals arise from deoxy photoproducts and negative signals from the CO-liganded reference state. (Data courtesy of Colleen M. Jones.)

discussion of SVD, it is worthwhile to compare briefly these two alternative approaches.

Suppose that we have collected a set of time-resolved spectra (e.g., the data in Fig. 1) measured at n_λ wavelengths and n_t times which we wish to analyze in terms of sums of exponential relaxations. That is, we wish to represent the measured data matrix in the form

$$A_{ij} = A(\lambda_i, t_j) = \sum_{n=1}^{n_k} a_n(\lambda_i) e^{-k_n t_j} \tag{1}$$

for each λ_i. An obvious approach to solving this problem is to use global analysis, in which all of the n_λ vectors of time-dependent amplitudes (i.e., all of the columns of the data matrix) are simultaneously fitted using the same set of n_k relaxation rates $\{k\}$.[6,7] The total number of parameters which must be varied in carrying out this fit is $(n_\lambda + 1) \times n_k$. Such a fit to the unsmoothed data represented in Fig. 1 would require fitting 91(480) = 43,680 data points to a total of (480 + 1)5 = 2405 parameters; reducing the data by averaging over the spectral bandwidth, pruning of regions where the signals are relatively small, and sampling at 101 wavelengths reduces this to fitting 91(101) = 9191 data points using (101 + 1)5 = 510 parameters.

To determine the number of relaxations necessary to fit the data, some statistical criterion of goodness-of-fit must be used to compare the fits obtained for different assumed values for n_k, the number of relaxations. The fitting of data will be discussed in more detail in Section IV,E on the application of physical models and is also discussed at length elsewhere in this volume. The value of n_k determined from the fitting procedure provides a lower limit for the number of kinetic intermediates which are present in the system under study.[8] Another piece of information which is useful in the analysis of such data is the number of spectrally distinguishable molecular species (n_s) which are required to describe the data set. It becomes difficult to determine this number from inspection of real experimental data when it exceeds two, in which case isosbestic points cannot be used as a criterion. In the case of global analysis, the only method for estimating n_s is indirectly (and ambiguously) from the number of relaxations, n_k.

---

[8] The simplest kinetic model for a system which contains n_s species is one in which species interconvert only via first-order reactions. Such a system may be described by an $n_s \times n_s$ matrix containing the elementary first-order rates. The kinetics of such a system may, in most cases, be completely described in terms of a set of exponential relaxations with rates given by the eigenvalues of the rate matrix. If the system comes to equilibrium, one of these eigenvalues is zero, leaving $n_s - 1$ nonzero relaxation rates. If the eigenvalues of the rate matrix are nondegenerate, all relaxations are resolved in the kinetic measurement, and all of the species in the system are spectrally distinguishable, then the number of relaxations is one less than the number of species, $n_k = n_s - 1$. Because the spectra of all of the kinetic intermediates may not be distinguishable, the number of relaxations often equals or exceeds the number of distinguishable spectra, that is, $n_k \geq n_s$. Under conditions where two or more species exchange so rapidly that the equilibration cannot be resolved by the experiment, both the number of relaxations and the number of observed species will be reduced.

We now turn to the SVD-based analysis of the same data. If the system under observation contains n_s species which are spectrally distinguishable, then Beer's law requires that the measured spectrum at time t_j can be described as a linear combination of the spectra of these species:

$$A_{ij} = A(\lambda_i, t_j) = \sum_{n=1}^{n_s} f_n(\lambda_i) c_n(t_j) \tag{2}$$

where A_{ij} is the element of measured spectrum A_j (the spectrum measured at time t_j) sampled at wavelength λ_i, $f_n(\lambda_i)$ is the molar absorbance of species n at wavelength λ_i multiplied by the sample pathlength, and $c_n(t_j)$ is the concentration of species n at time t_j. This result does not depend on the number of species present in the system or the size of the data matrix (i.e., how many spectra are measured and the number of wavelengths on which the spectra are sampled). One of the most useful and remarkable properties of an analysis based on SVD is that it provides a determination of n_s which is independent of any kinetic analysis. In the absence of measurement errors this number is the rank of the data matrix.[1-3] For real data, SVD provides information which can be used to determine the effective rank of the data matrix (i.e., the number of species which are distinguishable given the uncertainty of the data) which provides a lower limit for n_s. This determination is discussed in more detail below in Sections IV,C and IV,D which describe the analysis of SVD output and the rotation procedure.

When SVD is used to process the data matrix prior to carrying out the fit, the output is a reduced representation of the data matrix in terms of a set of n_s basis spectra and an associated set of n_s time-dependent amplitude vectors. A second important property of SVD is that if the set of output components (pairs of basis spectra and amplitude vectors) is ordered by decreasing size, each subset consisting of the first n components provides the best n-component approximation to the data matrix in the least-squares sense.[1-3] It is therefore usually possible to select a subset containing only n_s of the output components which describe the data matrix $\mathbf{A}$ to within experimental precision. Once n_s has been determined, fitting the data requires modeling the amplitudes for only n_s time-dependent amplitude vectors instead of the n_λ vectors required by global analysis. The total number of parameters which must be varied in carrying out the fit is, therefore, $(n_s + 1) \times n_k$. The determination of the number of relaxations required to best fit the data is accomplished using a weighted fitting procedure which is directly comparable to that used for the global analysis of the data, except that it requires fitting of a much smaller set of time-dependent amplitude vectors.

The effectiveness of this procedure is illustrated by the SVD of the data in Fig. 1, the first six components of which are presented in Fig. 2. The spectra and time-dependent amplitude vectors which describe the first two components clearly exhibit signals which are present in the data. Note, however, the progressive decrease in the singular values, s_i, and the signal-to-noise ratios of the subsequent amplitude vectors. Given this result, if n_s were chosen based on a visual inspection of Fig. 2, one might estimate n_s to be only 2; that is, nearly all of the information in the data can be described in terms of only the first two basis spectra and their associated amplitudes. Fitting the first two amplitude vectors from the SVD to five exponential relaxations would require fitting only 91(2) = 182 data points using only (2 + 1)5 = 15 parameters, as compared with the 9191 data points and 510 parameters required by global analysis of the data in Fig. 1.

This brief discussion and the example point out the advantages of using SVD in carrying out such an analysis when the number of wavelengths on which the data are sampled is large (i.e., $n_\lambda \gg n_s$). The use of SVD as an intermediate filter of the data matrix not only provides a rigorous and model-independent determination of n_s, but also enormously simplifies the fitting problem. If the data set includes experiments at only a small number of wavelengths, so that the number of wavelengths is smaller than the number of species in the system which exhibit distinguishable spectra, then $n_s \cong n_\lambda$ and SVD offers no clear advantage in the analysis. This brief discussion also points out why the use of SVD proliferated in the 1980s. Earlier experiments usually consisted of measuring time traces at a small set of selected wavelengths. Only the availability of array detectors and efficient data acquisition computers has made it possible to analyze sets of data sampled on a sufficiently dense array of wavelengths to demand the increases in processing efficiency which result from the use of SVD.

The matrix of data can be derived from a wide variety of experiments. Examples include sets of time-resolved optical spectra, obtained using either a rapid-scanning stopped-flow spectrometer[9] or a pulse-probe laser spectrometer,[10-13] and equilibrium spectra obtained during potentiomet-

[9] R. N. Cochran, F. H. Horne, J. L. Dye, J. Ceraso, and C. H. Suetler, *J. Phys. Chem.* **84,** 2567 (1980).

[10] J. Hofrichter, J. H. Sommer, E. R. Henry, and W. A. Eaton, *Proc. Natl. Acad. Sci. U.S.A.* **80,** 2235 (1983).

[11] J. Hofrichter, E. R. Henry, J. H. Sommer, and W. A. Eaton, *Biochemistry* **24,** 2667 (1985).

[12] L. P. Murray, J. Hofrichter, E. R. Henry, M. Ikeda-Saito, K. Kitagishi, T. Yonetani, and W. A. Eaton, *Proc. Natl. Acad. Sci. U.S.A.* **85,** 2151 (1988).

[13] S. J. Milder, T. E. Thorgeirsson, L. J. W. Miercke, R. M. Stroud, and D. S. Kliger, *Biochemistry* **30,** 1751 (1991).

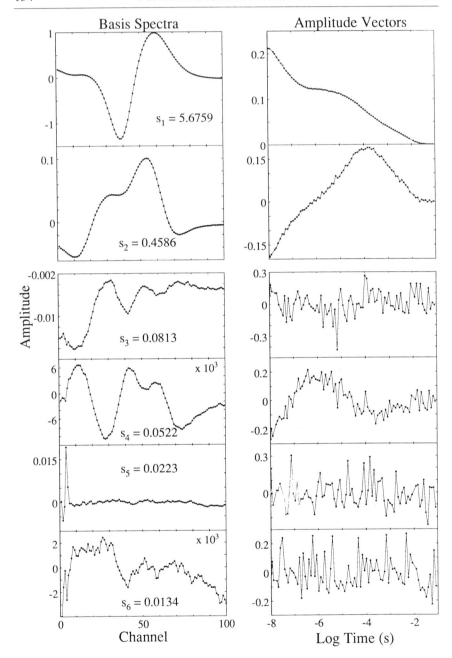

ric[14,15] or pH[16] titrations. This analysis has also been applied to other types of spectra, such as circular dichroism[17,18] and optical rotatory dispersion spectra.[19] The only constraint imposed by the analysis presented here is that the measured signal be linear in the concentrations of the chemical species. The data matrix can then be described by an expression analogous to Eq. (2). In general, the index j runs over the set of experimental conditions which are varied in measuring the spectra. In the case of time-resolved spectroscopy this index includes, but is not necessarily limited to, the time variable, whereas in pH or potentiometric titrations it would include the solution pH or voltage, respectively.

If all of the spectra, $f_n(\lambda)$, are known with sufficient accuracy, then the problem of determining the sample composition from the spectra is easily solved by linear regression. More often, however, the spectra of only a subset of the species are known, or the accuracy with which the reference spectra are known is insufficient to permit the analysis of the data to be carried out to within instrumental precision. Under these conditions one is interested in determining both the number and the shapes of a minimal set of basis spectra which describe all of the spectra in the data matrix. Because the information contained in the data matrix almost always over-determines the set of basis spectra, the algorithm must be robust when faced with rank-deficient matrices. SVD is optimally suited to this purpose. Two alternative procedures can be used to calculate the decomposition. One is to calculate it directly using an algorithm which is also called

[14] R. W. Hendler, K. V. Subba Reddy, R. I. Shrager, and W. S. Caughey, *Biophys. J.* **49,** 717 (1986).

[15] K. V. Subba Reddy, R. W. Hendler, and B. Bunow, *Biophys. J.* **49,** 705 (1986).

[16] S. D. Frans and J. M. Harris, *Anal. Chem.* **57,** 1718 (1985).

[17] J. P. Hennessey, Jr., and W. C. Johnson, Jr., *Biochemistry* **20,** 1085 (1981).

[18] W. C. Johnson, Jr., *Annu. Rev. Biophys. Biophys. Chem.* **17,** 145 (1988).

[19] D. W. McMullen, S. R. Jaskunas, and I. Tinoco, Jr., *Biopolymers* **5,** 589 (1965).

FIG. 2. Singular value decomposition of the data in Fig. 1. The basis spectra (columns of U · S) are plotted on the left, and the corresponding time-dependent amplitudes (columns of V) are plotted on the right. The first 10 singular values were as follows: $s_1 = 5.68$; $s_2 = 0.459$; $s_3 = 0.0813$; $s_4 = 0.0522$; $s_5 = 0.0223$; $s_6 = 0.0134$; $s_7 = 0.0109$; $s_8 = 0.0072$; $s_9 = 0.0047$; $s_{10} = 0.0043$. The data produce two significant basis spectra for which the time-dependent amplitudes have large signal-to-noise ratios. The first, which has a signal-to-noise ratio of about 250, results primarily from a decrease in the amplitude of the deoxy–CO difference spectrum, and its amplitude monitors the extent of ligand rebinding. The second, which has a signal-to-noise ratio of about 30, arises from changes in the spectra of the deoxy photoprod-uct and hence reflects changes in the structure of the molecule in the vicinity of the heme chromophore. The amplitudes of the SVD components are plotted as the points connected by solid lines.

singular value decomposition (SVD), and the other is to use a procedure called principal component analysis (PCA).[20-24] PCA was used in most of the early applications of rank-reduction algorithms to experimental data.[22,25-27] The output of the decomposition provides a set of basis spectra in terms of which all of the spectra in the data set can be represented to within any prescribed accuracy. These spectra are not the spectra of molecular species, but are determined by the mathematical properties of the SVD itself, most significantly by the least-squares property mentioned above. These spectra and their corresponding amplitudes can be used in a variety of ways to extend the analysis and thereby obtain the spectra of the molecular species. This problem is discussed in detail in Section IV,E. A historical summary of the approaches which have been brought to bear on this problem has been presented by Shrager.[23]

Practical applications of SVD to data analysis followed only after the development of an efficient computer algorithm for computing the SVD[28,29] and the experimental advances discussed above. Much of the existing literature which addresses the application of SVD to spectroscopic data has focused on describing specific algorithms for extracting the number of spectral components which are necessary to describe the data and for determining the concentrations of molecular intermediates from the basis spectra.[22,26,27,30] Since beginning to collect this type of data almost a decade ago, we have made extensive use of SVD in the analysis of time-resolved spectroscopic data. In addition to the utility of SVD in the quantitative analysis of data, we have found that a truncated SVD representation of the data also provides an ideal "chart paper" for array spectroscopy, in that it allows data to be compared both qualitatively and quantitatively at a range of levels of precision and also to be stored in a compact and uniquely calculable format. This application of SVD is extremely important to the experimental spectroscopist, since it is very difficult to compare directly raw data sets which may contain as many as several

[20] In this chapter we use the abbreviation SVD to refer both to the decomposition itself and to the SVD algorithm and the abbreviation PCA to refer specifically to the calculation of the SVD by the eigenvalue–eigenvector algorithm (see below).

[21] T. W. Anderson, *Ann. Math. Stat.* **34,** 122 (1963).

[22] R. N. Cochran and F. H. Horne, *J. Phys. Chem.* **84,** 2561 (1980).

[23] R. I. Shrager, *Chemom. Intell. Lab. Syst.* **1,** 59 (1986).

[24] R. I. Shrager and R. W. Hendler, *Anal. Chem.* **54,** 1147 (1982).

[25] J. J. Kankare, *Anal. Chem.* **42,** 1322 (1970).

[26] E. A. Sylvestre, W. H. Lawton, and M. S. Maggio, *Technometrics* **16,** 353 (1974).

[27] R. N. Cochran and F. H. Horne, *Anal. Chem.* **49,** 846 (1977).

[28] G. Golub and W. Kahan, *SIAM J. Numer. Anal. Ser. B* **2,** 205 (1965).

[29] G. H. Golub and C. Reinsch, *Numer. Math.* **14,** 403 (1970).

[30] R. I. Shrager, *SIAM J. Alg. Disc. Methods* **5,** 351 (1984).

hundred thousand data points. Moreover, because no assumptions are required to carry out the SVD portion of the analysis, it provides a simple intermediate screen of the relative quality of "identical" data sets which permits the selection of both representative and optimal data for further analysis.

We begin this chapter with a brief summary of the properties of the singular value decomposition which are relevant to data analysis. We then describe how the SVD of a noise-free data set for which the spectra, f, and concentration, c, vectors [Eq. (2)] are known can be calculated from consideration of the integrated overlaps[31] of these components. Because data analysis necessarily begins with matrices which are "noisy" at some level of precision, we next consider some of the properties of the SVD of matrices which contain noise. This section begins with a brief description of the SVD of random matrices (i.e., matrices which contain *only* noise). We then use perturbation theory to explore how the random amplitudes are distributed in the SVD output when noise is added to a data matrix which has a rank of one, a simple example which enables a quantitative analysis of the noise-averaging properties of SVD. The discussion of noisy matrices continues by describing an asymptotic treatment which permits the best estimate of the noise-free matrix to be calculated in the presence of noise, the details of which are presented elsewhere,[32] and concludes with a brief discussion of a special case in which the noise amplitudes are not random over all of the data matrix, but are highly correlated along either the rows or columns of **A**.

With this theoretical background, we proceed to a step-by-step description of how SVD-based analysis is carried out on real data. The steps include preparation and preprocessing of the data, the calculation of the SVD itself, and a discussion of how the SVD output is analyzed to determine the effective rank of the data matrix. This discussion includes the

---

[31] The integrated overlaps of two continuous spectra, $f_1(\lambda)$ and $f_2(\lambda)$, and of two sets of concentrations defined as continuous functions of conditions x, $c_1(x)$ and $c_2(x)$, are defined, respectively, as

$$\int_0^\infty f_1(\lambda)f_2(\lambda) \, d\lambda; \qquad \int_0^\infty c_1(x)c_2(x) \, dx$$

If f_1 and f_2 are vectors which represent the spectra $f_1(\lambda)$ and $f_2(\lambda)$ sampled on a discrete set of wavelengths $\{\lambda_i\}$, and c_1 and c_2 are vectors which consist of the concentrations $c_1(x)$ and $c_2(x)$ sampled on a discrete set of x values $\{x_i\}$, then the overlaps defined above are closely approximated by either $f_1 \cdot f_2$ or $c_1 \cdot c_2$ multiplied by the size of the appropriate sampling interval. We will conventionally ignore the sampling interval, which appears as a scale factor when comparing the overlaps of vectors sampled on the same points, and use the dot product as the definition of the "overlap" between two vectors.

[32] E. R. Henry, in preparation.

description of a "rotation" procedure which can be used to distinguish condition-correlated amplitude information from randomly varying amplitudes of nonrandom noise sources in the data matrix, the mathematical treatment of which is presented in the Appendix. The analysis of real data necessarily includes the use of molecular models as a means of obtaining from the data information about the system under study. We next describe how the output of the SVD procedures is used as input data for fitting to models and the weighting of the SVD output which optimizes the accuracy with which the fit describes the original data. In Section V we present simulations of the SVD-based analysis of sets of time-resolved spectra for the kinetic system $A \rightarrow B \rightarrow C$. These simulations address in some detail the effects of both random and nonrandom noise on data where the information content is known *a priori*, and they explore the range of noise amplitudes for which the rotation algorithm results in useful improvement of the retained SVD components.

II. Definition and Properties

The existence of the SVD of a general rectangular matrix has been known for over 50 years.[33] For an $m \times n$ matrix $\mathbf{A}$ of real elements ($m \geq n$) the SVD is defined by

$$\mathbf{A} = \mathbf{USV^T} \tag{3}$$

where $\mathbf{U}$ is an $m \times n$ matrix having the property that $\mathbf{U^TU} = \mathbf{I}_n$, where $\mathbf{I}_n$ is the $n \times n$ identity matrix, $\mathbf{V}$ is an $n \times n$ matrix such that $\mathbf{V^TV} = \mathbf{I}_n$, and $\mathbf{S}$ is a diagonal $n \times n$ matrix of nonnegative elements.[34] The diagonal elements of $\mathbf{S}$ are called the singular values of $\mathbf{A}$ and will be denoted by $s_k, k \in \{1, \ldots, n\}$. The columns of $\mathbf{U}$ and $\mathbf{V}$ are called the left and right singular vectors of $\mathbf{A}$, respectively.[1-3] The singular values may be ordered (along with the corresponding columns of $\mathbf{U}$ and $\mathbf{V}$) so that $s_1 \geq s_2 \geq \ldots \geq s_n \geq 0$. With this ordering, the largest index r such that $s_r > 0$ is the rank of $\mathbf{A}$, and the first r columns of $\mathbf{U}$ comprise an orthonormal basis of the space spanned by the columns of $\mathbf{A}$. An important property of the SVD is that for all $k \leq r$, the first k columns of $\mathbf{U}$, along with the correspond-

[33] C. Eckhart and G. Young, *Bull. Am. Math. Soc.* **45**, 118 (1939).

[34] There is some variability in the precise representation of the SVD. The definition given by Lawson and Hanson,[3] for example, differs from that given here in that both $\mathbf{U}$ and $\mathbf{V}$ are square matrices ($m \times m$ and $n \times n$, respectively), and $\mathbf{S}$ is defined to be $m \times n$, with the lower $(m - n) \times n$ block identically zero. The definition given here has advantages in terms of storage required to hold the matrices $\mathbf{U}$ and $\mathbf{S}$. The SVD is similarly defined for an arbitrary matrix of complex numbers. We assume, without loss of generality, that all of the matrices appearing in this chapter consist of real numbers.

ing columns of $\mathbf{V}$ and rows and columns of $\mathbf{S}$, provide the best least-squares approximation to the matrix $\mathbf{A}$ having a rank of k. More precisely, among all $m \times n$ matrices $\mathbf{B}$ having rank k, the matrix $\mathbf{B} = \mathbf{A}_k \equiv \mathbf{U}_k\mathbf{S}_k\mathbf{V}_k^T$, where $\mathbf{U}_k$ and $\mathbf{V}_k$ consist of the first k columns of $\mathbf{U}$ and $\mathbf{V}$, respectively, and $\mathbf{S}_k$ consists of the first k rows and columns of $\mathbf{S}$, yields the smallest value of $\|\mathbf{A} - \mathbf{B}\|$.[35] Furthermore, the magnitude of the difference $\|\mathbf{A} - \mathbf{A}_k\| = (s_{k+1}^2 + \ldots + s_n^2)^{1/2}$.[1-3]

The relationship between SVD and principal component analysis (PCA)[20] may be seen in the following way. Given the matrix $\mathbf{A}$ with the decomposition shown in Eq. (3), the matrix product $\mathbf{A}^T\mathbf{A}$ may be expressed as

$$\begin{aligned}\mathbf{A}^T\mathbf{A} &= (\mathbf{USV}^T)^T\mathbf{USV}^T \\ &= \mathbf{VSU}^T\mathbf{USV}^T \\ &= \mathbf{VS}^2\mathbf{V}^T\end{aligned} \qquad (4)$$

The diagonal elements of $\mathbf{S}^2$ (i.e., the squares of the singular values of $\mathbf{A}$) are the eigenvalues, and the columns of $\mathbf{V}$ are the corresponding eigenvectors, of the matrix $\mathbf{A}^T\mathbf{A}$. A principal component analysis of a data matrix $\mathbf{A}$ has traditionally derived the singular values and the columns of $\mathbf{V}$ from an eigenvalue–eigenvector analysis of the real symmetric matrix $\mathbf{A}^T\mathbf{A}$, and the columns of $\mathbf{U}$ either from the eigenvectors corresponding to the n largest eigenvalues of the reverse product $\mathbf{AA}^T$ [$= \mathbf{US}^2\mathbf{U}^T$, by a derivation similar to that shown in Eq. (4)], or by calculating $\mathbf{U} = \mathbf{AVS}^{-1}$. Although obtaining the matrices $\mathbf{U}$, $\mathbf{S}$, and $\mathbf{V}$ via this procedure is mathematically equivalent to using the direct SVD algorithm,[29] the latter procedure is more robust and numerically stable and is preferred in most practical situations.[23,29]

A. Singular Value Decomposition of Known Data Matrix

To understand how SVD sorts the information contained in a noise-free data matrix it is instructive to consider the SVD of matrices having the form of Eq. (2). To generalize Eq. (2), the $m \times n$ matrix $\mathbf{A}$ may be written

$$\mathbf{A} = \mathbf{FC}^T \qquad (5)$$

where the $m \times r$ matrix $\mathbf{F}$ consists of a set of r column vectors $\{F_i\}$ which

---

[35] The matrix norm used here is the so-called Frobenius norm, defined for an $m \times n$ matrix $\mathbf{M}$ as

$$\|\mathbf{M}\| = \left(\sum_{i=1}^{m}\sum_{j=1}^{n} M_{ij}^2\right)^{1/2}$$

are the spectra of r individual species and the $n \times r$ matrix $\mathbf{C}$ is a set of corresponding amplitude vectors $\{C_i\}$, describing the condition-dependent concentrations of these r species. The vectors $\{F_i\}$ and $\{C_i\}$ are both assumed to be linearly independent. The matrix $\mathbf{A}$ will then have rank r. We now consider the $r \times r$ matrices $\mathbf{F^T F}$ and $\mathbf{C^T C}$ which consist of the overlaps of all possible pairs of vectors in $\{F_i\}$ and $\{C_i\}$, respectively:

$$\begin{aligned} (\mathbf{F^T F})_{ij} &= F_i \cdot F_j \\ (\mathbf{C^T C})_{ij} &= C_i \cdot C_j \end{aligned} \tag{6}$$

The eigenvalues and eigenvectors of the $r \times r$ product of these two matrices, $\mathbf{F^T F C^T C}$, have a simple relationship to the SVD of $\mathbf{A}$. If ν is an eigenvector of this matrix with eigenvalue λ, then

$$\mathbf{F^T F C^T C}\nu = \lambda\nu \tag{7}$$

Premultiplying Eq. (7) by $\mathbf{C}$ yields

$$\begin{aligned} \mathbf{C F^T F C^T C}\nu &= \mathbf{C}\lambda\nu \\ (\mathbf{C F^T F C^T})\mathbf{C}\nu &= \lambda(\mathbf{C}\nu) \\ \mathbf{A^T A}(\mathbf{C}\nu) &= \lambda(\mathbf{C}\nu) \end{aligned} \tag{8}$$

The vector $\mathbf{C}\nu$ is therefore an eigenvector of the matrix $\mathbf{A^T A}$ with the same eigenvalue.

Because the eigenvalues of $\mathbf{A^T A}$ are the squared singular values of $\mathbf{A}$, and the normalized eigenvectors are the columns of the matrix $\mathbf{V}$ in Eq. (4), it follows that the r eigenvalues of $\mathbf{F^T F C^T C}$ are the squares of the r nonzero singular values of $\mathbf{A}$. Multiplying each corresponding eigenvector by $\mathbf{C}$ yields (to within a normalization factor) the corresponding column of $\mathbf{V}$. Note that the transpose of the overlap product matrix $(\mathbf{F^T F C^T C})^T = \mathbf{C^T C F^T F}$ has the same set of eigenvalues but a different set of eigenvectors. If ω is the eigenvector corresponding to eigenvalue λ, then by a derivation similar to the above:

$$\begin{aligned} \mathbf{C^T C F^T F}\omega &= \lambda\omega \\ \mathbf{F C^T C F^T F}\omega &= \lambda(\mathbf{F}\omega) \\ \mathbf{A A^T}(\mathbf{F}\omega) &= \lambda(\mathbf{F}\omega) \end{aligned} \tag{9}$$

$\mathbf{F}\omega$ is therefore an eigenvector of the matrix $\mathbf{A A^T}$. Normalization of $\mathbf{F}\omega$ yields the column of the matrix $\mathbf{U}$ corresponding to the singular value given by $\lambda^{1/2}$. The remaining $n - r$ columns of $\mathbf{U}$ and $\mathbf{V}$, corresponding to singular values which are equal to zero, may be made up of arbitrary orthonormal sets of vectors which are also orthogonal to the first r column vectors constructed as described here.

A useful result of this analysis is that, because the columns of **V** and **U** may be formed simply by normalizing the sets of vectors {**C**ν} and {**F**ω}, respectively, the individual elements of the eigenvectors ν and ω are the coefficients with which the various columns of **C** and **F** are mixed to produce the columns of **V** and **U**. This analysis of the overlap product matrix thus allows us to understand quantitatively how, in the absence of noise, SVD constructs the output matrices from the spectra and concentrations of the species which generate the input matrix.

III. Singular Value Decomposition of Matrices Which Contain Noise

To this point we have discussed the SVD of hypothetical data constructed from spectra and concentrations of a set of species which are known with arbitrary accuracy. In the analysis of experimental data, one is confronted with matrices which contain noise in addition to the desired information. One objective of the experimental spectroscopist is to extract the data from the noise using the smallest possible number of *a priori* assumptions. To take full advantage of SVD in accomplishing this task it is important to understand how SVD deals with matrices in which the individual elements include random as well as nonrandom contributions. Although some insight into this problem can be obtained from algebraic analysis, the problems encountered in the analysis of data are generally too complex to solve analytically, and simulations are required. In this section we use both algebraic analysis and simulations in treating some relatively simple examples which we have selected to illustrate the general principles involved in dealing with noisy matrices.

We begin with a description of the SVD of random matrices. We have carried out simulations to obtain distributions of singular values for a set of $m \times n$ matrices and for square matrices of finite size, and we compare these results with the known analytical result in the asymptotic limit of infinite matrix size. Next, we illustrate the noise-averaging properties of SVD by asking how random noise is partitioned among the SVD components in the case where the noise-free data matrix has a rank of one. We then present a procedure which generalizes an earlier treatment by Cochran and Horne[22,27] which specifies a weighting of the data matrix which can be used to obtain the best estimate of the noise-free data matrix from noisy data if the matrix of variances of the noise component is known. Finally, we consider the problem of noise which can be described as the outer product of two vectors (i.e., the noise amplitude matrix has a rank of 1).

A. Random Matrices

To explore the effects of noise in the data matrix, $\mathbf{A}$, on the SVD of $\mathbf{A}$ we begin by considering matrices which contain only random elements. Figure 3a depicts the distributions of the singular values for matrices of dimensions 10×10, 100×100, and 1000×1000. The distributions were calculated from simulations in which a total of $2(10^5)$ singular values were generated from the SVD of matrices having the specified size, each element of which was a normally distributed random variable having mean value zero and variance σ^2. Note that the rank of the $n \times n$ noise matrix is always close to n. This result can be readily understood, since it is not generally possible to write any one random vector of length n as a linear combination of the remaining $n - 1$ random vectors. This distribution is described in the limit as $n \to \infty$ by the so-called quarter-circle law.[36,37]

$$P(x) = \frac{1}{\pi}(4 - x^2)^{1/2}; \qquad x \equiv s/\sigma n^{1/2} \tag{10}$$

The distribution function describes the quarter-circle on the interval [0,2], also shown in Fig. 3a. The simulations show that the distribution of singular values closely approximates the quarter-circle distribution, even for relatively small matrices. Characteristic distortions, which are largest when n is small (10×10), are present in the regions of the maximum and minimum singular values but the asymptotic limit becomes a very good first-order description of the distribution of singular values for matrices larger than 100×100, a size which is often exceeded in the collection of real experimental data.

There is no analytical theory to describe the distribution of singular values for an $m \times n$ matrix where $m \neq n$. If $m > n$, then it is almost always possible to write $m - n$ of the rows of the matrix as linear combinations of a subset of n rows which are linearly independent. If the singular values of an $m \times n$ matrix are compared with those of an $n \times n$ matrix, where both are composed of random elements having the same variance, one intuitively expects that each singular value of the $m \times n$ matrix will, on the average, be larger than the corresponding singular value of the $n \times n$ matrix. This expectation is confirmed by the results of simulations which were carried out to determine the distribution of singular values for matrices varying from 200×200 to 1000×200 which are presented in Fig. 3b. The results show that the entire distribution of singular values shifts to higher values, with the magnitude of the shift being correlated with $m - n$. The results in Fig. 3b suggest that the largest singular value from the

[36] E. Wigner, *SIAM Rev.* **9**, 1 (1967).
[37] H. F. Trotter, *Adv. Math.* **54**, 67 (1984).

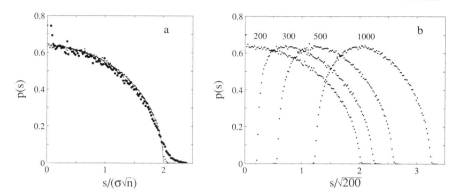

FIG. 3. Distributions of singular values of matrices of normally distributed random numbers having zero mean. (a) Calculated and asymptotic distributions for square matrices. The distribution predicted in the limit of infinite matrix size, described by the quarter circle law [Eq. (10)], is shown as the solid line. The average distributions obtained from calculation of a total of $2(10^5)$ singular values for matrices of the following sizes are shown for comparison: (•) 10×10; (·) 100×100; (·) 1000×1000. (b) Calculated distributions for $m \times n$ matrices where $m \geq n$. The average distributions obtained by calculating a total of $2(10^5)$ singular values for matrices of the following sizes are shown: 200×200; 300×200; 500×200; 1000×200. The number of rows is indicated above each distribution.

distribution increases roughly as $m^{0.3}$. It is important to note that when $m > n$, the entire set of singular values is effectively bounded away from zero, so a random matrix which is not square can be confidently assumed to have full rank (i.e., rank $= \min\{m,n\}$). In the simulations in Section V,A, we shall see that this conclusion can be generalized to matrices which contain nonrandom as well as random amplitudes.

B. Noise-Averaging by Singular Value Decomposition

As discussed above, the first component of the SVD of the matrix **A** provides the best one-component least-squares approximation to **A**. For a data set which consists of n spectra that are identical except for the admixture of random noise, the first singular vector (U_1) is, to within a scale factor, very nearly identical to the average of all of the spectra in the data matrix. This example illustrates the averaging properties of SVD. In this section we use perturbation theory to examine these properties in more detail for a particularly simple case. We consider a data matrix $\mathbf{A}_0$ which has a rank of 1 (i.e., $\mathbf{A}_0$ can be described as the outer product of two column vectors a and b, $\mathbf{A}_0 = ab^T$). We now add to $\mathbf{A}_0$ a random noise matrix, ε, each element of which is a normally distributed random variable having a mean value zero and variance σ^2, that is,

$$\mathbf{A} = \mathbf{A}_0 + \varepsilon = ab^T + \varepsilon \qquad (11)$$

One would like to know how the noise represented by ε alters the singular values and vectors of the matrix $\mathbf{A}$. If we consider for the moment the error-free data matrix $\mathbf{A}_0$, we can write

$$\mathbf{A}_0 = s_0 U_0 V_0^T \qquad (12)$$

where $U_0 = a/\|a\|$, $V_0 = b/\|b\|$, and the singular value $s_0 = \|a\|\,\|b\|$. Second-order perturbation theory can be used to determine how these quantities are modified by the addition of the random matrix ε.

We begin by calculating the eigenvalues and eigenvectors of the matrices $\mathbf{A}\mathbf{A}^T$ and $\mathbf{A}^T\mathbf{A}$.

$$\begin{aligned} \mathbf{A}\mathbf{A}^T &= \mathbf{A}_0\mathbf{A}_0^T + \mathbf{A}_0\varepsilon^T + \varepsilon\mathbf{A}_0^T + \varepsilon\varepsilon^T \\ &= \mathbf{A}_0\mathbf{A}_0^T + \mathbf{W} \end{aligned} \qquad (13)$$

The perturbed values of the largest eigenvalue, s^2, and the corresponding eigenvector, U, of this matrix can then be written

$$s^2 = s_0^2 + U_0^T\mathbf{W}U_0 + \sum_{n\neq0} \frac{(U_0^T\mathbf{W}U_n)^2}{s_0^2} \qquad (14)$$

$$U = U_0 + \sum_{n\neq0} \frac{(U_0^T\mathbf{W}U_n)}{s_0^2} U_n$$

where the $\{U_n\}$ are a set of normalized basis spectra which are orthogonal to U_0. We proceed by calculating the matrix elements in Eq. (14) and then calculating the expected values and variances of the resulting expressions for s^2 and U. The results, in which only terms that are first order in σ^2 have been retained, may be summarized by

$$s^2 \cong s_0^2 + (m + n - 1)\sigma^2 + 2s_0\varepsilon_s$$

$$s \cong s_0 + \frac{(m + n - 1)\sigma^2}{2s_0} + \varepsilon_s; \qquad [var(\varepsilon_s) = \sigma^2] \qquad (15)$$

$$U_i \cong U_{0i} + \frac{\varepsilon_{U_i}}{s_0}; \qquad [var(\varepsilon_{U_i}) = (1 - U_{0i}^2)\sigma^2]$$

where ε_s and the ε_{U_i} are random variables having zero mean and the indicated variances.[38] A similar calculation for the matrix $\mathbf{A}^{\mathrm{T}}\mathbf{A}$ yields the result

$$V_j = V_{0j} + \frac{\varepsilon_{V_j}}{s_0}; \qquad [var(\varepsilon_{V_j}) = (1 - V_{0j}^2)\sigma^2] \qquad (16)$$

The results in Eqs. (15) and (16) show that, while each element of the input matrix, $\mathbf{A}$, has variance σ^2, each element of the U and V vectors of the perturbed data matrix has a variance which is somewhat less than $(\sigma/s_0)^2$, and the variance of the singular value s is simply σ^2. As the matrix, $\mathbf{A}$, becomes large the squares of the individual elements of the normalized vectors U_0 and V_0 will, in most cases, become small compared to 1, and the variance of each of the individual elements of U and V will approach $(\sigma/s_0)^2$.

We expect *a priori* that the averaging of multiple determinations of a variable, each of which is characterized by random error of variance σ^2, decreases the error in the average value by a factor of $d^{1/2}$, where d is the number of determinations. It is interesting to consider an example for which the above results may be easily compared with this expectation. If we choose the matrix $\mathbf{A}_0$ to be an $m \times n$ matrix of ones, U_0 and V_0 are constant vectors having values of $1/m^{1/2}$ and $1/n^{1/2}$, respectively, and $s_0 = (mn)^{1/2}$. Equations (15) and (16) then become

$$s \cong (mn)^{1/2} + \frac{(m + n - 1)\sigma^2}{2(mn)^{1/2}} + \varepsilon_s; \qquad [var(\varepsilon_s) = \sigma^2]$$

$$U_i \cong \frac{1}{m^{1/2}} + \frac{\varepsilon_{U_i}}{(mn)^{1/2}}; \qquad \left[var(\varepsilon_{U_i}) = \left(1 - \frac{1}{m}\right)\sigma^2\right] \qquad (17)$$

$$V_j \cong \frac{1}{n^{1/2}} + \frac{\varepsilon_{V_j}}{(mn)^{1/2}}; \qquad \left[var(\varepsilon_{V_j}) = \left(1 - \frac{1}{n}\right)\sigma^2\right]$$

If the elements of the basis spectrum, U, were obtained by simply fitting the noisy data, $\mathbf{A}$, with the V_0 vector from the noise-free data which

---

[38] We have used the following properties of random variables in this derivation and in the discussion which follows. First, if X is any fixed vector and Y is a vector of random variables of mean zero and variance σ_Y^2, then

$$\langle X \cdot Y \rangle = 0; \qquad var(X \cdot Y) = |X|^2 \sigma_Y^2$$

Furthermore, if the individual elements of Y are independent and normally distributed, then $X \cdot Y$ is also normally distributed. Second, if Z is also a vector of random variables of variance σ_Z^2, which are independent of those in Y, then

$$\langle Y \cdot Z \rangle = \sum_{i=1}^{n} \langle Y_i \rangle \langle Z_i \rangle; \qquad var(Y \cdot Z) = \sum_{i=1}^{n} (\sigma_Y^2 \langle Z_i \rangle + \sigma_Z^2 \langle Y_i \rangle + \sigma_Y^2 \sigma_Z^2)$$

has n identical elements, one would expect a relative error in the fitted "U" vector of $\sigma/n^{1/2}$. Use of the corresponding procedure to obtain the amplitude vector, V, should produce a relative error of $\sigma/m^{1/2}$. The predictions of Eq. (17) are very close to these expected results: the variances of both the U and the V vectors are slightly less than would be obtained from the fits. This can be rationalized by the fact that one degree of freedom, that is, variations in the sum of the squares of the entries of the data matrix, is incorporated into the variations of s. Each element of the filtered matrix, reconstructed from the first singular value and vectors, sUV^{T}, can now be calculated

$$A_{ij} \cong 1 + \frac{(m + n - 1)}{mn}\sigma^2 + \frac{\varepsilon_{U_i}}{n^{1/2}} + \frac{\varepsilon_{V_j}}{m^{1/2}} + \frac{\varepsilon_s}{(mn)^{1/2}}; \qquad (n, m \gg 1) \quad (18)$$

where terms of higher order than $1/(mn)^{1/2}$ have been neglected. The amplitude of the noise in the reconstructed matrix is thus also significantly reduced from that of the input matrix if both n and m are large. This reduction results from discarding the amplitudes of the higher SVD components which are derived almost exclusively from the random amplitudes of ε.

These results point out a number of useful features in designing experiments to maximize signal-to-noise in the SVD-reduced representation of the data. Increasing the size of the data matrix in either dimension improves the signal-to-noise ratio in the singular vectors if it increases the magnitude of s_0. For the additional data to contribute to s_0, the added points must contain meaningful amplitude information and hence cannot include regions in which there is little or no absorbance by the sample. Increasing the size of the data matrix also does not help if it can only be accomplished by simultaneously increasing the standard deviation in the measurement for each data point. In most cases, the size of the data matrix must be determined by compromises. For example, increasing the value of m (i.e., increasing the wavelength resolution of the experiment) reduces the number of photons detected per resolution element of the detector. At the point where the noise in the measured parameter is dominated by statistical fluctuations in the number of photons detected (shot noise), further increasing the resolution will increase σ as $m^{1/2}$, so no improvement in signal-to-noise in the SVD output will result from accumulating more densely spaced data. Increasing the size of the data set by using a greater number of conditions necessarily increases the time required for data acquisition. In this case, reduction in the quality of the data matrix, perhaps by spectrometer drift or long-term laser drifts, may offset the improvements expected from increasing the number of conditions sampled.

C. Statistical Treatment of Noise in Singular Value Decomposition Analysis

We have seen in Section III,A that a matrix which includes random noise nearly always has full rank (i.e., rank = $\min\{m,n\}$). The presence of measurement noise in a data matrix thus complicates not only the best estimate of the error-free data contained therein but even the determination of the effective rank of the matrix. Two attempts have been made to treat quantitatively the statistical problems of measurement errors in the principal component analysis of data matrices. Based on a series of simulations using sets of synthetic optical absorption spectra having a rank of 2 in the presence of noise of uniform variance, Sylvestre et al.[26] proposed that an unbiased estimate of the variance could be obtained by dividing the sum of squared residuals obtained after subtraction of the rank r representation of a $p \times n$ data matrix by the quantity $(n - r)(p - r)$. This result is useful as a criterion in determining the rank of a matrix if the noise is uniform (see Section IV,C). This analysis was generalized by Cochran and Horne[27] to the case where the matrix of variances σ_{ij}^2 of the elements of the data matrix, A_{ij}, is any matrix having a rank of 1, rather than a constant matrix. Cochran and Horne[27] also introduced a scheme for statistical weighting of the data matrix prior to PCA so that the effective rank, r, is more easily determined and the rank-r representation of the data is optimized. In this section we discuss this analysis and its extension to the case where the matrix of variances, σ_{ij}^2, is arbitrary[32] and establish a connection between such a weighting scheme and SVD-based analysis.

Consider a single set of measurements, arranged as an $m \times n$ data matrix **A**. Successive determinations of **A** will differ because of measurement errors owing to noise and other factors. If multiple determinations of **A** were carried out, its expected value, $\langle \mathbf{A} \rangle$, could be calculated by averaging the individual elements. In the limit of a very large number of determinations, the matrix $\langle \mathbf{A} \rangle$ will become the best estimate of the error-free matrix. In the following discussion we make constant use of the fact that the SVD of **A** is closely related to the eigenvector–eigenvalue analyses of the matrices $\mathbf{AA}^T$ and $\mathbf{A}^T\mathbf{A}$. We consider the expected values $\langle \mathbf{AA}^T \rangle$ and $\langle \mathbf{A}^T\mathbf{A} \rangle$ that would be generated by making an infinite number of determinations of **A** and accumulating the averages of the resulting two product matrices. If we assume that individual elements of $\mathbf{A}$ may be treated as independent variables uncorrelated with other elements, the components of the average matrix $\langle \mathbf{AA^T} \rangle$ may be written

$$\langle \mathbf{AA^T} \rangle_{ij} = \left\langle \sum_k A_{ik} A_{jk} \right\rangle$$

$$= \sum_k \langle A_{ik} A_{jk} \rangle$$

$$= \sum_k (\langle A_{ik} \rangle \langle A_{jk} \rangle (1 - \delta_{ij}) + \langle A_{ik}^2 \rangle \delta_{ij}) \tag{19}$$

where δ_{ij} is the Kronecker delta. If we now define the elements of the variance matrix as

$$\sigma_{ij}^2 = \langle A_{ij}^2 \rangle - \langle A_{ij} \rangle^2 \tag{20}$$

Eq. (19) can be rewritten as

$$\langle \mathbf{AA^T} \rangle_{ij} = \sum_k (\langle A_{ik} \rangle \langle A_{jk} \rangle + \sigma_{ik}^2 \delta_{ij}) \tag{21}$$

Similarly, the elements of the average matrix $\langle \mathbf{A^TA} \rangle$ may be written

$$\langle \mathbf{A^TA} \rangle_{ij} = \sum_k (\langle A_{ki} \rangle \langle A_{kj} \rangle + \sigma_{ki}^2 \delta_{ij}) \tag{22}$$

These two results may be recast in matrix notation as

$$\begin{aligned} \langle \mathbf{AA^T} \rangle &= \langle \mathbf{A} \rangle \langle \mathbf{A} \rangle^T + \mathbf{X} \\ \langle \mathbf{A^TA} \rangle &= \langle \mathbf{A} \rangle^T \langle \mathbf{A} \rangle + \mathbf{Y} \end{aligned} \tag{23}$$

$\mathbf{X}$ and $\mathbf{Y}$ are diagonal matrices whose diagonal elements consist of sums of rows and columns of the matrix of variances, respectively, that is,

$$X_{ij} = \left(\sum_k \sigma_{ik}^2 \right) \delta_{ij}$$

$$Y_{ij} = \left(\sum_k \sigma_{ki}^2 \right) \delta_{ij} \tag{24}$$

　　　In Eq. (23) the effects of measurement errors on the expectation values of $\mathbf{AA^T}$ and $\mathbf{A^TA}$ have been isolated in the matrices $\mathbf{X}$ and $\mathbf{Y}$, respectively. In general, these matrices are not simple multiples of the identity matrices of the appropriate size, so there is no simple relationship between the eigenvectors of $\langle \mathbf{AA^T} \rangle$ and those of $\langle \mathbf{A} \rangle \langle \mathbf{A} \rangle^T$ or between the eigenvectors of $\langle \mathbf{A^TA} \rangle$ and those of $\langle \mathbf{A} \rangle^T \langle \mathbf{A} \rangle$. In the special case in which the matrix of variances σ_{ij}^2 has a rank of 1, Cochran and Horne[27] showed that it is possible to obtain diagonal matrices $\mathbf{L}$ and $\mathbf{T}$ such that the weighted or transformed matrix $\mathbf{A_W} = \mathbf{LAT}$ produces an expected value of the first product matrix of the form

$$\langle \mathbf{A_W A_W^T} \rangle = \langle \mathbf{A_W} \rangle \langle \mathbf{A_W} \rangle^T + c\mathbf{I_m} \tag{25}$$

where c is an arbitrary constant and $\mathbf{I_m}$ is the $m \times m$ identity matrix. Although not discussed by Cochran and Horne,[27] it may also be shown that the same choices of $\mathbf{L}$ and $\mathbf{T}$ produce an expected value of the reverse product matrix $\langle \mathbf{A_W^T A_W} \rangle$ which has a similar form. Equation (25) is significant because it shows that the eigenvectors of the "noise-free" product $\langle \mathbf{A_W} \rangle \langle \mathbf{A_W} \rangle^T$ are now identical to those of the average of "noisy" matrices $\mathbf{A_W A_W^T}$, with eigenvalues offset by the constant c; a similar description holds for the reverse products.

We show elsewhere[32] that for an arbitrary matrix of variances σ_{ij}^2 it is possible to construct diagonal matrices $\mathbf{L}$ and $\mathbf{T}$ such that the transformed matrix $\mathbf{A_W} = \mathbf{LAT}$ satisfies the following conditions:

$$\begin{aligned} \langle \mathbf{A_W A_W^T} \rangle &= \langle \mathbf{A_W} \rangle \langle \mathbf{A_W} \rangle^T + a\mathbf{I_m} \\ \langle \mathbf{A_W^T A_W} \rangle &= \langle \mathbf{A_W} \rangle^T \langle \mathbf{A_W} \rangle + b\mathbf{I_n} \end{aligned} \tag{26}$$

where a and b are constants such that $a/b = n/m$. This analysis shows that, by using the matrices $\mathbf{L}$ and $\mathbf{T}$, which can be determined from the matrix of variances σ_{ij}^2, it is possible to produce indirectly the singular value decomposition of the weighted noise-free matrix $\langle \mathbf{A_W} \rangle$ from the averages of the noisy products $\mathbf{A_W A_W^T}$ and $\mathbf{A_W^T A_W}$. It should be emphasized that this result is only rigorous in the limit of a large number of determinations of the data matrix $\mathbf{A}$ (and hence $\mathbf{A_W}$). The efficacy of such weighting schemes in improving the estimate of the noise-free data obtained from the analysis of a single determination of $\mathbf{A}$ can only be established by numerical simulations which incorporate the known characteristics of both the signal and the noise. Because the noise distribution in our experiments (e.g., Fig. 1) is nearly uniform, our experience with such schemes is severely limited. For this reason we will not discuss this issue in any detail in this chapter. One can argue intuitively, however, that the utility of such procedures for individual data matrices should depend both on the size of the data matrix and on the detailed distribution of the noise. That is, as the data matrix becomes large, a single data set should be able to sample accurately the noise distribution if the distribution of the variances is smoothly varying. On the other hand, the noise distribution might never be accurately sampled if the variances are large for only a very small number of elements of the data matrix.

Implementation of this general statistical weighting scheme requires solving a system of $m + n$ simultaneous nonlinear equations[32] and using the resulting diagonal matrices $\mathbf{L}$ and $\mathbf{T}$ to calculate the weighted data matrix $\mathbf{A_W} = \mathbf{LAT}$. The SVD of this matrix is then calculated, and the output screened and/or postprocessed by any of the methods discussed in this chapter (see below), yielding a set of r basis spectra $\mathbf{U'_W}$ and amplitudes $\mathbf{V'_W}$ for which $\mathbf{A_W} \cong \mathbf{U'_W V'_W T}$. A set of basis spectra and amplitudes of the unweighted matrix $\mathbf{A}$ which are consistent with those of $\mathbf{A_W}$ may then be constructed by simply "undoing" the weighting separately on $\mathbf{U'_W}$ (using $\mathbf{L}^{-1}$) and on $\mathbf{V'_W}$ (using $\mathbf{T}^{-1}$), that is,

$$\mathbf{A} \cong \mathbf{U'V^T}$$
$$\mathbf{U'} = \mathbf{L}^{-1}\mathbf{U'_W} \qquad (27)$$
$$\mathbf{V'} = \mathbf{T}^{-1}\mathbf{V'_W}$$

It is important to note that the final basis spectra and amplitudes are generally neither normalized nor orthogonal, but these mathematical properties are not usually critical for the subsequent steps in data analysis (see below). As discussed by Cochran and Horne[22,27] one of the advantages of producing a weighted matrix satisfying Eq. (26) is that, if $\mathbf{A_W}$ has rank r, then the last $m - r$ eigenvalues of $\langle \mathbf{A_W A_W^T} \rangle$ will equal a. This is equivalent to having only the first m singular values of $\langle \mathbf{A_W} \rangle$ nonzero. This suggests that one measure of the success in applying the procedure to a finite data set might be the extent to which it pushes one set of singular values toward zero and away from the remaining set.

D. Singular Value Decomposition of Matrices Containing Rank-1 Noise

In addition to the random noise which we have discussed to this point, data may contain signals which have random amplitudes when examined along either the rows or the columns of the data matrix, but nonrandom amplitudes when examined along the other set of variables. One example of a situation in which noise has these characteristics arises in single-beam optical spectroscopy using array detectors, where changes in the output energy of the source or in the sensitivity of the detector result in constant offsets across the entire measured spectrum. The amplitude of these offsets is highly correlated along the wavelength direction of the data matrix, but uncorrelated along the conditions dimension. Another example arises in the measurement of condition-dependent absorbances at a single wavelength, such as kinetic traces or titration curves, where the limits of the absorbance changes can often only be obtained by extrapolation of the data to conditions where precise measurement is not possible (e.g., infinite or zero time; complete saturation with substrate). Uncertainty in the value of the extrapolated absorbance can generate errors which are present with equal amplitude in all of the data measured at a single wavelength, but which vary from wavelength to wavelength.

The influence of this type of noise on the SVD output may be addressed using the formalism developed in Section II,A. We consider a noise-free $m \times n$ data matrix which can be written as the sum of the outer products of a small set of basis m-vectors $\{F_{0i}\}$ and corresponding amplitude n-vectors $\{C_{0i}\}$, namely, $\mathbf{A_0} = \mathbf{F_0 C_0^T}$. We consider the situation in which the noise $\mathbf{N}$ may also be written as the product of two vectors: X, an m-vector which describes the noise amplitudes as a function of the isolated variable and Y, an n-vector which describes the noise amplitudes as a function of the remaining variables. In other words, $\mathbf{N} = XY^T$ is rank-1.[39] We can then write the full data matrix as $\mathbf{A} = \mathbf{A_0} + \mathbf{N} = \mathbf{FC^T}$, where $\mathbf{F}$ and $\mathbf{C}$ are formed by simply appending the column vectors X and Y to the matrices $\mathbf{F_0}$ and $\mathbf{C_0}$, respectively. As discussed in Section II,A, the SVD of a matrix of this form is completely determined by eigenvalue–eigenvector analyses of the overlap product matrix $\mathbf{F^TFC^TC}$ and its transpose.

Either the vector X or the vector Y may contain the random amplitudes. For purposes of discussion, we will assume that the randomness appears only in the amplitude vector Y, which we assume to be a set of independent, normally distributed random variables. We will also assume for simplicity that the "noise-free" matrices $\mathbf{F_0}$ and $\mathbf{C_0}$ each have a single column; these column vectors will also be called F_0 and C_0, respectively. The analyses for situations in which X is the random vector, and in which the noise-free data matrix consists of more than one component, proceed in a similar fashion. We will assume further that both X and F_0 are normalized vectors, so that the overall amplitude information is contained in the vectors Y and C_0. Then the overlap matrix $\mathbf{F^TF}$ may be written simply as

$$\mathbf{F^TF} = \begin{bmatrix} 1 & \Delta \\ \Delta & 1 \end{bmatrix} \tag{28}$$

where Δ is the overlap of the normalized vectors F_0 and X.

The statistical properties of the overlap product matrix $\mathbf{F^TFC^TC}$ and its transpose are now determined by the statistical properties of the random

---

[39] This situation in which the noise *amplitude* matrix is rank-1 must be distinguished from the case in which the matrix of *variances* of the noise is rank-1, which was discussed in a different context in Section III,C.

overlap matrix $\mathbf{C}^T\mathbf{C}$. Using the results of Note 38, the expected value and variance of $C_0 \cdot Y$ become

$$\langle C_0 \cdot Y \rangle = 0$$
$$Var(C_0 \cdot Y) = \sum_i \sigma_{yi}^2 \langle (C_0)_i \rangle^2 \qquad (29)$$
$$= \sigma_Y^2 |C_0|^2$$

and $C_0 \cdot Y$ is normally distributed.

The expected value and variance of $Y \cdot Y$ may be determined in a similar fashion. The results, quoted here without proof, are

$$\langle Y \cdot Y \rangle = n\sigma_Y^2$$
$$Var(Y \cdot Y) = 2n(\sigma_Y^2)^2 \qquad (30)$$

However, in this case the resulting values of $Y \cdot Y$ are not normally distributed, but are characterized by the skewed distribution

$$p(x)\, dx = \frac{2^{n/2}\sigma_Y^n}{\Gamma(n/2)} x^{n/2-1}\, e^{-x/(2\sigma_Y^2)}\, dx \qquad (31)$$

where $x = Y \cdot Y$, $\Gamma(\ldots)$ is the gamma function, and n is the number of elements in Y.

With these results, the overlap matrix $\mathbf{C}^T\mathbf{C}$ in the case of two vectors may be written

$$\mathbf{C}^T\mathbf{C} = \begin{bmatrix} |C_0|^2 & 0 \pm \sigma_Y \\ 0 \pm \sigma_Y & n\sigma_Y \pm (2n)^{1/2}\sigma_Y^2 \end{bmatrix} \qquad (32)$$

where the notation $a \pm b$ denotes a random variable with expected value a and variance b^2. Because every instance of $\mathbf{C}^T\mathbf{C}$ in an ensemble is symmetric, the two off-diagonal elements are in fact the same normally distributed random variable (derived from the inner product of the random vector Y with C_0). This simplifying feature is offset by the fact that the lower right diagonal element in Eq. (32) is a random variable (derived from the inner product of the random vector Y with itself) which is *not* normally distributed and is neither independent of nor representable in terms of the off-diagonal elements. If the variance of this element (which is second order in $\sigma_Y^2/|C_0|^2$) is neglected, the analysis simplifies to the diagonalization of overlap product matrices which are functions of a single normally distributed random variable. Even in this approximation analytical expressions for the distributions of eigenvalues and eigenvectors of such matrices are unmanageably complex.

It is, however, possible to determine the statistical properties of the SVD of the perturbed data matrix by explicit simulation. The aim of such simulations is to produce an ensemble of noisy data sets, the mean of which corresponds to some prescribed, noise-free data set, and use this ensemble to calculate the statistical properties (means and variances) of the singular values and vectors. In most cases it is necessary to explicitly calculate the SVD of a large number of matrices synthesized by adding random noise having specified characteristics to the elements of the noise-free data matrix. In the present situation, however, the simulation procedure is greatly simplified because it is only necessary to create ensembles of overlap product matrices $\mathbf{F^T F C^T C}$. Because $\mathbf{F^T F}$ is determined by the overlaps of the (normalized) basis vectors, it can be specified by simply specifying the magnitude of the off-diagonal elements, Δ. An ensemble of the elements of $\mathbf{C^T C}$ which involve the random amplitudes can then be constructed by calculating the overlaps of an ensemble of random amplitude vectors with the various fixed amplitude vectors and with themselves. The results of a set of such simulations are presented in Fig. 4.

Figure 4 shows the extent of mixing of the random amplitudes, Y, with C_0 and the spectrum, X, with F_0 as a function of the spectral overlap, Δ, at a number of different values of the root-mean-square (RMS) noise amplitude, σ_Y. Let us start by examining the results in Fig. 4b, which describe the mixing of the basis spectrum X with F_0. When the RMS amplitude of the "noise" spectrum is significantly smaller than the noise-free data, SVD effectively discriminates against the "noise" spectrum. Figure 4b shows that the effectiveness of this discrimination depends on both Δ and the noise amplitude, σ_Y. For low to moderate noise levels the mixing coefficient increases linearly with Δ, and the inset to Fig. 4b shows that the initial slope of the curves increases roughly as σ_Y^2. In interpreting these results, we must remember that the amplitude of the "noise" spectrum is a random variable, so SVD is able to average and thereby substantially reduce these amplitudes in producing the output signal spectrum. This cancellation, however, can only be as effective as the ability of SVD to distinguish between the "signal" and "noise" spectra. The extent to which these spectra are distinguishable is determined by Δ. When Δ is zero, there is no mixing of noise with signal at any noise amplitude. When Δ is nonzero, SVD mixes the signal and noise spectra to produce orthogonal basis spectra, and the extent of mixing increases roughly as $\Delta\sigma_Y^2$. At high noise amplitudes the curves become nonlinear, and appear to saturate at a value of $2^{-1/2}$.

We now examine the mixing of the random amplitudes, Y, with the unperturbed composition vector, C_0, described by Fig. 4c. The extent of the mixing is essentially independent of the noise amplitude when the noise is small. As Δ increases, the mixing coefficient increases monotonically to a value of $2^{-1/2}$. Recall that, by design, *all* of the information on the

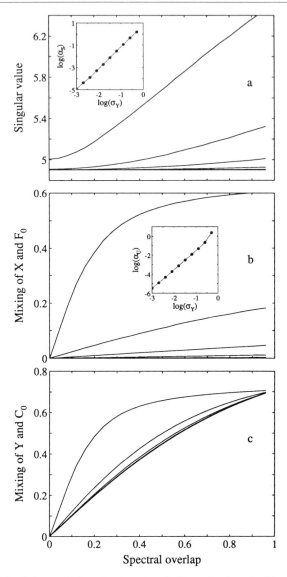

FIG. 4. Mixing between data and noise when both can be described by matrices having a rank of 1. Simulations were carried out at each of 10 noise amplitudes, σ_Y, spaced by factors of 2 from 0.001 to 0.512 and at each of 20 values of the overlap, Δ, ranging from 0

amplitude of the perturbation is contained in the norm of the vector Y. As a result, the amplitude-independent mixing coefficient in Fig. 4c actually reflects the fact that the noise content of the first amplitude vector increases in *direct* proportion to the amplitude of the perturbation. The only operative discrimination against the random amplitudes in deriving the first amplitude vector is the overlap. At the largest noise amplitudes the mixing coefficient approaches its saturating value at smaller values of Δ.

The contribution of the random amplitudes to the first amplitude vector is proportional to σ_Y, and their contribution to the first spectrum is proportional to σ_Y^2. This is a direct consequence of the averaging of this contribution over the random amplitudes by SVD. Given these results it becomes possible to rationalize the dependence of the singular values on the noise amplitudes shown in Fig. 4a. At low noise amplitudes, the singular value increases quadratically with increasing Δ. The second derivative at $\Delta = 0$ increases in proportion to σ_Y^2 as shown in the inset to Fig. 4a. We have seen in Section III,B that addition of a small random noise amplitude to a nonrandom matrix increases the singular value in proportion to its variance, σ^2. The observed results from the simulations parallel this behavior. The first-order effect of adding the noise at low noise amplitudes is to

---

to 0.96 in increments of 0.04. Each simulation was performed as follows. The matrix $\mathbf{F}$ was first formed as in Eq. (28) for a prescribed value of the overlap Δ. A noise-free amplitude vector, C_0, having elements $(C_0)_i = 0.5[\exp(-k_1 t_i) + \exp(-k_2 t_i)]$, for a set of 71 values of t_i, uniformly spaced on a logarithmic grid from 10^{-8} to 10^{-1} sec, was first calculated using $k_1 = 10^6$ sec^{-1} and $k_2 = 10^3$ sec^{-1}. An ensemble of 10^4 vectors, Y, each consisting of 71 normally distributed random numbers with variance σ_Y^2 and a mean value of zero was used to construct an ensemble of overlap product matrices $\mathbf{F}^T\mathbf{F}\mathbf{C}^T\mathbf{C}$ where $\mathbf{C}^T\mathbf{C}$ has the form $\begin{bmatrix} C_0 \cdot C_0 & C_0 \cdot Y \\ C_0 \cdot Y & Y \cdot Y \end{bmatrix}$. The eigenvalues and eigenvectors of these matrices and their transposes were then used to construct an ensemble of singular values and mixing coefficients as described in the text. (a) Averaged singular values, s_1. The inset in (a) describes the dependence of α_S on σ_Y (the square root of the variance of Y) where α_S is determined by fitting the initial portion of the curve to $\alpha_S \Delta^2$. (b) Mixing coefficients which describe the singular vectors U_1. These coefficients describe the mean amplitude of X, the normalized spectrum associated with the random amplitudes which is mixed with F_0 to generate U_1 under each set of conditions. The coefficients for U_1 depend on both the overlap, Δ, and the noise amplitude, σ_Y. The dependence of the mixing coefficient on the overlap is approximately linear for values of the mixing coefficient less than about 0.2. The inset in (b) describes the dependence of the initial slope of the curves in (b), α_U, on σ_Y. The second derivative of the curves in (a), α_S, and the slopes in (b), α_U, can both be approximately represented by the relation $\alpha = A\sigma_Y^2$ for $\sigma_Y < 0.1$. (c) Mixing coefficients which describe the singular vectors V_1. These coefficients describe the mean amplitude of the random amplitude vector, Y, which are mixed with C_0 to generate V_1. These coefficients depend primarily on the overlap, Δ, and are nearly independent of the noise amplitude, σ_Y. In each graph, the uppermost line represents the results of the calculations for $\sigma_Y = 0.512$, and each lower line represents the results for a value of σ_Y which is successively smaller by a factor of 2.

increase the random component in the amplitude vector in direct proportion to $\Delta\sigma_Y$. Without the normalization imposed by SVD, these random amplitudes would be expected to increase the norm of this vector by an amount proportional to $(\Delta\sigma_Y)^2$. This increase then appears in the singular value when the amplitude vector is normalized. No contribution to the singular value is expected from the mixing of F_0 and X, since both spectra are normalized prior to the mixing. At the highest noise amplitudes, σ_Y becomes comparable to the mean value of C_0, and an additional small contribution of Y to the singular value can be perceived as an offset in the value of the singular value s at $\Delta = 0$. This probably results from the fact that the random amplitudes have, at this point, become comparable to the "signal" amplitudes, and the "noise" component can no longer be treated as a perturbation.

These simulations provide considerable insight into the performance of SVD for data sets which contain one or more component spectra together with noise described by a well-defined spectrum having random amplitudes. The results show that when a perturbation having these characteristics is present in a data set, it will have a much larger effect on the amplitude vectors than on the spectra. Our observation that the degree of mixing with the signal spectrum increases as σ_Y^2 suggests that any steps taken to minimize such contributions will be particularly helpful in improving the quality of the resulting component spectra. The noise contribution to the amplitude vectors increases only in direct proportion to σ_Y, so reduction of the noise amplitude will be less effective in improving these output vectors. There are other analytical methods which can be used to supplement the ability of SVD to discriminate against such contributions. One such method, the so-called rotation algorithm, is discussed in Section IV,D. Because the mixing of the random amplitudes, Y, with the "signal" component, C_0, is directly proportional to the overlap between the spectra associated with these amplitudes, Δ, the results further argue that, in some cases, it may be advantageous to select a form for the data which minimizes this overlap. If, for example, the "noise" arises primarily from baseline offsets mentioned above, then the overlap can be minimized by arranging the collection and preprocessing of the data so that the spectra which are analyzed by SVD are difference spectra rather than absolute spectra. The spectral signature of such random components in a specific experiment (corresponding to X) can usually be determined by analysis of a data set which contains no "signal" but only experimentally random contributions. We shall return to this point when discussing the simulations presented in Section V,B below in which random noise comparable to that discussed in Sections III,A and III,B has also been included in the data matrix.

IV. Application of Singular Value Decomposition to Analysis of
    Experimental Data

Having considered some of the properties of the SVD of noise-free and
noisy matrices, we now turn to the problem of applying SVD to the analysis
of experimental data. The actual calculation of the SVD of a data matrix
is only one of a series of steps required to reduce and interpret a large data
set. For the purposes of this discussion we shall break the procedure
into four steps. The first step is the organization of the experimental
measurements into the matrix form required by SVD. In addition to the
processing of the measured signals to produce the relevant experimental
parameter (e.g., absorbance, linear dichroism, corrected fluorescence intensity) this step might include some preconditioning (i.e., truncation or
weighting) of the data. The second step is the calculation of the SVD of
the data matrix. The third step is the selection of a subset of the singular
values and vectors produced by the SVD that are judged sufficient to
represent the original data to within experimental error (i.e., the determination of the effective rank of $\mathbf{A}$). In some cases this step may be preceded
or followed by some additional processing of the matrices produced by
the SVD. We describe one such procedure, a rotation of subsets of the
left and right singular vectors which optimizes the signal-to-noise ratio of
the retained components. The effects of this rotation procedure are explored in more detail by the simulations described below. The final step is
the description of the reduced representation of the original data set that
is produced in the first three steps in terms of a physical model. This step
most often involves least-squares fitting.

A. Preparation of Data Matrix

To carry out the SVD of a set of data, the data must be assembled as
a matrix $\mathbf{A}$ which is arranged so that each column contains a set of
measurements for which a single isolated variable ranges over the same
set of values for each column, the values of all of the other variables
remaining fixed.[40] Different columns of $\mathbf{A}$ then correspond to different sets
of values for the remaining variables. For example, the data in Fig. 1

---

[40] The first step in any analysis is the reduction of the raw data to produce values for
the desired experimental parameter. This operation usually includes adjustment of the
measured data for offsets, instrument response, and instrument background, as well as
correction for baselines and other experimental characteristics. We assume that all such
calculations which are specific to a given experimental technique and instrument have
been carried out and tested by appropriate control experiments which demonstrate, for
example, the applicability of Eq. (2) to data collected and analyzed by these procedures.

consist of optical absorption difference spectra (i.e., a difference in optical densities between the photoproduct and the unphotolyzed sample measured as a function of wavelength) obtained at different times after photo-dissociation. To reduce these data using SVD, we create a data matrix $\mathbf{A}$, each column of which contains a single spectrum (i.e., varies only with wavelength). The matrix $\mathbf{A}$ is then built up from such column vectors (spectra) measured under different conditions [in this case, times as described by Eq. (2)]. In a properly constructed matrix each row then corresponds to a single wavelength.

Three types of preprocessing of the data matrix, $\mathbf{A}$, might be contemplated prior to calculation of the SVD. We shall refer to them as truncation, smoothing, and weighting. Truncation refers to the reduction of the size of the data matrix by selection of some subset of its elements; smoothing refers to any procedure in which noise is reduced by averaging of adjacent points; weighting refers to scaling of the data matrix to alter systematically the relative values of selected elements. Truncation of the data set, the first of these operations, may always be carried out. The effect of truncation is to reduce the size of the data matrix and thereby delimit the range of the experimental conditions. Truncation is clearly desirable if some artifact, such as leakage of light from a laser source into the spectrograph, preferentially reduces the quality of data on a subset of the data points. Smoothing of the data could, in principle, be performed either with respect to the isolated variable (i.e., "down the columns" of $\mathbf{A}$) or with respect to the remaining variables (i.e., "across the rows" of $\mathbf{A}$). As we have seen in the discussion of the noise-averaging properties presented in the previous section, SVD itself acts as an efficient filter to suppress random measurement noise in the most significant components. A data set reconstructed from the SVD components is therefore effectively noise-filtered without the artifacts that may arise when some of the more popular smoothing algorithms are used. For this reason, there is no clear advantage to presmoothing a data set across either variable, unless such an operation is to take place in conjunction with a sampling operation in order to reduce the data matrix to a size determined by limits on either computational speed or computer memory.

The statistical discussion of noise in Section III,C suggests that it would be advantageous to weight the data matrix in accordance with the measured variances of its individual elements. A detailed discussion of the desirability of and strategies for statistical weighting is beyond the scope of this chapter and will be addressed in more detail elsewhere.[32] It would appear, however, from the discussion of Cochran and Horne[22,27] that a weighting procedure should probably be incorporated into the analysis both in cases where the variances of the data set have been very well characterized and in cases where the variances range over values which differ by a significant factor. In the latter cases, it is likely that any reasonable weighting scheme will produce better results than no weighting at all. It is difficult to judge *a priori* whether weighting will significantly improve the accuracy of the SVD analysis. The only unambiguous method for determining the effects of weighting for a given type of data appears to be to carry out statistical simulations that incorporate the known properties of the data as well as the variances characteristic of the measurement system.

B. Calculation of Singular Value Decomposition

The computation of the SVD of a data matrix is the most clear-cut of all the analytical steps in the treatment of experimental data. The input matrix can be either A or A_W, depending on whether the weighting procedure has been used. When the SVD of the data matrix A, assembled as described above, is calculated [Eq. (3)], the left singular vectors of A (the columns of U) are an orthonormal set of basis spectra which describe the wavelength dependencies of the data, and the corresponding right singular vectors (the columns of V) are normalized sets of time-dependent amplitudes for each basis spectrum (see Fig. 2). The singular values, $\{s_i\}$, are the normalization factors for each basis spectrum U_i and amplitude vector V_i. Thoroughly tested FORTRAN subroutines for computing the SVD, based on the work of Golub and co-workers,[28,29] are generally available as part of the LINPACK[41] and *Numerical Recipes*[42] subroutine packages. The reader is referred to the original work for a discussion of the computational details of the SVD algorithm, which are outside the scope of this chapter.[1,28,29,42]

C. Analysis of Singular Value Decomposition Output

The SVD provides a complete representation of the matrix A as the product of three matrices U, S, and V having well-defined mathematical properties. Equation (3) represents the $m \times n$ elements of A in terms of $m \times n$ (elements of U) + $n \times n$ (elements of V) + n (diagonal elements

[41] "Linpack Users Guide" (J. J. Dongarra, J. R. Bunch, C. B. Moler, and G. W. Stewart, eds.). SIAM, Philadelphia, Pennsylvania, 1979.

[42] W. H. Press, B. P. Flannery, S. A. Teukolsky, and W. T. Vetterling, "Numerical Recipes: The Art of Scientific Computing." Cambridge Univ. Press, Cambridge, 1986.

of **S**) = $(m + n + 1)n$ numbers.[43] The effective use of SVD as a data reduction tool therefore requires some method for selecting subsets of the columns of **U** and **V** and corresponding singular values which provide an essentially complete representation of the data set. This selection then specifies an "effective rank" of the matrix **A**. In practice, a reasonable selection procedure produces an effective rank which is much less than the actual number of columns of **A**, effecting a drastic reduction in the number of parameters required to describe the original data set.

A first criterion for the selection of usable components is the magnitude of the singular values, since the ordered singular values provide a quantitative measure of the accuracy of the representation of the original data matrix **A** in terms of any subset of the columns of **U** and **V**. In the absence of measurement noise and other perturbations, the number of nonzero singular values is the number of linearly independent component spectra required to describe the data set. In experimental data, however, the presence of noise results in all nonzero singular values (see Section III,A). Despite this complexity, it is still possible to use the singular values, together with an estimate of the measurement uncertainties, to determine how many component spectra are sufficient to describe the data set to within experimental error. If the data have not been weighted, and the variance, σ^2, is identical for all elements of the data matrix, it is reasonable to argue that a component $k + 1$ may be considered negligible if the condition

$$\|\mathbf{A} - \mathbf{U}_k\mathbf{S}_k\mathbf{V}_k^T\| = \sum_{i=k+1}^{n} s_i^2 \leq \mu\nu\sigma^2 \tag{33}$$

is satisfied. $\mathbf{U}_k$, $\mathbf{S}_k$, and $\mathbf{V}_k$ are the representation of **A** in terms of k basis vectors and their corresponding amplitudes, as defined in Section II,A, and μ and ν are related to the size of the data matrix. This expression simply states that the neglect of this and all subsequent components should yield a reconstructed data matrix that differs from the original by an amount that is less than the noise. The choice of μ and ν rests on the determination of the number of degrees of freedom for the representation which remain after the selection of k basis vectors. Shrager has suggested that $\mu = m$ and $\nu = n$.[23,24,30] The results of Sylvestre et al. mentioned in Section III,C suggest that a better choice may be $\mu = m - k$ and $\nu =$

---

[43] The number of independent parameters required to specify **U**, **S**, and **V** is reduced because these numbers are constrained by the mathematical properties of the matrices **U** and **V**. A total of $n(n + 1)/2$ constraints arise from the orthonormality conditions on the columns of each matrix, giving a total of $n(n + 1)$ constraints. The total number of independent parameters is therefore $(m + n + 1)n - n(n + 1) = mn$, which is the number of independent parameters in the matrix **A**.

$n - k.$[26] The index r of the least significant component that does not satisfy this condition is then an estimate of the effective rank of $\mathbf{A}$, and the first through the rth components are retained for further consideration.

Some guidance in selecting significant components from the SVD of a weighted data matrix is obtained from the work of Cochran and Horne.[22,27] Weighting of the data using the procedure described in Section III,C produces a weighted matrix $\mathbf{A_W}$ such that, if $\mathbf{A_W}$ has rank r, then the last $m - r$ eigenvalues of $\langle \mathbf{A_W A_W^T} \rangle$ will have the same value, a [Eq. (26)]. This is equivalent to having only the first m singular values of $\langle \mathbf{A_W} \rangle$ nonzero. Successful application of the weighting algorithm thus produces a set of singular values which are pushed toward zero away from the remaining set. If such a bifurcation is found, the point at which the singular values separate can be used to estimate of the rank of the matrix.

Another reasonable criterion for the selection of usable components from the SVD is the signal-to-noise ratio of the left and right singular vectors (columns of $\mathbf{U}$ and $\mathbf{V}$). Under some experimental conditions, particularly when noise is present which is random only along one dimension of the data matrix (see Sections III,D and V,B), selection of usable components from the SVD based on singular values alone can produce a representation of the data matrix that approximates the original to within experimental error, but in which some of the selected components do not contain enough signal to lend themselves to further analysis (e.g., by least-squares fitting with a physical model). An example of such behavior is seen in the SVD presented in Fig. 2, where the amplitude of the third basis spectrum exhibits almost no time-correlated "signal," but the fourth component, which is only about half as large, clearly does. Under these circumstances additional criteria may be required to select those components for which the signal-to-noise ratios are sufficiently large to be candidates for further processing.

A useful measure of the signal-to-noise ratio of given columns of $\mathbf{U}$ (U_i) and $\mathbf{V}$ (V_i), introduced by Shrager and co-workers,[23,24,30] are their autocorrelations defined by

$$C(U_i) = \sum_{j=1}^{m-1} U_{j,i} U_{j+1,i} \tag{34}$$

$$C(V_i) = \sum_{j=1}^{n-1} V_{j,i} V_{j+1,i} \tag{35}$$

where $U_{j,i}$ and $V_{j,i}$ represent the jth elements of the ith columns of $\mathbf{U}$ and $\mathbf{V}$, respectively. Because the column vectors are all normalized to unity, those vectors which exhibit slow variations from row to row ("signal") will have values of the autocorrelation that are close to but less than one. Rapid row-to-row variations ("noise") will result in autocorrelations which are much less than one, and possibly negative. (The smallest possible value is -1.) For column vectors with many elements (>100 rows) that are subjectively "smooth," autocorrelation values may exceed 0.99, whereas values less than about 0.8 indicate signal-to-noise ratios approaching 1. Components which have been selected based on singular value can be further screened by evaluating the autocorrelations of the corresponding columns of U and V and rejecting the component if either autocorrelation falls below some threshold value. A proper choice of this threshold depends on the number of elements in the columns being considered and other experimental details.

D. Rotation Procedure

The presence of measurement noise and other random components in the data matrix decreases the effectiveness with which SVD extracts useful information into the rank-ordered singular values and vectors. As we have seen in Section III,D, when the magnitudes of signal and noise components of the data become comparable, they may be mixed in the SVD. The signal amplitude is "spread" by this mixing over two or more of the singular values and vectors. In some cases, the columns of U and V ordered by decreasing singular value do not exhibit monotonically decreasing signal-to-noise ratios (see Fig. 2). A component which is primarily "noise" may actually be sufficiently large to supersede a signal component in the hierarchy. If this problem is addressed by simply discarding the "noise" component from the data, one effectively introduces "holes" in the set of retained components where components having large amplitudes are ignored and those having small amplitudes are retained. In other cases one encounters a set of components which satisfy the condition in Eq. (33) and contain some signal, but are not individually of sufficient quality to pass a signal-to-noise test such as the autocorrelation criterion just described. Because such small signals are almost always of interest, some procedure for "concentrating" the signal content from a number of such noisy components into one or a very small number of vectors to be retained for further analysis can be extremely useful.

One such optimization procedure transforms a selected set of such noisy components by finding normalized linear combinations for which the autocorrelations [Eqs. (34) and (35)] are maximized. The autocorrelations may be optimized either for the columns of U [Eq. (34)] or for the columns of V [Eq. (35)]. The choice depends on whether the signal-to-noise ratio of the determinations as a function of the isolated variable (e.g., wavelength), or as a function of the remaining variables (e.g., time, pH), is considered more important. For purposes of discussion, the transformations will be applied to a set of p columns of $\mathbf{V}$ to be denoted by $\{V_k\}$, where the indices k are taken from the set $\{k1, k2, \ldots, kp\}$. Clearly, blocks of consecutive columns of either matrix are the most obvious candidates for transformation, because they correspond to blocks of consecutively ordered singular values, but this choice is not required by the algorithm. It is only necessary that the processing of the columns of one matrix be accompanied by the compensatory processing of the corresponding columns of the other matrix so that the contribution of the product of the two matrices to the decomposition in Eq. (3) is preserved. The problem then is to determine coefficients $\{r_i\}$, for $i = 1, \ldots, p$, such that the autocorrelation of the normalized vector

$$V' = r_1 V_{k1} + \ldots + r_p V_{kp} \tag{36}$$

is a maximum. Because the set of vectors $\{V_k\}$ is an orthonormal set, the requirement that V' be normalized is enforced by the constraint $r_1^2 + \ldots + r_p^2 = 1$. The solution of this problem is described in the Appendix. The procedure yields p distinct sets of coefficients $\{r_i\}$ for which the autocorrelations of the transformed vectors given by Eq. (36) have zero derivatives (yielding some maxima, some minima, and some saddle points) with respect to the coefficients. The transformed vectors with the largest autocorrelations may then be inspected individually to determine whether they should be retained for subsequent analysis.

To represent the effect of this transformation on the entire matrix $\mathbf{V}$, the p sets of coefficients $\{r_i\}$ provided by the transformation procedure may be arrayed as columns of an orthogonal matrix $\mathbf{R}_{\{k\}}$ (see Appendix). This matrix may be viewed as describing a rotation of the ordered set of orthonormal vectors $\{V_k\}$ onto a transformed set of orthonormal vectors $\{V_k'\}$. We can define an $n \times n$ matrix $\mathbf{R}$ by

$$\begin{aligned} R_{ij} &= \delta_{ij} \qquad \text{if } i \text{ or } j \notin \{k\} \\ R_{ki,kj} &= (R_{\{k\}})_{ij} \end{aligned} \tag{37}$$

that is, by embedding the matrix $\mathbf{R}_{\{k\}}$ into an identity matrix using the indices $\{k\}$. It is easily verified that $\mathbf{R}$ is also an orthogonal matrix. We can then define a transformed matrix $\mathbf{V^R}$ in terms of the entire original matrix $\mathbf{V}$ by

$$\mathbf{V^R} = \mathbf{VR} \tag{38}$$

The columns of $\mathbf{V}$ that are in the set $\{V_k\}$ are transformed in $\mathbf{V^R}$ to the corresponding vectors in the set $\{V_k'\}$, and columns of $\mathbf{V}$ that are not in $\{V_k\}$ are carried over to $\mathbf{V^R}$ unchanged. If we define a transformed $\mathbf{U^R}$ matrix by

$$\mathbf{U^R} = \mathbf{USR} \tag{39}$$

then the decomposition in Eq. (3) may be written

$$\begin{aligned}
\mathbf{A} &= \mathbf{USV^T} \\
&= \mathbf{USRR^TV^T} \\
&= \mathbf{(USR)(VR)^T} \\
&= \mathbf{U^R(V^R)^T}
\end{aligned} \tag{40}$$

where we have exploited the orthogonality of $\mathbf{R}$ (i.e., $\mathbf{RR^T} = \mathbf{I}_n$) on the second line. The matrices $\mathbf{U^R}$ and $\mathbf{V^R}$ contain new "basis vectors" and amplitudes, respectively, in terms of which the data matrix $\mathbf{A}$ may be represented.[44] It is important to point out that, while the columns of $\mathbf{V^R}$ still comprise an orthonormal set of vectors, the columns of $\mathbf{U^R}$ are neither normalized nor orthogonal. Furthermore, the mixing of different components results in the loss of the optimal least-squares property (see Section II) when the data matrix is described in terms of any but the complete set of transformed vectors produced by this procedure.

The set of column vectors produced by the rotation procedure (columns of $\mathbf{V^R}$) are mutually "uncorrelated" (in the sense that the symmetrized cross-correlation matrix defined in the Appendix is now diagonal). One consequence of this fact is that variations which are correlated in the original columns of $\mathbf{V}$ (the "signal" distributed by the SVD over several components) will tend to be isolated in single vectors after the rotation. Another consequence is that columns of $\mathbf{V}$ which are uncorrelated will not be mixed by the rotation procedure. Therefore, one anticipates that components having totally random amplitudes (i.e., those which result from random noise in the data matrix) which are included in the rotation will not be significantly altered by the rotation procedure and will subsequently be eliminated on screening of the transformed vectors based on the autocorrelation criterion. Extension of this line of reasoning suggests that including in the set of rotated vectors additional vectors beyond those that might be expected to contain usable signal will not significantly alter the characteristics of the transformed vectors which contain signal and will be retained after rotation.

The question of which components to include in the rotation procedure has no simple answer. It is clear that even two components which have very high signal-to-noise ratios (i.e., autocorrelations which are close to 1) may be mixed in the transformation if their variations are correlated in

---

[44] In practice, of course, only those columns of $\mathbf{U}$ and $\mathbf{V}$ whose indices are in the set $\{k\}$ need be transformed by postmultiplication by $\mathbf{R}_{\{k\}}$; the remaining columns of $\mathbf{V}$ are simply carried over unchanged, and the remaining columns of $\mathbf{U}$ are multiplied by the corresponding singular values to produce the transformed basis vectors.

the sense defined above. As a result, any component of the SVD output which is interesting or useful in its present form, either for mathematical reasons or based on experience, should be excluded from the rotation procedure.[10,11] Furthermore, although the discussion in the previous paragraph suggests that it is "safe" to include more components than are clearly required, the set of included components should be kept to some small fraction of the total set.[45] A procedure that we have used with some success with data matrices of about 100 columns is to select as candidates roughly 10% of the components which have the largest singular values, exclude any of these which either should not be mixed with other components for some reason or will not be significantly improved by such mixing, and apply the rotation procedure to the rest.[11,12,46]

An example which demonstrates the effectiveness of rotation in reordering and improving the autocorrelations of the amplitude vectors is shown in Fig. 5. Columns 3 through 10 of the SVD shown in Fig. 2 were included in the rotation, which was carried out with the expectation of removing random contributions to the small signal observed in V_4 of the SVD. Columns 1 and 2 were excluded because their singular values were, respectively, 70 and 5.6 times larger than that of the "noise" component 3 and the signal-to-noise ratios of these components were already about 250 and 30, respectively. The first effect of rotation is that which was anticipated: the signal-to-noise ratio in the third amplitude vector is improved from about 2 to more than 7 and the derivative-shaped "signal" in channels 1 through 40 is concentrated in the third basis spectrum.[47] The autocorrelation of the rotated V_3^R is 0.933, whereas that of V_4 is only 0.742. The second effect is to suppress the random offset amplitudes represented by the third component in the original SVD (Fig. 2) to the point that they do not even appear in the first six components after rotation. The bulk of

[45] Additional constraints placed on the elements of the transformed vectors by the rotation procedure tend to determine the individual elements as the size of the included set approaches that of the complete set. Specifically, if p vectors out of a total of n columns of **V** are included, the $p \times n$ elements of the resulting transformed vectors are subject to p^2 constraints—$p(p-1)/2$ from the fact that the symmetrized cross-correlation matrix (see Appendix) has all off-diagonal elements equal to zero, $p(p-1)/2$ from the orthogonality of all the transformed vectors, and p from the normalization of each vector. As p approaches n, these constraints obscure any relationship between the shapes of the untransformed and the transformed vectors, and the set of vectors required to represent the signal content of the original data matrix will actually increase rather than decrease.

[46] J. Hofrichter, E. R. Henry, A. Szabo, L. P. Murray, A. Ansari, C. M. Jones, M. Coletta, G. Falcioni, M. Brunori, and W. A. Eaton, *Biochemistry* **30,** 6583 (1991).

[47] This "signal" arises from the perturbation of the absorption spectra of the cobalt porphyrins in the α chains of the hybrid hemoglobin tetramer. The time course for this spectral change is distinctively different from that of the second component.[11]

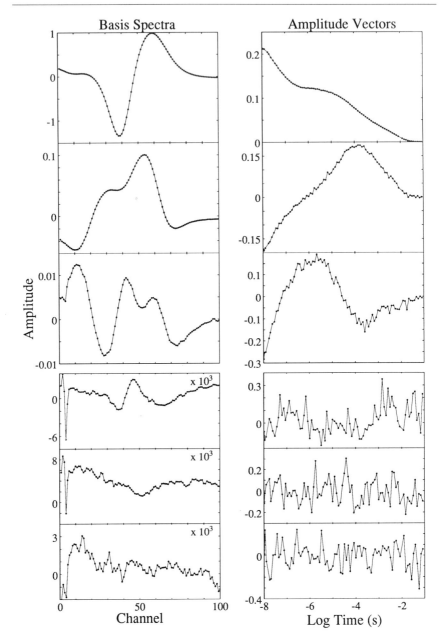

FIG. 5. Rotated SVD of the data in Fig. 1. Components 3 through 10 of the SVD for which the first six components are shown in Fig. 2 were rotated using the algorithm discussed in the text and derived in the Appendix. The autocorrelations of the components included in

the offset amplitude actually appears as component 8 of the rotated SVD, and the autocorrelation of V_8^R is slightly less than zero (-0.12).

E. Application of Physical Models to Processed Singular Value Decomposition Output

The discussion to this point has dealt with the purely mathematical problem of using SVD and associated processing to produce a minimal but faithful representation of a data matrix in terms of basis vectors and amplitudes. The next step in the analysis of the data is to describe this representation of the data matrix in terms of the concentrations and spectra of molecular species. This step requires that some physical model be invoked to describe the system and an optimization procedure be carried out to adjust the parameters of the model so that the differences between the data and the model description are minimized. Several assumptions are inherent in such a procedure. First, a set of pure states or species which are accessible to the system must be enumerated. The measured spectra are then assumed to be linear combinations of the spectra of the various pure species, weighted by their populations under each set of conditions [e.g., Eq. (2)]. The dependence of the populations of these species [the $\{c_n\}$ of Eq. (2)] on the conditions is further assumed to be quantitatively described by a kinetic or thermodynamic model. If the model provides for r distinct species, then the first two of these assumptions permit the (m wavelengths) $\times$ (n conditions) matrix $\mathbf{A}$ to be written in the form of Eq. (5), that is,

$$\mathbf{A} = \mathbf{FC^T} \qquad (41)$$

where the columns of the $m \times r$ matrix $\mathbf{F}$ contain the spectra of the individual species, and the corresponding columns of the $n \times r$ matrix $\mathbf{C}$ contain the populations of the species as a function of conditions.

The most common means for reducing a representation of a set of experimental data to a description in terms of a physical model is through the use of least-squares fitting. Using this approach, the amplitudes of all of the vectors which describe the data matrix would be simultaneously

---

the transformation (3–10) were 0.149, 0.746, 0.062, −0.089, 0.337, −0.010, 0.031, and 0.099 before rotation. The signal-to-noise ratio for the component with the highest autocorrelation (V_4) evaluated by comparing a smoothed version of this component with the original is approximately 2. The autocorrelations of transformed components 3 through 10 were 0.932, 0.473, 0.277, 0.191, 0.001, −0.115, −0.165, and −0.268, and their normalized amplitudes were 0.058, 0.015, 0.041, 0.011, 0.023, 0.057, 0.014, and 0.029. The signal-to-noise ratio for the most highly correlated component (V_3^R) is about 7.

fitted to the model to produce a set of coefficients which describe the spectra of each of the species in the model as well as the dependence of the species concentrations on experimental conditions.[23] A common alternative to using molecular or physical models to directly fit the data is to assume functional forms which result from analysis of generalized or simplified models of the system and to use these forms to fit the data. For example, if the kinetics of a system can be described by a set of first-order or pseudo-first-order processes, then the kinetics of the changes in system composition can be described by sums of exponentials, with relaxation rates which are the eigenvalues of the first-order rate matrix.[8] Under these circumstances, the time-dependent vectors which describe the changes in the spectra can be empirically described by sums of exponential relaxations, and fitting can be carried out using functions of this form. Similarly, pH titration curves can be assumed to be sums of simple Henderson–Hasselbach curves describing the uncoupled titration of individual groups, and the measured dependence of the spectra on pH can be fitted to sums of these curves.[16,23] Because use of this approach requires the assumption of some functional form, it is therefore less rigorous than the use of an explicit model. It also does not permit direct determination of the spectra of the species in the model. As pointed out in Section I, the advantage of using the output of SVD in any fitting procedure is that the number of basis spectra required to describe the data matrix, and hence the number of amplitudes which must be fitted, is minimized by the rank reduction which has been accomplished by SVD.

Suppose that a population matrix $\mathbf{C}'$ is derived from a specific set of model parameters. If $\mathbf{C}'$ has rank r so that $(\mathbf{C}'^{\mathbf{T}}\mathbf{C}')^{-1}$ exists, the generalized inverse of $\mathbf{C}'^{\mathbf{T}}$ can be written as $\mathbf{C}'(\mathbf{C}'^{\mathbf{T}}\mathbf{C}')^{-1}$,[3] and the corresponding matrix $\mathbf{F}'$ of species spectra which minimizes the difference $\|\mathbf{A} - \mathbf{F}'\mathbf{C}'^{\mathbf{T}}\|$ may be written[3,22]

$$\mathbf{F}' = \mathbf{A}\mathbf{C}'(\mathbf{C}'^{\mathbf{T}}\mathbf{C}')^{-1} \tag{42}$$

Least-squares fitting of the matrix $\mathbf{A}$ with the model then requires varying the parameters of the model in some systematic way so that the population matrix $\mathbf{C}'$ calculated from the parameters, and the matrix $\mathbf{F}'$ of spectra calculated using Eq. (42), result in the smallest possible value of the difference $\|\mathbf{A} - \mathbf{F}'\mathbf{C}'^{\mathbf{T}}\|$. The suitability of the model as a description of the measurements would then be assessed on the basis of how well the final matrices $\mathbf{F}'$ and $\mathbf{C}'$ describe the original data.

This approach of least-squares fitting the entire data matrix, commonly referred to as global analysis, has been applied in a large number of studies. Examples include the analysis of sets of spectra obtained from pH titrations of multicomponent mixtures,[48] analysis of fluorescence decay curves,[7] and analysis of flash photolysis data on the bacteriorhodopsin photocycle.[6,49-51] In principle it provides the most complete possible description of a data matrix in terms of a postulated model; however, it has certain features that make it difficult to use in many cases. The most obvious difficulties are associated with the matrix $\mathbf{F}'$, which specifies the spectra of the species in the model. If the number of wavelengths (m) sampled in collecting the data matrix is large, this matrix, which contains the extinction coefficient of each of the r species at each of m wavelengths, is also large, containing a total of $m \times r$ adjustable parameters. The fitting procedure then tends to become computationally cumbersome, in that every iteration of a search algorithm in parameter space requires at least one recalculation of $\mathbf{F}'$ using Eq. (42) or the equivalent. It should be noted that in most of the applications cited above the number of wavelengths included in the analysis was 15 or less. Furthermore, numerical instabilities may arise in the direct application of Eq. (42) if $\mathbf{C}'$ is rank-deficient, or nearly so, because calculation of the inverse of $\mathbf{C}'^T\mathbf{C}$ then becomes problematic.

SVD provides a reduced representation of a data matrix that is especially convenient for a simplified least-squares fitting process. In the most general terms, after SVD and postprocessing have been performed, an essentially complete representation of the data matrix $\mathbf{A}$ in terms of k components may be written

$$\mathbf{A} \cong \mathbf{U}'\mathbf{V}'^T \tag{43}$$

where $\mathbf{U}'$ is a matrix of k basis spectra, and $\mathbf{V}'$ contains the amplitudes of the basis spectra as a function of conditions. If only the SVD has been performed, then $\mathbf{U}'$ consists of the k most significant columns of $\mathbf{US}$, and $\mathbf{V}'$ the corresponding columns of $\mathbf{V}$; if a rotation or similar procedure has been performed as well, then $\mathbf{U}'$ consists of the k most significant columns of the matrix $\mathbf{U}^R$ [Eq. (39)] and $\mathbf{V}'$ the corresponding columns of the matrix $\mathbf{V}^R$ [Eq. (38)]. If the data have been weighted prior to SVD, then $\mathbf{U}'$ consists of the k most significant columns of $\mathbf{U}'$ and $\mathbf{V}'$ the corresponding columns of $\mathbf{V}'$ as calculated from Eq. (27). The assumed completeness of the representations of $\mathbf{A}$ in Eqs. (41) and (43) suggests the *ansatz* that the columns of any matrix $\mathbf{C}'$ of condition-dependent model populations may be written as linear combinations of the columns of $\mathbf{V}'$. This linear relation-

[48] S. D. Frans and J. M. Harris, *Anal. Chem.* **56,** 466 (1984).
[49] R. Mauer, J. Vogel, and S. Schneider, *Photochem. Photobiol.* **46,** 247 (1987).
[50] R. Mauer, J. Vogel, and S. Schneider, *Photochem. Photobiol.* **46,** 255 (1987).
[51] A. H. Xie, J. F. Nagle, and R. H. Lozier, *Biophys. J.* **51,** 627 (1987).

ship between $\mathbf{C}'$ and $\mathbf{V}'$ may be inverted, at least in the generalized or least-squares sense, so that we can write formally[52]

$$\mathbf{V}' \cong \mathbf{C}'\mathbf{P} \tag{44}$$

In the least-squares fit, the model parameters used to calculate $\mathbf{C}'$ and the set of linear parameters $\mathbf{P}$ are varied to produce a population matrix $\mathbf{C}' = \hat{\mathbf{C}}$ and a parameter matrix $\mathbf{P} = \hat{\mathbf{P}}$ such that the difference $\|\mathbf{V}' - \mathbf{C}'\mathbf{P}\|$ is minimized. The optimal approximation to $\mathbf{V}'$ will be denoted $\hat{\mathbf{V}}(\equiv\hat{\mathbf{C}}\hat{\mathbf{P}})$. This then yields the further approximation

$$\mathbf{A}' = \mathbf{U}'\mathbf{V}'^{\mathbf{T}} \cong \mathbf{U}'\hat{\mathbf{V}}^{\mathbf{T}} = \mathbf{U}'\hat{\mathbf{P}}^{\mathbf{T}}\hat{\mathbf{C}}^{\mathbf{T}} \equiv \hat{\mathbf{F}}\hat{\mathbf{C}}^{\mathbf{T}} \tag{45}$$

where the matrix $\hat{\mathbf{F}}$ is the set of corresponding "least-squares" species spectra. Equation (45) permits the identification of $\hat{\mathbf{F}}$ in terms of the basis spectra:

$$\hat{\mathbf{F}} = \mathbf{U}'\hat{\mathbf{P}}^{\mathbf{T}} \tag{46}$$

It is important to note that, because all of the species spectra must be represented in terms of the set of basis spectra which comprise $\mathbf{U}'$, the matrix $\hat{\mathbf{P}}$ is much smaller than the matrix $\hat{\mathbf{F}}$. Accordingly, the number of adjustable parameters which must be specified in fitting the SVD representation of the data is significantly reduced relative to the number required to fit the original data matrix.

This somewhat formal discussion may be made clearer by considering an example from the field of time-resolved optical absorption spectroscopy, which is similar to the example described in Section I. Recall that the data consist of a set of absorption spectra measured at various time delays following photodissociation of bound ligands from a heme protein by laser pulses.[10,11] Each column of the data matrix $\mathbf{A}$ describes the absorbances at each wavelength measured at a given time delay. After the SVD and postprocessing, we are left with a minimal set of basis spectra $\mathbf{U}'$ and time-dependent amplitudes $\mathbf{V}'$ (see Fig. 5). Suppose that we now postulate a "model" which states that the system contains r "species," the populations of which each decay exponentially with time with a characteristic rate:

$$C_{ij} = e^{-\kappa_j t_i}, \quad j = 1, \ldots, r \text{ and } i = 1, \ldots, m \tag{47}$$

where the set $\{t_i\}$ represents the times at which the spectra (columns of $\mathbf{A}$)

---

[52] The formal inversion [Eq. (44)] may optionally be used to facilitate the fitting procedure.[4] When the model parameters which produce the population matrix $\mathbf{C}'$ are varied in each step of the optimization, the generalized inverse may be used to produce the matrix of linear parameters, $\mathbf{P}$, which produces the best approximation to $\mathbf{V}'$ corresponding to the specified set of model parameters.

are measured and the set $\{\kappa_j\}$ represents the characteristic decay rates of the populations of the various "species." The fit in Eq. (45) optimizes the relation

$$V'_{ij} \cong \sum_{q=1}^{r} P_{qj}\, e^{-\kappa_q t_i} \tag{48}$$

Producing an optimal least-squares approximation to $\mathbf{V}'(\equiv \hat{\mathbf{V}})$ clearly involves simultaneously fitting all the columns of $\mathbf{V}'$ using linear combinations of exponential decays, with the same set of rates $\{\kappa_q\}$, but with distinct sets of coefficients $\{P_{qj}\}$, for each column j. The resulting best-fit rates $\{\hat{\kappa}_q\}$ then produce a best-fit set of "model" populations $\hat{\mathbf{C}}$ and best-fit coefficients $\{\hat{P}_{qj}\}$ [Eq. (48)]. The set of "species" spectra $\hat{\mathbf{F}}$ which produce a best fit to the matrix $\mathbf{U}'$ are then obtained from Eq. (46), that is,

$$\hat{F}_{iq} = \sum_{j=1}^{k} \hat{P}_{qj} U'_{ij} \tag{49}$$

Although this "model" is admittedly highly contrived, in that descriptive kinetic models involving interconverting species will not produce species populations that all decay to zero as simple exponentials, it illustrates the general fitting problem.

Least-squares fitting the columns of $\mathbf{V}'$ obtained from SVD, when the residuals from each column of $\mathbf{V}$ are correctly weighted, is mathematically equivalent to least-squares fitting the entire data matrix using the global analysis procedure described in Section I. Shrager[23] has shown that for SVD alone (no postprocessing) the two procedures in fact yield the same square deviations for any set of parameters if the sum of squared residuals from each of the columns of $\mathbf{V}$ is weighted by the square of the respective singular value. In other words, the function to be minimized in the simultaneous fit to all the columns of $\mathbf{V}$ should be

$$\phi^2 = \sum_{i=1}^{m} s_i^2 \| V'_i - (\mathbf{C}'\mathbf{P})_i \|^2 \tag{50}$$

where V'_i is the ith column vector of $\mathbf{V}'$. It is shown in the Appendix that, if a rotation has been performed as described in Eqs. (36)–(40), approximately the same squared deviations will be obtained if all the columns of $\mathbf{V}^R$ are fit with the ith column weighted by the ith diagonal element W_{ii} of the matrix $\mathbf{W} = \mathbf{R}^T\mathbf{S}^2\mathbf{R}$.

In practice the SVD is truncated to generate $\mathbf{V}'$, and only a small fraction of the columns of $\mathbf{V}$ are included in the fitting procedure. This is equivalent to setting the weighting factors of the remaining columns to zero. If the truncation is well designed, then the columns of V which are discarded either have small weighting factors, s_i^2, or have autocorrelations which are small enough to suggest that they contain minimal condition-correlated signal content. If a rotation procedure described in Eqs. (36)–(40) has been performed prior to selecting V', then singular values of very different magnitudes may be mixed in producing the retained and discarded columns of V^R and their corresponding weighting factors (see Appendix). Because the rotation procedure is designed to accumulate the condition-correlated amplitudes into the retained components, the discarded components, while not necessarily small, also have little or no signal content. In both cases the neglected components clearly contribute to the sum of squared residuals, ϕ^2. Because their condition-correlated amplitudes are small, however, their contribution to ϕ^2 should be nearly independent of the choice of fitting parameters. To the extent that this is true, parameters optimized with respect to either truncated representation of the data should closely approximate those which would have been obtained from fitting to the entire data set.

In summary, an SVD-based analysis almost always simplifies the process of least-squares fitting a data matrix with a physical model by reducing the problem to that of fitting a few selected columns of V'. Reducing the rank of the data matrix also minimizes the number of parameters which must be varied to describe the absorption spectra of the molecular species [the elements of the matrix $\hat{P}$ in Eq. (46)]. Attention must be paid to the proper choice of weighting factors in order to produce a result which faithfully minimizes the deviations between the fit and the full data matrix, but the increase in the efficiency of fitting afforded by this approach argues strongly for its use under all conditions where the rank of the data matrix is significantly smaller than the number of rows and/or columns (i.e., rank $\ll \min\{m,n\}$).

V. Simulations for a Simple Example: The Reaction A → B → C

To explore in more detail the effects on the SVD output of introducing noise into data sets we have carried out simulations of noisy data for the simple kinetic system

$$A \rightarrow B \rightarrow C \tag{51}$$

This model was used to generate sets of data consisting of sample absorption spectra and difference spectra (with difference spectra calculated as sample spectrum $-$ C) using rates $k_{AB} = 10^6 \, sec^{-1}$ and $k_{BC} = 10^3 \, sec^{-1}$. The spectra of A, B, and C were represented as peaks having Gaussian bandshapes centered about wavelengths $\lambda_A = 455$ nm, $\lambda_B = 450$ nm, and

λ_C = 445 nm. The bandwidths (half-widths at $1/e$ of maximum) and peak absorbances for the three species were chosen to be Δ_A = 20 nm, Δ_B = 18 nm, and Δ_C = 16 nm and $\varepsilon_A c_t l$ = 0.9 OD, $\varepsilon_B c_t l$ = 1.0 OD, and $\varepsilon_C c_t l$ = 1.1 OD ($c_t l$ is the product of the total sample concentration and path length). These spectra were selected so that the ordered nonzero singular values of the noise-free data successively decreased by a factor of between 5 and 10. These data thus represent many of the problems encountered in the processing of real data in which some processes may produce changes in absorbance as large as 1 OD, whereas other processes produce changes as small as a few thousandths of an optical density unit. To derive reliable kinetic information under such unfavorable circumstances, careful consideration must be given to the effects of measurement noise on the data analysis.

Two different types of noise were added to the data. The first noise component, which we refer to as random noise, was selected independently for each element of the data matrix from a Gaussian distribution having an expectation value of zero and variance σ_r^2. Random noise simulates shot noise or instrumental noise in the determination of each experimental absorbance. The assumption that the amplitude of the random noise component is constant over the entire data matrix is certainly an oversimplification for real data: shot noise results from random deviations in the number of photons measured for each data point, and therefore depends on a number of factors, including the intensity of the source and the optical density of the sample. Moreover, the photon and electronic noise actually appears in the measured intensities, not in the absorbance, which is calculated from the logarithm of the ratio of two intensities. The second noise source consists of a spectrum having an identical absorbance at each wavelength, but having a different amplitude for every measured spectrum, selected from a Gaussian distribution with mean value zero and variance σ_λ^2. We shall refer to noise having these characteristics as wavelength-correlated noise. This noise approximates changes in the DC baseline of the spectrometer. In single-beam spectroscopy, such noise can arise from changes in the output energy of the lamp or changes in the sensitivity of the detector. In double-beam spectroscopy, it can result from electronic drift. In both cases, however, σ_λ can be significantly larger than the error inherent in the determination of the dependence of a given spectrum on wavelength, which is characterized by σ_r.

We have already seen that independent addition of these two kinds of noise to noise-free data has qualitatively different effects on the SVD. Based on the results of Sections III,A and III,B, random noise is expected to introduce a spectrum of singular values similar in magnitude to those obtained for a random matrix and to perturb the singular values and vectors of the noise-free data as discussed in Section III,B. The effects of adding wavelength-correlated noise have been explored in Fig. 4 for the case where the noise-free data matrix is rank-1. As shown there, the SVD contains only a single extra component, which arises primarily from the constant spectrum assumed as the noise source.

To examine the statistical properties of the SVD of data sets having specified amplitudes for random (σ_r) and wavelength-correlated (σ_λ) noise, the SVD of each of 5000 independently generated data matrices was calculated. Each matrix contained the identical "signal," which consisted of absorbances at 101 wavelengths and 71 time points evenly spaced on a logarithmic time grid, as well as randomly selected noise. For each set of 5000 trials, the means and standard deviations of the individual singular values and of the individual elements of the appropriate singular vectors were calculated. In calculating the statistical properties of the SVD, one is confronted with the problem of choosing the sign of each of the SVD components. Because SVD only determines unambiguously the sign of each of the products $U_i \cdot V_i$, some independent criterion must be used to choose the sign of each of the U_i or V_i. The algorithm chosen in these simulations was to require that the inner product of the left singular vector with the corresponding left singular vector obtained from the SVD of the noise-free data matrix be positive. To present the results of the simulations in a compact form we have chosen to display the singular values, together with the square root of the mean of the variances of the relevant singular vectors (noise amplitudes). The singular values facilitate comparison of the magnitude of the noise contributions with those of the signals which result from the noise-free data. The noise amplitude provides a compact characterization of the signal-to-noise ratio for a given parameter or vector.

A. Effects of Random Noise

The first set of simulations was carried out to explore the consequences of adding random noise of arbitrary amplitude to the data. Based on the discussion of noise presented earlier, addition of random noise to a noise-free data matrix would be expected to have two effects on the SVD output. First, the random amplitudes will generate a set of additional singular values having amplitudes comparable to those of a random matrix having the same size and noise amplitude [see Eq. (10) and Fig. 3] in addition to those which derive from the noise-free data. When the noise amplitude is small, all of these values should be significantly smaller than the smallest singular value of the noise-free data, and the noise should not interfere with the ability of SVD to extract the spectral information from the data.

Second, the noise should perturb the singular values and vectors which derive from the noise-free data by the addition of random amplitudes as shown in Eqs. (15) and (16) for the case in which the noise-free data matrix has a rank of 1. One objective of the simulations was to extend this analysis to explore both data sets which had a rank higher than 1 and the effects of larger noise amplitudes. In particular, we were interested in determining the random noise amplitudes at which signals became unrecoverable. It is intuitively expected that the noise amplitudes must become large compared to the signal for this to occur, so information on this point cannot be obtained by treating the noise as a perturbation.

An example of the input data at a relatively low noise amplitude is shown in Fig. 6b, and the results of the simulations are summarized in Fig. 7. The averages of the first three singular values are shown in Fig. 7a, and the square roots of the variances of the first three singular values and singular vectors are plotted as a function of σ_r in Fig. 7b–d. The results in Fig. 7a show that, for small σ_r ($<3 \times 10^{-2}$), the average values of s_1, s_2, and s_3 are essentially unperturbed from the values obtained from the noise-free data. Figure 7b shows that the presence of the noise in the data matrix at these noise amplitudes is observable as an increase in the variances of the first three singular values and vectors. The square roots of the variances of s_1, s_2, and s_3 are each very nearly equal to σ_r. Figures 7b and 7c show that the square roots of the variances of the singular vectors, σ_U and σ_V, are very nearly equal to σ_r/s_j, where s_j is the relevant singular value from the noise-free data. Because these noise amplitudes are small compared to all of the SVD components of the noise-free data, these results can be compared directly to the results predicted by the perturbation treatment in Section III,B.

The observed variances suggest that it may be possible to generalize Eqs. (15) and (16) which state that, for a data matrix having a rank of 1, the square root of the variance of s_1 is equal to σ_r, whereas the square roots of the variances of the vectors U_1 and V_1 approximate σ_r/s_1 for large matrices. The proposed generalization of these equations would predict that the jth SVD component is described by

$$s_j \cong s_{j0} + \frac{(m + n - 1)\sigma^2}{2s_{j0}} + \varepsilon_{s_j}; \qquad [var(\varepsilon_{s_j}) = \sigma_r^2]$$

$$U_{ji} \cong U_{ji0} + \frac{\varepsilon_{U_{ji}}}{s_{j0}}; \qquad [var(\varepsilon_{U_{ji}}) = (1 - U_{ji0}^2)\sigma_r^2] \qquad (52)$$

$$V_{ji} \cong V_{ji0} + \frac{\varepsilon_{V_{ji}}}{s_{j0}}; \qquad [var(\varepsilon_{V_{ji}}) = (1 + V_{ji0}^2)\sigma_r^2]$$

A more careful examination of the variances of the singular values and

a

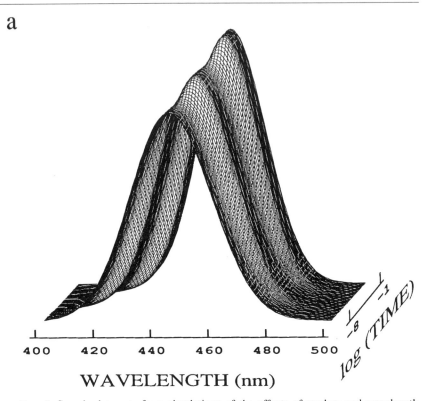

WAVELENGTH (nm)

FIG. 6. Sample data sets from simulations of the effects of random and wavelength-correlated noise on singular values and vectors. (a) The noise-free data. The spectra are calculated as the sums of the spectra of the three species A, B, and C times the concentrations calculated for a population which is 100% A at $t = 0$, using the rates $k_{AB} = 10^6$ sec^{-1} and $k_{BC} = 10^3$ sec^{-1}. The spectra are described in the text. The time points were chosen on a logarithmic grid with 10 points per decade, beginning at 10^{-8} and ending at 10^{-1} sec. Random and wavelength-correlated noise having amplitudes σ_r and σ_λ, respectively, were added to the noise-free data. (b) Spectra with random noise only: $\sigma_r = 0.016$ OD; $\sigma_\lambda = 0$. (c) Spectra with both random and wavelength-correlated noise: $\sigma_r = 0.016$ OD; $\sigma_\lambda = 0.016$ OD. (d) Difference spectra with both random and wavelength-correlated noise. The difference spectra were calculated by subtracting the spectrum of pure C from all of the calculated spectra. The noise amplitudes in (d) are identical to those in (c).

vectors obtained from these simulations shows that these approximations are quite accurate, describing the variance of the first SVD component to within 1%, and the second and third components to within 2–3%. We can conclude from this analysis that if the noise in the data is small and purely random, then the variances of the singular values and vectors should be related as described by Eq. (52). Failure of the variances to meet this

b

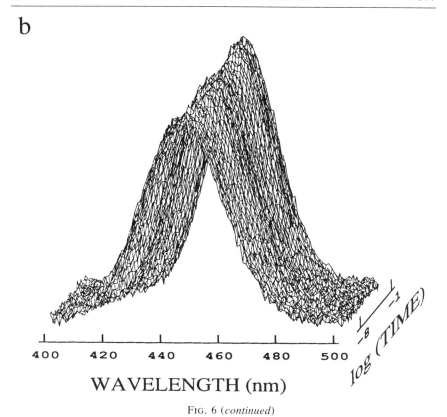

WAVELENGTH (nm)

FIG. 6 (*continued*)

criterion argues strongly for the presence of other sources of noise in the data.

As σ_r rises above 0.03 OD, s_3 begins to increase. The noise amplitudes for all the singular values and vectors continue to increase in direct proportion to σ_r. Once σ_r exceeds 0.1 OD, the square roots of the variances of the third singular vectors, U_3 and V_3, saturate at the values expected from completely random vectors ($m^{-1/2}$ and $n^{-1/2}$, respectively), showing that the elements of these vectors have become completely uncorrelated. This result implies that the third SVD component of the noise-free data has been replaced by a component which is generated by the random noise. Further increasing σ_r produces proportional increases in s_3, the singular value associated with this pair of random basis vectors. This behavior is analogous to that observed for random matrices, where the U and V vectors are random and the variance of the random variable appears only as a scale factor multiplying the singular values [Fig. 3; Eq. (10)]. At the

c

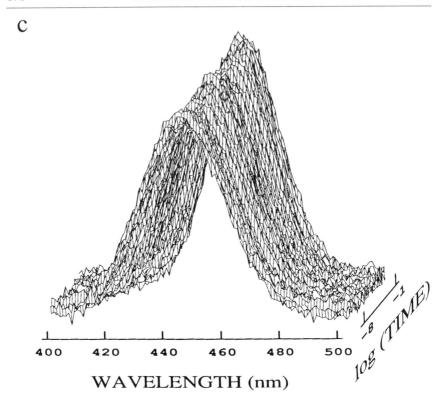

400   420   440   460   480   500   log (TIME)

WAVELENGTH (nm)

d

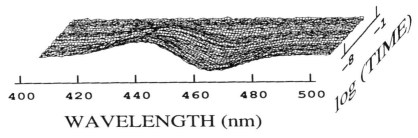

400   420   440   460   480   500   log (TIME)

WAVELENGTH (nm)

FIG. 6 (*continued*)

maximum values of σ_r, similar behavior is observed in the second singular value and vectors as the noise becomes sufficiently large that its contribution to the spectrum of singular values begins to dominate s_2.

The dependence of the standard deviations of the singular values and vectors suggests that a smooth connection can be made between the perturbation treatment for low noise amplitudes and the results obtained from the simulations of random matrices described in Fig. 3 at higher noise amplitudes. To quantitatively compare the results of the simulations with those in Fig. 3 we first calculate from Eq. (10) that, for an $n \times n$ square matrix, the largest singular value resulting from the random amplitudes is equal to $s_{max} = 2\sigma_r n^{1/2}$. Inspection of the results in Fig. 3b for $m/n < 2$ suggests that the corresponding first approximation for an $m \times n$ matrix is $s_{max} \cong 2\sigma_r(mn)^{1/4}$. The second and third singular vectors in Fig. 7, which have singular values $s_2 = 7.7$ and $s_3 = 1.04$, become dominated by random noise when the noise amplitudes exceed 1.0 and 0.12 OD, respectively. The corresponding values of s_{max} calculated from the noise amplitudes are 9.2 and 1.1, very close to s_2 and s_3, respectively. If this result can be generalized, it suggests that the random noise dominates a signal component k when $s_{max} \geq s_k$. To restate this conclusion, the signal described by a given singular value, s_j, and its associated singular vectors, U_j and V_j, becomes totally obscured by noise when the random noise amplitude, σ_r, exceeds $s_j/2(mn)^{1/4}$. Attempts to improve the quality of signal components in the presence of random noise by rotation or other postprocessing algorithms were uniformly unsuccessful (a result which is expected because the noise components generated by SVD are almost completely uncorrelated).

B. Combined Effects of Random and Wavelength-Correlated Noise

The rotation procedure described in Section IV,D and derived in the Appendix was designed to discriminate between signals which are correlated in one dimension of the data matrix and those which are not. The simulations just described show that this procedure has little or no effect when confronted with random noise, which produces SVD components which are uncorrelated in both dimensions of the data matrix. We anticipate, however, that this algorithm will be more successful in discriminating against mixing of wavelength-correlated noise with the noise-free data. To explore the effectiveness of this procedure we performed two sets of simulations in which both wavelength-correlated noise and random noise were added to the data. In the first set of simulations the spectra were calculated as in Section V,A above, and in the second set the same spectra were used but the spectrum of pure C was subtracted from the sample spectrum at each time point to produce a difference spectrum. Sample

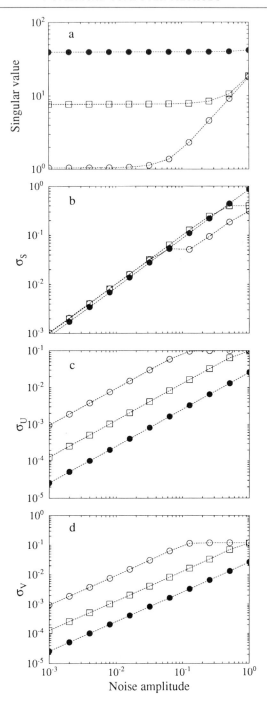

data sets at the same noise amplitudes are shown in Fig. 6c,d. Recall that in the absence of random noise, addition of wavelength-correlated noise increases the rank of the matrix of the absolute spectra from 3 to 4 and the rank of the matrix of difference spectra from 2 to 3.

At each of three amplitudes for the random noise (σ_r = 0.016, 0.032, and 0.064 OD), σ_λ was increased systematically from a point where the noise amplitudes were much smaller than the smallest SVD component of the noise-free data to a point where the contribution of this component to the resulting data matrix was larger than that of the second component. At each set of noise amplitudes 5000 data sets were generated and analyzed by SVD. For the absorption spectra components 3 through 10 were then rotated as described above to optimize the autocorrelation of the retained V_3^R component. For the difference spectra components 2 through 10 were rotated to optimize the retained V_2^R component.

Figure 8 presents the first three singular values and the noise amplitudes of the smallest singular vectors (U_3 and V_3) of the noise-free data in the first case where the data are absorption spectra. The random amplitudes of the wavelength-correlated noise would produce a singular value, $s_\lambda = (mn\sigma_\lambda^2)^{1/2}$ if the noise were isolated in a single SVD component. The noise amplitudes thus become comparable to the signal represented by the third SVD component of the noise-free data when σ_λ is slightly larger than 0.01 OD. Figure 8a permits us to track the magnitude of this component once it exceeds s_3 from the noise-free data. Note that s_3 doubles when the noise amplitude reaches a value of about 0.03 OD and then climbs steadily until it reaches a value of about 0.1 OD, at which point it levels off. This is the region in which we are primarily interested.

If we now examine the results in Fig. 8c, we find that the square root of the variance of V_3, $\sigma(V_3)$, begins to increase detectably at values of σ_λ as small as 0.004 OD. When σ_λ becomes about 10 times larger ($\sim$0.05

FIG. 7. Effects of the amplitude of random noise on the SVD of simulated data for the reaction A → B → C. The spectra were calculated as described in the text. (a) The first three singular values; (b) square root of the variance of the first three singular values; (c, d) square root of the average variance of the first three left (c) and right (d) singular vectors. For each noise amplitude, σ_r, the SVD of each of 5000 independently generated data matrices was calculated. For each set of 5000 trials the means and variances of each of the singular values, s_1, s_2, and s_3, and of each element of the three most significant singular vectors (columns 1 to 3 of the matrices U and V) were calculated. The value plotted for each singular vector and noise amplitude in (c) and (d) is the square root of the mean value of the variances calculated for all of the elements of that singular vector. (Preliminary calculations in which the dispersions of the variances were examined showed that variations were too small to be analyzed from the results of these simulations.)

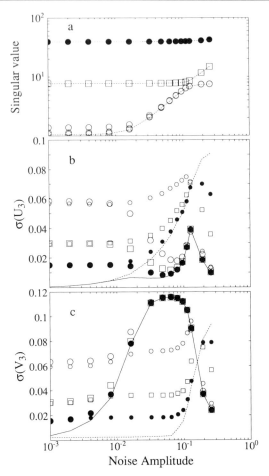

FIG. 8. Effects of the RMS amplitude of an offset spectrum on the singular values and variances of the third singular vectors of simulated absorption spectra for the reaction A → B → C in the presence of random noise. Simulations were carried out using the spectra described in the text and analysis procedures identical to those described in the legend to Fig. 7, both in the absence of random noise and at random noise amplitudes (σ_r) of 0.016, 0.032, and 0.064 OD. (a) The first three singular values. The dashed lines are the result in the absence of random noise; the results for all values of σ_r are plotted using the identical symbol: s_1, filled circles; s_2, open squares; s_3, open circles. (b, c) Square root of the average variance of the third singular vectors, U_3 (b) and V_3 (c). The results are shown both before (large symbols and solid lines) and after (small symbols and dashed lines) rotation to optimize the autocorrelation of the retained right singular vector, V_3^R.

OD), $\sigma(V_3)$ approaches the value expected for a random variable $n_t^{-1/2} \cong$ 0.119], strongly suggesting that the third SVD component results almost exclusively from the wavelength-correlated noise at these noise amplitudes. This conclusion is supported by examining U_3 of the individual SVDs under these conditions, which shows that the U_3 of the noise-free data has been replaced by a nearly constant spectrum arising from the offsets. The variance of V_3 increases monotonically as σ_λ increases between these two values. This result argues that most of the information originally contained in component 3 of the noise-free data has been displaced into component 4 by the wavelength-correlated noise when σ_λ is larger than 0.05 OD.

These results show that the sorting of wavelength-correlated noise from the data by SVD is moderately efficient as long as s_λ is about a factor of 3 smaller than the singular value of a given component ($\sigma_\lambda < 0.004$ OD). What is even more interesting is that SVD is able to sort efficiently when s_λ is more than a factor of 3 *larger* than this singular value ($\sigma_\lambda > 0.04$ OD). There is a "mixing zone," delimited by the requirement that the amplitude of the wavelength-correlated noise, measured by s_λ, be within a factor of about 3 of that of the signal, measured by s_3, in which significant mixing occurs. At all values of σ_λ, rotation of the SVD output has a dramatic effect on $\sigma(V_3)$, reducing it to a value which is almost identical to that obtained in the presence of only the random noise component. This improvement persists until σ_λ exceeds 0.1, where the wavelength-correlated noise is almost an order of magnitude larger than the third component of the noise-free data.

If we now examine the noise amplitude for U_3 shown in Fig. 8b we find that it remains small throughout this mixing zone and even decreases as σ_λ becomes large enough to dominate the third SVD component. This result is consistent with the results of the simulations shown in Fig. 7, which showed that the noise amplitude in each of the left singular vectors produced by mixing of random noise amplitudes with the data is inversely proportional to its singular value. Because U_3 of the noise-free data and the spectral signature associated with the wavelength-correlated noise are smooth, the contribution of random noise to the noise amplitude of U_3 should be determined by the magnitude of s_3. When the SVD output is rotated to optimize $\sigma(V_3)$, the variance of U_3^R increases systematically as σ_λ increases, but never reaches the value expected if the spectrum were completely uncorrelated. It is also interesting to note that the random noise component acts almost only as a "background" to the wavelength-correlated noise in all of the output vectors. The noise amplitude contributed by the wavelength-correlated noise appears to be simply superimposed on this background.

A second set of simulations was carried out to explore to what extent the mixing of wavelength-correlated noise was dependent on the preprocessing of the data. The noise-free data used in these simulations were identical to those used above except the spectrum of pure C was subtracted at each time point. The data are therefore representative of data processed to produce difference absorption spectra at a given time point with the equilibrium sample used as a reference. The sample data set in Fig. 6d shows that the calculation of difference spectra removes much of the signal from the data matrix. From the point of view of SVD, the major consequence of this change is effectively to remove the first SVD component, which corresponds to the average absorption spectrum of the sample, from the analysis. The first SVD component now corresponds to the average *difference* spectrum observed in the simulated experiment.

The first two singular values and the noise amplitudes obtained for the second singular vectors (U_2 and V_2) are shown in Fig. 9. Note that the values of s_1 and s_2 of the noise-free data in these simulations are only slightly larger than s_2 and s_3 in Fig. 8. The dependence of the variance of V_2 on σ_λ, shown in Fig. 9c, is qualitatively similar to that found for V_3 in Fig. 8c. The ability of the rotation algorithm to reduce the variance of V_2 is also qualitatively similar to that shown in Fig. 8c, but the reduction in $\sigma(V_2)$ which results from rotation in Fig. 9c is even larger than that for $\sigma(V_3)$ in Fig. 8c. As in the previous simulation, rotation is able to "rescue" the original signal-to-noise ratio in V_2^R up to the point where the noise becomes as large as the first component of the noise-free data. At this point, significant noise amplitudes begin to mix into V_1. Comparable simulations which included rotation of V_1 show that the original signals in both V_1 and V_2 can be recovered even at these very large noise amplitudes.

Closer comparison of the results in Figs. 8 and 9 shows that there are significant differences between the analyses of the absorption spectra and the difference spectra. One major difference is the width of the "mixing zone," or transition region in which the relevant component (V_3 in Fig. 8 or V_2 in Fig. 9) is a combination of signal and wavelength-correlated noise. For the difference spectra (Fig. 9c) this zone is significantly narrower, covering about a factor of 3 change in σ_λ. (An intermediate value of $\sigma(V_2)$ is obtained for only a single value of σ_λ.) In Fig. 8c, on the other hand, this transition zone extends over at least a factor of 10 in σ_λ. The RMS deviations of V_3 in Fig. 8c are also somewhat greater than the corresponding deviations in V_2 in Fig. 9c at low noise amplitudes ($\sigma_\lambda <$ 0.02). These results suggest that SVD alone sorts the wavelength-correlated noise from the signal more efficiently when the input data are in the form of difference spectra.

Another significant difference in the two simulations is that the variance

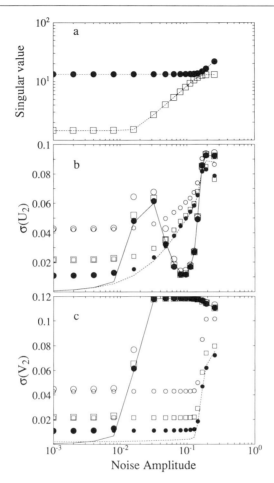

FIG. 9. Effects of the RMS amplitude of an offset spectrum on the singular values and variances of the second singular vectors of simulated difference spectra for the reaction A → B → C in the presence of random noise. Simulations were carried out using the spectra described in the text except that the spectrum of pure C was subtracted from the calculated sample spectrum at each time point and using analysis procedures identical to those described in the legend to Fig. 7. (a) The first two singular values. The dashed lines are the result in the absence of random noise, and the results for all values of σ_r are plotted using the same symbol: s_1, filled circles; s_2, open squares. (b, c) Square root of the average variance of the second singular vectors, U_2 (b) and V_2 (c). The results are shown both before (large symbols and solid lines) and after (small symbols and dashed lines) rotation to optimize the autocorrelation of the retained right singular vector, V_2^R. It should be noted that in the absence of random noise the SVD consists of only three singular values and vectors, so rotation only involves the mixing of V_2 and V_3.

of U_2 from the difference spectra also becomes very large in the transition region in Fig. 9b. There is essentially no evidence of such an effect in Fig. 8b. Because the offset spectrum has no wavelength dependence, this cannot arise from high-frequency contributions to U_2 in the SVD of a single data set. Rather, it must result from variations in the sign and magnitude of the offsets which are mixed with the U_2 of the noise-free data in different simulated data sets. Because the relative sign of the signal and offset contributions to U_2 is presumably determined by statistical fluctuations in the offset amplitudes, it is not unreasonable that the offset contribution varies both in magnitude and sign from data set to data set. The major difference between the two simulations is the absence of the largest SVD component in the difference spectra simulations. Examination of individual SVD outputs shows mixing of significant offset amplitudes with this component. Our tentative interpretation of this difference, then, is that significant mixing of the offsets with the average absorption spectrum can occur when the absorption spectra are used as data (Fig. 8b), but that this mixing cannot occur when difference spectra are used (Fig. 9b).[53]

The results of these simulations can be summarized as follows. As the amplitude of the wavelength-correlated noise increases, it first mixes with the components of the noise-free data, systematically increasing the random component of V_3 (V_2) and altering U_3 (U_2) (components in parentheses refer to the simulations of the difference spectra). As the noise amplitude increases further, the third (second) SVD component obtained from the noisy data results primarily from the wavelength-correlated offset, and its singular value, s_3 (s_2), increases in proportion to σ_λ. At these noise amplitudes V_3 (V_2) is essentialy a random vector. For the absorption spectra, the increased value of s_3 reduces $\sigma(U_3)$, suggesting that nearly all of the noise in U_3 results from random noise contributions, but more complex behavior is observed for the U_2 obtained from the difference spectra. Under these conditions most of the information contained in the third (second) component of the SVD of the noise-free data has been displaced into the fourth (third) SVD component.

The rotation procedure is extremely effective in reclaiming the signals represented by V_3 (V_2) under all conditions where σ_λ is less than about 0.1. Processing of components 3 through 10 (2 through 10) by this algorithm dramatically reduces the noise of V_3 (V_2). The cost of this decrease in the noise of the rotated V_3 is a significant increase in the noise amplitude of

---

[53] The possibility cannot be ruled out that a remaining ambiguity in the determination of the sign of U_2 as calculated in these simulations contributes to the variance of U_2 at higher noise levels.

the rotated $U_3(U_2)$. The noise amplitude of the rotated $V_3(V_2)$ only increases at the point where the wavelength-correlated noise becomes comparable in magnitude to the second (first) SVD component of the noise-free data. At this point the wavelength-correlated noise begins to mix significantly with the second (first) SVD component, and a different processing algorithm (specifically, a rotation which includes $V_2(V_1)$ in the rotated set) would be required to extract the signal from the noise. The ineffectiveness of the rotation algorithm when confronted with random noise amplitudes, noted earlier, is strongly reinforced by these simulations. We therefore conclude that the rotation procedure can be of significant benefit in extracting correlated noise contributions to the data matrix which are comparable in magnitude to a given signal component. When the signal is significantly larger than the noise, rotations appear to be of questionable benefit, sometimes resulting in a less efficient extraction of the signal by reducing the difference in the norms of the mixed components.

VI. Summary

In writing this chapter we have attempted to point out the significant advantages which result from the application of SVD and complementary processing techniques to the analysis of large sets of experimental data. Although SVD performs no "magic," it does efficiently extract the information contained in such data sets with a minimum number of input assumptions. Moreover, SVD is the *right* approach to use for such a reduction. The least-squares property which we have described both in the introduction (Section I) and in Section II, together with the theorem which shows that fitting of appropriately weighted SVD output is equivalent to fitting of the entire data set, argues that there are few, if any, additional risks which accompany the use of SVD as part of the analysis procedure.

The simple examples of the SVD of noisy matrices (Section III) together with the simulations in Section V teach several lessons in regard to the choice of an algorithm for processing data. First, when random noise is mixed with data the results are totally predictable. SVD very effectively averages over random contributions to the data matrix. However, when the random noise amplitude becomes sufficiently large to completely submerge the averaged signal amplitude from a given SVD component, that component becomes unresolvable. This result is a straightforward extension of conventional signal-averaging wisdom. The only possible approaches to extracting additional information when all of the higher SVD components are uncorrelated (i.e., they result from random noise) are either to increase the size of the data matrix and hence increase the number of observations which are averaged or to improve the quality of the data matrix by decreasing the noise level in the experiment. Second, when the data contain random amplitudes of a correlated noise source (e.g., the wavelength-correlated noise in our simulations) the mixing of the "signal," represented by a given SVD component, k, with the noise becomes significant when σ_λ becomes sufficiently large that the contribution of the noise component to the data matrix, $s_\lambda = (mn\sigma_\lambda^2)^{1/2}$, becomes comparable in magnitude to the "signal" represented by that SVD component, s_k. When the noise is small, it does not seriously degrade the component, but is efficiently sorted by SVD into a separate component, appearing as a larger than random contribution to the spectrum of singular values which result from the noise. When the noise becomes significantly larger than a given SVD component, the ability of SVD to sort by signal magnitude again separates such noise from the desired signal relatively cleanly, but the efficiency with which the two components are separated appears to depend on the specific data being analyzed. When such noise is mixed with the signal component in the SVD, dramatic improvement in signal-to-noise ratios can be produced by rotation of the mixed and highly cross-correlated SVD components to improve the autocorrelation of the retained component(s).

In conclusion, it must be noted that in the examples and simulations which we have presented we know the "signal" *a priori* and are also using noise having well-defined characteristics. As a result, it is considerably easier to understand the behavior observed in these cases than when one is confronted with real data. The results of the SVD analysis either with or without the rotation procedure depend on both the spectra which are used as input and on the detailed characteristics of the noise which is added to the data. Furthermore, the distribution of correlated noise in the SVD output can be altered by preprocessing as simple as the calculation of difference spectra. Extrapolation of these results would suggest that, in order to optimize the processing algorithm for any specific data, the experiment and its processing should be tested by simulations which use spectra that closely match those measured and noise that closely approximates that measured for data sets in which there is no signal. The only obvious alternative to using simulations appears to be to process the same experimental data using a variety of different procedures and to compare the results. Either of these approaches may pay dividends in the analysis of real experimental data.

APPENDIX: Transformation of SVD Vectors to Optimize
    Autocorrelations

Suppose that we have an orthonormal set of column vectors $\{V_i\}$, for example, the set of right singular vectors of an $m \times n$ matrix. The autocorrelation of column vector V_i is defined by

$$C(V_i) = \sum_{j=1}^{n-1} V_{ji} V_{j+1,i} \tag{A.1}$$

where V_{ji} is the jth element of the vector V_i and n is the number of elements in each vector. The rotation procedure produces a linear transformation $\mathbf{R}_{\{k\}}$ which takes a given subset consisting of the p basis vectors $\{V_k\}$ ($k \in \{k1, \ldots, kp\}$) to a new set of vectors $\{V'_k\}$ for which the autocorrelations as defined in Eq. (A.1) are optimized. The problem is to find coefficients r_{ji}, the elements of the matrix $\mathbf{R}_{\{k\}}$, such that the autocorrelations of the vectors

$$V'_j = \sum_{i \in \{k\}} r_{ji} V_i \tag{A.2}$$

are optimized. From Eq. (A.1) we have

$$C(V'_i) = \sum_{j=1}^{n-1} V'_{ji} V'_{j+1,i}$$

$$= \sum_j \left(\sum_k r_{ik} V_{jk} \right) \left(\sum_p r_{ip} V_{j+1,p} \right)$$

$$= \sum_{k,p} r_{ik} r_{ip} \left(\sum_j V_{jk} V_{j+1,p} \right) \tag{A.3}$$

If we define the $p \times p$ cross-correlation matrix $\mathbf{X}$ by

$$X_{ij} = \sum_k V_{ki} V_{k+1,j} \tag{A.4}$$

then Eq. (A.3) may be written in the compact form

$$C(V') = \sum_{i,p} r_i r_p X_{ip} \tag{A.5}$$

where we have also dropped the vector index j for brevity.

We now require that the coefficients r_i produce extremum values of the autocorrelation in Eq. (A.5), subject to the constraint that the normalized and orthogonal vectors V_i are transformed into normalized vectors V'_j, that is, $\sum r_i^2 = 1$. These requirements are easily formulated using the method of undetermined multipliers:

$$\frac{\partial F}{\partial r_i} = 0, \quad i = 1, \ldots, m \tag{A.6}$$

where

$$F = C(V') + \lambda \sum_i r_i^2 \tag{A.7}$$

Using Eq. (A.5), Eq. (A.6) becomes

$$\sum_p (X_{ip} + X_{pi}) - 2\lambda r_i = 0 \tag{A.8}$$

This may be rewritten in matrix–vector notation as

$$\mathbf{X}^S r = \lambda r \tag{A.9}$$

where the matrix $\mathbf{X}^S = (\mathbf{X} + \mathbf{X}^T)/2$ is the symmetrized cross-correlation matrix and r is the vector of coefficients r_i.

Equation (A.9) represents a simple eigenvalue problem for the real symmetric matrix $\mathbf{X}^S$. The individual eigenvectors r_j ($j = 1, \ldots , p$) of $\mathbf{X}^S$ are sets of coefficients r_{ji} that produce distinct transformed vectors V_j' from the starting vectors V_i according to Eq. (A.2). It is easily shown that the corresponding eigenvalues λ_j are in fact the autocorrelations of the transformed vectors V_j'. The matrix $\mathbf{R}_{\{k\}}$ of the linear transformation that we seek is just the matrix of column vectors r_j, and the transformation in Eq. (A.2) may be rewritten

$$\mathbf{V}^R = \mathbf{V}\mathbf{R}_{\{k\}} \tag{A.10}$$

where $\mathbf{V}$ and $\mathbf{V}^R$ are the matrices of untransformed and transformed vectors, respectively. The matrix $\mathbf{R}_{\{k\}}$ is clearly orthogonal, that is, $\mathbf{R}_{\{k\}}^{-1} = \mathbf{R}_{\{k\}}^T$, allowing the transformation of Eq. (A.10) to be identified as a rotation. The eigenvectors r_j that make up the matrix $\mathbf{R}_{\{k\}}$ are conventionally arranged in order of decreasing eigenvalues, so that the columns of $\mathbf{V}^R$ will be arranged in order of decreasing autocorrelations of the transformed vectors.

It should be pointed out that the conditions of Eq. (A.6) only ensure extremum values of the autocorrelation with respect to the coefficients r_i, not necessarily maximum values. In general, the procedure produces a set of p vectors V_j' some of which have autocorrelations that are maxima, some of which have autocorrelations that are minima, and some of which have autocorrelations that represent saddle points in the space of coefficients r_i. For this reason, the procedure may be viewed as producing a new set of autocorrelations, some of which are improved (with respect to the original set) at the expense of others. It should also be noted that this straightforward procedure for reducing the optimization problem to an eigenvalue problem for determining the transformation matrix $\mathbf{R}_{\{k\}}$ is easily generalized to optimization of a much broader class of autocorrelation and other bilinear functions than has been considered here.

One final topic of discussion is the proper choice of weighting factors to be used when performing least-squares fits using transformed columns of $\mathbf{V}$. As discussed in the text, fits using the untransformed columns of $\mathbf{V}$ require that the fit to each column be weighted by the square of the corresponding singular value. In this case, for any set of model parameters the weighted squared deviation between $\mathbf{V}$ and the matrix $\tilde{\mathbf{V}}$ ($\tilde{\mathbf{V}} \equiv \mathbf{C'P}$ in Section IV,E) of fitted columns produced by explicit evaluation of the model may be written

$$\phi^2 = \|\mathbf{S}(\mathbf{V^T} - \tilde{\mathbf{V}}^\mathbf{T})\|^2 \tag{A.11}$$

Inserting the product $\mathbf{RR^T} = \mathbf{I}_n$, where $\mathbf{R}$ is the full orthogonal transformation matrix defined in Eq. (37), and using the definition of Eq. (A.10),

$$\phi^2 = \|\mathbf{SRR^T}(\mathbf{V^T} - \tilde{\mathbf{V}}^\mathbf{T})\|^2$$
$$= \|\mathbf{SR}[(\mathbf{V^R})^\mathbf{T} - (\tilde{\mathbf{V}}^\mathbf{R})^\mathbf{T}]\|^2 \tag{A.12}$$

We define the matrix $\mathbf{\Delta}$ as

$$\mathbf{\Delta} = \mathbf{V^R} - \tilde{\mathbf{V}}^\mathbf{R} \tag{A.13}$$

Using the identity

$$\|\mathbf{M}\|^2 = Tr(\mathbf{M^T M}) \tag{A.14}$$

where $Tr(. . .)$ signifies the matrix trace operation, we can write

$$\phi^2 = Tr(\mathbf{\Delta R^T S^2 R \Delta^T})$$
$$= \sum_i \sum_{kl} \Delta_{ik}(\mathbf{R^T S^2 R})_{kl}\Delta_{il} \tag{A.15}$$

If the model being used is adequate to describe the data matrix to within some tolerance, then within some neighborhood of the minimum of Eq. (A.11) in parameter space it is reasonable to expect that the deviations represented by the different columns of $\mathbf{\Delta}$ will be uncorrelated. If this is the case, we can write

$$\sum_i \Delta_{ik}\Delta_{il} = \sum_i \delta_{kl}\Delta_{ik}^2 \tag{A.16}$$

where δ_{kl} is the Kronecker delta. This allows us to simplify Eq. (A.15) to

$$\phi^2 \cong \sum_k (\mathbf{R^T S^2 R})_{kk} \sum_i \Delta_{ik}^2$$
$$= \sum_k W_{kk}\|\Delta_k\|^2 \tag{A.17}$$

where the matrix $\mathbf{W} = \mathbf{R^T S^2 R}$ and Δ_k is the kth column of the matrix $\mathbf{\Delta}$.

Thus, subject to caveats concerning the assumptions leading to Eq. (A.16), it is reasonable to choose the squares of the corresponding diagonal elements of the matrix $\mathbf{R}^T\mathbf{S}^2\mathbf{R}$ as weighting factors in fits to columns of $\mathbf{V}^R$.

Another way to estimate the weighting factors is via the amplitudes of the columns of $\mathbf{U}^R$ corresponding to the normalized columns of $\mathbf{V}^R$ being fit. In the absence of rotations, the weighting factors for fitting the columns of $\mathbf{V}$ are simply the squared amplitudes of the corresponding columns of $\mathbf{US}$. By analogy, the weighting factors for fitting the columns of $\mathbf{V}^R$ could be chosen as the squared amplitudes of the corresponding columns of $\mathbf{U}^R$. These squared amplitudes are given by the diagonal elements of the product $(\mathbf{U}^R)^T\mathbf{U}^R$. By Eq. (39) this product may be written

$$(\mathbf{U}^R)^T\mathbf{U}^R = \mathbf{R}^T\mathbf{S}\mathbf{U}^T\mathbf{U}\mathbf{S}\mathbf{R}$$
$$= \mathbf{R}^T\mathbf{S}^2\mathbf{R} \qquad (A.18)$$

Thus, this approach to estimating the weighting factors yields the same result as was produced by the first method.

Acknowledgments

We are grateful to Attila Szabo for numerous discussions and suggestions, as well as for critical comments on the manuscript. We also thank our colleagues in using SVD, William A. Eaton, Anjum Ansari, and Colleen M. Jones, for helpful comments and suggestions. We especially thank Colleen M. Jones for providing the data used as the example.

[9] Fourier Resolution Enhancement of Infrared Spectral Data

By Douglas J. Moffatt and Henry H. Mantsch

Introduction

Infrared spectroscopy has become an important tool in the study of biological structures and processes. The vast improvement of infrared spectroscopic instrumentation (three orders of magnitude in sensitivity in the past 15 years) has made possible many new applications for this technique, most notably in biology, with the routine acquisition of high-quality spectra of aqueous solutions. At the same time, the use of digital data acquisition and storage has greatly facilitated the application of numerical computer methods as an aid in data interpretation. One area of biology in which infrared spectroscopy has become particularly useful is the structural analysis of proteins and, in particular, the determination of the secondary structure of soluble and membrane-bound proteins.[1-3]

Herein we present a critical discussion of the computational methods of Fourier resolution enhancement (band narrowing) and their role in the analysis of the infrared absorption spectra of proteins. Although, strictly speaking, the use of the expression "resolution enhancement" applied to the process of band narrowing is ambiguous, it is widely used in the literature, and therefore we shall use it here in that context. This type of numerical data processing does not affect the instrumental resolution, but changes the shape of the infrared bands and, by reducing their width, allows a better visual separation (i.e., resolution) of individual component bands.

General Theory

There are two common methods of resolution enhancement via band narrowing that use the Fourier transform: Fourier self-deconvolution (FSD) and Fourier derivation (FD).

Fourier Self-Deconvolution

Fourier self-deconvolution, which is based on well-established data processing methods in communication technology, was introduced to infrared spectroscopy in 1981.[4] It assumes a spectrum to be composed of a number of bands with different positions and intensities, but with the same band shape, $W(\nu)$. This input band shape, $W(\nu)$, is then replaced with a narrower output band shape, $W'(\nu)$. Let $I(x)$ be the Fourier transform, $\mathscr{F}\{\ \}$, of a spectrum $E(\nu)$ and apply the standard equation for the Fourier transform operator

$$I(x) = \mathscr{F}\{E(\nu)\} = \int_{-\infty}^{\infty} E(\nu) \exp(i2\pi x\nu) \, d\nu \tag{1}$$

where, for an infrared spectrum, the independent variables x and ν will have the reciprocal units of cm and cm^{-1}, respectively. If one now defines $D(x) = \mathscr{F}\{W(\nu)\}$ and $D'(x) = \mathscr{F}\{W'(\nu)\}$, the resolution-enhanced output spectrum, $E'(\nu)$, may be calculated as

[1] H. Susi and D. M. Byler, this series, Vol. 130, p. 290.
[2] W. K. Surewicz and H. H. Mantsch, *Biochim. Biophys. Acta* **952**, 115 (1988).
[3] E. Goormaghtigh, V. Cabiaux, and J.-M. Ruysschaert, *Eur. J. Biochem.* **193**, 409 (1990).
[4] J. K. Kauppinen, D. J. Moffatt, H. H. Mantsch, and D. G. Cameron, *Appl. Spectrosc.* **35**, 271 (1981).

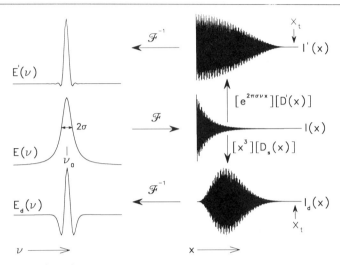

FIG. 1. Illustration of the steps of the FSD and FD procedures applied to a Lorentzian band $E(\nu)$, centered at ν_0. $I(x)$, the Fourier transform of $E(\nu)$, is an exponentially decaying cosine wave. In FSD, $I(x)$ is multiplied by an exponentially increasing function for band narrowing and also by an apodization function, $D'(x/x_t)$, for band shape, to yield a new function $I'(x)$. An inverse Fourier transform of $I'(x)$ then yields the narrower output band $E'(\nu)$. In FD, $I(x)$ is multiplied by $x^3 D_s(x/x_t)$, to give a new function $I_d(x)$, the inverse Fourier transform of which yields a new band $E_d(\nu)$, which also has a narrower half-width but significantly larger side lobes.

$$E'(\nu) \; = \; \mathscr{F}\{I(x) \cdot [D'(x)/D(x)]\} \tag{2}$$

In practice, the spectrum $E(\nu)$ is known, the band shape $W(\nu)$ is estimated, and $D'(x)$ is a weighting function, also known as an apodization function, that is unity at $x = 0$, decays monotonically to zero at some point x_t (see Fig. 1), and is zero for $x > x_t$. This weighting function is usually chosen as either Bessel, $[1 - (x/x_t)^2]^2$, or triangular squared, $[1 - (x/x_t)]^2$. In FSD applications, a useful parameter is K, the band narrowing factor, defined as the ratio of the bandwidth before to that after band narrowing

$$K \; = \; (\text{FWHH of } W)/(\text{FWHH of } W') \tag{3}$$

where FWHH is the full width at half-height of the band. In most applications, the maximum value for K, $K_{\max}$, is limited by the signal-to-noise ratio, (S/N), in the spectrum.[5] When applied to a condensed phase infrared absorption spectrum, $W(\nu)$ is usually a Lorentzian band shape with a

[5] J. K. Kauppinen, D. J. Moffatt, H. H. Mantsch, and D. G. Cameron, *Appl. Opt.* **20,** 1866 (1981).

FWHH, (2σ), in the range 5 to 25 cm^{-1}, in which case $K_{max} \cong \log_{10}(S/N)$ and $D(x) = e^{-2\pi\sigma x}$. To be a candidate for FSD, a spectrum should typically have a signal-to-noise ratio exceeding 100 and an instrumental resolution of at least 4 cm^{-1}, preferably higher.

Fourier Derivation

Compared to FSD, Fourier derivation is a simpler procedure in which $I(x)$, the Fourier transform of the spectrum, is multiplied by a function x^p and also by a weighting function $D_s(x)$. An inverse Fourier transform then yields the derivative spectrum

$$E_d(\nu) = \mathcal{F}\{I(x)x^p D_s(x)\} \qquad (4)$$

In practice, p, which is roughly analogous to K in FSD, is commonly set to 3.0 (lower values degrade the resolution enhancement while higher values can result in an inordinate number of side lobes in the output spectrum).[6] $D_s(x)$ is the first lobe of a sinc squared function, $[\sin(\pi x/x_t)/(x/x_t)]^2$, with x_t as the only adjustable parameter. When p is an even integer, the output is congruent with the corresponding analytical derivative of the input spectrum smoothed by $D_s(x)$. A distinct advantage of Fourier derivation over conventional derivation procedures is that p is not restricted to integral values, thus allowing the degree of band narrowing to be adapted to the quality (i.e., S/N ratio) of the input spectrum.

Comparison of Fourier Self-Deconvolution with Fourier Derivation

Figure 1 details the application of the two procedures to a synthetic absorbance spectrum consisting of a single Lorentzian band. Although both procedures produce essentially the same result, namely, narrowing of the original input band, there are significant practical differences. FSD has a major advantage of retaining the integrated intensity of the component bands of a profile and, when properly applied, producing a desirable output band shape with a minimum of negative side lobes. However, proper application requires a spectrum composed of bands with fairly uniform shapes and widths and at least an approximate knowledge of their shape. The derivation procedure, on the other hand, has the advantage that no band shape information is required, rendering it simpler and more reliable when applied to unknown spectra. Yet, there are two drawbacks with Fourier derivation: the loss of integrated intensity information (which precludes many quantitative studies) and a poor output band shape with side lobes that can mask weak component bands (see Fig. 4). Also, in

---

[6] D. G. Cameron and D. J. Moffatt, *Appl. Spectrosc.* **41**, 539 (1987).

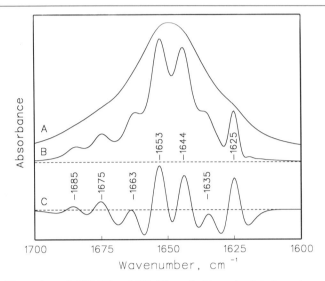

FIG. 2. Illustration of FSD (B) and FD (C) applied to a simulated spectrum of a typical protein in the region of the conformation-sensitive amide C=O stretching bands (A). It can be seen that although derivation leads to a better separation of the seven bands, in the latter case one has to deal with many negative lobes.

cases where band widths differ significantly, the results can be misleading, with narrow bands being greatly enhanced relative to the broader bands while extremely broad features are totally suppressed.

Figure 2 compares the results of FSD and FD as applied to a simulated protein spectrum in the region of the amide C=O stretching bands. The C=O stretching band envelope consists of seven components with different intensities. The band at 1625 cm^{-1} is 30% narrower than the other bands, and its peak height is considerably enhanced relative to that of the other bands. This shows that the peak height after band narrowing can be deceptive. In FSD, however, the integrated intensity of each of the seven bands is preserved. For a more detailed description of these Fourier transform methods, the interested reader is referred to Refs. 7–10. A set of computer programs for FSD and FD was developed at the National

[7] J. K. Kauppinen, in "Spectrometric Techniques" (G. Vanasse, ed.), Vol. 3, p. 199. Academic Press, New York, 1983.

[8] H. H. Mantsch, H. L. Casal, and R. N. Jones, *Adv. Spectrosc.* (*Chichester, U.K.*) **13**, 1 (1986).

[9] P. R. Griffiths and G. L. Pariente, *Trends Anal. Chem.* **5**, 209 (1987).

Research Council of Canada[11] and is available on request from the authors free of charge.

Practical Considerations

The following are a number of practical considerations in the application of these numerical computer methods for band narrowing.

Estimation of Input Band Shape

The input band shape is the most important parameter in the FSD procedure. The input band shape and bandwidth should at least approximate those of the component bands comprising the spectrum. In practice, the band shape can usually be derived from a theoretical model with the correct width determined by trial and error. For condensed phase infrared spectra, one usually employs a Lorentzian band shape and starts with an estimated bandwidth. Most commercial FT-IR instruments include software that allows quick viewing of FSD results, usually with an option to vary one of the parameters, bandwidth or K, and iterate.

Figure 3 illustrates the significance of a correct bandwidth estimate. The composite contour in trace A consists of two Lorentzian bands that have the same width (20 cm^{-1}) and intensity. Deconvolution with a band narrowing factor K of 2.5 and the correct half-width (FWHH = 20 cm^{-1}) leads to trace B, with FWHH = 10 cm^{-1} to trace C, and with FWHH = 30 cm^{-1} to trace D. Clearly, the correct estimate of the half-width produces the best result, whereas the use of too large a half-width (also referred to as "overdeconvolution") results in significant negative side lobes. One approach in optimizing the half-width is to use a fixed, low K (e.g., 1.5) and start with a low width, then increase it until just before side lobes (see Fig. 3D) start appearing in the output. At this point, K may be increased as explained below. Note that the optimum half-width will be smaller, sometimes much smaller than the half-width of the band contour. In Fig. 2 the half-width of the composite band is about 50 cm^{-1}, whereas the half-width used in the deconvolution is 13 cm^{-1}.

[10] H. H. Mantsch, D. J. Moffatt, and H. L. Casal, *J. Mol. Struct.* **173**, 285 (1988).

[11] D. J. Moffatt, J. K. Kauppinen, D. G. Cameron, H. H. Mantsch, and R. N. Jones, "Computer Programs for Infrared Spectrophotometry," National Research Council of Canada Bulletin No. 18, p. 1. NRCC, Ottawa, 1986.

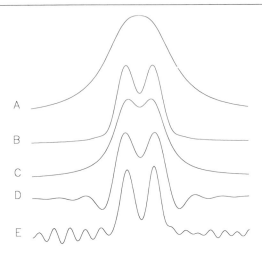

FIG. 3. Illustration of the correct choice of bandwidth with FSD. (A) Composite band contour consisting of two equally intense Lorentzian bands 10 cm^{-1} apart, each with a FWHH of 20 cm^{-1}. The effect of FSD [with $K = 2.5$ and $D'(x) = $ triangular squared], by use of a FWHH of 20 cm^{-1} (B), 10 cm^{-1} (C), and 30 cm^{-1} (D). (E) Morphology of noise after FSD. The input spectrum was (A) with noise added to degrade the S/N ratio to a value of 1500. FSD parameters: width, 20 cm^{-1}; K, 3.8.

Signal-to-Noise Ratio and K

All experimental data will contain a certain amount of noise which FSD enhances by approximately 10^{K-1} (for Lorentzian bands). Moreover, the noise attains a morphology very similar to that of the output bands (see Fig. 3E). Hence, to aid in interpretation of the FSD output, one should (1) extend the deconvolved region to include a portion of the spectrum known to be devoid of peaks, (2) run replicate spectra and compare deconvolved results, or (3) estimate or measure the S/N and keep K below $\log_{10}(S/N)$. Also, because the noise in an absorbance spectrum increases exponentially with absorbance, samples should be prepared so that peak absorbances do not exceed 1.5 absorbance units.

Water Vapor

A common experimental problem associated with FT-IR absorption spectroscopy, especially of aqueous samples, is the presence of weak water vapor absorption lines caused by a slight change in the atmospheric moisture in the spectrometer in the time between the collection of reference and sample data. To the FSD procedure, these lines are essentially "noise" and have the same effect as noise—reducing the maximum value

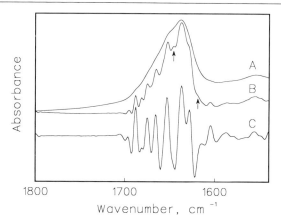

FIG. 4. Fourier resolution enhancement applied to the infrared spectrum of α-chymotrypsinogen in aqueous buffer. Original spectrum (A), resolution-enhanced spectrum after FSD by use of $K = 3.5$ and FWHH $= 13$ cm^{-1} (B), and resolution-enhanced spectrum after FD by use of $p = 3.0$ and $x_t = 0.25$ cm^{-1} (C).

for K. This is especially troublesome for the study of proteins because this interference occurs between 1200 and 1800 cm^{-1}, covering the conformation-sensitive region of the amide C=O stretching bands (1600–1700 cm^{-1}). This source of error may be greatly reduced by careful attention to spectrometer purging, combined, if necessary, with digital subtraction of residual water vapor lines.

Edge Effects

If the input spectrum has a significant slope near the edges of the region subjected to band narrowing, large distortions will appear near the edges of the output spectrum. This problem may be circumvented in two ways: by baseline correcting the spectrum before band narrowing or by choosing a wider than desired region and ignoring the output data within 4σ of the edge.

Applications to Protein Analysis

The sensitivity of infrared spectra (particularly that of the amide C=O stretching band) of proteins to the secondary structure of the polypeptide backbone has been recognized for some time. Yet the original infrared spectra are difficult to interpret because the separation between bands originating from the different protein segments is generally smaller than the half-width of the bands. This results in a broad overall band contour consisting of many overlapping bands. Thus, the primary role of the Fourier methods of band narrowing is to identify the individual component bands in a given protein. This is illustrated by the example in Fig. 4 which shows the infrared spectrum, in the amide $C=O$ stretching region, of an aqueous solution of the commonly used model protein α-chymotrypsinogen. The original spectrum (Fig. 4A) exhibits only a broad band profile with a few shoulders. The spectrum after FSD (Fig. 4B) reveals the presence of at least eleven components in the spectral region between 1600 and 1700 cm^{-1}. The spectrum after FD also reveals these components; however, the side lobes in FD mask two weak features evidenced by FSD (indicated by arrows). Once the complex band contour is resolved, the different component bands can be submitted to curve fitting for quantitative analysis as the area of the different components of the amide $C=O$ bands is directly related to the amount of secondary structure present in that particular protein. A first comprehensive study, based on this methodology, of the secondary structure of a number of globular proteins was published in 1986,[12] and we estimate that between 1986 and 1990 over 100 reports were published on this topic.

Concluding Remarks

Although we have only illustrated the utility of Fourier resolution enhancement in applications to condensed phase infrared spectra, it may be employed to advantage on any data set that satisfies two criteria: (1) high S/N ratio and (2) composite bands with intrinsic widths that are large compared to the instrumental resolution. As the quality of instrumentation, and that of the spectroscopic data produced, improves with time, these techniques will be called on increasingly to extract as much information as possible from experimentally recorded data.

[12] M. Byler and H. Susi, *Biopolymers* **25**, 469 (1986).

[10] Maximum Likelihood Analysis of Fluorescence Data

By Željko Bajzer and Franklyn G. Prendergast

I. Introduction

In principle, fluorescence spectroscopy provides a powerful tool for the investigation of structure and function in biological macromolecules. There are two fundamental considerations in the use of fluorescence data for such purposes. First, the fluorescence may be used simply as a signal of an event. Under such circumstances only a change in the signal, in response to some perturbation, needs be measured, and such measurements can be conducted sensitively and accurately especially if steady-state fluorescence intensity is the property being quantified. Second, fluorescence data may be interpreted in terms of the physicochemical character, the structure, and the dynamics of the environs of the fluorophore. This latter use derives from the well-documented sensitivity of the fluorescence process to the physicochemical properties of the environment.[1,2] However, use of fluorescence data in this second manner is much more problematic, especially when one is studying typical biologically derived fluorophores such as tryptophan or tyrosine. The photophysics of these fluorophores is intrinsically so complex, and their fluorescence so sensitive to a wide variety of environmental factors, that inferences regarding structure drawn from fluorescence intensity decay data are difficult at best, impossible at worst to justify. Still, there are several cases where careful measurements of the fluorescence parameters can yield quantitative data which are interpretable in terms of structure within the bounds of well-constructed physical (conceptual) models.

Of all the measurable fluorescence properties, the determination of the fluorescence intensity decay is potentially one of the most useful. Thus, intensity decay data may be employed either for the measurement of fluorescence lifetimes or for the determination of the time-dependent fluorescence anisotropy decay—from the decay of the polarized intensities—which yields assessments of the dynamics of the fluorophore. The parameters determined from such measurements are valuable both in the study of structure per se and for the study of time-dependent processes such as molecular motion. However, it is now clear that even in systems bearing a single fluorophore, such as a protein with a single tryptophan residue, the fluorescence intensity decay is often apparently multiexponential, evoking physical models implying multiple conformers of the molecule being studied and speculations regarding the physical origin of each fluorescence lifetime detected. Obviously, confidence in such inferences, especially where they involve interpretation of the physical basis for particular parameters, must be influenced markedly by one's confidence in the accuracy of the data analysis. And, it matters not at all which method is used in the determination of the intensity decay profile; either time-correlated single photon counting or phase/modulation fluorometry may

[1] J. R. Lakowicz, "Principles of Fluorescence Spectroscopy." Plenum, New York and London, 1983.
[2] J. M. Beechem and L. Brand, *Annu. Rev. Biochem.* **54,** 54 (1985).

be used confidently for data acquisition. Rather, the key consideration lies in the method(s) used for data analysis.

In this chapter we focus entirely on the analysis of fluorescence intensity decay data gathered with time-correlated single photon counting instrumentation. The method of analysis considered obviously can be applied generally, but our principal interest is in its use for the study of protein fluorescence. To some extent this realization biases the description we provide, but does so only marginally, as our main issue is a consideration of the merits of the process of analysis we describe, discuss, and finally recommend.

The extraction of fluorescence decay lifetimes from time-correlated single photon counting (TCSPC) measurements is a problem with a long history, because of the difficulties involved. The issue centers on two numerically ill-conditioned problems: deconvolution with respect to the instrument response function (IRF) and parameter estimation of multiexponential functions. Favorable characteristics of the problem relate to the possibility that the level of noise can be controlled, and to the fact that one can obtain dense sampling of the fluorescence intensity decay curve.

The essential property of any method for the analysis of multicomponent decay curves determined experimentally, and contaminated by noise, is the ability to separate the individual components (separability, or resolvability). For a given level of noise, a given number of sampled data, and a given algorithm, there is a critical lifetime ratio and a critical fraction ratio beyond which the corresponding components cannot be resolved. Thus, if the lifetime ratio is smaller than the critical lifetime ratio, or if the fraction ratio is greater than the critical fraction ratio, the corresponding two components are likely to be detected by the algorithm as one component. It is also essential for the method to be accurate, that is, the values obtained for the parameters must be sufficiently close to the "true" values. In ideal circumstances the latter can be known from experiments in which the contribution of each component has been determined prior to creation of a mixture of components. More realistically, one assumes that the true values of parameters are those obtained by the analysis of measured data, and then, using these values, one can generate synthetic data. Applying the same analysis to these data, one can *a posteriori* conclude whether the method is sufficiently accurate.

In the present chapter we consider use of the maximum likelihood method for the analysis of TCSPC fluorescence intensity decay data. This method was introduced for such analysis by Hall and Selinger,[3,4] but only

[3] P. Hall and B. K. Selinger, *J. Phys. Chem.* **85,** 2941 (1981).
[4] P. Hall and B. K. Selinger, *Z. Phys. Chem. Neue Folge* **141,** 77 (1984).

recently has it been fully developed.[5] The emphasis here is on computational and numerical aspects. Section II is devoted to a detailed mathematical presentation of the method. First, we provide a description of the maximum likelihood principle (Section II,A) and the mathematical model describing TCSCP data of fluorescence intensity decay, with inclusion of the effects of zero-time shift and with correction for scattered light (Section II,B). Different approaches to the discretization of integrals involved in convolution of the instrument response function and the intensity decay model function are presented and discussed from a rather general point of view in Section II,C. The required minimization of Poisson deviance is considered in conjunction with the elimination of the background and light scattering parameters (Section II,D). To determine the number of exponentially decaying components we invoke the likelihood ratio test (Section II,E). Some general and one specific test are used to assess the goodness-of-fit to the data and the adequacy of the model employed (Section II,F). In Section II,G we describe "separability" and "detectability" indices which provide measures on how accurately a particular decay component can be detected. Finally, Section 2 is concluded with a brief summary of theoretical considerations. The reader less interested in mathematical details basically can skip the whole of Section II, except this summary, and proceed to the rest of the chapter.

The effectiveness and accuracy of the maximum likelihood method is investigated by use of simulations (Section III) and illustrated on particular measured fluorescence intensity decays of tryptophan-bearing proteins (Section IV). In Section V we briefly describe the available software. In conclusion (Section VI) we provide some recommendations on how the complex problem of deconvolution and parameter estimation of multiexponential functions appearing in fluorescence intensity decay analysis might be approached.

II. Theory

A. Maximum Likelihood Method

In this section we describe the maximum likelihood (ML) method[6–8] within the context of parameter estimation of the fluorescence decay process as measured by time-correlated single photon counting.

[5] Ž. Bajzer, T. M. Therneau, J. C. Sharp, and F. G. Prendergast, *Eur. Biophys. J.* **20,** in press (1991).

[6] Y. Bard, "Nonlinear Parameter Estimation." Academic Press, New York, 1974.

[7] P. McCullagh and J. A. Nelder, "Generalized Linear Models." Chapman & Hall, London, 1983.

[8] S. L. Meyer, "Data Analysis for Scientists and Engineers." Wiley, New York, 1975.

The observed counts $c_1, \ldots, c_n$ in n channels which represent the fluorescence intensity decay curve can be considered as n independent observations with associated probabilities $p(c_i; \boldsymbol{\theta})$, $i = 1, \ldots, n$, where $\boldsymbol{\theta} = (\theta_1, \ldots, \theta_m)$ is a parameter vector which characterizes the decay process. This vector will be determined by maximizing the joint probability,

$$L(c_1, \ldots, c_n; \boldsymbol{\theta}) = \prod_{i=1}^{n} p(c_i; \boldsymbol{\theta}) \tag{1}$$

called the likelihood function. To actually perform the maximization of the likelihood function it is necessary to assume the form of the function $p(c_i; \boldsymbol{\theta})$. This can be done in two steps. First, the observed counts c_i can be assumed to follow the Poisson distribution:

$$p(c_i; \boldsymbol{\theta}) = e^{-\langle c_i \rangle} \langle c_i \rangle^{c_i}/c_i! \tag{2}$$

where $\langle c_i \rangle$ is the expected value of the number of counts in the ith channel. Second, $\langle c_i \rangle$ is assumed to be modeled by the function $g_i(\boldsymbol{\theta})$ which incorporates our theoretical knowledge about the measured fluorescence intensity decay process:

$$\langle c_i \rangle = g_i(\boldsymbol{\theta}), \qquad i = 1, \ldots, n \tag{3}$$

The explicit expression for $g_i(\boldsymbol{\theta})$ will be presented and discussed in Section II,B. The parameter vector $\hat{\boldsymbol{\theta}}$ which maximizes the likelihood function (1) represents the "most probable parameter vector" in the sense that for $\hat{\boldsymbol{\theta}}$, the probability of observing counts $c_1, \ldots, c_n$ is the highest providing Eqs. (2) and (3) are valid. Usually $\hat{\boldsymbol{\theta}}$ is referred to as the maximum likelihood estimator of the parameter vector $\boldsymbol{\theta}$.

The log-likelihood function, $\ln L$, attains its maximum for the same value of $\boldsymbol{\theta}$ as the likelihood function, L. It is customary to determine $\hat{\boldsymbol{\theta}}$ by minimizing $-\ln L$, which, in the case of Poisson distribution, is equivalent to minimizing the Poisson deviance,[7]

$$D(\boldsymbol{\theta}) = 2 \sum_{i=1}^{n} \{c_i \ln[c_i/g_i(\boldsymbol{\theta})] - [c_i - g_i(\boldsymbol{\theta})]\} \tag{4}$$

In the least-squares (LS) approach customarily used for fluorescence decay analysis, the function which is minimized is analytically quite different:

$$\chi^2(\boldsymbol{\theta}) = \sum_{i=1}^{n} [c_i - g_i(\boldsymbol{\theta})]^2/\sigma_i^2 \tag{5}$$

This is the deviance for a Gaussian distribution of the observed counts, with a standard deviation σ_i. In the customary approach used for fluorescence decay analysis, σ_i is estimated as $c_i^{1/2}$ yielding Neyman's χ^2. Statistically more appropriate, but computationally more complex, is Pearson's χ^2, where $\sigma_i = [g_i(\boldsymbol{\theta})]^{1/2}$ This form of χ^2 is not standardly used for fluorescence decay analysis, but it has been considered by Hall and Selinger[3] and has been implemented in some software packages.

It can be shown that the $\boldsymbol{\theta}$ corresponding to the minimum of $\chi^2(\boldsymbol{\theta})$ is the maximum likelihood estimator when the Gaussian distribution of c_i is assumed. At the same time it is also the weighted least-squares estimator. When the counts c_i are large, the deviances (4) and (5) are numerically close. This is a consequence of the central limit theorem by which the Poisson distribution can be approximated by a Gaussian distribution.[8] Thus, from a theoretical point of view, the maximum likelihood estimator will differ from the least-squares estimator by the extent to which the low-count region of the fluorescence decay curve determines the values of the parameters.

The necessary condition for the minimum of the deviance (4) expressed as $\partial D(\boldsymbol{\theta})/\partial \theta_j = 0$, $j = 1, \ldots, m$, leads to the likelihood equations:

$$\sum_{i=1}^{n} \left[1 - \frac{c_i}{g_i(\boldsymbol{\theta})} \right] \frac{\partial g_i(\boldsymbol{\theta})}{\partial \theta_j} = 0 \qquad (6)$$

In the case of minimization of Gaussian deviance (5) with $\sigma_i = c_i^{1/2}$, the corresponding equations are similar:

$$\sum_{i=1}^{n} \left[1 - \frac{g_i(\boldsymbol{\theta})}{c_i} \right] \frac{\partial g_i(\boldsymbol{\theta})}{\partial \theta_j} = 0 \qquad (7)$$

and one may expect their solution to be numerically close to the solution of Eq. (6) if c_i is well approximated by $g_i(\boldsymbol{\theta})$. As discussed in our recent paper,[5] the difference between the Gaussian and Poisson deviances is of the order of $\sum_{i=1}^{n} [1 - c_i/g_i(\boldsymbol{\theta})]^3$, which is negligible for $c_i/g_i(\boldsymbol{\theta}) \cong 1$.

B. Fluorescence Decay Model

As stated in the previous section the theoretical knowledge about a fluorescence decay process and its measurement is supposed to be built into the function $g_i(\boldsymbol{\theta})$ which is standardly given by[9]

$$g_i(\boldsymbol{\theta}) = \int_{(i-1)h}^{ih} I_0(t) \, dt + b, \qquad i = 1, \ldots, n \qquad (8)$$

where h denotes channel width (time calibration) and b denotes the con-

[9] D. V. O'Connor and D. Phillips, "Time-Correlated Single Photon Counting." Academic Press, New York, 1984.

stant background counts per channel. $I_0(t)$ is the "observed" fluorescence decay function expressed as a convolution of the instrument response function, $R(t)$, and the actual fluorescence intensity decay law $I(t)$:

$$I_0(t) = \int_0^t R(u + \delta)I(t - u)\,du + \xi R(t) \tag{9}$$

$$R_i = \langle R_i \rangle + \nu_i, \qquad \langle R_i \rangle = \int_{(i-1)h}^{ih} R(t)\,dt, \qquad i = 1, \ldots, n \tag{10}$$

$$I(t) = \sum_{k=1}^N A_k\, e^{-t/\tau_k}, \qquad A_k = f_k/\tau_k, \qquad \sum_{k=1}^N f_k = f, \qquad N = (m - 3)/2 \tag{11}$$

Here ξ is the parameter determining the light scattering correction.[10] This parameter measures how much scattered exciting light is present. R_i is the observed number of counts in channel i obtained by measurement of the instrument response function.[11] The quantities $\langle R_i \rangle$ and ν_i are the corresponding expected value and noise, respectively. Taking into account Eqs. (9) and (10), expression (8) can be rewritten in a form more convenient for further considerations,

$$g_i(\boldsymbol{\theta}) = F_i + \xi\langle R_i \rangle + b, \qquad F_i = \int_{(i-1)h}^{ih} dt \int_0^t du\, R(u + \delta)I(t - u),$$

$$i = 1, \ldots, n \tag{12}$$

Here, for the decay law (11) the parameter vector $\boldsymbol{\theta}$ is given by decay lifetimes and the corresponding fractions, $\theta_{2k-1} = \tau_k$, $\theta_{2k} = f_k$, $k = 1, \ldots, N$, by the zero-time shift, $\theta_{m-2} = \delta$, by the light scatter parameter, $\theta_{m-1} = \xi$, and by the constant background per channel, $\theta_m = b$. The number of exponential components, N, is either a small number corresponding, for example, in a protein-bound fluorophore to distinct conformational states of the fluorophore, or may be arbitrarily large. In the latter case, A_k not only could describe a quasi-continuous distribution of lifetimes which may correspond to a distribution of states, but it also may represent a mathematical approximation of a nonexponential decay.[12] When N is large the lifetimes are usually given by a fixed sequence (e.g.,

[10] I. Isenberg, *J. Chem. Phys.* **59**, 5696 (1973).

[11] It is assumed that in R_i the constant background is already subtracted. This background contribution, which is small, is usually determined as an average dark count per channel in the same way as the background parameter in the model function (8) (see Ref. 9). Such a method of background determination agrees with the maximum likelihood approach as will be shown in Section II,D.

[12] A. Siemiarczuk, B. D. Wagner, and W. R. Ware, *J. Phys. Chem.* **94**, 1661 (1990).

in the exponential series method[12] and in the maximum entropy method[13]), and only the amplitudes A_k constitute the adjustable parameter vector $\boldsymbol{\theta}$.

C. Discretization of Integrals

The model function (12) is still given in an implicit form with respect to $\boldsymbol{\theta}$, and this is not practical for actual minimization of the Poisson deviance (4). We devote this section to the numerically and computationally important question of how to evaluate the integrals involved in Eqs. (10) and (12). In the most of the literature on fluorescence data analysis, this question has not been treated with particular care, and so we provide here a more complete and general presentation. The important aspect of the problem is to obtain a formula which is sufficiently accurate, but at the same time can be rapidly computed. The basic approach to evaluation of the integrals will be valid for any form of decay law, $I(t)$, but we specifically consider the commonly used multiexponential form (11) and shall present three different computationally convenient expressions developed in the literature.

First we approximate the integral F_i, using the first mean value theorem of integral calculus[14]:

$$\int_a^{a+h} f(x)\, dx = hf(a + \eta), \qquad 0 \le \eta \le h \tag{13}$$

where $f(x)$ is a bounded, continuous function in the interval $[a, a + h]$. If the function $f(x)$ is linear in this interval, then $\eta = h/2$. Now, from Eqs. (12) and (13) it follows, for $\delta = 0$,

$$F_i = h \int_0^{(i-1)h+\eta_i} \psi(u)\, du = h \sum_{j=1}^{i-1} \int_{(j-1)h}^{jh} \psi(u)\, du + h \int_{(-1)h}^{(i-1)h+\eta_i} \psi(u)\, du \tag{14}$$

$$\psi(u) = R(u)I[(i - 1)h + \eta_i - u] \tag{15}$$

Here, according to the mentioned theorem

$$0 \le \eta_i \le h, \qquad i = 1, \ldots, n \tag{16}$$

By systematic application of this theorem to the integrals of (14) one obtains

[13] A. K. Livesey and J. C. Brochon, *Biophys. J.* **52**, 693 (1987).
[14] I. N. Bronshtein and K. A. Semendyayev, "Handbook of Mathematics." Van Nostrand-Reinhold, New York, 1985.

$$F_i = h^2 \sum_{j=1}^{i-1} R[(j-1)h + \eta'_j] I[(i-j)h + \eta_i - \eta'_j]$$

$$+ h\eta_i R[(i-1)h + \eta''_i] I(\eta_i - \eta''_i), \qquad 0 \le \eta'_j \le h, \qquad 0 \le \eta''_i \le \eta_i \quad (17)$$

This expression is exact, but it is of little practical importance unless the unknown quantities η_i, η'_j, and η''_i are not specified. By choosing

$$\eta_i = \eta'_j = \eta''_i = h/2, \qquad i = 1, \ldots, n; \qquad j = 1, \ldots, i-1 \quad (18)$$

it is implied that the integrands in Eqs. (12) and (14) are approximated by linear functions within the intervals of integrations. Such piecewise linear approximation ultimately leads to the common formula for F_i obtained by Grinvald and Steinberg.[15] Thus, inserting Eq. (18) into Eq. (17) and implying piecewise linear approximation of $R(t)$ [cf. Eqs. (10) and (13)],

$$\langle R_i \rangle = \int_{(i-1)h}^{ih} R(t)\, dt \cong hR[(i-1)h + h/2] \quad (19)$$

F_i is simplified and expressed in terms of $\langle R_j \rangle$, $j = 1, \ldots, i$, which subsequently have to be replaced by the measured quantities R_j to yield

$$F_i = h \left\{ \sum_{j=1}^{i-1} R_j I[(i-j)h] + I(0)R_i/2 \right\} + \varepsilon_i \quad (20)$$

Here ε_i is the combined approximation error arising from: (a) the replacement of $\langle R_j \rangle$ by R_j, (b) the piecewise linear approximation [Eq. (18)], and (c) the approximation in Eq. (19). This formula is obviously valid for any decay model function $I(t)$.

The recursive formula of Grinvald and Steinberg for F_i can be obtained from Eq. (20) by specifying $I(t)$ as a multiexponential function [Eq. (11)] and introducing the notation

$$F_i = \sum_{k=1}^{N} F_i^k + \varepsilon_i, \qquad F_i^k = hA_k \sum_{j=1}^{i-1} e^{-(i-j)h/\tau_k} R_j + hA_k R_i/2 \quad (21)$$

Then, it follows

$$F_i^k = (F_{i-1}^k + hA_k R_{i-1}/2) e^{-h/\tau_k} + hA_k R_i/2, \qquad F_1^k = hA_k R_1/2 \quad (22)$$

This simple formula is usually implemented in codes for fluorescence data analysis by the least-squares method.

To obtain the recursive formula of McKinnon et al.[16] one has to start from Eq. (14) choosing

[15] A. Grinvald and I. Z. Steinberg, Anal. Biochem. **59**, 583 (1974).
[16] A. E. McKinnon, A. G. Szabo, and D. R. Miller, J. Phys. Chem. **81**, 1564 (1977); note that numerical approaches are described in the supplementary microfilm material.

$$\eta_i = 0 \tag{23}$$

and assuming piecewise linear approximation of $R(t)$:

$$R(t) = a_i t + b_i, \qquad (i - 1)h \le t \le ih \tag{24}$$

$$a_i = (\rho_{i+1} - \rho_i)/h, \qquad b_i = \rho_{i+1} - a_i ih, \qquad \rho_i \equiv R[(i - 1)h] \tag{25}$$

The choice of Eq. (23) is equivalent to piecewise constant approximation of the integrand. With the assumption of multiexponential decay, the integrations in Eq. (14) can be performed analytically and after straightforward algebra yield the following relation for F_i:

$$F_i = \sum_{k=1}^{N} F_i^k + \varepsilon_i \tag{26}$$

$$F_i^k = F_{i-1}^k e^{-h/\tau_k} + hA_k\tau_k[\rho_i - \rho_{i-1} e^{-h/\tau_k} - (\rho_i - \rho_{i-1})(1 - e^{-h/\tau_k})\tau_k/h] \tag{27}$$

where ε_i is an approximation error. In order to relate ρ_i to the measured counts R_i, at this point, a piecewise constant approximation of $R(t)$ is assumed, yielding

$$\langle R_i \rangle \cong hR[(i - 1)h] = h\rho_i \tag{28}$$

Replacing $\langle R_i \rangle$ by R_i as before, Eq. (27) leads to the formula of McKinnon *et al.*:

$$F_i^k = F_{i-1}^k e^{-h/\tau_k} + A_k\tau_k[R_i - R_{i-1} e^{-h/\tau_k} - (R_i - R_{i-1})(1 - e^{-h/\tau_k})\tau_k/h], \qquad F_1^k = 0 \tag{29}$$

Better approximation for F_1^k can be obtained by performing analytical integration in Eq. (12) with $\delta = 0$, $R(u) = u\rho_2/h$, $\rho_2 \equiv R(h) = 2R_1/h$. This results in

$$F_1^k = A_k R_1[\tau_k - 2\tau_k^2/h + 2(1 - e^{-h/\tau_k})\tau_k^3/h^2] \tag{30}$$

The approximation error ε_i is a combination of errors arising from (a) the replacement of $\langle R_i \rangle$ by R_i, (b) the choice of Eq. (23), (c) the piecewise linear approximation of $R(t)$ [Eq. (24)], and (d) the piecewise constant approximation $R(t)$ [Eq. (28)].

From the above derivations it is clear that in this formula piecewise constant and piecewise linear approximations are used interchangeably without consistent justification. On the contrary, in deriving the Grinvald–Steinberg formula [Eq. (22)], piecewise linear approximation *is* consistently used. Nevertheless, according to McKinnon *et al.*,[16] the analytical integration implied in their approach yields more accurate approximations when the lifetimes τ_k are of the order of h. This was recently apparently confirmed by Periasamy,[17] who performed numerous simulations to compare the numerical accuracy of discretization formulas (22) and (29) within the least-squares method. However, it is important to note that in these latter simulations the counts of the instrument response function, R_i, were inappropriately simulated with $R[(i - 1)h]$ [cf. Eq. (10)]. This means that the effects of the approximation given in Eq. (29) were not included in this study. It is not clear *a priori* what would be the result of such studies had these effects been included.

When comparing two of the formulas [Eqs. (22) and (29)] from a computational point of view, it is obvious that for a given i formula (29) requires two more additions and one more multiplication than does formula (22). Additional computation of i independent factors is negligible because i usually loops from 1 to 512 and k at most from 1 to 5. (In the case of quasi-continuous distributions of lifetimes this is not so because the number of components may be large.) Periasamy[17] indicated an average increase of approximately 10% of total computational time for estimating lifetimes and fractions by the least-squares method if formula (29) is used instead of Eq. (22).

Recently, we have developed another approach to the evaluation of the integrals in Eqs. (8)–(10). In our scheme we avoid the approximation of the integral (8), which in the approach of Grinvald and Steinberg is performed by choosing $\eta_i = h/2$ and in the approach of McKinnon *et al.* is accomplished by choosing $\eta_i = 0$. We also avoid approximations of the type in Eqs. (19) and (28). We approximate only one integral, and all others are evaluated analytically. Here we briefly indicate the main ideas of our scheme; more detailed derivations can be found elsewhere.[5]

For the multiexponential model given by Eq. (11), expression (12) for F_i (with $\delta = 0$) becomes

$$F_i = \sum_{k=1}^{N} A_k \int_{(i-1)h}^{ih} e^{-t/\tau_k} \int_0^t R(u) \, e^{u/\tau_k} \, du \, dt \tag{31}$$

Performing the partial integration with respect to t, this expression reduces to a form which involves only the integral

$$J_{ik} = \int_{(i-1)h}^{ih} R(u) \, e^{u/\tau_k} \, du \tag{32}$$

This integral is now approximated by using the generalized first mean value theorem of integral calculus which states

[17] N. Periasamy, *Biophys. J.* **54**, 961 (1988).

$$\int_a^{a+h} f(x)g(x)\,dx = f(a + \eta) \int_a^{a+h} g(x)\,dx, \qquad 0 \le \eta \le h \qquad (33)$$

where $f(x)$ is a bounded, continuous function in the interval $[a, a + h]$, and $g(x)$ is a function everywhere positive (or negative) and integrable in the same interval. Thus, taking into account Eq. (10) the integral J_{ik} can be expressed in terms of $\langle R_i \rangle$:

$$J_{ik} = e^{[(i-1)h + \eta_i]/\tau_k} \langle R_i \rangle, \qquad 0 \le \eta_i \le h \qquad (34)$$

The value of η_i is now chosen according to the minimax principle for the unknown factor $\exp(\eta_i/\tau_k)$, namely, we choose such a value of η_i that the largest possible error in approximating η_i is minimized. This is

$$\eta_i = h/2, \qquad i = 1, \ldots, n \qquad (35)$$

After this approximation, and assuming the form (26) for F_i, the straightforward algebra leads to the following expression for F_i^k:

$$F_i^k = A_k \tau_k \left[B_k \sum_{j=1}^{i-1} e^{-(i-j)h/\tau_k} R_j + B_k' R_i \right],$$

$$B_k = 2 \sinh(h/2\tau_k), \qquad B_k' = 1 - e^{-h/(2\tau_k)} \qquad (36)$$

For $h/\tau_k \ll 1$ this expression reduces to formula (23) of Grinvald and Steinberg $[B_k \cong h/\tau_k,\ B_k' \cong h/(2\tau_k)]$. A computationally convenient recursive form of Eq. (36),

$$F_i^k = F_{i-1}^k e^{-h/\tau_k} + A_k \tau_k B_k' (R_{i-1} e^{-h/(2\tau_k)} + R_i), \qquad F_1^k = A_k \tau_k B_k' R_1 \qquad (37)$$

requires the same number of additions and multiplications with respect to i dependent terms as the standard Grinvald–Steinberg formula. Elsewhere[5] we have also derived an explicit expression for F_i when $\delta \ne 0$. For this case $R_j, j = 1, \ldots, i$ in formulas (36) and (37) have to be replaced by

$$P_j = R_{j+1} + \varepsilon(R_{j+l+1} - R_{j+l})/h, \qquad \delta = lh + \varepsilon \qquad (38)$$

where $-h/2 \le \varepsilon < h/2$. This is, of course, an approximation which includes only the linear term in ε.

Comparing now the three presented discretization formulas [Eqs. (22), (29), and (37)] and underlying approximations, one notes that the last one seems to be most advantageous from a theoretical and computational point of view. It is based on only one approximation step in which the maximal possible approximation error is minimized. Other formulas include several approximation steps, and the largest possible approximation error cannot then be easily estimated. Computationally Eq. (37) is practically equivalent to the simplest formula, Eq. (22). From a numerical point of view, it is not clear which of these formulas is the most accurate. For $h/\tau_k \ll 1$ one can reasonably expect that all three approaches will yield very similar results. However, in real applications the case $h/\tau_k \cong 1$ can also appear. We have performed preliminary numerical studies of the three discretization schemes within the maximum likelihood approach. The results indicated[5] that the application of formula (37) is a good compromise, yielding equal or better accuracy than the Grinvald–Steinberg formula Eq. (22) or the formula of McKinnon *et al.* for $h/\tau_k \gg 1$ [Eq. (29)]. In the case of $h/\tau_k \cong 1$, formula (37) performs better than formula (22) but not as good as Eq. (29). However, the formula of McKinnon *et al.* is less accurate than both formulas (37) and (22) when information about parameters decreases (smaller lifetime ratio). The results of our numerical studies related to formula (29) were obtained using a more accurate approximation of F_1^k given by Eq. (30).

Our numerical studies of the three discretization schemes are still preliminary, and we intend to perform more complete investigation which would also include nonnegligible zero-time shift and the light scatter effects.

D. Minimization of Poisson Deviance

The difficulties inherent to numerical minimization of the nonlinear function of several variables rapidly increase with the number of variables. Therefore, it is beneficial when the specific form of the function allows the number of variables to be reduced by analytical procedures. In the case of the Poisson deviance (4), this can be done if the model function $g_i(\boldsymbol{\theta})$ depends linearly on some of the components of the parameter vector $\boldsymbol{\theta}$. As can be seen from the preceding two sections, the model function for the fluorescence decay process is linear in parameters ξ and b, and also in fractions f_k if the multiexponential decay law in Eq. (11) is used. Indeed, from Eqs. (8)–(12) and (26) it follows

$$g_i(\boldsymbol{\theta}) = \sum_{k=1}^{N} f_k H_i(\tau_k) + \xi R_i + b + \varepsilon_i, \qquad H_i(\tau_k) = F_i^k/(A_k\tau_k) \quad (39)$$

where F_i^k and ε_i depend on the discretization scheme employed, as discussed in the preceding section. We first eliminate the constant background level b using the common measurement procedure in which n_1 channels (usually between 10 and 20) are separately used for the background counts.[9] The model function is then $g_i(\boldsymbol{\theta}) = b$, $i = 1, \ldots, n_1$, and the likelihood Eq. (6) simplifies to

$$\sum_{i=1}^{n_1} (1 - c_i/b) = 0 \qquad (40)$$

yielding for the maximum likelihood estimator, $\hat{b}$, the known result

$$\hat{b} = \frac{1}{n_1} \sum_{i=1}^{n_1} c_i \qquad (41)$$

Next, we eliminate the light scatter parameter ξ. To do so we use Eq. (39) and express the likelihood Eqs. (6) for parameters f_k, ξ, and b in the form

$$\sum_{i=1}^{n} H_i(\tau_k) = \sum_{i=1}^{n} c_i H_i(\tau_k)/g_i(\boldsymbol{\theta}) \qquad (42)$$

$$\sum_{i=1}^{n} R_i = \sum_{i=1}^{n} c_i R_i/g_i(\boldsymbol{\theta}) \qquad (43)$$

$$n = \sum_{i=1}^{n} c_i/g_i(\boldsymbol{\theta}) \qquad (44)$$

Simple summation of Eqs. (42)–(45) multiplied by appropriate factors eliminates $g_i(\boldsymbol{\theta})$ on the right-hand sides of these equations, yielding (the approximation error ε_i is not taken into account here)

$$\sum_{k=1}^{N} f_k \sum_{i=1}^{n} H_i(\tau_k) + \xi \sum_{i=1}^{n} R_i + bn = \sum_{i=1}^{n} c_i \qquad (45)$$

From here we can express ξ as a function of the other parameters:

$$\xi = \sum_{i=1}^{n} \left[c_i - \sum_{k=1}^{N} f_k H_i(\tau_k) - b \right] \Big/ \sum_{j=1}^{n} R_j \qquad (46)$$

and insert it into the model function (39), which then becomes a function only of fractions, lifetimes, and zero-time shift, if the background level b is fixed by its maximum likelihood estimate $\hat{b}$. Furthermore, the insertion of $g_i(\boldsymbol{\theta})$ with ξ given by Eq. (46) also simplifies the Poisson deviance to the form

$$D(\boldsymbol{\theta}) = 2 \sum_{i=1}^{n} c_i \ln \frac{c_i}{g_i(\boldsymbol{\theta})} \qquad (47)$$

From the above considerations it follows that elimination of ξ required parameter b to be estimated by simultaneous minimization of the Poisson deviance with $g_i(\boldsymbol{\theta}) = b$ and of the Poisson deviance with $g_i(\boldsymbol{\theta})$ given by

Eq. (39).[18] In this way we have reduced the problem of minimization of the Poisson deviance (4) with $2N + 3$ variables to the minimization of a nonlinear function [Eq. (48)] with $2N + 1$ variables (N being the number of exponential components).

In the very simplified model of Hall and Selinger,[4] where the effects of the instrument response function, as well as the light scatter correction and the background level, are neglected, it is possible to eliminate two fractions out of $2N$ parameters. Here we briefly present this specific case in a compact form. The model function is now simplified to

$$g_i(\boldsymbol{\theta}) = \int_{(i-1)h}^{ih} I(t) \, dt = \sum_{i=1}^{n} f_k H_i^s(\tau_k) \tag{48}$$

$$H_i^s(\tau_k) = e^{-(i-1)h/\tau_k}(1 - e^{-h/\tau_k}) \tag{49}$$

and the likelihood equations become

$$\sum_{i=1}^{n} \left[1 - \frac{c_i}{g_i(\boldsymbol{\theta})} \right] H_i^s(\tau_k) = 0 \tag{50}$$

$$\sum_{i=1}^{n} \left[1 - \frac{c_i}{g_i(\boldsymbol{\theta})} \right] f_k \frac{\partial H_i^s(\tau_k)}{\partial \tau_k} = 0 \tag{51}$$

The derivatives of H_i^s satisfy a special relation

$$\frac{\partial H_i^s(\tau_k)}{\partial \tau_k} = h H_i^s(\tau_k)[i - (1 - e^{-h/\tau_k})^{-1}] \tag{52}$$

Multiplying Eq. (50) by f_k, and summing with respect to $k = 1, \ldots, N$, one obtains

$$\sum_{k=1}^{N} f_k \sum_{i=1}^{n} H_i^s(\tau_k) = \sum_{i=1}^{n} c_i \tag{53}$$

In a similar way, multiplying Eq. (51) by f_k, summing with respect to k, and using Eq. (52) one obtains

$$\sum_{k=1}^{N} f_k \sum_{i=1}^{n} i H_i^s(\tau_k) = \sum_{i=1}^{n} i c_i \tag{54}$$

Now, Eqs. (53) and (54) represent a linear system of equations for fractions

[18] This requirement is reasonable as long as we can assume that the background level does not change during measurements of actual fluorescence intensity decay.

f_k, so that two of fractions can be expressed in terms of other fractions and lifetimes. Thus, the number of parameters in the Poisson deviance for this simplified model is reduced to $2N - 2$.

In their first paper on the maximum likelihood method Hall and Selinger[3] discussed the one-component case ($N = 1$) whereby from Eq. (53) it follows immediately that f_1 can be expressed as a function of τ_1 and the minimization of the Poisson deviance is reduced to solving a nonlinear equation

$$\sum_{i=1}^{n} [1 - c_i H(\tau_1)/c] H_i^s(\tau_1)[i - (1 - e^{-h/\tau_1})^{-1}] = 0,$$

$$H(\tau_1) = \sum_{i=1}^{n} H_i^s(\tau_1), \qquad c = \sum_{i=1}^{n} c_i \qquad (55)$$

which can be further simplified.[3] In their second paper[4] Hall and Selinger considered the Spencer–Weber model for time-resolved anisotropy measurements which can be expressed as a multiexponential function with two components. In this case two fractions can be eliminated as indicated above, and the Poisson deviance is only a function of the two lifetimes. However, some of the statements in that paper[4] may be misleading because of an implicit allusion that for any number of components all fractions can be eliminated by likelihood equations, and that minimization of Poisson deviance can be reduced to minimization with respect to only N lifetimes variables. Unfortunately this is not true as can be seen by careful inspection of the likelihood equations for this model.

Another comment related to the second paper of Hall and Selinger[4] is also appropriate here. In their introduction they state that finding the least-squares estimator for a two-exponential model cannot be reduced to a two-parameter problem, in contrast to finding the maximum likelihood estimator where this reduction can be done. This statement is true as long as Pearson's χ^2 is used for the least-squares estimator. However, more often than not, Neyman's χ^2 is used and then the elimination of *all* fractions can be performed, even for the complex model discussed in the beginning of this section. Such an approach has been considered only recently,[19] although the mathematical basis for it has been known for a long time.[20]

The rest of this section we devote to some numerical aspects of the minimization of the deviance, Eq. (47). In the process of numerical minimization one must pay attention to the admissible values of parame-

[19] A. L. Wong and J. M. Harris, *Anal. Chem.* **61**, 2310 (1989).
[20] M. R. Osborne, *SIAM J. Numer. Anal.* **12**, 571 (1975).

ters. Thus, the fractions and the lifetimes in the multiexponential fluorescence decay model usually should be positive. To assure their positivity, minimization actually can be performed with respect to new variables, u_j, which can assume any real value. We define these variables conveniently (but otherwise mostly arbitrarily) as $f_k = u_{2k}^2$, $\tau_k = |u_{2k-1}|$, $\delta = u_{2k+1}$. The light scatter parameter given by Eq. (46) should, by its physical meaning, assume positive values or zero. If at the minimum it assumes a negative value, this fit may be rejected and the given model considered as an inadequate description of the data. However, a small negative ξ ($|\xi| < 0.01$) may be attributed to noise in the data and to the discretization error.

It is well known that nonlinear minimization is a difficult numerical problem, and, together with theoretical knowledge about different algorithms, a fair amount of experience within a particular application is needed to effect successful minimization. We have used the SIMPLEX algorithm of Nelder and Mead[21] for minimization of the Poisson deviance, in particular an implementation from *Numerical Recipes*.[22] This algorithm utilizes an iterative procedure which moves a simplex,[23] defined by $m +$ 1 function values as vertices, toward the region of the function minimum. Only the function evaluations are necessary, which is generally the main advantage of the SIMPLEX algorithm over other algorithms (Newton, Levenberg–Marquardt) in which the evaluation of function derivatives is also required.

Usually the iterations are stopped when the simplex is confined within a region determined by the prescribed tolerance in function values. This means that the function values determining the simplex must not differ more than tolerance allows. However, in applications where the Poisson or Gaussian deviance with model function (39) must be minimized, sometimes the minimum is surrounded by a very shallow m-dimensional valley and the volume of the final simplex is large (e.g., in the two-dimensional case the corresponding triangle has large area). This means that numerically the minimum is not well defined if only the tolerance criterion for function values is used. For this reason we have also introduced a tolerance criterion for parameter vector values, which means that in this scheme the final simplex is shrunken to a "point," defining the minimum within given tolerances.

[21] J. A. Nelder and R. Mead, *Comp. J.* **7**, 308 (1965).

[22] W. H. Press, B. P. Flannery, S. A. Teukolsky, and W. T. Vetterling, "Numerical Recipes." Cambridge Univ. Press, Cambridge, 1986.

[23] The simplex is a geometric figure having plane faces and $m + 1$ vertices in m-dimensional space; it is a triangle in two-dimensional space and a tetrahedron in three-dimensional space.

E. Determination of Number of Decay Components by Likelihood
   Ratio Test

For a given measured fluorescence intensity decay profile a priori one
does not know the number (N) of multiexponential components present.
To obtain this information we could use some other method of analysis
which intrinsically provides this number (e.g., the generalized Padé–Laplace method,[24] but we can also rely on the likelihood ratio test.[25] This
test is based on the ratio of likelihood functions at the maximum for two
corresponding nested models (e.g., two- and one-component models). The
logarithm of this ratio equals the difference of related deviances, and the
following applies:

Assume that a sequence of models with 3, 5, 7, . . . parameters has
been fit to a given data set and that D_3, D_5, D_7, . . . are the resulting
deviances. If the true model has r parameters then $D_r - D_{r+2}$, the apparent
improvement in fit, will be distributed as a χ^2 distribution on 2 degrees of
freedom, χ_2^2. The difference $D_{r-2} - D_r$ will be distributed as a random
variable which is stochastically larger than a χ_2^2; how much larger depends
both on the lack of fit and the amount of information.

The practical test of whether a particular fluorescence decay profile can be described by $N - 1$ or N components may then be formulated as follows: First, an arbitrary cut point is chosen, for example,
$Pr(X > 9.2) = 0.01$ when X is chi-squared on 2 degrees of freedom. Then,
if for the difference of deviances the inequality

$$\Delta_N \equiv D_{2N-1} - D_{2N+1} \geq 9.2 \tag{56}$$

is valid, the model with $N - 1$ components is rejected as unacceptable.
The probability that a model with $N - 1$ components will be erroneously
rejected, namely, that too many components are chosen, is then 0.01.

By using simulations we have investigated[5] how the difference of devi-
ances varies with the lifetime ratio $r = \tau_2/\tau_1 = \tau_3/\tau_2 = \tau_4/\tau_3$ for 2, 3, and
4 exponential components with equal fractions. The following facts were
revealed: (1) Δ_N is a very steep function of the lifetime ratio (approximately
the power law); (2) the more components present the less steep is the
function; and (3) the threshold of the lifetime ratio for detection of the
exact number of components increases with the number of components
(it is approximately 1.1 for two components, 1.5 for three components,
and 2.1 for four components in case of 512 data points with 40,000 counts

[24] Ž. Bajzer, J. C. Sharp, S. S. Sedarous, and F. G. Prendergast, Eur. Biophys. J. **18,** 101
   (1990).
[25] M. G. Kendall and A. Stuart, "The Advanced Theory of Statistics," Vol. 2, Hafner, New
   York, 1961.

in a peak of the fluorescence intensity decay function, with channel width of 0.025 nsec and $\tau_1 = 1$ nsec). The latter two findings reflect the known fact that the problem of fitting data with multiexponential functions becomes increasingly more difficult as the number of exponentials increases.

An important practical question now arises, namely, when should one stop adding a new component, if the above criterion force us to go beyond, say, $N = 5$. Obviously, that depends on the lifetime ratios, the signal-to-noise ratio, and the number of data points involved. In general, going beyond $N = 5$ is not recommendable, because the problem is likely to be numerically too ill-conditioned. Consequently the estimated values of parameters become unreasonably dependent on the initial values of parameters used to start the minimization. Then, a better approach would be to employ a method of analysis which offers the possibility of detecting quasi-continuous lifetime distributions.[12,13] The most simple of such approaches is the exponential series method[12] in which many lifetime values (of the order of 100) over a certain range are given, and minimization of the Gaussian deviance is performed only with respect to fractions. The latter would then define a quasi-continuous distribution of lifetimes sufficient to describe the intensity decay. By using such an approach one can, at least in principle, rule out the possibility that a genuine lifetime distribution has been "replaced" by a relatively small number of distinct exponential decay components which appear to fit the data "well."

F. Goodness-of-Fit and Adequacy of Model

We have shown how the number of exponential decay components can be determined by the likelihood ratio test. By this procedure we actually choose the most adequate of a series of nested models. However, the question remains whether such a model is adequate; for example, are there any effects supported by the data which are not included in the model? In other words, is there enough statistical evidence that a given model describes the data? And, therefore, is the interpretation of parameters is reliable? Answers to these questions are based on goodness-of-fit criteria which are described in detail in the chapter by Straume and Johnson[26] in this volume. Here we shall discuss some of the most commonly employed statistical criteria, and we introduce a criterion for the adequacy of the model which is based on specific properties of the multiexponential model being considered.

Once the minimization of deviance is performed we know the value of the deviance at the minimum. From a statistical point of view, the minimal

[26] M. Straume and M. L. Johnson, this volume [5].

deviance should be a random variable distributed as a χ^2 distribution with $\nu = n - m$ degrees of freedom. Therefore, its mean value should be equal to ν, which implies that $D(\hat{\boldsymbol{\theta}})/\nu \cong 1$. This is the most commonly employed goodness-of-fit criterion. A more useful test is based on the standard normal variate defined as

$$Z = [D(\hat{\boldsymbol{\theta}}) - \nu]/(2\nu)^{1/2} \tag{57}$$

Catterall and Duddell[27] discussed the application of Z within the framework of fluorescence intensity decay analysis and recommended that acceptable values of Z should lie between -3 and $+3$. It is also useful to calculate the probability, Q, that the obtained deviance occurred by chance:

$$Q = \Gamma(\nu/2, D/2)/\Gamma(\nu/2), \qquad \Gamma(\alpha, x) \equiv \int_x^\infty e^{-t} t^{a-1} \, dt, \qquad D \equiv D(\hat{\boldsymbol{\theta}}) \tag{58}$$

According to Press et al.,[22] if Q is larger than 0.1 one can believe that the model describes the data properly; for Q larger than 0.001 the fit may still be considered acceptable if one may assume that the errors in data are underestimated, possibly owing to the presence of a minor nonrandom error component.

Nonrandom (systematic) errors can be detected by inspection of standard Pearson's residuals defined as[7]

$$r_i = [c_i - g_i(\hat{\boldsymbol{\theta}})/[g_i(\hat{\boldsymbol{\theta}})]^{1/2} \tag{59}$$

Visual inspection of the plot of residuals may reveal larger nonrandom deviations. More objective indication of nonrandomness can be obtained by statistical tests of randomness such as the runs test, the difference sign test for linear trend, the rank correlation test, Noether's cyclic test for periodicity of residuals, and some others, all systematically described by Catterall and Duddell.[28] We found that the most sensitive is the runs test for signs of residuals which are supposed to occur in random order. The standardized statistic z is defined as follows: A run is a sequence of residuals with the same sign. The number of runs, r, in the set of n residuals is counted and compared with the number expected, μ_r, for a set of random numbers:

$$\mu_r = 1 + 2n_p n_n/n_s, \qquad n_s = n_n + n_p \tag{60}$$

[27] T. Catterall and J. Duddell, in "Deconvolution and Reconvolution of Analytical Signals" (M. Bouchy, ed.), p. 445. Ecole Nationale Supérieure des Industries Chimique de l'Institute National Polytechnique de Lorraine, Nancy, France, 1982.

[28] T. Catterall and J. Duddell, in "Time-Resolved Fluorescence Spectroscopy in Biochemistry and Biology" (R. B. Cundall and R. E. Dale, eds.), p. 173. Plenum, New York and London, 1983.

where n_p and n_n are the numbers of positive and negative residuals in the set, respectively. The statistic z is given by

$$z = (r - \mu_r)/\sigma_r, \qquad \sigma_r = \left[\frac{2n_p n_n(2n_p n_n - n_s)}{n_s^2(n_s - 1)}\right]^{1/2} \qquad (61)$$

and is normally distributed. Thus, the ideal value of z should be zero, and values between -3 and 3 may be acceptable. Higher or lower values indicate that nonrandomness is very probable.

Now, after discussing general and common goodness-of-fit criteria, we introduce an index which reflects the specific features of our model and which is useful in assessing how adequately the model describes the data. By applying the Laplace transformation to Eq. (9) and using Eq. (11), one can show that

$$\int_0^\infty I_o(t) \, dt = f \int_0^\infty R(u + \delta) \, du + \xi \int_0^\infty R(u) \, du \qquad (62)$$

The integral on the left-hand side should closely approximate the total number of counts c, and the integrals on the right-hand side are closely related to the total number of counts, R, in the IRF:

$$c = \sum_{i=1}^n c_i, \qquad R = \sum_{i=1}^n R_i \qquad (63)$$

If δ and ξ in Eq. (62) are neglected, then the value of the sum of fractions f can be approximately estimated by using the measured quantities c_i and r_i:

$$f \cong f_{exp} \equiv c/R \qquad (64)$$

Using this we can normalize the fractions and minimize the deviance with respect to f_k/f_{exp}. If the model correctly describes the data, the sum of the estimated fractions, $\sum_{k=1}^N \hat{f}_k/f_{exp}$, should be approximately equal to one. More precisely, starting from Eq. (62) one can show by relatively straightforward calculation that

$$\phi \equiv 1 - \sum_{k=1}^N \frac{\hat{f}_k}{f_{exp}} = \frac{\rho}{R - \rho} - \frac{RI_{tail}}{(R - \rho)c} + \frac{R\Delta f}{c} + \frac{(v - 2R)\Delta g + bn}{(R - \rho)c} \qquad (65)$$

where

$$\Delta f = \sum_{k=1}^N (\hat{f}_k - f_k), \qquad \Delta g = \sum_{i=1}^n [g_i(\hat{\theta}) - c_i], \qquad v = \sum_{i=1}^n v_i \qquad (66)$$

By definition $\rho = 0$ for $\delta \leq 0$, and

$$\rho = \int_0^\delta R(t)\, dt, \qquad \delta > 0 \tag{67}$$

Because in practice the zero time shift is small, ρ is very small compared to R, so that the first term of expression (65) will always be small if not zero. The second term can be relatively large if

$$I_{tail} = \int_{nh}^\infty I_0(t)\, dt \tag{68}$$

is large. The third term specifically depends on how well we have estimated the fractions. The main contribution of the last term comes from Δg which should be small if the model function $g_i(\boldsymbol{\theta})$ describes the data c_i well.

In general, we found by simulations that the index ϕ is of the order of 0.01 or -0.01 if enough of the tail of the fluorescence intensity decay curve was taken into account (nh is approximately equal to or greater than $3\tau_{max}$, where τ_{max} is the maximal lifetime). When ϕ is negative and of the order of -0.1, this represents a warning that a large part of the tail of the decay profile is neglected and/or the model is inadequate. The simulations based on estimated parameters and measured IRF can then reveal whether such a ϕ value is to be expected as a consequence of the too "short" tail or is also a consequence of the model inadequacy. As we shall see in Section III, if the tail is too short the estimated parameters can be very inaccurate, depending on the lifetime ratio. If ϕ is positive and of the order of or is larger than the smallest fraction detected, this is a clear sign that the model is inadequate, and/or the estimates of the fractions are considerably inaccurate. Such a conclusion is consequence of expression (65) and the fact that $I_{tail} \geq 0$.

A particular inadequacy of the model may be expected if the IRF changes its shape during the fluorescence measurement. This is known as the excitation light drift.[10] Such changes are usually rather slow, but their effect may still be detectable. Let us assume that the IRF is generally represented by

$$R(t',t) = R(t') + U(t',t) \tag{69}$$

where $U(t',t)$ is the correction at the time of measurement t, and $U(t',0) = 0$ so that $R(t',0) = R(t')$ is the IRF we usually measure. The "observed" corrected fluorescence decay function $I_0^c(t)$ is then given by

$$I_0^c(t) = I_0(t) + \int_0^t U(t',t) I(t - t')\, dt' + \xi U(t,t) \tag{70}$$

where $I_0(t)$ is defined in Eq. (9). The two correction terms on the right-hand side of Eq. (70) will introduce two more contributions to the expression (65) for the index ϕ [cf. Eq. (62) with I_o replaced by I_o^c]. Thus, the substantial inadequacy of the model owing to the excitation light drift generally will be detected by the index ϕ.

G. Properties of Maximum Likelihood Estimator and Information Matrix

The important aspect of parameter estimation is how "close" the maximum likelihood estimators are to the truth. When applied to fluorescence decay analysis, we may ask how accurate are the estimated lifetimes and fractions? The properties of the maximum likelihood estimator related to these aspects are summarized in basic statistical texts.[8,25,29]

The uncertainty in the maximum likelihood estimator of $\boldsymbol{\theta}$ is related to the concept of information about the parameter vector $\boldsymbol{\theta}$. For a given model function $g_i(\boldsymbol{\theta})$, information about $\boldsymbol{\theta}$ contributed by the ith number of counts c_i (with an error distributed according to the Poisson distribution) is the matrix $\mathbf{I}_i(\boldsymbol{\theta})$ defined by the following elements:

$$I_{jk} = \left\langle \frac{\partial}{\partial \theta_j} \ln p(c_i;\boldsymbol{\theta}) \frac{\partial}{\partial \theta_k} \ln p(c_i;\boldsymbol{\theta}) \right\rangle = \frac{1}{g_i(\boldsymbol{\theta})} \frac{\partial g_i(\boldsymbol{\theta})}{\partial \theta_j} \frac{\partial g_i(\boldsymbol{\theta})}{\partial \theta_k} \tag{71}$$

The total information about $\boldsymbol{\theta}$ contributed by the set of counts $\{c_1, \ldots, c_n\}$ is

$$\mathbf{I}(\boldsymbol{\theta}) = \sum_{i=1}^{n} \mathbf{I}_i(\boldsymbol{\theta}) \tag{72}$$

For any nonbiased estimator $\boldsymbol{\theta}_e$ of $\boldsymbol{\theta}$, and any vector $\mathbf{a} = (a_1, \ldots, a_m)$ of constants, irrespective of the observed data set and the estimation method, the following inequality is valid:

$$var(\mathbf{a}\boldsymbol{\theta}_e) \geq \mathbf{a}[\mathbf{I}(\boldsymbol{\theta})]^{-1}\mathbf{a}^T \tag{73}$$

This is a special case of "information (or Cramér–Rao) inequality." A biased estimator can have smaller variance.

We found this inequality especially useful for obtaining some information, *a priori*, on the possible separability of two components. Sandor *et al.*[30] defined the index of separability $\bar{S}_{kl}$ for components with lifetimes τ_k and τ_l as

[29] E. L. Lehmann, "Theory of Point Estimation." Wiley, New York, 1983.
[30] T. Sandor, M. F. Conroy, and N. K. Hollenberg, *Math. Biosci.* **9**, 149 (1970).

$$\bar{S}_{kl} = \frac{|\tau_k - \tau_l|}{[var(\tau_k) + var(\tau_l)]^{1/2}} \tag{74}$$

The higher the value of this index the higher is the probability that the components can be separated. Now, by using inequality (73), we can calculate the upper bound of the index of separability:

$$\bar{S}_{kl} \leq \frac{|\tau_k - \tau_l|}{(I'_{kk} + I'_{ll})^{1/2}} \equiv s_{kl}, \qquad I'_{kk} = [\mathbf{I}^{-1}]_{2k-1,2k-1}, \qquad k = 1, \ldots, N \tag{75}$$

Thus, s_{kl} can be calculated for a given instrument response function and assumed values of parameters (including the value of the background per channel), and it provides information on whether two components can be separated before estimation of the parameters is attempted. It is clear that if $s_{kl} < 1$ the two components are likely to be nonseparable.

In a similar way we introduced[5] the index of detectability, $\bar{D}_k$, for the fraction f_k:

$$\bar{D}_k = \frac{f_k}{[var(f_k)]^{1/2}} \leq \frac{f_k}{(I''_{kk})^{1/2}} \equiv d_k,$$

$$I''_{kk} = [\mathbf{I}^{-1}]_{2k,2k}, \qquad k = 1, \ldots, N \tag{76}$$

Again, the upper bound of the index of detectability d_k can be calculated before estimation of the parameters is attempted, and thus can provide information, *a priori,* on whether a component is detectable. For $d_k < 1$ a considered component is likely to be undetectable. We have found by simulations that if any of d_k and s_{kl} values are smaller than 2 the estimates of parameters may be unacceptably inaccurate.

The maximum likelihood estimate is related to the information matrix in a key way: As information about $\boldsymbol{\theta}$ increases (Trace $\mathbf{I}^{-1} \to 0$), the bias of $\hat{\boldsymbol{\theta}}$ tends to zero and the covariance matrix $\mathbf{V}$ of $\hat{\boldsymbol{\theta}}$ tends to $\mathbf{I}^{-1}$. By information inequality (74), $\hat{\boldsymbol{\theta}}$ is thus asymptotically fully efficient, that is, no other unbiased estimator can have a smaller variance.

As the information about each parameter increases, it is also true that the maximum likelihood estimates will more and more assume a Gaussian distribution, and confidence limits for each parameter can be computed as $\pm 1.96\sigma(\theta_j)$, where the standard deviation $\sigma(\theta_j)$ in parameter θ_j is given by

$$\sigma(\theta_j) = V_{jj}^{1/2}, \qquad \mathbf{V} \cong [\mathbf{I}(\hat{\boldsymbol{\theta}})]^{-1} \tag{77}$$

H. Summary of Theoretical Considerations

We wish now to summarize the most salient theoretical and computational aspects of the maximum likelihood method as applied to the analysis of fluorescence decay.

1. The maximum likelihood method makes it possible to determine the most probable lifetimes of fluorescence intensity decay and their corresponding fractional contributions.
2. The estimation of lifetimes and fractions is performed by minimizing the Poisson deviance given by Eq. (4).
3. The maximum likelihood estimation of decay parameters will differ from the least-squares estimation by the extent to which the low-count region of the fluorescence decay curve determines the values of the parameters.
4. The mathematical model of fluorescence intensity decay is given by Eq. (12). The integrals involved can be described in several numerically and computationally different ways. We recommend the method based on the generalized first mean value theorem of integral calculus and the minimax principle, resulting in the recursive formula Eq. (37).
5. The correction of the model for the possible presence of scattered light can be performed analytically, yielding expression (46) for the related parameter and simplifying the Poisson deviance into the form of Eq. (47). When minimizing this deviance the positivity of lifetimes and fractions is introduced by transformations of variables. If by minimization the light scatter parameter [Eq. (46)] becomes negative, the fit must be generally considered as unacceptable.
6. Numerically, the minimization of the Poisson deviance can be conveniently performed by using the SIMPLEX algorithm of Nelder and Mead.[21] We recommend the termination of SIMPLEX when both the function and independent variable values are determined within prescribed tolerances. When dealing with three or more components the SIMPLEX algorithm may become nonrobust with respect to the starting parameter values.[5] In such cases it is useful to repeat the minimization for several different starting values of parameters and accept the result which corresponds to the lowest minimum.
7. The number of exponential decay components can be determined by using the likelihood ratio test defined by Eq. (56).
8. There are many statistical criteria to judge the goodness-of-fit. Based on this judgment one may accept or reject a given model as an adequate description of an intensity decay curve. We specifically used tests based on a standard normal variate for deviance defined by Eq. (57), based on the probability that the obtained deviance occurred by chance [Eq. (58)], and on the runs test for signs of residuals which are supposed to occur randomly [Eq. (61)]. In addition, we established a criterion based on the deviation ϕ of "normal-

ized'' fractions from 1 [defined by Eqs. (65)–(68)]. The inadequacy of the model is indicated when ϕ is positive and of the order of the smallest fraction.

9. The uncertainty in estimated parameters is related to the information matrix [defined by Eqs. (71) and (72)]. Namely, the variance matrix is approximately equal to the inverse of the information matrix. In addition, by using the information matrix, one can estimate upper bounds to indices which indicate whether for given decay data, two components (or more) can be resolved and whether a component can be detected [see Eqs. (75) and (76)]. If these upper bounds are smaller than 2 it is likely that the parameters are estimated with unacceptable accuracy.

III. Testing the Maximum Likelihood Method by Simulations

Tests by simulations offer much more flexibility than can tests on real data with the result that some hidden features of the method can be revealed. In this section we shall present some simulations which reveal how important the ''tail'' of the fluorescence intensity decay profile is for an accurate estimation of lifetimes and fractions. We shall also summarize briefly the results of simulations presented in our recent paper.[5]

The synthetic data for fluorescence intensity decay curves were generated by using Eqs. (27) and (37) with various measured instrument response functions standardly defined on 512 channels. The number of counts was normalized so that the counts in the peak channel (CPC) are equal to the counts in the peak channel of the measured instrument response function. According to observations in measurements on our single photon counting system, the background level was set to be 0.01% of CPC. For each set of lifetimes, fractions, and instrument response function, 101 synthetic data sets were generated corresponding to different realizations of the Poisson noise (a noise generator from Press *et al.*[22] was used).

As an error measure in estimating the lifetimes and fractions, we have used the average relative error E, defined as

$$E = \frac{1}{2N} \sum_{k=1}^{N} \left(\left| \frac{\tau_k - \tau_l'}{\tau_k} \right| + \left| \frac{f_k - f_l'}{f_k} \right| \right) \tag{78}$$

where τ_l' are estimated lifetimes and f_l' are corresponding estimated fractions. The index l is such that for a given k, $|\tau_k - \tau_l'|$ is minimal. This error measure successfully characterizes the overall accuracy of parameter estimation when the true values of parameters are assumed.[24,31] The overall

[31] J. Eisenfeld and C. C. Ford, *Biophys. J.* **26**, 73 (1979).

TABLE I

ACCURACY OF MAXIMUM LIKELIHOOD AND LEAST-SQUARES ESTIMATES FOR SELECTED
CHANNEL WIDTHS AND LIFETIME RATIOS

IRF[a]			mean$(E)^c$		Information indices		
h	FWHM	$\tau_1 = 1, \tau_2{}^b$	ML	LS	s_{12}	d_1	T^d
10	96	1.5	0.15(1)[e]	0.15(1)	3.6	2.8	0.08
25	130	1.5	0.043(4)	0.105(5)	14.6	10.6	0.006
40	135	1.5	0.062(5)	0.173(6)	13.0	9.4	0.007
10	96	3.0	0.022(2)	0.022(2)	15.9	26.3	0.016
25	130	3.0	0.0061(4)	0.0062(4)	117	113	0.0003
40	135	3.0	0.0054(4)	0.0088(5)	152	129	0.0002

[a] Three measured IRFs are characterized by the number of counts in the peak channel being 20,000, by the channel width h expressed in picoseconds, and by the full width at half of the maximum (FWHH) expressed in picoseconds.

[b] The lifetimes τ_1 and τ_2 are expressed in nanoseconds. The fractions were $f_1 = f_2 = 0.5$.

[c] ML, Maximum likelihood; LS, least-squares.

[d] T, Trace $\mathbf{I}^{-1}$.

[e] The number in parentheses is the estimated uncertainty (standard deviation) in the last significant digit.

accuracy of estimated lifetimes and fractions obtained from a given set of 101 synthetic data was characterized by the mean value of the corresponding average relative errors E_i, $i = 1, \ldots, 101$, or by the median value of E_i, E_{med}.

The accuracy of estimation of lifetimes and fractions is influenced by several factors: the signal-to-noise ratio (characterized by the CPC), the ratios of lifetimes and fractions, the channel width, and the length of the tail of the fluorescence intensity decay curve. The importance of two last factors is illustrated in Table I, which shows the results of simulations with three measured IRFs basically different only in channel width. For comparison we have also included the results obtained by application of the standard least-squares method to the same data. The synthetic fluorescence intensity profile based on the IRF with an h of 40 psec includes a tail of the curve that is 4 times longer than the tail of the profile based on an IRF with an h of 10 psec. The mean value of the error in estimating lifetimes and fractions is substantially smaller for the medium tail ($h = 25$ psec) curve than for the short tail ($h = 10$ psec) curve for both methods. For the decay with the longest tail ($h = 40$ psec) the accuracy of the maximum likelihood estimates decreases with respect to the accuracy obtained for the medium tail curve for the smaller lifetime ratio τ_2/τ_1 but is still very acceptable. The accuracy of the least-squares estimates for the longest tail of decay curves is significantly worse than that for the medium tail curves. Such findings indicate that there is an optimal channel width for given lifetimes, fractions, and signal-to-noise ratio, as Hall and Selinger[3] suggested in the case of a one-component model. In general, the Table I reveals that the maximum likelihood method is much more accurate than the least-squares method when the lifetime ratio is small and when the tail is longer. This is simply due to the fact that the tail of the decay curves carries a considerable amount of information about parameters and that the maximum likelihood method adequately takes into account Poisson noise in the low-count region of the tail.

The lower bounds of the index of separability and the index of detectability basically reflect the behavior of the mean(E) for the maximum likelihood method. Along the same lines, the decrease of the trace of $\mathbf{I}^{-1}$ generally reflects the increase of accuracy although a small deviation to such behavior is noted in Table I. The results from Table I also indicate how the lifetime ratio affects the accuracy of the estimated lifetimes and fractions. The behavior of E_{med} as a function of lifetimes is more comprehensively studied elsewhere.[5] Here we quote the main results of simulation studies presented there. First, the maximum likelihood estimates are always at least as good as those from the least-squares approach and, in most cases, are better. Second, the accuracy with which the lifetimes and fractions are estimated behave similarly for both methods as lifetime or fraction ratios are changed. In two-component models as the lifetime ratio increases (with constant fractions), the accuracy increases (E_{med} drops more than an order magnitude as the lifetime ratio changes from 1.2 to 2). E_{med} depends on the fraction $f_1(f_1 + f_2 = 1)$, approximately as a quadratic function with the minimum at $f_1 = 0.5$. In models with three or four components whose lifetimes are chosen so that their ratios can be defined as $\tau_2/\tau_1 = \tau_2/\tau_2 = \tau_4/\tau_3$, the accuracy increases with increasing lifetime ratio and fixed equal fractions. Third, the accuracy with which the lifetimes and fractions are estimated increases for both the maximum likelihood and the least-squares method as the level of the signal-to-noise ratio increases.

Fourth, the tail of the fluorescence intensity decay curves may carry considerable information on fractions and lifetimes. The maximum likelihood method is more suitable than the least-squares method for extracting this information (i.e., the maximum likelihood method provides more accurate estimates). There is an optimal choice of the channel width (which determines the length of the tail) for which the maximum likelihood and least-squares estimates are most accurate. Fifth, the lower bounds of separability and detectability indices are useful for assessing how accurately and whether under the given conditions (of signal-to-noise ratio, number of channels, channel width, and measured IRF) one can resolve two components with close lifetimes. The lower bound of the detectability index is specifically a useful determinant of whether, and how accurately, one can detect a component with a small fraction. Finally, the likelihood ratio test is useful in deciding whether an additional decay component is needed for "best fit." If the test implies values just above the threshold, then the corresponding parameter estimates in most cases will be strongly biased.

IV. Illustrations

It is our intention in this section to focus on the analysis per se rather than on attempts to interpret the data in terms of the biochemical significance of the derived parameters.

A. Analysis of Fluorescence in Scorpion Neurotoxin

In this subsection we apply the maximum likelihood method to the analysis of the fluorescence decay of the single Trp residue in scorpion neurotoxin. The particular data chosen comprised 512 data points with 20,000 counts at peak and channel width of 0.0139 nsec.

First we performed the analysis with the generalized Padé–Laplace method[24] which revealed that at least a three-component multiexponential model has to be assumed. The maximum likelihood analysis with three components led to a minimal deviance of 662 $[D(\hat{\theta})/\nu = 1.36]$. Assuming four components the minimal deviance decreased to 559 $[D(\hat{\theta})/\nu = 1.16]$ so that the difference Δ_4 is 103, much greater than the threshold value of 9.2 (see Section II,E). This means that at least four components are required to describe the data. The minimal deviance obtained when five components were assumed was 542 so that $\Delta_5 = 17$. On the basis of the likelihood ratio test we may still consider accepting the five-component model as a better descriptor of the data than the four-component model. However, in the case of the five-component model the light-scattering parameter was negative ($\xi = -0.23$) and approximately 3 times larger in absolute value than the light-scattering parameter in a four-component analysis ($\xi = 0.025$). This situation suggests clearly that the fifth component is compensated by the negative light-scatter parameter, which, by its nature, is not an admissible solution. We may therefore reject the five-component model and accept the four-component model as the most adequate to describe the particular data on the fluorescence intensity decay of the Trp residue in scorpion neurotoxin.

The common criteria on goodness-of-fit (see Section II,F) for a four-component fit are fairly well satisfied: $Z = 2.4$ and $Q = 0.01$. The residuals

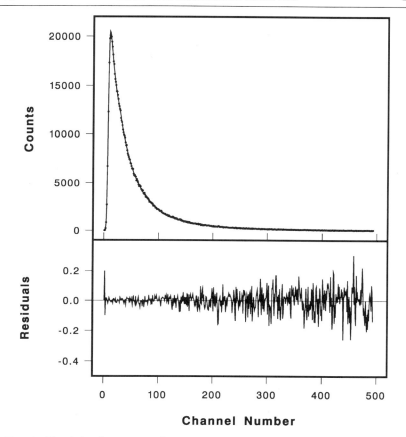

Fig. 1. Fit of the fluorescence intensity decay curve of the Trp residue in scorpion neurotoxin by use of the maximum likelihood method. The measured counts are depicted by dots and the fitted curve by the solid line. Below are the residuals calculated using Eq. (59).

appear to be random (see Fig. 1), and the runs test we described in Section 2.6 yielded $z = -2.1$ which is acceptable, although the existence of minor nonrandom errors in the data cannot be denied. The next question is, Is there enough information about the parameters in the data to resolve the given components, and how accurate are the estimated lifetimes and fractions (see Table II)?

All of the calculated indices of separability and detectability are considerably greater than 2, so we can be confident that the given components are resolvable and detectable under the amount of information provided by the data. The accuracy of estimated parameters is assessed by the standard deviations calculated from the inverse of the information matrix

TABLE II
ANALYSIS OF SCORPION NEUROTOXIN

Index:	τ_1 s_{12}	τ_2 s_{23}	τ_3 s_{34}	τ_4 E_{med}	f_1 d_1	f_2 d_2	f_3 d_3	f_4 d_4	δ ϕ	ξ	$D(\hat{\theta})/\nu$
	0.035	0.25	0.66	3.1	0.021	0.26	0.58	0.139	0.0012	0.025	1.16
S.D.[a]	0.005	0.01	0.01	0.1	0.005	0.02	0.01	0.004	0.0002		
Index	19	23	22		9	16	43	33	-0.014		
Simulation with 101 noise realizations[b]											
	0.042	0.26	0.66	3.1	0.023	0.26	0.58	0.141	0.0008	0.037	0.98
S.D.[c]	0.006	0.01	0.02	0.1	0.003	0.02	0.02	0.005			
Index				0.046					-0.013		

[a] The standard deviations (S.D.) in parameters are estimated from the inverse information matrix [see Eq. (78)].
[b] Displayed parameters for a given method correspond to that particular simulation (among 101) for which the average relative error [Eq. (78)] achieves its median value.
[c] The standard deviations (S.D.) in parameters are calculated as deviations of estimated parameters (in performed analyses of 101 synthetic decay curves) from their assumed "true" values given in the first row of the table.

(see Section II,G). As can be seen, these are always smaller than 15% of the values of the estimated lifetimes and fractions, except for the small fraction of shortest lifetime where it is 25%. The obtained index ϕ (describing the deviation of sum of fractions from its theoretical value, see Section II,F) was -0.0139. According to the discussion in Section II,F the fact that this index is negative does not allow us to conclude that the model is adequate, but it also does not preclude such a conclusion as it would happen if the parameter had been positive and larger than the smallest fraction.

To obtain more information about the adequacy of the model and the accuracy of estimated parameters we performed simulations. Assuming that the values of estimated parameters are true values we synthesized 101 fluorescence intensity decay curves (with given measured IRF) and analyzed these curves by the maximum likelihood method assuming four-component decay. The mean value and the standard deviation of index ϕ was -0.0139 ± 0.0002, which agrees perfectly with the value of ϕ obtained in the real data case, indicating that the model represents an adequate description of the data. However, one should note that some (undetermined) inadequacy of model to which the index ϕ is not sensitive may still exist.

The median of the average relative error in lifetimes and fractions [Eq.

TABLE III
ANALYSIS OF RIBONUCLEASE T_1 AT pH 5.5^a

Index:	τ_1 s_{12}	τ_2 E_{med}	f_1 d_1	f_2 d_2	δ ϕ	ξ	$D(\hat{\theta})/\nu$
	0.9	3.990	0.010	0.990	−0.0012	0.061	1.55
S.D.	0.1	0.008	0.002	0.002	0.0006		
Index	21		5	552	−0.03		
			Simulation with 101 noise realizations				
	1.0	4.00	0.011	0.989	−0.0013	0.067	0.92
S.D.	0.2	0.01	0.03	0.03			
Index		0.062			−0.03		

^{a} The symbols and comments are the same as for Table II.

(78)], E_{med}, was 0.046, and the mean with standard deviation was 0.053 ± 0.003. This means that the error in estimated parameters on average is about 5%. In Table II we have also displayed standard deviations for parameters defined as deviations from the assumed true values of parameters (used to generate the simulated data). These standard deviations are in good agreement with the standard deviations estimated for real data on the basis of the inverse of the information matrix (compare second and fifth rows in Table II). Thus, in this case we can be confident that the estimation of standard errors in the real data case is reasonably accurate. This can be seen also by comparing the estimates of lifetimes and fractions in the real data case to their estimates in simulation displayed in Table II (row 4). This particular simulation represents the whole set of 101 performed simulations in a sense that its average relative error is the median of obtained, E_i, $i = 1, \ldots, 101$.

In conclusion, the analysis we performed using the maximum likelihood method reveals that the fluorescence intensity decay of Trp in scorpion neurotoxin (SN3) is best described by four-component model.

B. Analysis of Fluorescence in Ribonuclease T_1

As a next illustration of the maximum likelihood method we have analyzed the fluorescence intensity decay of the Trp residue in ribonuclease T_1 in two cases: for pH 5.5 and for pH 7.5. In both cases the data comprised 512 data points with 15,000 counts at peak and a channel width of 0.030 nsec.

We first applied the GPL method, which revealed only one component in both cases. Subsequent application of the maximum likelihood method led to accepting two components (see Tables III and IV) in both cases: for

TABLE IV
ANALYSIS OF RIBONUCLEASE T_1 AT pH 7.5

Index:	τ_1 s_{12}	τ_2 E_{med}	f_1 d_1	f_2 d_2	δ ϕ	ξ	$D(\hat{\theta})/\nu$
	1.6	3.66	0.06	0.94	−0.0013	0.062	1.56
S.D.	0.1	0.02	0.01	0.01	0.0004		
Index	19		6.4	95	−0.02		
	Simulation with 101 noise realizations						
	1.6	3.66	0.07	0.93	−0.0012	0.061	0.94
S.D.	0.2	0.03	0.01	0.01			
Index		0.035			−0.02		

pH 5.5 the likelihood ratio test gave $\Delta_2 = 99$, and for pH 7.5, $\Delta_2 = 587$. A three-component model was not required by any of these data: for pH 5.5, $\Delta_3 = 0.09$, and for pH 7.5, $\Delta_3 = 103$, but with a large negative light-scatter parameter $\xi = -1.7$.

The common criteria for the goodness-of-fit show that there are some nonrandom errors in the data or that the multiexponential model is not most adequate. The reduced deviances are almost equal (see Table III). The corresponding standard normal variate Z is 8.6, thus outside the acceptable interval $(-3,3)$. The probability that obtained minimal deviances occurred by chance is smaller than 0.001. The residuals also show some nonrandomness. This is more emphasized in case of data with pH 5.5 where the runs test for signs of residuals yielded a standard normal variate z of -5.0, whereas for data with pH 7.5 the runs test yielded $z = -2.8$. It is interesting that in both cases a noticeable nonrandom error appears between channels 100 and 180 as the plots of residuals show in Figs. 2 and 3.

The question also arises whether such a small component as the one in the pH 5.5 data with fraction 0.01 (see Table III) is detectable. The magnitude of the detectability index is high enough that one can assume detectability of this small component. Furthermore, the estimated standard deviation in the obtained fraction is only 20% of its value. We also performed simulations as in Section 3. The results of these simulations are displayed in Table III. It is clear that assuming the multiexponential model as adequate the parameters can be obtained with a fair accuracy ($E_{med} = 0.06$ for pH 5.5 data and $E_{med} = 0.04$ for pH 7.5 data). Standard deviations in parameters obtained by simulations agree very well with those estimated from data. The index ϕ is negative and therefore inconclusive, but its

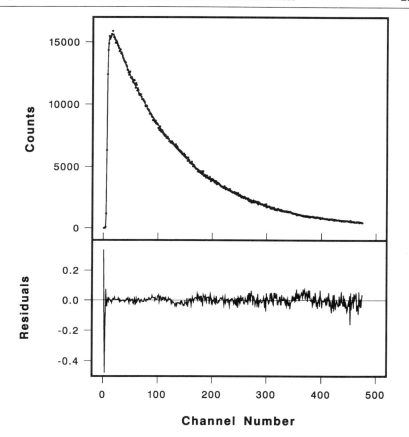

FIG. 2. Fit of the fluorescence intensity decay curve of the Trp residue in ribonuclease T_1 at pH 5.5 by the maximum likelihood method. The measured counts are depicted by dots and the fitted curve by the solid line. Below are the residuals calculated using Eq. (59).

values agree perfectly with corresponding mean values obtained in simulations: $\phi = 0.03$ (with standard deviation smaller than 10^{-4}) in the case of the pH 5.5 data and $\phi = 0.02$ (with standard deviation smaller than 10^{-4}) in the case of the pH 7.5 data. This means that inadequacy of the model suggested by the goodness-of-fit criteria does not affect the index ϕ.

V. Available Software

The results of analyses we have presented here as an illustration were obtained by the computer program FDA (written in FORTRAN 77) which we have developed in the last few years. The program is interactive, but

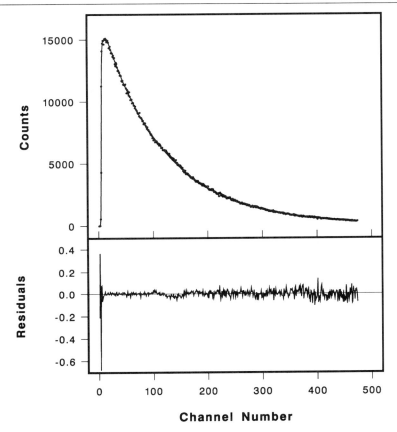

FIG. 3. Fit of the fluorescence intensity decay curve of the Trp residue in ribonuclease T_1 at pH 7.5 by the maximum likelihood method. The measured counts are depicted by dots and the fitted curve by the solid line. Below are the residuals calculated using Eq. (59).

it can also run in batch mode. It incorporates analysis by the GPL method[24] and by the maximum likelihood and least-squares methods as described in this chapter. The program offers the possibility of starting the minimization in the maximum likelihood (or least-squares) method using the parameters obtained in a previous analysis by the GPL method. One can also easily restart the minimization from any previously obtained parameters and, if necessary, change the tolerances and/or size of the initial simplex. The uncertainties in estimated parameters are determined using Eq. (77), but also using two other approaches, which we shall not discuss here because all three approaches most often give numerically the same result. However, in some cases, any one of these approaches may fail due to ill-conditioned matrix inversion, but the other two may still give a reasonable result.

The graphics capability of the FDA program is based on "grap," the UNIX V supported DOCUMENTER'S WORKBENCH software tool by AT&T. Thus FDA produces files for graphs of experimental and theoretical fluorescence intensity decay curves and of corresponding residuals. By a single command these graphs can be viewed on the screen, or a hard copy can be obtained on laser printers supporting POSTSCRIPT. The analysis of residuals by different tests (as described in Section II) is provided by a separate program.

The program FDA is capable of performing simulations based on synthetic data generated from given lifetimes and fractions and measured IRF as described in Section III. This and other features mentioned above make the FDA program rather versatile. However, having been developed for our own use it is not yet well documented, it is not yet structurally and computationally optimized, and it is not particularly user friendly. The main disadvantage of FDA is its inability to perform simultaneous fits to several data sets in manner of the global analysis. For this reason, in collaboration with Dr. J. Beechem, we have extracted the part including the maximum likelihood and the least-squares methods from the FDA program and have implemented it in the environment of the known software package GLOBALS UNLIMITED.[32]

As a separate project in collaboration with Dr. P. Axelsen we are developing an efficient and user-friendly program which incorporates the features not included in GLOBALS UNLIMITED software. In particular this new program offers the possibility of linking both the lifetimes and the fractions as well as to link ratios of fractions corresponding to different experiments. In its final version this program will allow the user to perform simulations and to analyze them similarly to how these are accomplished in the FDA program.

VI. Conclusions

We have presented the numerical and computational aspects of fluorescence intensity decay analysis by the maximum likelihood method, specifically paying attention to those details which are usually omitted in the literature. The analysis of fluorescence intensity decay profiles belongs to those complex problems which involve the art of scientific computing.

[32] J. M. Beechem, E. Gratton, and W. W. Mantulin, in "GLOBALS UNLIMITED—Technical Reference Manual" (J. K. Butzow, ed.). Laboratory for Fluorescence Dynamics, Univ. of Illinois, Urbana, Illinois, 1990.

With the present state of knowledge this problem cannot be reduced to a precise algorithm. Rather, one needs to employ a roughly defined strategy based on experience and involving the use of several approaches. The experience we have gained from numerous simulations and real data analyses allow us to make some recommendations on how to approach this problem.

It is useful to estimate the number of exponential decay components and the corresponding parameters by using the GPL method (or another similar technique such as the method of moments). In this way one has an excellent starting point for subsequent application of the maximum likelihood method, namely, the expected number of components and the values of parameters for the initial guess.

If analysis by the maximum likelihood method is performed, and if there is evidence (based on the likelihood ratio test) that two or more decay components are present, we recommend calculation of the lower bounds to the separability and detectability indices for the lifetimes and fractions obtained. These calculations require but a small computational effort and provide information on how resolvable and detectable the assumed components are. (If any of these indices is smaller than 2, the parameter estimates may be very inaccurate.) The next step is to do simulations based on the preliminary obtained parameters and on a *measured* instrument response function. These are computationally much more intensive, but they should nonetheless be performed. Then, by using E_{med}, one should be able to assess the global accuracy achieved for this set of parameters. The same simulations can be used to estimate the variances for particular lifetimes or fractions and can thus provide further insight into how accurate the estimated parameters might be.

If, by this additional analysis, it becomes clear that the estimation of parameters is not satisfactorily accurate, one should proceed with simulations of fluorescence intensity decay data containing greater amounts of information about the desired parameters. The amount of information can be enhanced by increasing the CPC (counts at the peak channel) and the number of channels, and/or by adding decay curves as done for global analysis.[33] The maximum likelihood analysis of these data should then yield more accurate estimates of parameters. Depending on the results of these simulations one should be able then to design and perform additional measurements, and by applying the maximum likelihood technique obtain more accurate estimates of both lifetimes and fractions.

The maximum likelihood method described above appears to be the most accurate approach if the assumed multiexponential decay model

[33] J. R. Knutson, J. M. Beechem, and L. Brand, *Chem. Phys. Lett.* **102**, 501 (1983).

is correct. To assess how adequate the multiexponential decay model is, one should use methods of analysis capable of detecting quasi-continuous distributions of lifetimes which may correspond to a nonexponential decay model. Several such methods have been studied recently,[13,34–38] but, with the exception of the extended method of moments,[35] they involve minimization of the Gaussian deviance for analysis of time-correlated single photon counting data. It would now be desirable to modify these methods to include the more appropriate Poisson deviance and to conform with the maximum likelihood approach. This is in principle a trivial modification, which requires only a change of the function to be minimized, and no additional computational effort need be expected.

Acknowledgments

We wish to thank A. Zelić who carefully checked all equations and their derivations. We also thank K. D. Peters for assistance with the preparation of figures. Supported by Grant GM 34847 from the U.S. Public Health Service.

[34] J. R. Alcala, E. Gratton, and F. G. Prendergast, *Biophys. J.* **51**, 925 (1987).

[35] E. W. Small, L. J. Libertini, D. W. Brown, and J. R. Small, *SPIE Proc. Fluorescence Detection III* **1054**, 36 (1989).

[36] K. J. Willis, A. G. Szabo, M. Zuker, J. M. Ridgeway, and B. Alpert, *Biochemistry* **29**, 5270 (1990).

[37] S. W. Provencher, *Comput. Phys. Commun.* **27**, 229 (1982).

[38] R. B. Gregory and Y. Zhu, *Nucl. Instrum. Methods Phys. Res., Sect. A* **290**, 172 (1990).

[11] Method of Moments and Treatment of Nonrandom Error

By Enoch W. Small

Introduction

The main mirror for the Hubble Space Telescope was painstakingly ground and polished with extremely high precision, yet its final dimensions were disastrously inaccurate! The reader should keep this fact in mind when reading this and other chapters in this volume. To obtain this precision of 1/65 of the 632.8 nm wavelength of a HeNe laser (~9.7 nm) over the surface of the 2.4 m mirror, a sophisticated new optical instrument was built capable of measuring the position of any point on the surface to nearly 1/1000 wavelength. Unfortunately, the accuracy of the determination was only about $\frac{1}{4}$ wavelength.[1]

It is important to remember that there are two kinds of errors in any measurement, random statistical errors and nonrandom errors, usually of instrumental origin. Modern laser-based monophoton decay fluorometers provide extremely high precision over a huge dynamic range. It is often, if not usually, found that the ability to extract relevant decay parameters is not limited by the small statistical fluctuations in the data, but by nonrandom errors put there by the instrument and sample.

The method of moments is a transform method for the deconvolution and analysis of decays. It will work on many kinds of data, but it was designed to work on fluorescence decay data measured by time-correlated single photon counting. Such data contain certain errors such as light scatter leakage, time origin shifts, lamp width errors, and impurity decays. Good instrumental procedures will keep these errors to a minimum, but they will always be present at some level and will often decrease the accuracy of recovered decay parameters or prevent the resolution of closely spaced decays. The unique feature of the method of moments is its ability to correct automatically for important nonrandom errors. It performs corrections for the first three errors mentioned here through a process we call moment index displacement, or MD. Contaminant decays may be dealt with by processes we call a Cheng–Eisenfeld filter or F/F deconvolution. Whether one is analyzing fluorescence decays or grinding large mirrors, great precision with respect to the random errors does little good if nonrandom errors destroy the accuracy of a determination.

Overview

Applications of moments for the analysis of decays began over 40 years ago.[2] The method of moments was first applied to fluorescence decays in 1967 by Wahl and Lami[3] and a little later by Isenberg and Dyson.[4] Most of the early development of the method was done by Isenberg and co-workers,[5–9] with important mathematical contributions by Eisenfeld and

[1] M. M. Waldrop, *Science* **249**, 735 (1990).

[2] Z. Bay, *Phys. Rev.* **77**, 419 (1950).

[3] P. Wahl and H. Lami, *Biochim. Biophys. Acta* **133**, 233 (1967).

[4] I. Isenberg and R. D. Dyson, *Biophys. J.* **9**, 1337 (1969).

[5] R. D. Dyson and I. Isenberg, *Biochemistry* **10**, 3233 (1971).

[6] I. Isenberg, R. D. Dyson, and R. Hanson, *Biophys. J.* **13**, 1090 (1973).

[7] I. Isenberg, *J. Chem. Phys.* **59**, 5696 (1973).

[8] I. Isenberg, *J. Chem. Phys.* **59**, 5708 (1973).

[9] I. Isenberg, *in* "Biochemical Fluorescence: Concepts, Volume 1" (R. F. Chen and H. Edelhoch, eds.), p. 43. Dekker, New York, 1975.

co-workers[10–17] and Lee.[18] An early paper by Isenberg on the treatment of nonrandom errors[7] led to much of the development of the method as it is used today in our laboratory. It is this later work that will be described here.

This chapter begins with a brief glossary of terms unique to the method of moments. The following two sections (How the Method Works and Treatment of Nonrandom Errors) explain how our current approach to the method of moments works and how it corrects for nonrandom errors. These sections aim at providing the necessary background for using the method intelligently. It is also hoped that the section on nonrandom errors might persuade the reader to consider the specific effects of nonrandom errors in other forms of data analysis. The reader who just wants to try the computer program might initially skip these two mathematical sections and go on to the subsequent four sections (Rules for Accepting Analysis, Fluorescence Shell Program, Structure of Program FLUOR, and Using Program FLUOR) which describe the program for deconvoluting and analyzing for sums of exponentials and provide an example of its use on a complex decay. The final section gives some advice on when to use and when not to use the method of moments. Data analysis by the method of moments has been recently summarized at a more technical level.[19]

Definition of Terms

Cheng–Eisenfeld filter: a method of fixing a lifetime in an analysis for a sum of exponential components. Because the method fixes the lifetime and not its associated amplitude, filtering can be used to estimate how much of a particular lifetime component is present in a multicomponent mixture.

Cutoff correction: integrals of fluorescence decays which extend past

[10] J. Eisenfeld and D. J. Mishelevich, *J. Chem. Phys.* **65**, 3384 (1976).

[11] J. Eisenfeld, S. R. Bernfeld, and S. W. Cheng, *Math. Biosci.* **36**, 199 (1977).

[12] J. Hallmark and J. Eisenfeld, *Appl. Nonlinear Anal.* **543**, 553 (1979).

[13] C. C. Ford and J. Eisenfeld, *Appl. Nonlinear Anal.* **531**, 542 (1979).

[14] J. Eisenfeld, *Math. Biosci.* **47**, 15 (1979).

[15] J. Eisenfeld and C. C. Ford, *Biophys. J.* **26**, 73 (1979).

[16] S. W. Cheng and J. Eisenfeld, *in* "Applied Nonlinear Analysis" (V. Lakshmikantham, ed.), p. 485. Academic Press, New York, 1979.

[17] J. Eisenfeld and S. W. Cheng, *Appl. Math. Comput.* **6**, 335 (1980).

[18] J. W. Lee, *J. Chem. Phys.* **77**, 2806 (1982).

[19] E. W. Small, L. J. Libertini, D. W. Brown, and J. R. Small, *Proc. SPIE Int. Soc. Opt. Eng.* **1054**, 36 (1989).

the last channel of measured data. An iterative technique is used to estimate the cutoff corrections.

Exponential depression: multiplication of the raw data, $E(t)$ and $F(t)$, by the function $e^{-\lambda t}$, where λ is a constant chosen by the user. Exponential depression increases the relative contribution that short-time data make to the moments.

F/F deconvolution: use of a single-component fluorescence decay to analyze another fluorescence decay. F/F deconvolution is distinguished from the normal F/F deconvolution in which a measured excitation is used for deconvoluting the decay. The analyzing fluorescence or measured excitation used for deconvolution is sometimes called the reference function.

λ-Invariance plot: a plot of the decay parameters (amplitudes and lifetimes) as a function of the exponential depression parameter, λ. One chooses values of the parameters from locally flat regions of the plot.

λ scan: a sequence of analyses performed at different values of λ. The resulting decay parameters are saved to generate the λ-invariance plot.

Moments: integrals of data multiplied by the factor t^k. The moments are just a set of a few numbers generated from the data and containing all the necessary information for recovering the decay parameters of interest. Because of the factor t^k, higher moments emphasize data at longer times.

Moment index displacement (MD): use of a different set of moments to calculate the decay parameters. Analyses done at different values of MD will not agree unless one is using the correct decay model for analyzing the data and nonrandom errors are not distorting the results. Higher MD uses higher moments and therefore emphasizes later data.

How the Method Works

To illustrate the method of moments, we will begin with a simple system. Let us assume that we have a perfect decay fluorometer capable of measuring fluorescence decays without any errors at all. We measure the decay of a solution containing two pure components with different lifetimes. We excite our sample with a brief flash of light, and we measure the response of the instrument, $F(t)$, to the emitted fluorescence. The excitation source is not infinitely narrow in time, and even our perfect instrument cannot respond infinitely fast to the emitted light. We therefore use a scatter sample to obtain $E(t)$, the measured excitation. Because we assume a perfect instrument, $E(t)$ and measured fluorescence, $F(t)$, are related by the following convolution[9]:

$$F(t) = E(t) * f(t) = \int_0^t E(t - u)f(u)\, du \tag{1}$$

where $f(t)$ is called the impulse response function. It is the fluorescence response that would have been obtained by an infinitely fast instrument using an infinitely fast excitation. Because we arbitrarily set the number of components, n, equal to two,

$$f(t) = \sum_{i=1}^n \alpha_i e^{-t/\tau_i} = \alpha_1 e^{-t/\tau_1} + \alpha_2 e^{-t/\tau_2} \tag{2}$$

The goal of the fluorescence decay experiment is to use $E(t)$ to deconvolute $F(t)$ so that we can estimate the $2n$ parameters, α_i and τ_i, which describe the physical system.

We begin by calculating at least $2n$ moments of both $E(t)$ and $F(t)$:

$$m_k = \int_0^\infty t^k E(t)\, dt \tag{3}$$

$$\left.\begin{array}{c} \\ \\ \\ \end{array}\right\} \text{for } k = 0, 1, 2, \ldots$$

$$\mu_k = \int_0^\infty t^k F(t)\, dt \tag{4}$$

This integration step can be thought of as a smoothing function, minimizing the effects of random errors. Note the t^k factors in the integrands of Eqs. (3) and (4). The t^k increases quickly with t and k, and therefore higher moments obtain their information from longer times.

We begin by defining a new set of moments:

$$G_{k+1} = \frac{1}{k!} \int_0^\infty t^k f(t)\, dt \tag{5}$$

These are called reduced moments of the impulse response function, since the factor $1/k!$ differentiates them from the moments of the data. The factor is included simply to make subsequent calculations easier. Our goal will be to use the moments of the data to derive these reduced moments of the impulse response.

By substituting Eq. (1) into Eq. (4), followed by a number of subsequent manipulations,[4] one can derive a recursion relationship relating the moments of the data with G_k:

$$G_1 = \frac{\mu_0}{m_0}$$

$$G_2 = [\mu_1 - G_1 m_1]/m_0$$

$$G_3 = \left[\frac{\mu_2}{2} - \frac{G_1 m_2}{2} - G_2 m_1 \right] / m_0$$

$$\vdots \tag{6}$$

$$G_{k+1} = \left[\frac{\mu_k}{k!} - \sum_{s=1}^{k} G_s \frac{m_{k+1-s}}{(k+1-s)!} \right] / m_0$$

By "recursion" we mean that we begin by calculating G_1. To calculate each higher G_k we must use the previously calculated lower G values. To a computer, such a recursion is trivial because of the use of nested loops.

The use of the recursion relationship is the deconvolution step in the method of moments. One begins with many channels of lifetime data, calculates a small set of simple integrals using Simpson's rule, and then combines them to form a set of moments of the deconvoluted function. The resulting reduced G moments are just a set of numbers. With the proper number of G values, one has all the required information to extract the desired decay parameters. Instead of fitting perhaps 1024 noisy channels of convoluted fluorescence data, one can instead fit a small set of moments to recover the parameters. As we shall see, fitting moments is very different from fitting raw data, because nonrandom errors affect different moments in different ways.

For the example presented here, we are assuming a double exponential decay. However, it is important to note that it is not necessary to assume any particular functional form for the impulse response function in order to perform deconvolution by this method.[20]

Putting our double exponential impulse response function [Eq. (2)] into Eq. (5) and integrating we find

$$G_1 = \alpha_1 \tau_1 + \alpha_2 \tau_2$$

$$G_2 = \alpha_1 \tau_1^2 + \alpha_2 \tau_2^2$$

$$G_3 = \alpha_1 \tau_1^3 + \alpha_2 \tau_2^3$$

$$\vdots \tag{7}$$

$$G_k = \sum_{i=1}^{n} \alpha_i \tau_i^k$$

Four of these equations provide a unique solution for the four unknown decay parameters. To find these parameters we being by solving a set of two simultaneous linear equations for d_0 and d_1:

[20] T. N. Solie, E. W. Small, and I. Isenberg, *Biophys. J.* **29**, 367 (1980).

$$G_1 d_0 + G_2 d_1 = G_3 \tag{8}$$
$$G_2 d_0 + G_3 d_1 = G_4$$

The d_0 and d_1 are the coefficients of a quadratic equation:

$$x^2 + d_1 x + d_0 = 0 \tag{9}$$

whose two roots are the lifetimes τ_1 and τ_2. To find the amplitudes, α_1 and α_2, we substitute the calculated lifetimes back into the expressions for G_1 and G_2 shown in Eq. (7). This yields two linear equations in two unknowns, which are then solved for the amplitudes. These simple computations have extracted the wanted four decay parameters.

For completeness we also show the result for an arbitrary number of components, n. First, one solves a set of n simultaneous linear equations for n values of d_i. In matrix form these equations are[15]

$$\begin{bmatrix} G_1 & G_2 & \cdots & G_n \\ G_2 & G_3 & \cdots & G_{n+1} \\ \vdots & \vdots & & \vdots \\ G_n & G_{n+1} & \cdots & G_{2n-1} \end{bmatrix} \begin{bmatrix} d_0 \\ d_1 \\ \vdots \\ d_{n-1} \end{bmatrix} = \begin{bmatrix} G_{n+1} \\ G_{n+2} \\ \vdots \\ G_{2n} \end{bmatrix} \tag{10}$$

The d_i are coefficients of an nth order polynomial:

$$P_n = x^n + d_{n-1} x^{n-1} + \cdots + d_0 = 0 \tag{11}$$

the roots of which are the n lifetimes. Substituting the lifetimes back into the first n expressions of Eq. (7) yields n linear equations for n unknowns, which can be readily inverted to give n values of α_i.

Moment Index Displacement

As we shall see in the section on nonrandom data errors, such errors are incorporated into the moments in very specific ways. MD provides a means of extracting the decay parameters from a different set of G values, thereby changing the sensitivity of the analysis to the errors.[7,21,22] For our two-component data there are an infinite number of equations like those shown in Eq. (8), which relate the quadratic coefficients d_0 and d_1. For example, we could have chosen the following set:

$$G_2 d_0 + G_3 d_1 = G_4 \tag{12}$$
$$G_3 d_0 + G_4 d_1 = G_5$$

where we have displaced the indices of the G values by one. We call use of this next set of G values an analysis with MD equal to 1. In order to

[21] E. W. Small and I. Isenberg, *Biopolymers* **15,** 1093 (1976).
[22] E. W. Small and I. Isenberg, *J. Chem. Phys.* **66,** 3347 (1977).

solve for d_0 and d_1, we are merely required to choose any two equations which are linearly independent. An MD2 analysis would use the moments G_3 through G_6. Higher MD values can also be used, but in practice, we rarely go above an MD of 4. After recovering the d_i, the quadratic Eq. (9) is solved for the lifetimes, which are then substituted back into Eq. (7) to find the amplitudes.

To find n values of d_i for an n-component analysis, we use the following set of linear equations to find the d_i:

$$
\begin{bmatrix}
G_{MD+1} & G_{MD+2} & \cdots & G_{MD+n} \\
G_{MD+2} & G_{MD+3} & \cdots & G_{MD+n+1} \\
\vdots & \vdots & & \vdots \\
G_{MD+n} & G_{MD+n+1} & \cdots & G_{MD+2n-1}
\end{bmatrix}
\begin{bmatrix}
d_0 \\
d_1 \\
\vdots \\
d_{n-1}
\end{bmatrix}
=
\begin{bmatrix}
G_{MD+n+1} \\
G_{MD+n+2} \\
\vdots \\
G_{MD+2n}
\end{bmatrix}
\tag{13}
$$

Again, the lifetimes are the roots of Eq. (11). The amplitudes are found by substituting the lifetimes into expressions for n of the G values shown in Eq. (7) beginning with G_{MD+1}.

Exponential Depression and λ Invariance

Multiply both sides of Eq. (1) by $e^{-\lambda t}$, where λ is an arbitrary constant we call the exponential depression parameter:

$$
\begin{aligned}
e^{-\lambda t}F(t) &= e^{-\lambda t}\int_0^t E(t-u)f(u)\,du \\
&= \int_0^t e^{-\lambda t}\,e^{\lambda u}\,e^{-\lambda u}E(t-u)f(u)\,du \\
&= \int_0^t e^{-\lambda(t-u)}E(t-u)\,e^{-\lambda u}f(u)\,du
\end{aligned}
\tag{14}
$$

Equation (14) is equivalent to

$$
F_\lambda(t) = E_\lambda(t)f_\lambda(t)
\tag{15}
$$

with

$$
E_\lambda(t) = e^{-\lambda t}E(t)
\tag{16}
$$
$$
F_\lambda(t) = e^{-\lambda t}F(t)
\tag{17}
$$
$$
f_\lambda(t) = e^{-\lambda t}f(t)
\tag{18}
$$

We call this process exponential depression.[6] We begin by multiplying our raw data, $E(t)$ and $F(t)$ by $e^{-\lambda t}$ and refer to the new functions $E_\lambda(t)$ and $F_\lambda(t)$ as the depressed excitation and emission, respectively. After deconvolution we recover the moments of the depressed impulse response function $f_\lambda(t)$. Exponential depression is independent of the functional form of $f(t)$. For a sum of exponentials the recovered depressed lifetimes are related to the true lifetimes by

$$\tau_{i,\lambda} = \tau_i/(1 + \lambda\tau_i) \tag{19}$$

The method of moments program simply prints out the true lifetimes calculated using Eq. (19).

The function $e^{-\lambda t}$ starts at 1 and decays off at a rate of λ. Multiplying data by this function depresses the noisy long-time tail of the data, counteracting the t^k factor in the moment calculations of Eqs. (3) and (4). High λ values therefore increase the relative amount of information obtained from short-time data.

λ Invariance

Method of Moments analyses, as performed in our laboratory, always require a λ-invariance test. According to the derivation shown in Eq. (14), the convolution should be invariant to multiplication by any value of λ. However, if one uses too high a value, then information will be lost from long-time data; if one uses too low a value, then the noisy tail of the data will dominate the moments, and information will be lost at short times. Only a certain range of values will be acceptable. To perform a λ-invariance test the program analyzes the data over a preselected range of λ values starting at the highest value and working back to lower values. The current convention is to analyze at 21 different values of λ. We call this a λ scan. Recovered decay parameters are plotted as a function of λ, and values are only accepted from flat or relatively flat regions of the plot.[23,24] A λ-invariance plot has two important functions[25]: (1) λ-Invariance plots can only be flat if one uses a decay model which is justified by the data. For example, if one has three-component data and analyzes for only two components, then the λ-invariance plots will slope downward with increasing λ. (2) λ-Invariance plots can only be flat if the recovered decay parameters have not been distorted by the presence of certain nonrandom data errors.

λ-Invariance can be very frustrating to the novice user who wants a

---

[23] I. Isenberg and E. W. Small, *J. Chem. Phys.* **77,** 2799 (1982).
[24] E. W. Small and I. Isenberg, *in* "Time-Resolved Fluorescence Spectroscopy in Biochemistry and Biology" (R. B. Cundall and R. E. Dale, eds.), p. 199. Plenum, New York, 1983.
[25] Acceptance of these two critically important statements rests partly on experience and partly on mathematical rigor. Although partial proofs exist,[18] their presentation is out of the scope of this chapter.

quick answer and does not care much whether the result is justified by the data. Use of "bad" data or the use of an incorrect decay model will often cause the method of moments to fail to give an acceptable answer, even when there may be any number of combinations of decay parameters which will give what would normally be regarded as an excellent "fit" to the data.

Nothing in the mathematics limits λ-invariance to the method of moments. In fact, it has also been used with a least-squares analysis.[26]

Cutoff Correction

The actual data can only be measured to a finite time, T, whereas the integrals shown in Eqs. (3) and (4) extend to infinity. The initially calculated depressed moments are therefore

$$m_{k,T} = \int_0^T t^k E_\lambda(t)\, dt \tag{20}$$

$$\mu_{k,T} = \int_0^T t^k F_\lambda(t)\, dt \tag{21}$$

One can assume, without approximation, that $E(t)$ drops instantaneously to zero at time T, and therefore the measured moments of the excitation are correct [i.e., $E(t)$ for $t > T$ cannot affect the measured $F(t)$]. However, a cutoff correction, $\Delta\mu_k$, must be calculated to correct the moments of the fluorescence.[6] Thus

$$\mu_k = \mu_{k,T} + \Delta\mu_k \tag{22}$$

where

$$\Delta\mu_k = \int_T^\infty t^k F_\lambda(t)\, dt \tag{23}$$

Substituting Eq. (15) into (23), assuming that $f_\lambda(t)$ is a sum of exponentials, and integrating, one can show[6,27]

[26] S. Hirayama, *J. Photochem.* **27**, 171 (1984).

[27] The equations shown here differ from those originally presented (see Ref. 6) in that β_i has been multiplied by e^{-T/τ_i} and $I_{i,k}$ by e^{+T/τ_i}. This redefinition of terms was recently done in order to avoid problems in calculation of β_i values experienced for very small values of τ_i.

$$\Delta\mu_k = \sum_{i=1}^{n} \beta_i I_{i,k} \tag{24}$$

with

$$\beta_i = \alpha_i \int_0^T E_\lambda(u) \, e^{-(t-T)/\tau_i} \, du \tag{25}$$

and $I_{i,k}$ calculated using the recursion relationship:

$$
\begin{aligned}
I_{i,0} &= \tau_i \\
I_{i,1} &= T\tau_i + \tau_i I_{i,0} \\
&\;\;\vdots \\
I_{i,k} &= T^k \tau_i + k\tau_i I_{i,k-1}
\end{aligned}
\tag{26}
$$

Iterative looping is used to approximate $\Delta\mu_k$. First, estimates are made of the decay parameters so that approximate cutoff corrections can be calculated. (The source of these estimates is described in the section on the structure of program FLUOR.) In the iteration procedure the cutoff corrections are added to the moments of the measured decay, after which new α_i and τ_i values are computed; these parameters are then used to calculate new approximations for the $\Delta\mu_k$. Convergence is accepted when the differences between the current and previous calculated decay parameters become very small. For succeeding values of λ the starting $\Delta\mu_k$ values are estimated using the decay parameters from the analysis at the previous value of λ. Although the requirement for a cutoff correction might be regarded as a disadvantage of the method, in practice, convergence only rarely requires more than a single loop to a few iterations. This is true not only because exponential depression greatly reduces the magnitude of the correction, particularly at high values of λ, but also because the initial estimate of decay parameters using the previous point on the λ-invariance plot is generally a very good one.

Treatment of Nonrandom Errors

Common nonrandom errors in fluorescence decay data include light *scatter* which leaks though emission monochromators. Scatter typically derives from optical reflections, particulate matter in the solution, or Raman scattering from the solvent, and it is always present at some level. Also, small *time origin shifts* between the fluorescence and the reference function used for its deconvolution are always present. In addition to time origin shifts, the fluorescence and measured excitation can be distorted relative to each other in other ways. The most obvious manifestation of such distortions is the difference in width of the instrumental response between the scatter and reference function. For historical reasons I shall refer to this last error as a *lamp width error*, although in our laboratory we no longer use flash lamps to excite the fluorescence. Time origin shifts and lamp width error normally result from the use of different wavelengths for the collection of the fluorescence and reference functions, but even if both are collected at the same wavelength, these nonrandom errors will result from artifacts in the constant fraction timing circuits. These three basic errors are treated using MD.[7,21,22] A more detailed description of these and other instrumental errors has been recently published.[28]

Another knid of nonrandom error always present at some level is the presence of impurity decays. These decays can derive from a number of different sources besides the obvious contamination of the sample with fluorescing impurities. The two methods discussed here for dealing with contamination errors are called Cheng–Eisenfeld filtering and F/F deconvolution.

In the discussion which follows, a superscripted asterisk will be used next to a quantity to denote that it is in error owing to the presence of nonrandom distortion. Only the effects of pure errors will be mentioned. In other words, scatter will be considered only in the absence of a time origin shift, and so forth. It is not usually necessary to do this, and in many cases the extension of the derivation to mixed errors is obvious. The effects of some combinations have been published.[22]

The following three sections describe how MD corrects for light scatter, time origin shifts, and lamp width error, respectively. It is important to note that none of the derivations shown assumes any functional form for the impulse response function, $f(t)$. A sum of exponentials is not required.

MD and Light Scatter

Suppose one makes a fluorescence decay measurement and a small amount of light scatter leaks through the emission monochromator. The measured incorrect fluorescence, $F^*(t)$, therefore contains a fraction of the measured excitation superimposed on it:

$$F^*(t) = F(t) + \xi E(t) \tag{27}$$

The scatter coefficient, ξ, measures the amount of contaminating scatter. When we use our measured excitation to deconvolute $F^*(t)$, we recover the wrong impulse response function, $f^*(t)$:

[28] E. W. Small, *in* "Topics in Fluorescence Spectroscopy, Volume 1: Techniques" (J. R. Lakowicz, ed.), p. 97. Plenum, New York, 1991.

$$F^*(t) = \int_0^t E(t - u)f^*(u)\, du \qquad (28)$$

Since

$$E(t) = \int_0^t E(t - u)\delta(u)\, du \qquad (29)$$

and $F(t)$ is given by Eq. (1), we have

$$F^*(t) = \int_0^t E(t - u)f(u)\, du + \xi \int_0^t E(t - u)\,\delta(u)\, du$$

$$= \int_0^t E(t - u)f^*(u)\, du \qquad (30)$$

where $\delta(t)$ is the Dirac delta function. Therefore

$$f^*(t) = f(t) + \xi\delta(t) \qquad (31)$$

When we perform a method of moments analysis on data contaminated by scatter, we recover distorted G^* values which are the reduced moments of this distorted impulse response function. Replacing $f(t)$ in Eq. (5) with $f^*(t)$,

$$G^*_{k+1} = \frac{1}{k!}\int_0^\infty t^k f(t)\, dt + \frac{1}{k!}\int_0^\infty t^k \xi\delta(t)\, dt \qquad k = 0, 1, 2, \ldots \qquad (32)$$

The first integral is simply our expression for G_{k+1}, and the second integral is zero for all $k > 0$:

$$\begin{aligned}
G_1^* &= G_1 + \xi \\
G_2^* &= G_2 \\
G_3^* &= G_3 \\
&\ \vdots \\
G_k^* &= G_k
\end{aligned} \qquad (33)$$

The only reduced moment which contains any information on the scatter is G_1^*. Because only MD0 uses this first moment, analyses using any higher values of MD will completely ignore the scatter.[29]

We are led to a surprising result. Regardless of how much scatter is present, any analysis with an MD greater than 0 will simply ignore its presence. If we wish to know how much scatter is present, we need only to calculate the decay parameters using an MD greater than 0, thus ignoring

---

[29] It is important to note that scatter in the presence of time origin shifts is not completely corrected by MD (see Ref. 22). In such cases it is corrected in a perturbation sense; scatter has decreasing effect on analyses performed at increasing MD.

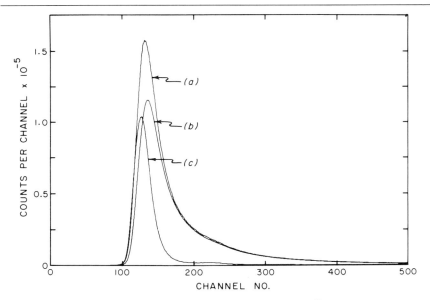

FIG. 1. Example of fluorescence decay data containing scatter.[21] Three compounds with lifetimes of 1.6, 7.8, and 18.7 nsec were mixed in solution and their total fluorescence decay measured. After a period of time, the instrument was rearranged so that scatter was collected superimposed on the fluorescence. An analysis with MD1 recovered the correct lifetimes of 1.6, 8.0, and 18.8 nsec with the expected amplitudes. (a) Measured fluorescence with scatter; (b) fluorescence decay calculated using the correct, recovered decay parameters; (c) measured excitation. Note that the calculated decay does not fit the data.

the scatter, and use Eq. (7) to calculate what G_1 should have been. G_1 is simply the integral of the fluorescence [see Eq. (5) with $k = 1$]. The fractional increase in the fluorescence due to the presence of scatter is given by[22]

$$\frac{\xi}{G_1} = \frac{G_1^*}{G_1} - 1 \qquad (34)$$

An example of a three-component fluorescence decay badly contaminated with scatter is shown in Fig. 1. In spite of the presence of the scatter as well as other serious distortions in this early flash-lamp excited fluorescence decay, an analysis with an MD of 1 was sufficient to recover the correct decay parameters. Figure 1 illustrates that the method of moments does not "fit" data. It also illustrates that MD does more than simply "weight" the data.

MD and Time Origin Shifts

By time origin shift we mean that the fluorescence is collected such that zero time is in a different position than the zero time for the measured excitation. We define a positive shift, s, to be how far the measured excitation is shifted to shorter times:

$$E^*(t) = E(t + s) \qquad (35)$$

It is straightforward to show that after deconvolution one recovers the moments of a distorted impulse response function,

$$f^*(t) = f(t - s) \qquad (36)$$

Substituting Eq. (36) into Eq. (5) and integrating, we find

$$G^*_{k+1} = \frac{1}{k!} \int_0^\infty t^k f(t - s) \, dt = \frac{1}{k!} \int_{-s}^\infty (t + s)^k f(t) \, dt \qquad (37)$$

$$
\begin{aligned}
G^*_1 &= G_1 \\
G^*_2 &= G_2 + sG_1 \\
G^*_3 &= G_3 + sG_2 + \text{terms in } s^2 \\
G^*_4 &= G_4 + sG_3 + \text{higher order terms in } s \qquad (38) \\
&\vdots \\
G^*_k &= G_k + sG_{k-1} + \text{higher order terms in } s
\end{aligned}
$$

Because we assume that s is a small number relative to the reduced moments, all higher order terms in s can be assumed to be zero. Equation (38) is not as easy to interpret as the simpler result for light scatter.

Isenberg has shown[7] that if one were to shift $E(t)$ relative to $F(t)$, then

$$\left(\frac{\partial \tau_i^*}{\partial s} \right)_{s=0} = 0 \qquad (39)$$

for MD > 0. The partial derivative of the recovered incorrect lifetime with respect to the applied shift equals zero when evaluated at $s = 0$. The restriction that MD > 0 is important.

When using the method of moments, it is trivial to introduce an artificial shift between $E(t)$ and $F(t)$, because all that is required is to start the integration for the moments of $E(t)$ at different times relative to $F(t)$. Figure 2A shows a plot of the recovered lifetimes for both MD0 and MD1 analyses of a three-component data set similar to the one shown in Fig. 1, except that no scatter was deliberately added. The analyses are plotted as a function of an artificially added shift. It is obvious from the plot that the MD0 analyses are sensitive to the shift, because they slope downward to

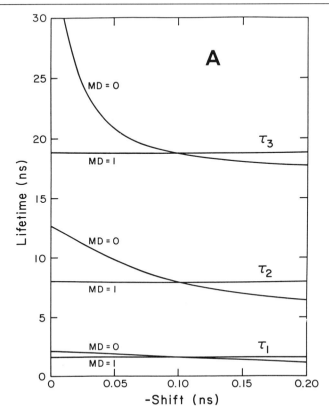

FIG. 2. Recovered lifetimes for MD0 and MD1 analyses as a function of applied shift.[24] Three-component data were collected as described in the caption to Fig. 1, except that no scatter was deliberately added. (A) For all three lifetimes the MD0 analysis is sensitive to the applied shift, but the MD1 analysis is not and recovers the correct lifetimes. (B) An enlargement of the plot near the lowest lifetime shows that the MD1 plot is in fact curved. According to Eq. (39), the amount of shift in the data (-0.105 nsec) can be recovered as the point at which the derivative equals zero. On the scale of this plot the sensitive MD0 result appears as an almost vertical line.

the right, but the MD1 analyses are not. MD has corrected the shift. Actually the MD1 analyses show curvature as well, but it is not apparent on the scale of the plot. A greatly expanded version of the lower region of the plot, showing only the shortest lifetime, is shown in Fig. 2B. Now the MD1 result for this lifetime is seen to be roughly parabolic. Assuming that the dominant error in these data is a time origin shift, then, according to Eq. (39), the slope of the MD1 plot becomes zero at the point at which the applied shift equals that present in the data. This occurs at an added shift

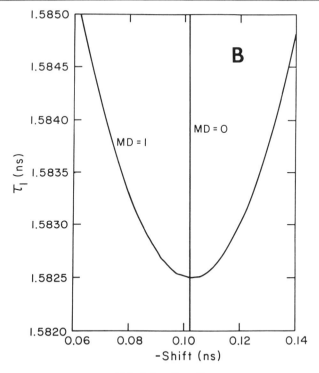

FIG. 2 (*continued*)

of -0.105 nsec. On the expanded scale of this plot the sensitive MD0 result is essentially a vertical line piercing the MD1 curve at this value of the shift. Similar plots can be drawn for the other two lifetimes.[24]

Isenberg has also shown that for MD > 1 higher orders of derivatives of τ_i evaluated at $s = 0$ are also zero.[7] In fact, the number of derivatives equal to zero is equal to the value of MD. He argues that this means that plots such as that shown in Fig. 2 become flatter and flatter as MD increases, and therefore MD corrects time origin shifts in a perturbation sense. That is, higher MD analyses are increasingly better at correcting shifts. This behavior is observed in practice.

Moment Index Displacement and Lamp Width Errors

Explanations for why MD corrects lamp width errors are intrinsically less satisfying than those for scatter and time origin shift, because one must make assumptions about the functional form of the lamp width error. For the purposes of this discussion I assume the simple form:

$$E^*(t) = \left[\frac{\delta(t)}{2} + \frac{\delta(t + s)}{2}\right] * E(t)$$

$$= \tfrac{1}{2}E(t) + \tfrac{1}{2}E(t + s) \tag{40}$$

This form of lamp width error is a convolution with a broadening function that contains half of its response at two positions in time. One might actually find this error if, during the measurement of $E(t)$, something changed in the timing circuits of the instrument and half of $E(t)$ were measured at a slightly different time.[30]

We recover the reduced moments of a distorted impulse response function:

$$f^*(t) = \tfrac{1}{2}f(t) + \tfrac{1}{2}f(t - s) \tag{41}$$

which after integration can be shown to yield the following:

$$
\begin{aligned}
G_1^* &= G_1 \\
G_2^* &= G_2 + \tfrac{1}{2}sG_1 \\
G_3^* &= G_3 + \tfrac{1}{2}sG_2 + \text{terms in } s^2 \\
G_4^* &= G_4 + \tfrac{1}{2}sG_3 + \text{higher order terms in } s \\
&\;\vdots \\
G_k^* &= G_k + \tfrac{1}{2}sG_{k-1} + \text{higher order terms in } s
\end{aligned} \tag{42}
$$

Note that these incorrect G values are identical in form to those presented in Eq. (38) for the time origin shift. The lamp width error described here behaves as though it were a time origin shift of $s/2$, and therefore MD will correct this error in the same manner.

Sources of Contaminant Decays

A particularly damaging kind of nonrandom error occurs when

$$f^*(t) = f(t) + f_c(t) \tag{43}$$

We would like to determine the decay parameters of $f(t)$, but there is contaminant decay, $f_c(t)$, present as well. If $f_c(t)$ were a single exponential decay of significant amplitude and lifetime well separated from any others in $f(t)$, then we might be able to simple resolve it in our analysis. A more likely situation is that $f_c(t)$ is unresolvable from the main decays. Such low-level contaminants will prevent the resolution of very closely spaced

---

[30] The description of a lamp width error given here is actually quite general. It could be readily extended to any simple convolution with a distortion consisting of an infinite number of little shifts, assuming that each individual shift is small enough.

decays.[31] They also represent a particularly insidious kind of error when one wishes to analyze for distributions of lifetimes, because they can lead one to conclude that a distribution is much broader than it really is.[32] Unless dealt with, the contaminant decays discussed here will distort an analysis (often severely) regardless of the analysis method being used.

There are several sources of low-level contaminant decays which can distort a deconvoluted impulse response function measured using time-correlated single photon counting. Five of them are (1) impurities in the sample or its solvent, (2) impurities in the scatter sample, (3) luminescence of optical components which is measured differently in $E(t)$ and $F(t)$, (4) multiple photon events, and (5) polarization errors, such as those which derive from errors in measuring sensitivity corrections.

It is obvious that extraneous fluorescence from a sample can lead to contamination of the form of Eq. (40), but it is less obvious that a contaminant in the measurement of $E(t)$ can have the same effect. Actually, if $E(t)$ contains an exponential decay of lifetime τ_c, then the recovered impulse response function will be contaminated by an exponential decay of the same lifetime. The amplitude corresponding to this spurious decay depends on the relationship between its own lifetime and those in $f(t)$. For example, if $f(t)$ is a sum of two exponentials, and τ_c is greater than either of their lifetimes, then the amplitude of $f_c(t)$ will be negative; if τ_c is less than either of their lifetimes, then the amplitude of $f_c(t)$ will be positive. Intermediate values of τ_c can contribute either negative or positive amplitude, depending on the relationship between τ_c and the two true lifetimes. (A rigorous derivation of these assertions can be found in the Appendix of Ref. 31.)

Luminescence of optical components can easily lead to serious contaminations. For example, use of the common RCA 8850 photomultiplier with a Pyrex window at wavelengths near 300 nm will result in a substantial decaying background luminescence.[31] Presumably this is also true when using other photomultiplier tubes with Pyrex or other windows at wavelengths where the window partially absorbs the light to be measured. If both $E(t)$ and $F(t)$ are measured at precisely the same wavelength, then this contamination would have no effect on recovered decays after deconvolution. However, $E(t)$ and $F(t)$ are usually measured at different wavelengths, and the contamination can be substantial.

When making a measurement using time-correlated single photon counting, one assumes that at most one photon is detected per flash of the light source. Nevertheless, according to statistical probability, multiple

[31] L. J. Libertini and E. W. Small, *Rev. Sci. Instrum.* **54**, 1458 (1983).
[32] L. J. Libertini and E. W. Small, *Biophys. Chem.* **34**, 269 (1989).

events will be detected as well. Because the electronics will trigger from the first detected event, later events will be ignored, creating a small bias of the data to shorter times. This error, which has been recognized for some time,[33-36] is sometimes called photon pileup. The effects of multiple events may be minimized by a number of different means. Most commonly the signal is attenuated until their level is below some arbitrarily chosen limit. They may also be minimized electronically by the use of energy windowing.[37,38]

To show the effect of multiple events, we begin by assuming that a particular sample has a single exponential fluorescence, that the instrument is very fast so that the data do not require deconvolution, and that triple and higher multiple events will not occur at an appreciable rate. We also assume that the arrival of the second event does not affect the timing of the first event. It can then be shown[39] that double photon events become a single contaminant decay with half the lifetime of the original decay:

$$f^*(t) = \alpha_1 e^{-t/\tau_1} + 2R_{2,1}\alpha_1 e^{-2t/\tau_1} \tag{44}$$

$R_{2,1}$ is the ratio of double to single events. If $f(t)$ is a sum of two exponentials, then there will be three $(1 + 2)$ contaminant decays; if $f(t)$ is a sum of three exponentials, then there will be six $(1 + 2 + 3)$; etc.[39] Relaxing our strict assumptions of no deconvolution and precise timing only adds to the complexity of the resulting contamination. Contaminant decays from multiple photon events will always occur at some level in lifetime data measured using time-correlated single photon counting. We have shown that they can have detrimental effects on difficult lifetime resolutions.[31,38]

The final source of contaminant decays listed here is a polarization error resulting from an incorrect measurement of a sensitivity correction. The parallel and perpendicular impulse response functions of the fluorescence decay are given by[40]

$$f_{\parallel}(t) = \tfrac{1}{3}p(t)[1 + 2r(t)] \tag{45}$$
$$f_{\perp}(t) = \tfrac{1}{3}p(t)[1 - r(t)] \tag{46}$$

[33] P. B. Coates, *J. Sci. Instrum.* **1**, 878 (1968).
[34] C. C. Davis and T. A. King, *J. Phys. A: Gen. Phys.* **3**, 101 (1970).
[35] P. B. Coates, *Rev. Sci. Instrum.* **43**, 1855 (1972).
[36] C. H. Harris and B. K. Selinger, *Aust. J. Chem.* **32**, 2111 (1979).
[37] R. Schuyler and I. Isenberg, *Rev. Sci. Instrum.* **42**, 813 (1971).
[38] E. W. Small, L. J. Libertini, and I. Isenberg, *Rev. Sci. Instrum.* **55**, 879 (1984).
[39] J. J. Hutchings and E. W. Small, *Proc. SPIE Int. Soc. Opt. Eng.* **1024**, 184 (1990).
[40] J. R. Lakowicz, "Principles of Fluorescence Spectroscopy." Plenum, New York, 1983.

where $p(t)$ is the total fluorescence of the fluorophore and $r(t)$ is the anisotropy decay. The goal is to find the decay parameters of $r(t)$, because these parameters supply information on the movement of the fluorophore. One of the best ways to get this information is to analyze the difference function[41]:

$$d(t) = f_{\parallel}(t) - f_{\perp}(t) = p(t)r(t) \tag{47}$$

Because $p(t)$ is relatively easy to measure, one can use various means to combine $d(t)$ and $p(t)$ to recover the parameters in $r(t)$.

Unfortunately, in order to obtain the difference function, one must measure a sensitivity correction so that the correct amount of the perpendicular component can be subtracted from the parallel component. Measuring accurate sensitivity corrections can be a difficult problem.[41] An error in the sensitivity would cause us to recover a difference function which is in error, containing a small fraction, Δ, of $p(t)$:

$$d^*(t) = p(t)r(t) + \Delta p(t) \tag{48}$$

Either of the two methods described here, Cheng–Eisenfeld filtering or F/F deconvolution, can be used to mitigate this error.

Cheng–Eisenfeld Filtering to Remove Contaminant Decays

Cheng and Eisenfeld generalized the method of moments by pointing out that any series of numbers, $s_1, s_2, \ldots, s_k$, can be regarded as a moment series if s_k can be written in the form[16,17]

$$s_k = \sum_{i=1}^{n} \beta_i \Theta_i^k \tag{49}$$

Obviously, the G_k of Eq. (7) are a special case of a moment series with Θ_i equal to τ_i and β_i equal to α_i. This generalization led Cheng and Eisenfeld to a number of important conclusions about moments, and even led them to devise a whole new method of moments, called the DCM or direct computational method, in which Θ_i is actually the reciprocal of the lifetime.[16] One of the most useful results deriving from their work is referred to as the Cheng–Eisenfeld filter.

If one has a decay which is a sum of exponentials, then it is possible to remove the contribution a lifetime makes to the fluorescence by a performing simple operation on the moment series. We call the lifetime to be removed the filter lifetime, τ_f. To filter its contribution, we make a new

---

[41] M. G. Badea and L. Brand, this series, Vol. 61, p. 378.

moment series, G_k', by subtracting the product of τ_f and the next higher moment. Using Eq. (7),

$$G_k' = G_{k+1} - \tau_f G_k$$

$$= \sum_{i=1}^{n} \alpha_i \tau_i^{k+1} - \tau_f \sum_{i=1}^{n} \alpha_i \tau_i^{k}$$

$$= \sum_{i=1}^{n-1} \alpha_i(\tau_i - \tau_f)\tau_i^{k}$$

$$= \sum_{i=1}^{n-1} \alpha_i' \tau_i^{k} \qquad \text{for} \quad k = 1, 2, 3, \ldots \tag{50}$$

Equation (50) is also a moment series, but it contains one fewer lifetime components. We proceed by solving for one less lifetime using these filtered moments. We then substitute the resulting $n - 1$ lifetimes as well as the filtered lifetime back into the unfiltered G values of Eq. (7) and recover all n amplitudes, including an amplitude for the filter lifetime. The filter lifetime has been essentially fixed by the analysis, except that we have been able to recover its amplitude. The process may be repeated to fix more than one lifetime.

The Cheng–Eisenfeld filter gives us a way of asking how much amplitude is associated with a particular lifetime, and it serves the basis for a method of analyzing distributions of lifetimes.[32] We have also found that fixing the value of a poorly resolved component of little interest will simplify an analysis and thereby improve resolution of the remaining decays. An example of this ability, taken from Ref. 31, is shown here.

Solutions of three compounds were measured individually and found to have lifetimes of 8.8, 10.1, and 12.8 nsec. They were mixed in pairs, and fluorescence decays of each mixture were measured. The two easiest pairs to resolve (8.8, 12.8 and 10.1, 12.8) gave lifetimes biased to lower values. The most difficult pair (8.8, 10.1) simply did not resolve. Three-component analyses of the mixtures indicated an additional component of very low amplitude with a lifetime slightly below 4 nsec. We took the best average value that we could for this component (3.85 nsec) and used it in a Cheng–Eisenfeld filter. λ-Invariance plots for the three-component filtered analyses are shown in Fig. 3. The lifetimes have been corrected, and the resolutions are now possible. The amplitude returned for the filtered lifetime indicated that the background fluorescence which prevented resolution amounted to between 0.8 and 1.6% of the total fluorescence.

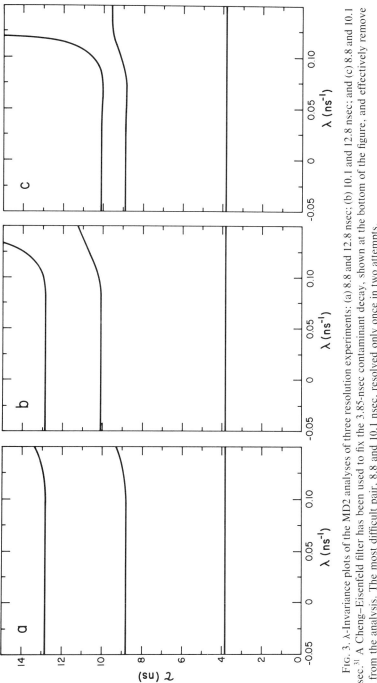

FIG. 3. λ-Invariance plots of the MD2 analyses of three resolution experiments: (a) 8.8 and 12.8 nsec; (b) 10.1 and 12.8 nsec; and (c) 8.8 and 10.1 nsec.[31] A Cheng–Eisenfeld filter has been used to fix the 3.85-nsec contaminant decay, shown at the bottom of the figure, and effectively remove it from the analysis. The most difficult pair, 8.8 and 10.1 nsec, resolved only once in two attempts.

F/F Deconvolution

In a typical fluorescence decay experiment, one uses $E(t)$ to deconvolute $F(t)$. We call this E/F deconvolution. It is also possible to deconvolute a measured multiexponential fluorescence instead of using another convoluted single component fluorescence. We call this F/F deconvolution.[42] F/F deconvolution is similar to the procedure used by van Resandt and co-workers[43] and by Lakowicz and Balter.[44] A later version has also been described.[45]

Instead of measuring the excitation function using a scatter sample, we measure a single-component fluorescence whose impulse response function is given by

$$f_a(t) = \phi \exp(-t/\theta) \tag{51}$$

where ϕ is the preexponential amplitude and θ is the single lifetime. The convoluted fluorescence of this standard is

$$F_a(t) = E(t) * f_a(t) \tag{52}$$

We call $F_a(t)$ the analyzing fluorescence, because we are going to use it instead of $E(t)$ to deconvolute a fluorescence of interest.

If one uses $F_a(t)$ to deconvolute a fluorescence decay, $F(t)$, then instead of the impulse response function, one instead obtains the function $f^*(t)$ in the expression:

$$F(t) = F_a(t) * f^*(t) \tag{53}$$

Strictly speaking, $f^*(t)$ is not in error, because we are performing our analysis this way on purpose. It is used here because we are simply treating this problem in a manner consistent with the errors described above.

If we assume that $F(t)$ is a sum of exponentials, then

$$\sum_{i=1}^{n} \alpha_i \exp(-t/\tau_i) = \phi \exp(-t/\theta) * f^*(t) \tag{54}$$

It is straightforward[42] to show that

$$f^*(t) = \sum_{i=1}^{n} \frac{\alpha_i}{\phi} \left\{ \delta(t) + \left[\frac{1}{\theta} - \frac{1}{\tau_i} \right] \exp(-t/\tau_i) \right\} \tag{55}$$

From Eq. (55) we conclude that, if one takes a measured single expo-

[42] L. J. Libertini and E. W. Small, *Anal. Biochem.* **138**, 314 (1984).
[43] R. W. Wijnaendts van Resandt, R. H. Vogel, and S. W. Provencher, *Rev. Sci. Instrum.* **53**, 1392 (1982).
[44] J. R. Lakowicz and A. Balter, *Biophys. Chem.* **15**, 353 (1982).
[45] M. Zuker, A. G. Szabo, L. Bramall, D. T. Krajcarski, and B. Selinger, *Rev. Sci. Instrum.* **56**, 14 (1985).

nential fluorescence and uses it to deconvolute a measured sum of exponentials, then one will obtain a constant times a delta function plus a sum of exponentials with the same set of lifetimes as the original sum. This is basically the same functional form for $f^*(t)$ that we encountered in Eq. (31) for the presence of scatter in the data and is treated the same way by the method of moments. Because the use of MD $\geq$ 1 does not include G_1, a method of moments F/F deconvolution is unaffected by the delta function term and directly recovers the desired lifetimes, τ_i. The α_i can be simply calculated from the value of β_i, since ϕ and θ can be determined by a simple deconvolution of $F_a(t)$.

Note from Eq. (55) that, if θ approaches one of the lifetimes to be resolved, its corresponding amplitude diminishes and becomes zero when the two lifetimes (θ and τ_i) are equal. The lifetime of the analyzing fluorescence can have any value as long as it is not too close to a lifetime that one wishes to resolve. Also, if one chooses the two lifetimes to be equal, then one of the lifetimes in the multicomponent sum will be cancelled, and the analysis will be simplified from an n to an $n - 1$ component analysis. This can greatly aid in the resolution of decays. An example of this simplification is given in Table I, which shows the F/F deconvolution analyses of a five-component data set.[19]

Five individual single-component decays were measured and summed in the computer to give a five-component decay. An analysis using a measured excitation to deconvolute this complex decay was unable to resolve the five components. F/F deconvolution was then attempted using the single component decay to see which ones could serve as an analyzing function. As can be seen in Table I, the first three decays with lifetimes of 1.06, 2.02, and 4.12 nsec each permitted correct resolution of the remaining four lifetimes.

We have also published an example of the use of F/F deconvolution to determine the fast rotational diffusion of small dye molecules in nonviscous solution.[46] As mentioned above, a small error in the measurement of the sensitivity correction causes the polarization difference function to be contaminated by a trace of the single lifetime of the dye. Using the total fluorescence of the dye to analyze the difference function will force the amplitude of the contaminating component to zero.

Precision versus Accuracy in Presence of Nonrandom Errors

The following simulated example illustrates two important points about precision and accuracy. More details can be found in Ref. 19.

Table II compares the performance of the method of moments (MM)

---

[46] E. W. Small, L. J. Libertini, D. W. Brown, and J. R. Small, *Opt. Eng.* **30**, 345 (1991).

TABLE I
RESULTS OF F/F DECONVOLUTION ANALYSES
OF FIVE-COMPONENT DATA[a]

| | Measured[c] | | | | |
θ	τ_1	τ_2	τ_3	τ_4	τ_5
1.06	—	1.7	3.5	7.8	16.9
2.02	1.1	—	3.9	8.6	17.6
4.12	0.9	1.8	—	8.8	27.7
Expected[b]	1.06	2.02	4.12	8.02	16.7

[a] Results were taken from Ref. 19.

[b] Five single-component decays with lifetimes of approximately 1, 2, 4, 8, and 16 nsec were measured
using four derivatives of 1-aminonaphthalene-n-sul-
fonic acid with n equal to 2, 5, 5, 4, and 7, respectively.
The compounds were dissolved in 10 mM Tris and
adjusted to pH 4.8 with glacial acetic acid, then
quenched with various concentrations of acrylamide
to give the lifetimes listed (obtained by analyzing the
individual decays using E/F deconvolution).

[c] The five decays were first summed and found to be
unresolvable by E/F deconvolution analysis. Results
listed were obtained by F/F deconvolution analyses
of the resultant five-component decay using the single-component decays (as indicated by θ) in place of
the measured excitation.

and least-square iterative reconvolution (LSIR) in analyzing two groups
of 20 simulated data sets.[19] To generate the data, a three-component
fluorescence decay with lifetimes of 0.50, 2.0, and 4.0 nsec was simulated
using a measured excitation obtained with an RCA 31000M photomultiplier
tube. In test (1) no nonrandom error was introduced. In test (2) 3% scatter
was added, the excitation was broadened by 20 psec in the manner described in the section on lamp width error, and a time origin shift of 10
psec was introduced. For each test, 20 pairs of data [$E(t)$ and $F(t)$] were
generated by adding multinomial noise to the original excitation and to the
simulated curve, bringing the total number of counts in each to 10^6. Table
II shows the average decay parameters, α_i and τ_i, recovered by the analyses. It also shows the average standard deviations of the results predicted
compared to the average standard deviation actually observed.

Both methods work fine on test (1), which was without nonrandom
error. The method of moments statistical algorithm[19] tends to overestimate

TABLE II
EFFECTS OF NONRANDOM ERROR ON STATISTICAL ANALYSES[a]

Method[b] (test #)	Parameter	α_1	τ_1	α_2	τ_2	α_3	τ_3
MM	avg[c]	0.404	0.50	0.253	2.01	0.072	4.07
(1)	σ_m[d]	0.0153	0.030	0.0089	0.109	0.0163	0.214
	σ_p[e]	0.0213	0.029	0.0072	0.077	0.0100	0.140
LSIR	avg	0.398	0.50	0.245	1.96	0.083	3.90
(1)	σ_m	0.0057	0.010	0.0074	0.073	0.0114	0.124
	σ_p	0.0058	0.010	0.0063	0.071	0.0106	0.120
MM	avg	0.410	0.50	0.250	2.00	0.076	4.00
(2)	σ_m	0.0146	0.028	0.0080	0.112	0.0147	0.173
	σ_p	0.0115	0.023	0.0063	0.076	0.0101	0.128
LSIR	avg	0.480	0.25	0.29	1.23	0.166	3.29
(2)	σ_m	0.0136	0.025	0.0137	0.105	0.0126	0.070
	σ_p	0.0062	0.0072	0.0048	0.029	0.0040	0.025
Expected values:		0.400	0.50	0.250	2.00	0.075	4.00

[a] Results were taken from Ref. 19. The particular example shown here for test (1) was chosen because it was the one example out of several combinations of lifetimes in which LSIR most outperformed the method of moments in terms of precision. For each LSIR analysis the expected values were used as starting guesses. Unless this is done, analyses will occasionally converge on incorrect results for difficult resolutions such as these, because the χ^2 surface is very flat. The method of moments does not use such prior estimates.

[b] MM and LSIR refer to the method of moments and least-squares iterative reconvolution, respectively. Test (1) is a three-component simulation with no nonrandom errors intentionally introduced. Test (2) has been distorted by the presence of scatter, time origin shift, and lamp width error as described in the text.

[c] avg, Average decay parameters obtained from single-curve analyses of 20 sets of simulated decay data.

[d] σ_m, Standard deviation of the mean actually observed on analysis of the 20 sets of simulated data.

[e] σ_p, Standard deviation predicted by the statistical algorithm. It was calculated as the average of the deviations predicted for the 20 sets of data.

slightly the actual precision of the results. The LSIR algorithm[47] predicts the precision quite well and delivers somewhat better precision in the decay parameters, especially on the shortest decay. The story is quite different for the second test containing nonrandom error. The presence of the errors makes essentially no difference in the performance of the method of moments, but LSIR fails to return the correct answer for any

[47] P. R. Bevington, "Data Reduction and Error Analysis for the Physical Sciences." McGraw-Hill, New York, 1969.

of the six decay parameters. LSIR predicts high precision, and indeed fits the data decently, but does not even deliver the expected precision.

Nonrandom errors can alter both the precision and accuracy of a determination. The two important points made by these analyses are (1) even though a method predicts excellent precision, it may not deliver it; and (2) great precision means nothing if there is no accuracy.

Those who work with LSIR are fully aware that analyses of data such as these containing nonrandom errors should not return the correct parameters.[48,49] In fact, the analyses in the last section of Table II give χ^2 values ranging from 1.2 to 1.4, and this small increase over the ideal of 1.0 should tell the user that something is amiss. Perhaps the user could devise a scheme of fitting these three errors. The method of moments solves the problem by simply being insensitive to them as long as MD is greater than zero.

Rules for Accepting Analysis

The method of moments computer program for solving multiexponential decays is set up to automatically scan through a series of analyses, varying the exponential depression parameter, λ. This series of analyses is then automatically repeated using different values of MD, a parameter which specifies which actual moments are used for the calculations. The array of resulting decay parameters are used to generate a λ-invariance plot. Such a plot typically consists of the fluorescence lifetimes plotted versus λ for each of the values of MD and must be visually interpreted according to a set of rules. Both the mathematics of the previous sections and considerable experience with the method of moments have led us to formulate the following[19]:

Rule 1. λ-Invariance: For a given MD and number of components, the parameter values (alpha's and tau's) should be taken from a region of the λ-invariance plot which is locally flat (i.e., $d\alpha_i/d\lambda \cong 0$, $d\tau_i/d\lambda \cong 0$). It is common to find that true flatness is not attained. In such a case it is necessary to estimate the parameters from the region of the scan which has minimum absolute slope. Confidence in that parameter should be inversely related to the slope at that point. Confidence should also be directly related to the range of λ over which the parameter value is relatively constant.

Rule 2. MD Agreement. Results obtained from scans at different MD values should agree. Perfect MD agreement is never observed with realis-

[48] A. Grinvald, *Anal. Biochem.* **75,** 260 (1976).
[49] M. L. Johnson and S. G. Frasier, this series, Vol. 117, p. 301.

tic data. Confidence in a result should increase with the number of MD values for which agreement is attained. It is not unusual to obtain agreement of only two MDs (in the usual MD range of 1 to 4) and still have a good result when data of known decay parameters are analyzed.

The rule of MD agreement does not require that the different MDs give exactly the same parameters. Confidence in the result should be roughly inversely related to the relative magnitudes of the differences.

Rule 3. Component Incrementation: An analysis for $n + 1$ components should indicate the same parameter values as the n-component analysis. The $(n + 1)$-component λ scans must also be judged by the λ-invariance and MD agreement rules. In a good analysis, n of the alpha's and tau's will agree with those obtained from the n-component scans, whereas the remaining parameters will be scored low by the λ-invariance and/or MD agreement criteria. The highly variable additional component from the $(n + 1)$-component analysis will almost always constitute a small percentage of the total decay. Lack of component incrementation, even when a relatively good n-component analysis has been obtained, is a clear indication that the latter result is not correct and should not be accepted.

It is important to emphasize that, even when an analysis satisfies these three rules, complete resolution of the decays may not have occurred. A satisfactory result merely says that the parameters derived describe the decay to the best of the method's ability to resolve exponentials, given the random and nonrandom errors in the data.[50]

Fluorescence Shell Program[51]

Computers have undergone impressive advances during the development of the method of moments program. The original program was written in FORTRAN and developed on a PDP 11/34 computer. Now, orders of magnitude more computing power can be purchased in the form of a 386-based IBM-compatible for a small fraction of the cost, and even these new computers will soon seem out of date. In recent updates of the method of moments programs our goal has been to maintain much of the original carefully optimized FORTRAN code and still have access to the power of the new inexpensive computers. Unfortunately the simple and powerful FORTRAN is no longer a well-supported language, and for this reason we

---

[50] What is meant here by the term "resolution" is discussed in detail in Ref. 23.

[51] F_Shell and the associated programs are available from the author. They should run on any IBM or IBM-compatible computer with an Intel 80286 or higher processer and appropriate math coprocessor. A color EGA or higher graphics adaptor is required. Source code for a basic FORTRAN version of FLUOR is also available.

have chosen to work with C. The Microsoft C 6.0 that we use permits mixed language calls to Microsoft FORTRAN 5.1. C is a complex "programmer's language" that is not user friendly to a biophysicist who spends very little time programming; nevertheless, the language is well-supported and powerful. The real power of C derives from the availability of other program packages, such as good graphics and windowing routines, from other vendors. We are using GraphiC, a set of powerful graphic functions, and Cscape, a package of windowing functions that permit pull-down menus, data entry screens, and popup menus with scrolling and mouse support.[52] The goal of this programming effort is to make the method of moments programs run on DOS or OS2 and easy to use.

One powerful feature of C is the ability to spawn child processes, permitting the user to switch at will from one program to another, keeping certain programs in memory. Each of the programs involved in the analysis and display of data can be run individually by a knowledgeable user, or the user can choose to coordinate program use by running a new program we call F_Shell. F_Shell resides in memory, manipulates data files, and spawns other programs for analysis and graphics. It acts as a shell program in that it insulates the user from the operating system.

To use and understand F_Shell and the associated programs, it is important to know the conventions we use for storing data. Data are stored in binary form, so that ASCII-to-binary conversion is not required every time data are read from disk. Each data file contains a header consisting of three identifying strings entered by the user, followed by the kind of file, the number of channels of data per data set, the nanoseconds per channel, the first channel of data corresponding to time equal to zero, and the background levels for each of the data sets in the file. Following the header are one, two, or three sequential data sets. The allowed combinations are E, F, EF, EFF, and FF, where E and F refer to individual measured excitations and fluorescences. For example, if one were to measure an anisotropy decay consisting of a measured excitation as well as parallel and perpendicular components of the fluorescence, then one would use an EFF file type.

Invoking F_Shell (by typing "FShell" at the DOS or OS2 prompt) presents the user with a main pull-down menu and a popup window displaying data files in the currently active directory. Programs which process or analyze data can access only the files in this active directory. The user can switch from the data window to the menu bar at the top of the screen and back by using the mouse or the keyboard. The data window is used

---

[52] GraphiC is available from Scientific Endeavors Corp., Kingston, TN; and Cscape from Oakland Group, Inc., Cambridge, MA.

for selecting files, and the menu bar is used for running programs. For convenience, "hot keys" (such as Ctrl_A, F3, etc.) can be used from the data window to directly invoke individual pull-down options, and the F1 key can be used at any time to provide rudimentary help. The pull-down options of the current version (1.03) of F_Shell are summarized in Table III.

The Program option of the menu bar displays the volume number of the program or assists the user to adapt the program to a particular computer. The Data File option provides a pull-down list of housekeeping operations which can be used on data files. The first pull-down option, "Select new files," simply returns the user to the data file selection window. The next five options, "Open other directory," "Copy to other directory," "Copy from other directory," "Delete in other directory," and "Close other directory," give the user access to other files stored on various other directories in the computer. The last two options "Format diskette" and "Write to diskette" can be used to back up files on the default diskette drive. Finally, "Delete files" deletes the selected files from the active directory.

The next menu bar option, "Utilities," provides a series of utility programs for manipulating data files. The first pull-down option, "BG calculation," runs a program to determine the background level for each data set in the file and store the information in the file heading. To determine a background, it uses an Andrews sine estimate of location for the channels preceding the first channel of data. Such an estimate is designed to be insensitive to the presence of outliers.[53] The next option, "Calculate anisotropy," runs a program to calculate an anisotropy decay from raw data. An "Edit data file" option is available which converts a file to ASCII and automatically invokes the system editor displaying the converted file. After the user exits the editor, another program automatically converts the edited version back to binary. This edit option can be used to alter entries in the file heading or just to examine the data. Finally, "Import data" runs a program which reads ASCII data from another source, prompts the user for heading information, and then creates an ordinary binary data file for use by the programs associated with F_Shell.

The Analyze option of the menu bar allows the user to run the analysis programs EXPO and FLUOR. FLUOR is a method of moments program for the deconvolution and analysis of sums of exponentials. EXPO is a very similar program except that it assumes that the impulse response function is measured directly and requires no deconvolution. It calculates

[53] D. F. Andrews, F. R. Bickel, P. J. Hampel, P. J. Huber, W. H. Rogers, and J. W. Tukey, "Robust Estimators of Location." Princeton Univ. Press, Princeton, New Jersey, 1972.

TABLE III
PULLDOWN MENU FOR F_SHELL

Program	Data file		Utilities		Analyze		Synthesize		Graphics	Quit (Ctrl_Q)
About F_Shell	Select new files	(Esc)	BG calculation	(Ctrl_B)	Expo (F2)		Simple exponential	(Shft_F2)	Plot (Ctrl_P)	
Set defaults	Open other directory	(Alt_F2)	Calculate anisotropy	(Ctrl_A)	Fluor (F3)		Add noise	(Shft_F3)		
	Copy to other directory	(Alt_F3)	Edit data file	(Ctrl_E)						
	Copy from other directory	(Alt_F4)	Import data	(Ctrl_I)						
	Delete in other directory	(Alt_F5)								
	Close other directory	(Alt_F6)								
	Format diskette	(Ctrl_F)								
	Write to diskette	(Ctrl_W)								
	Delete files	(Ctrl_D)								

the reduced moments of Eq. (5) directly from the data. To run EXPO, the user selects a single data file for analysis and then selects the EXPO menu option. F_Shell displays header information from the data file and then runs program EXPO, using the file name in a command line argument. To run FLUOR, the user selects one or two data files for analysis and then selects the FLUOR option. F_Shell displays header information and makes a reasonable guess as to which data in the selected file or files the user probably plans to use for $E(t)$ and $F(t)$. This choice is usually trivial. For example, if one chooses an EF file, F_Shell will select the E for $E(t)$ and the F for $F(t)$. The choice becomes less obvious if the user wishes to perform F/F deconvolution, because F_Shell must decide which F is to be used in the place of an $E(t)$. To give the user the necessary control, a popup window appears showing the automatic choices made by F_Shell and permitting the user to change them. Again, F_Shell runs FLUOR using the final choice of data in a command line argument. After existing EXPO or FLUOR, F_Shell returns to the data selection window.

The Graphics option of the menu bar runs a general graphics program for plotting data and the results of analyses. Quit returns to the DOS shell.

Structure of Program FLUOR

Program FLUOR deconvolutes and analyzes a fluorescence for a sum of exponential components. The following discussion describes the program in its current form. The author hopes that a more advanced form of the program will soon be available with menus, help screens, and data entry windows, but it will be structured and will function essentially the same as the current version described here.

The basic structure of the program FLUOR is illustrated in Fig. 4. The main program provides little more than a main menu which calls subroutines to provide all of the necessary functions of the program. On entry into the program subroutine START is called to initialize variables and arrays. It also calls INPUT to read data into memory and calculates a default range of channels to be used for the calculation of the moments. The main program calls MOMS to generate a set of depressed moments. The program then goes to the main menu where the user must enter an option code to tell the program what to do. Except for the code GON (where N is an integer from 1 to 6), all of the codes require the entry of only the first two letters. Either uppercase or lowercase will work. In the current version of the program the available codes are as follows:

1. LSearch allows the operator to estimate the number of components present and to choose a range of λ values for a λ scan.

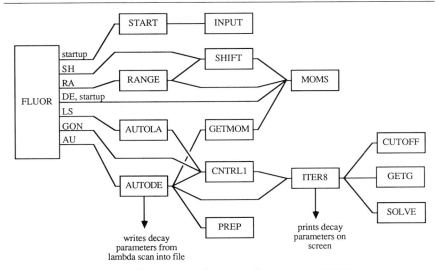

Fig. 4. Subroutine call diagram for program FLUOR.

2. AUto automatically scans through 21 different values of λ, performs analyses for specified number of components and range of MD, and saves the resulting decay parameters in a file (with the extension .FLU). The scan begins at the high λ end of the range and proceeds to low λ, each time using the previous decay parameters to estimate the cutoff correction to begin the new analysis. The data from the λ scan is used to generate the λ-invariance plot.

3. GON performs a "cold start" analysis for N components (e.g., GO2, GO5). N is an integer between 1 and 6. The program will perform a one-component analysis and use the resulting α and τ to begin a two-component analysis, and the two-component result to begin a three, etc. Analyses are performed at each value of the default MD range. A λ scan is necessary for obtaining and interpreting the final results, but the GON option can be useful for getting one's bearings.

4. MD permits the operator to change the default range of MD for which analyses will be performed. The initial default range is 1 to 4. The operator might wish to use higher MD to better resolve long lifetime components or lower MD to better resolve short lifetime components, although analyses at lower MD will tend to be more sensitive to nonrandom data errors.

5. DEpress lets the user change the value of λ and calculate a new set of depressed moments.

6. SHift lets the operator change the relative time position of $E(t)$ relative to $F(t)$. Positive shift shifts $E(t)$ forward in time relative to $F(t)$.

7. RAnge displays the current range of data used to calculate the moments and permits the user to change it.

8. FIlter permits the entry of a lifetime which will be held constant using a Cheng–Eisenfeld filter during the next analysis or λ scan. This option can be used more than once in succession so that more than one lifetime will be held constant.

9. SCatter uses Eq. (34) to calculate a scatter coefficient. Scatter coefficients are displayed on the screen and are saved in the file with the λ-scan data.

10. BG allows the user to change the background level on $E(t)$ or $F(t)$. This could also be done using the ''Edit data file'' option of F_Shell.

11. EPsilon allows the user to change a parameter used to determine whether an individual analysis has converged. The default value is 0.001. Decreasing this value makes convergence more difficult.

12. OP presents a list of the code options available.

13. EXit is used to return to F_Shell.

One can begin analyses by typing either GON, LS, or AU. Suppose the user types ''GO3.'' Program FLUOR then calls subroutine CNTRL1, which controls a ''cold start'' analysis. First, it chooses the first MD value of the MD range, calculates a crude estimate of α_1 and τ_1 from the G values without any cutoff corrections, and calls ITER8 to start a one-component analysis. ITER8 must have a first guess of decay parameters, because it begins by estimating cutoff corrections using Eqs. (24)–(26). After the one-component analysis is complete, the resulting decay parameters are then used as a first guess to call ITER8 to begin a two-component analysis, and the resulting two-component decay parameters are used to begin a three-component analysis. At the completion of the three-component analysis, CNTR1 increments MD and repeats the procedure. The process continues until MD reaches its maximum value.

The iteration procedure is controlled by subroutine ITER8. First, ITER8 begins with the guess given to it by CNTRL1. It calls CUTOFF to calculate cutoff corrections; it then calls GETG, which uses the cutoff corrections to correct the moments of $F(t)$ in order to calculate new G values. Finally it calls SOLVE3 to take the G values and solve for the appropriate number of decay parameters, α_i and τ_i. ITER8 then continues to cycle through calls to CUTOFF, GETG, and SOLVE in sequence. Each time, the previously determined decay parameters are used to estimate the cutoff correction. As soon as the newly determined values of α_i and τ_i

agree within preset limits with the previous values, ITER8 prints them out and returns to CNTR1.

Depending on how many components are sought, SOLVE calls a number of other subroutines to find the decay parameters. One- and two-component analyses are relatively trivial and are handled within SOLVE. For a three-component analysis the lifetimes are found as the three roots to a third order polynomial using Eqs. (13) and (11). Equation (7) is inverted to find the amplitudes. For four-, five-, and six-component analyses, QDROOT uses a QD (quotient-differnce) algorithm to estimate the roots.[54]

Suppose the user types "LS" for a λ search. FLUOR then calls AUTOLA which uses CNTRL1 to perform a sequence of cold start analyses with the goal of finding a value of λ for which the maximum value of the cutoff correction is about 0.1%. How to use this routine is discussed with the example in the next section.

Finally, suppose the user types "AU" to perform a λ scan. FLUOR then calls subroutine AUTODE, which begins by calling GETMOM. GETMOM repeatedly calls MOMS to generate an array of depressed moments that will be needed for all of the analyses of the λ scan. This is necessary so that all of the moments will not need to be recalculated for each new MD. After getting the necessary moments, AUTODE prompts the user for a range of λ, and a range of the number of components desired. For example, three- and four-component analyses might be wanted. In this case AUTODE starts with the lowest MD, the highest λ, and three-component analyses. First it calls CNTRL1 to perform a cold start analysis for three components, and then it uses this first analysis as the first guess to call ITER8 directly. Each subsequent analysis is used as the first guess for the analysis at the next lower value of λ. When analyses are completed for 21 different values of λ, AUTODE increments MD, repeating the procedure until analyses at the maximum MD are finished. When all of the three-component analyses are complete, AUTODE repeats the whole process for four components.

Using Program FLUOR

The goal of this section is to describe a particular general way in which the program can be used. This procedure will aim at obtaining a set of λ-invariance plots that will tell the user (1) whether the decay can be adequately described as a sum of exponentials, (2) how many resolvable components are present in the decay, and (3) what the decay parameters (α_i and τ_i) are.

[54] H. Rutihauser, Z. Angew. Math. Phys. 5, 496 (1954).

Entering Program and Finding Right λ Range

On entry into FLUOR, the data are read into memory and some useful information about the data is printed on the screen. FLUOR then requests a value for λ so that it can generate a set of moments. The user can enter zero to skip this calculation, because the way we are about to use the program will not need this initial set. The program then prompts for an option code. This is the main menu of the program. The user should type "LS" in order to perform a λ search.

A λ search provides the operator with a partially automated procedure for estimating the number of components in the decay and the range of λ values to be used for the λ-invariance plot. The λ search begins with a one-component analysis and then proceeds to higher numbers as directed by the user. The object is to find a value of λ such that the cutoff correction for the highest used moment of $F(t)$ is about 0.1% of the moment. Once the proper value of λ has been estimated, the results of MD2 to MD4 analyses are printed.

At this point the user must examine the result and make a choice of having the above procedure repeated for one more component or of returning to the main menu. If the number of components is insufficient, then the lifetimes will increase progressively with MD (or may decrease progressively if the decay contains a negative amplitude). If the lifetimes are essentially independent of the MD or change erratically with MD, then it is a good guess that more components will not be needed. The source of this behavior will become more apparent below, where we examine an actual λ-invariance plot.

Before quitting to the main menu, the user should note the final λ value as well as the final number of components. This latter number is probably the correct number of decay components in the data. As a first estimate of the appropriate range of λ for the λ scan, the user should take the final value of λ from the search, round it off to the nearest number easily divisible by 3 (e.g., round 0.86 to 0.90). This will be the maximum λ. Divide it by 3 to get the minimum λ.

Generating λ Scans

From the main menu the user should next enter the code "AU" in order to generate a λ scan. The program will ask for the minimum and maximum number of components to be resolved. Because one probably knows the number of components in the data from the λ search, it should be used for the minimum. To check for component incrementation, enter one more for the maximum. The program will then ask for the minimum and maximum values of λ. Enter the values just determined in the λ search.

After the λ scan analyses are complete the program will return to the main menu. Exit the program to return to F_Shell. Choose graphics and plot the λ-invariance plot.

Interpreting λ-Invariance Plots

As an example, monoexponential decays were measured on four individual samples, including tyrosine in ethanol and tyrosine quenched by different concentrations of potassium iodide. These decays are shown in Fig. 5A. The four decays were summed in the computer to give the four-component decay of Fig. 5B.[55]

A set of five λ-invariance plots is shown in Fig. 6 for one- through five-component analyses of the four-component data. The first three show clear lack of resolution, and in fact would not have been plotted if the user had followed the instructions given above for generating λ scans. Note that throughout the range of λ, higher MD returns higher lifetimes, and a λ search would have clearly indicated that the data contain more than three components. The four- and five-component analyses indicate flat regions and thus satisfy Rule 1 on λ-invariance.

MD0 analyses were not performed, because we knew that the presence of nonrandom errors would prevent them from resolving the decays. MD1 was performed, but it was omitted from the plot because it did not agree with the other MD values due to nonrandom errors. However, in the four-component plot MD3 and MD4 agree quite well. Even better MD agreement is seen in the five-component plot, where MD2, MD3, and MD4 agree quite well. Thus Rule 2, MD agreement, is satisfied.

Finally, it is obvious from Fig. 6E that the five-component analysis returns the same four components found in Fig. 6D. The fifth lifetime has a small amplitude and never even appears on the plot. Therefore Rule 3, component incrementation, is satisfied as well.

Summary

If one has a convoluted fluorescence decay and wishes to analyze it for a sum of exponentials, then one can begin by asking either of two questions: (1) What sum of exponentials best fits the data? (2) What physical decay parameters gave rise to the data? At first these two questions may sound equivalent; in fact, they represent different philosophical approaches to data analysis. In resolving the first question, one adjusts the decay parameters until a calculated curve agrees within arbitrarily

---

[55] L. J. Libertini and E. W. Small, *Anal. Biochem.* **163,** 500 (1987).

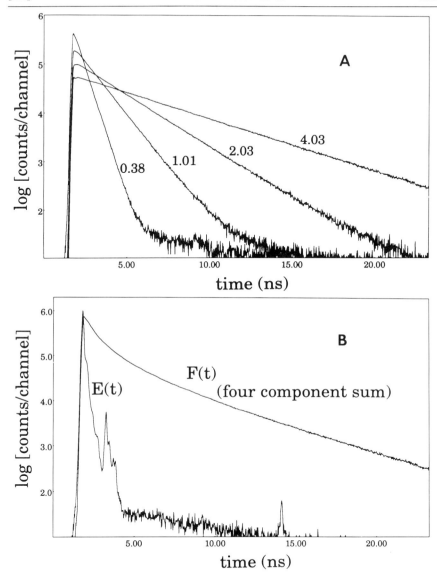

FIG. 5. Four-component fluorescence data.[55] Four single-component fluorescence decays with lifetimes of 0.38, 1.01, 2.03, and 4.03 nsec are shown in (A). These data were summed to give the four-component decay in (B). The measured $E(t)$ is also shown. The data were measured using a picosecond laser source and a triple-microchannel plate photomultiplier.

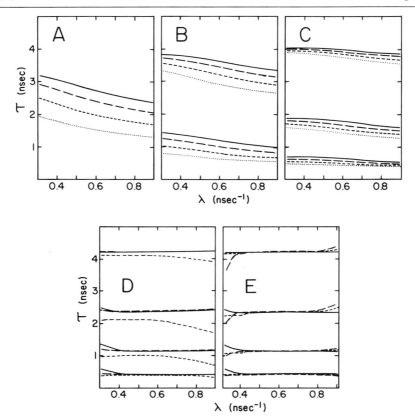

FIG. 6. Method of moments analysis of four-component data.[55] λ-Invariance plots are shown for one- to five-component analyses (A to E, respectively) of the four-component data shown in Fig. 5B. Results for MD1, MD2, MD3, and MD4 are represented by dotted, short-dashed, long-dashed, and solid lines, respectively. Lifetimes recovered are 0.43, 1.13, 2.33, and 4.21 nsec.

chosen limits to the original data. This is what we did in the fourth section of Table II. The fit obtained was decent, but the resulting parameters were wrong.

A more difficult approach is to design a method of data analysis which is intrinsically insensitive to the presence of anticipated errors, aiming directly at recovering the decay parameters without regard to the fit. This is what we have done with the method of moments with MD. If particular errors do not have much effect on the recovered parameters, then such a method of data analysis is said to be robust with respect to those errors.[56] Robust methods are widely used in engineering but have not seen much introduction yet to biophysics. Least-squares, the basis of the commonly used data fitting methods for pulse fluorometry, is nonrobust with respect to underlying noise distributions.[53,57,58] Isenberg has shown that least-squares is nonrobust with respect to the nonrandom light scatter, time origin shift, and lamp width errors as well.[59] As shown in Isenberg's paper, as well as here, the method of moments with MD is quite robust with respect to these nonrandom errors. Perhaps question (1) could be modified to include all of the errors that might be present in the data; but then, how would one decide which errors to include and whether an error is present? What fitting criterion would tell one this? Why choose a method which depends so strongly on this information when robust alternatives exist?

As a rule, fitting should not be used as a criterion for correct decay parameters, unless all of the significant nonrandom errors have been included in the fit. If one fits the data but has not incorporated an important error, then the best fit will necesssarily give the wrong answer. The method of moments provides clear criteria for accepting or rejecting an analysis. λ-Invariance plots will not be flat if the result is distorted by the presence of a nonrandom error, or if one is using the wrong decay model to fit the data.

It is important to remember that, even if the method of moments is intrinsically insensitive to certain nonrandom errors, it is still possible to use the method to obtain a sensitive quantitative measure of errors. For example, Eq. (34) can be used to determine the amount of scatter in data, or a plot such as that shown in Fig. 2 can be used to determine a time origin shift.

For relatively easy decay resolutions, using high-quality time-resolved fluorescence data, any method of analysis will probably give the same correct answer. The method of moments will excel for data which contain significant nonrandom errors such as those we have shown to be corrected by MD. The method of moments will also excel for difficult resolutions of sums of exponentials, because such resolutions are easily disrupted by small amounts of nonrandom error. Besides the sum of exponentials analy-

---

[56] In general, a robust method of estimating parameters from data is one that is insensitive to small deviations from the assumptions underlying the choice of the estimator. We assume a convoluted sum of exponentials.

[57] J. W. Tukey, *in* "Contributions to Probability and Statistics" (I. Olkin, ed.), p. 448. Stanford Univ. Press, Stanford, California, 1960.

[58] P. J. Huber, "Robust Statistics." Wiley, New York, 1981.

[59] I. Isenberg, *Biophys. J.* **43**, 141 (1983).

ses shown here, the method of moments can be used to analyze data for which the impulse response function is given by $\alpha \exp(-at - bt^{1/2})$.[20] All of the examples shown in this chapter have been single-curve analyses and have not taken advantage of the extreme resolving power of global analyses. Global analysis, however, is available for the method of moments.[15,19] It is also possible to perform analyses for distributions of lifetimes.[32,60] These methods have not been discussed here owing to space limitations.

Other than the examples mentioned in the previous paragraph, the method of moments cannot readily be used for the analysis of complex nonexponential decay functions. Also, it is not yet possible to link decay parameters in a complex manner, although Cheng–Eisenfeld filtering will hold any number of lifetimes constant. There is no intrinsic reason why other complex decay models or linked parameters could not be solved using the method of moments. For example, one could fit the reduced deconvoluted moments to any necessary function, but, to my knowledge, this has not been attempted yet.

Programs FLUOR and EXPO were written for the analysis of fluorescence decays, but they can be easily modified for use with other techniques, for example, the oscillating waveforms of photoacoustics.[19,61] It is easy to learn how to use the method of moments. No starting guesses are required, and one is quickly presented with a visual result, a λ-invariance plot, which is straightforward to interpret.

Try using the method of moments. Remember the Hubble telescope. It is wise not to launch into orbit any theory dependent on complex decay analyses before first performing some simple tests on the ground. Find out whether accuracy has been compromised by nonrandom errors.

Acknowledgments

The late Dr. Irvin Isenberg was the strongest proponent of the method of moments. One of his main stated reasons for developing the method of moments was to introduce robust data analysis methods to biophysics. Anyone can write another fitting program and then conclude that the program works because the data are fit well. Developing a robust method requires knowledge of the types of errors likely to be present in the data and significant efforts to circumvent their effects. Isenberg believed that it is merely a matter of time before data analysis methods robust with respect to both random and nonrandom errors are developed and commonly used throughout science.

[60] Unfortunately at this time, owing to the finite time and resources of the author, copies of method of moments programs other than FLUOR and EXPO cannot be supplied.
[61] J. R. Small, S. H. Watkins, B. J. Marks, and E. W. Small, *Proc. SPIE Int. Soc. Opt. Eng.* **1204**, 231 (1990).

Dr. Isenberg taught me the method of moments. Dr. Louis Libertini measured the data shown in Figs. 3 and 5 and generated and analyzed the data of Table I. This work was supported by National Institutes of Health Grant GM25663. A grant from the Medical Research Foundation of Oregon played a crucial role in keeping this effort active in 1986. A critical reading of the manuscript by Dr. Jeanne Rudzki Small was greatly appreciated.

[12] Laplace Deconvolution of Fluorescence Decay Surfaces

By MARCEL AMELOOT

Introduction

Fluorescence used as a probing technique in the life sciences is very popular because of its sensitivity to the environment and the proper time scale of the phenomenon. Fluorescence experiments conducted under constant illumination can reveal interesting information. However, in many cases time-resolved techniques are required to completely unravel the various processes in the excited state which influence the fluorescence characteristics of the probe. The fluorescence decay of the probe can be measured in the time or in the frequency domain.[1-4] In this chapter the focus is on measurements in the time domain. Among the various experimental methods used to collect the data in the time domain, the time-correlated single photon counting technique is very popular because of the high signal-to-noise ratio and the high data density.

Data analysis has received much attention in the field of time-resolved fluorescence spectroscopy. The high quality of the collected data justifies an intensive study. There are two main problems encountered in the analysis of time-resolved decay data. The first is due to the finite width of the excitation pulse. In a first approximation the measuring system and the fluorescent sample are linear systems. This means that the collected curves are a convolution product of the instrumental response function and the delta response function of the sample under study. In many cases the impulse response function of the sample can be written as a sum of exponentially decaying functions. The nonorthogonality of the exponential functions is the basis of the second and most important problem.

[1] J. N. Demas, "Excited State Lifetime Measurements." Academic Press, New York, 1983.
[2] R. B. Cundall and R. E. Dale, eds., "Time-Resolved Fluorescence Spectroscopy in Biochemistry and Biology." Plenum, New York, 1983.
[3] J. R. Lakowicz, "Principles of Fluorescence Spectroscopy." Plenum, New York, 1983.
[4] D. V. O'Connor and D. Phillips, "Time-Correlated Single Photon Counting." Academic Press, New York, 1984.

Because of this nonorthogonality, the estimation of the number of exponentially decaying functions and of the preexponential and exponential factors is severely complicated.

The methods for data analysis have been extensively reviewed and compared.[1,2,4–8] The various approaches can be divided in two groups mainly. Data analysis can be performed either in the time domain, using nonlinear least-squares methods (NLLS),[9,10] or in transform domains, using Fourier transforms,[11–13] the method of moments,[14,15] Laplace transforms,[11,16–21] modulating functions,[22] or the phase plane method.[23] A procedure which specifically analyzes the data from pulse fluorimetry in terms of distributions of relaxation times has been developed by using the maximum entropy formalism.[24]

The fluorescence decay traces can be measured under various experimental conditions so that a decay data surface is obtained. It is important to note that current technology allows fluorescence decays to be collected with a high signal-to-noise ratio in a matter of minutes. This means that it is feasible to collect complete fluorescence decay surfaces. Simultaneous analysis of related fluorescence decay experiments has been suggested for

[5] A. E. McKinnon, A. G. Szabo, and D. R. Miller, *J. Phys. Chem.* **81,** 1564 (1977).

[6] D. V. O'Connor, W. R. Ware, and J. C. André, *J. Phys. Chem.* **83,** 1333 (1979).

[7] M. G. Badea and L. Brand, this series, Vol. 61, p. 387.

[8] M. Bouchy, ed., "Deconvolution–Reconvolution." Conference Proceedings, Ecole Nationale Supérieure des Industries Chimique de l'Institut National Polytechnique de Lorraine, Nancy, France, 1982.

[9] A. E. W. Knight and B. K. Selinger, *Spectrochim. Acta Part A* **27A,** 1223 (1971).

[10] A. Grinvald and I. Z. Steinberg, *Anal. Biochem.* **59,** 583 (1974).

[11] W. P. Helman, *Int. J. Radiat. Phys. Chem.* **3,** 283 (1971).

[12] U. P. Wild, A. R. Holzwarth, and H. P. Good, *Rev. Sci. Instrum.* **48,** 1621 (1977).

[13] J. C. André, L. M. Vincent, D. V. O'Connor, and W. R. Ware, *J. Phys. Chem.* **83,** 2285 (1979).

[14] I. Isenberg and R. D. Dyson, *Biophys. J.* **9,** 1337 (1969).

[15] E. W. Small, L. J. Libertini, D. W. Brown, and J. R. Small, *Proc. SPIE Int. Soc. Opt. Eng.* **1054,** 36 (1989).

[16] A. Gafni, R. L. Modlin, and L. Brand, *Biophys. J.* **15,** 263 (1975).

[17] A. Gafni, in "Time-Resolved Fluorescence Spectroscopy in Biochemistry and Biology" (R. B. Cundall and R. E. Dale, eds.), p. 259. Plenum, New York, 1983.

[18] M. Ameloot and H. Hendrickx, *Biophys. J.* **44,** 27 (1983).

[19] P. S. N. Dixit, A. J. Waring, K. O. Wells, P. S. Wong, G. V. Woodrow, and J. M. Vanderkooi, *Eur. J. Biochem.* **126,** 1 (1982).

[20] M. Ameloot, J. M. Beechem, and L. Brand, *Biophys. Chem.* **23,** 155 (1986).

[21] Bajzer, J. C. Sharp, S. S. Sedarous, and F. G. Prendergast, *Eur. Biophys. J.* **18,** 101 (1990).

[22] B. Valeur and J. Moirez, *J. Chim. Phys. Chim. Biol.* **70,** 500 (1973).

[23] J. N. Demas and A. W. Adamson, *J. Phys. Chem.* **75,** 2463 (1971).

[24] A. K. Livesey and J. C. Brochon, *Biophys. J.* **52,** 693 (1987).

the resolution of fluorescence spectra[25,26] and to improve the accuracy of the recovered parameters.[20,27–31] This simultaneous analysis has been denoted as the global analysis approach, and its various aspects are discussed in several chapters in this volume. The underlying principle of this methodology is to utilize the relationships between related decay curves by linking the parameters which are common to the various fitting functions. It has been shown that global analysis helps both in discerning competing models and in the recovery of model parameters. The global procedure has been applied to the analysis of total luminescence decay curves in both real time[28–31] and transform domains,[20,25,27] or features from both domains may be used.[32] Similarly, the polarized intensity decays in a fluorescence polarization experiment can be analyzed simultaneously by nonlinear least squares[33–37] or by Laplace transforms.[18] It has to be emphasized that the linkage between parameters is not restricted to relaxation times and preexponentials. On the contrary, by fitting directly for the parameters of the assumed model (e.g., the rate constants of an excited state reaction), the set of experimental conditions over which parameters can be linked is extended. When excited-state reactions are analyzed in terms of the rate constants, data collected at various concentrations can be globally analyzed. This would not be possible if the fitting parameters were preexponentials and relaxation times. The procedure for analyzing decay data surfaces in terms of rate constants is called the global compartmental analysis and is discussed in [14] in this volume.

This chapter describes the use of Laplace transforms in the analysis of decay data obtained by a pulse method. The original Laplace transform method described for the deconvolution of decay traces required an iterative procedure.[11,16,17] However, the modified Laplace transform

[25] P. Wahl and J. C. Auchet, *Biochim. Biophys. Acta* **285,** 99 (1972).

[26] F. J. Knorr and J. M. Harris, *Anal. Chem.* **53,** 272 (1981).

[27] J. Eisenfeld and C. C. Ford, *Biophys. J.* **26,** 73 (1979).

[28] J. R. Knutson, J. M. Beechem, and L. Brand, *Chem. Phys. Lett.* **102,** 501 (1983).

[29] J. M. Beechem, M. Ameloot, and L. Brand, *Anal. Instrum. (N.Y.)* **14,** 379 (1985).

[30] J.-E. Löfroth, *Eur. Biophys. J.* **13,** 45 (1985).

[31] L. D. Janssens, N. Boens, M. Ameloot, and F. C. De Schryver, *J. Phys. Chem.* **94,** 3564 (1990).

[32] J. P. Privat, P. Wahl, J. C. Auchet, and R. H. Pain, *Biophys. Chem.* **11,** 239 (1980).

[33] C. W. Gilbert, *in* "Time-Resolved Fluorescence Spectroscopy in Biochemistry and Biology" (R. B. Cundall and R. E. Dale, eds.), p. 605. Plenum, New York, 1983.

[34] A. J. Cross and G. R. Fleming, *Biophys. J.* **46,** 45 (1984).

[35] M. Ameloot, H. Hendrickx, W. Herreman, H. Pottel, H. Van Cauwelaert, and W. van der Meer, *Biophys. J.* **45,** 525 (1984).

[36] A. Arcioni and C. Zannoni, *Chem. Phys.* **88,** 113 (1984).

[37] S. R. Flom and J. H. Fendler, *J. Phys. Chem.* **92,** 5908 (1988).

method[18,20] is noniterative and does not require initial guesses to start the analysis. Because of its speed, the modified Laplace transform method is ideal for the analysis of fluorescence decay data surfaces.

Theory

Iterative Laplace Deconvolution Method (LAP1)

In the following, it will be assumed that the fluorescent sample and the measuring chain behave as a linear system. In a pulsed experiment and under ideal conditions, the fluorescence relaxation of the sample, $g(t)$, can be written as a convolution product of the measured excitation function, $l(t)$, and of the impulse response function $f(t)$ of the fluorescent system,

$$g(t) = \int_0^t l(u)f(t - u)\, du = l(t) * f(t) \tag{1}$$

The function $f(t)$ has to be determined from the measurement of $g(t)$ and $l(t)$. In many cases $f(t)$ can be described adequately by a sum of exponential decaying functions,

$$f(t) = \sum_{i=1}^n a_i e^{-t/\tau_i} \tag{2}$$

The problem is then reduced to the estimation of n and the parameters $a = (a_1, \ldots, a_n)$ and $\tau = (\tau_1, \ldots, \tau_n)$.

Helman[11] introduced the use of the Laplace transform for the determination of the preexponential factor and the relaxation time of a monoexponential decay. The method has been extended by Gafni et al.[16,17] to analyze bi- and triexponential decays. However, separate expressions were required for $n = 1, 2,$ and 3. A more general treatment has been presented by Ameloot and Hendrickx.[18] The latter implementation is described below and is referred to as LAP1. The definition of the Laplace transform will be given first.

Let $p(t)$ be a function of t specified for $t > 0$. The Laplace transform of $p(t)$, denoted by $\mathscr{L}(p)$, is defined by

$$\mathscr{L}(p) = P(s) = \int_0^\infty p(t) e^{-st}\, dt \tag{3}$$

where s is the transform parameter. In the following, s is taken to be real and nonnegative.

The Laplace transform of a convolution product of two functions is the product of the Laplace transforms of the functions. The transformation with parameter s of Eq. (1) yields $G(s) = L(s)F(s)$, where $G(s)$, $L(s)$, and

$F(s)$ denote the Laplace transforms of $g(t)$, $l(t)$, and $f(t)$, respectively. It will be supposed further on that $f(t)$ is described by Eq. (2). Because of the linearity of the Laplace operator, $F(s)$ can be written as

$$F(s) = \sum_{i=1}^{n} \frac{a_i}{s + (1/\tau_i)} \tag{4}$$

providing the equality

$$\frac{G(s)}{L(s)} = \sum_{i=1}^{n} \frac{a_i}{s + (1/\tau_i)} \tag{5}$$

By calculating $2n$ Laplace transforms (s, real and positive) of $g(t)$ and $l(t)$, $2n$ nonlinear equations like Eq. (5) are obtained from which a_i and τ_i are solved.

However, the experimental curves are only defined for a finite time window $[0,T]$ and, in general, do not vanish in the last data channel at time T. This means that Eq. (5) must be corrected for this cutoff error. One may write

$$G(s) = \int_0^T g(t)\, e^{-st}\, dt + \int_T^\infty g(t)\, e^{-st}\, dt \tag{6a}$$

$$= G^T(s) + G^\infty(s) \tag{6b}$$

$G^T(s)$ denotes the transform calculated from the actual data. The cutoff correction $G^\infty(s)$ is completely determined by the assumed extension of the excitation profile $l(t)$ on $[T,\infty]$. Although several extensions are valid, an obvious choice is to set $l(t)$ equal to zero for $t > T$. This means for $l(t)$ that $L(s) = L^T(s)$. The cutoff correction $G^\infty(s)$ is then given by

$$G^\infty(s) = e^{-st} \sum_{i=1}^{n} \frac{a_i c_i}{s + (1/\tau_i)} \tag{7}$$

in which $a_i c_i$ is the contribution of the ith component to the last channel of the measured decay,

$$c_i = \int_0^T l(u)\, e^{-(T-u)/\tau_i}\, du \tag{8}$$

Gafni et al.[16,17] proposed an iterative procedure for calculating the corrected transforms. The solution of a set of Eq. (5) using $G^T(s)$ and $L^T(s)$ leads to a first approximation of a_i and τ_i, which in turn are used in calculating $G^\infty(s)$ according to Eqs. (7) and (8). The transforms of $g(t)$ can then be corrected, which will lead to a new set of estimates for the parameters. This procedure is repeated until some convergence criterion is satisfied.

The parameters a_i and τ_i in each iteration step can be determined as follows. For each value s_j of the transform parameter, Eq. (5) can be rewritten as

$$L(s_j) \sum_{i=1}^{n} s_j^{i-1} E_i(a,\tau) - G(s_j) \sum_{i=1}^{n} s_j^i D_i(\tau) = G(s_j) \tag{9}$$

where for $1 \leq i \leq n$,

$$E_i(a,\tau) = \sum_{1 \leq k_1 < k_2 < \ldots < k_1 \leq n} \tau_{k_1} \tau_{k_2} \cdots \tau_{k_i}(a_{k_1} + a_{k_2} + \ldots + a_{k_i}) \tag{10}$$

$$D_i(\tau) = \sum_{1 \leq k_1 < k_2 < \ldots < k_i \leq n} \tau_{k_1} \tau_{k_2} \cdots \tau_{k_i} \tag{11}$$

and

$$D_0(\tau) \equiv 1 \tag{12}$$

The functions D_i and E_i can be determined from a system of linear equations of the type of Eq. (9) using $2n$ values of s_j. The relaxation times τ_i are the roots of the polynomial of degree n,

$$\sum_{i=0}^{n} (-1)^i D_i(\tau) \tau^{n-i} = 0 \tag{13}$$

Common numerical techniques can be used to determine the roots of this polynomial. Once $\tau = (\tau_1, \ldots, \tau_n)$ is determined, $a = (a_1, \ldots, a_n)$ can be readily obtained from the values obtained for E_i by solving the appropriate linear system.

When unacceptable negative parameter values result from the first step in the analysis, another choice for the transformation parameters can be advised. However, it is found that in many cases the first set of transform parameters s_j can be maintained, provided that the decay component with wrong parameter values is ignored in the calculation of the cutoff correction.[18]

Instrumental artifacts, such as scatter and time shift, can be corrected with this deconvolution method by modifying Eq. (5). The amount of scattered lamp light, $S_c l(t)$, can be calculated by considering

$$\frac{G(s)}{L(s)} = \sum_{i=1}^{n} \frac{a_i}{s + (1/\tau_i)} + S_c \tag{14}$$

$$L(s_j) s_j^n D_n(\tau) S_c + L(s_j) \sum_{i=1}^{n} s_j^{i-1} K_i(a,\tau,S_c) - G(s_j) \sum_{i=1}^{n} s_j^i D_i(\tau) = G(s_j) \tag{15}$$

with

$$K_i(a,\tau,S_c) = E_i(a,\tau) + S_c D_{i-1}(\tau) \tag{16}$$

Determination of $D_n(\tau)S_c$, $K_i(a,\tau,S_c)$, and $D_i(\tau)$ from a system of equations using $2n + 1$ different values for s_j leads to the final solution for S_c, $a = (a_1, \ldots, a_n)$ and $\tau = (\tau_1, \ldots, \tau_n)$.

The excitation profile can usually not be recorded at the same wavelength settings as used for the sample. This may imply that the convolution product in Eq. (1) is not correct. When the decay curve of the sample is shifted to longer times due to effects in the detection photomultiplier, the time shift $Q, Q > 0$, may be estimated by solving[16,17]

$$\frac{G(s)}{L(s)} = e^{-Qs} \sum_{i=1}^{n} \frac{a_i}{s + (1/\tau_i)} \tag{17}$$

The time-shift Q can be determined from measuring the decay of a standard with a single relaxation time under the same experimental conditions as the sample of interest. Once the value for Q is obtained one can define

$$G_Q(s_j) = G(s_j)\, e^{Qs_j} \tag{18}$$

The unknown parameters a and τ can now be determined by replacing $G(s_j)$ by $G_Q(s_j)$ in Eq. (9).

The wavelength dependence of the instrumentation may not lead to a shift effect only. The shape of the functions may change as well. An alternative method, called the reference convolution, remediates for other effects as well. The Laplace implementation of this method is described in a separate section below.

In the case of a large cutoff error, LAP1 may not yield a valid first approximation for $a = (a_1, \ldots, a_n)$ and $\tau = (\tau_1, \ldots, \tau_n)$ for multiexponential decays, or the iteration procedure will take a long computing time. Another implementation of the Laplace transform method which circumvents the problem of the cutoff correction is described in the next section.

Noniterative Laplace Deconvolution Method (LAP2)

The shortcomings of LAP1 can be remediated by using a modified Laplace transform, leading to an algorithm denoted as LAP2.[18,20] For a function $p(t)$ on the interval $[0,T]$, the modified transformation $\mathscr{L}^T$ is defined by

$$\mathscr{L}^T(p) \equiv P^T(s) = \int_0^T p(t)\, e^{-st}\, dt \tag{19}$$

The result of applying the transformation $\mathcal{L}^T$ to the convolution product in Eq. (1) can be calculated as follows. For the model function $f(t)$ in Eq. (2) the convolution product $g(t)$ can be written as

$$g(t) = \sum_{i=1}^{n} z_i(t) \tag{20}$$

where

$$z_i(t) = \int_0^t l(u)a_i e^{-(t-u)/\tau_i} du \tag{21}$$

is the contribution of the ith component in $f(t)$ to the convolution product $g(t)$. Note that $z_i(t)$ is purely mathematical and does not have to be associated with a particular excited species. Applying the transformation $\mathcal{L}^T$ to $g(t)$ gives

$$G^T(s) = \sum_{i=1}^{n} Z_i^T(s) \tag{22}$$

Evaluating $Z_i^T(s)$ by partial integration and taking into account that

$$dz_i(t)/dt = -(1/\tau_i)z_i(t) + a_i l(t) \tag{23}$$

gives

$$Z_i^T(s) = -(1/s) e^{-sT} z_i(T) - (1/s\tau_i) Z_i^T(s) + (a_i/s)L^T(s) \tag{24}$$

Rearranging Eq. (24), performing the summation in Eq. (22), and using $z_i(T) = a_i c_i$ yields

$$\frac{G^T(s)}{L^T(s)} = \sum_{i=1}^{n} \frac{a_i}{s + (1/\tau_i)} \left[1 - \frac{e^{-sT} c_i}{L^T(s)} \right] \tag{25}$$

Equation (25) is exact. No extrapolation of $l(t)$ beyond the actual time window of the experiment is required. The functions D_i and E_i can be obtained as follows. For the transform parameter s_j, Eq. (25) can be rewritten as[20]

$$G^T(s_j) \sum_{i=1}^{n} s_j^i D_i(\tau) = L^T(s_j) \sum_{i=1}^{n} s_j^{i-1} E_i(a,\tau) - e^{-s_j T} \sum_{i=1}^{n} s_j^{i-1} H_i(a,\tau,c) \tag{26}$$

with

$$H_i(a,\tau,c) = \sum_{1 \le k_1 < k_2 < \ldots < k_i \le n} \tau_{k_1} \tau_{k_2} \ldots \tau_{k_i} (a_{k_1} c_{k_1} + a_{k_2} c_{k_2} + \ldots + a_{k_i} c_{k_i}) \tag{27}$$

To make the method noniterative, the functions $H_i(a,\tau,c)$ have to be eliminated by considering different values of s_j. Using equidistant values of the transform parameters s_j, namely, $s_{j+k} = s_j + k\Delta$, $\Delta > 0$, it has been shown[20] that the resulting expression is given by

$$\sum_{i=1}^{n} M_{i,n}[G^T(s_j)]D_i(\tau) = \sum_{i=1}^{n} M_{i-1,n}[L^T(s_j)]E_i(a,\tau) \tag{28}$$

The operation $M_{i,n}$ on a transform $P^T(s)$ is defined by

$$M_{i,n}[P^T(s_j)] \equiv \sum_{k=0}^{n} (-1)^k \frac{n!}{j!(n-j)!} P^T(s_{j+k})s_{j+k}^i e^{k\Delta T} \tag{29}$$

The variables $E_i(a,\tau)$ and $D_i(\tau)$ can be determined from $2n$ equations of the type of Eq. (28). Once their values are obtained, the decay parameters are recovered as indicated for LAP1.

LAP2 also allows correction for several instrumental artifacts. The scatter conttibution S_c can be estimated from $2n + 1$ equations of the following type

$$\sum_{i=0}^{n} M_{i,n}[G^T(s_j)]D_i(\tau) = M_{n,n}[L^T(s_j)]S_cD_n(\tau) + \sum_{i=1}^{n} M_{i-1,n}[L^T(s_j)]K_i(a,\tau,S_c) \tag{30}$$

The time-shift error Q can be taken into account by replacing $G^T(s_j)$ by $G_Q^T(s_j)$, analogously to Eq. (18).

Simultaneous Analysis of Multiple Total Intensity Decay Curves

We now indicate how LAP2 can be used in the simultaneous analysis of fluorescence decay data surfaces by linking preexponentials and relaxation times. Consider a set of q experiments. For each experiment k consisting of the instrumental response function $l_k(t)$ and the recorded fluorescence relaxation $g_k(t)$ one has

$$\sum_{i=1}^{n_k} \mathbf{D}_{k,i}(s_j)D_{k,i} - \sum_{i=1}^{n_k} \mathbf{E}_{k,i}(s_j)E_{k,i} = -D_{k,0}(s_j) \tag{31}$$

with

$$\mathbf{E}_{k,i}(s_j) = M_{i-1,n}[L_k^T(s_j)] \tag{32}$$

$$\mathbf{D}_{k,i}(s_j) = M_{i,n}[G_k^T(s_j)] \tag{33}$$

and where $D_{k,i}$ and $E_{k,i}$ are the equivalents of the definitions in Eqs. (10) and (11); n_k denotes the number of exponentially decaying functions required in the kth experiment.

For many systems of interest decay curves obtained at several excitation/emission wavelengths will exhibit common decay times and varying preexponential components. Because LAP2 eliminates the cutoff error, decay experiments performed with different time windows may be combined to obtain a high accuracy in resolving multiexponentials with extreme relaxation times. Experiments of these types give rise to different functions $E_{k,i}$ and to equal functions of the relaxation times, that is, $D_{k,i} = D_i$ for $i = 1, \ldots, n; k = 1, \ldots, q$. In this case a system of linear equations has to be solved for $n(q + 1)$ unknowns, $\{E_{k,i} \mid k = 1, \ldots, q; i = 1, \ldots, n\}$ and $\{D_i \mid i = 1, \ldots, n\}$. The minimum number of equations that has to be constructed from the experimental decay surface depends on the number of experiments and on the number of relaxation times. It is not required that the same number of equations of the type of Eq. (28) be generated from each individual experiment. When in total more than $n(q + 1)$ equations are calculated, the overdetermined system can be solved by applying least-squares. Once the values of the functions D_i are obtained, the relaxation times τ_j are determined as indicated in the previous section. Using the recovered values for τ_j and $E_{k,i}$, the preexponentials for experiment k can be found. For the described linkage, each curve is analyzed for the same lifetimes. The preexponentials then indicate whether a particular lifetime contributes to the observed decay for the considered experiment.

The global analysis approach using NLLS as described by Knutson *et al.*[28] is more general because any desired linkage between the parameters is possible. To achieve this with LAP2, one has to solve the system of equations directly for the preexponentials and the relaxation times instead of solving for the functions $E_{k,i}$ and $D_{k,i}$. Nonlinear techniques must then be used. This version of the program will be referred to as the nonlinear LAP2. Although the nonlinear LAP2 is iterative, it is still a very fast method because the numerous data points in the time domain are replaced by a few equations in the transform domain. Furthermore, because the calculation of the transforms of the observed data is the most time-consuming part of the program, storage of the calculated transforms in computer memory allows for efficient investigation of alternative linkages. The mapping of the parameters of each experiment into the global parameter vector for the total decay data surface is more complicated in this general case. A scheme similar to that used for NLLS[28,29] can be used.

Simultaneous Analysis of Multiple Polarized Intensity Decays

In a time-resolved anisotropy experiment, the time course of the fluorescence emission is analyzed into the parallel and perpendicular components, $i_\parallel(t)$ and $i_\perp(t)$, with respect to the polarization of the excitation light. The emission anisotropy is defined as

$$r(t) = \frac{i_\parallel(t) - i_\perp(t)}{i_\parallel(t) + 2i_\perp(t)} \tag{34}$$

In general, the decay of the emission anisotropy for a rotating molecule in an isotropic environment is described by a sum of exponentials.[38] If the rotational motion of the fluorescent molecule is restricted, a limiting anisotropy r_∞ has to be considered,[39,40]

$$r(t) = \sum_{j}^{m} \beta_j e^{-t/\phi_j} + r_\infty \tag{35}$$

The two polarization components can be expressed as

$$i_\parallel(t) = \frac{f(t)}{3}[1 + 2r(t)] \tag{36}$$

$$i_\perp(t) = \frac{f(t)}{3}[1 - r(t)] \tag{37}$$

Both components are linear combinations of $n(m + 1)$ exponential terms with the same relaxation times.

In practice, the number of exponentials, m, in the anisotropy $r(t)$ is restricted to one or two. In isotropic solvents, this may correspond to an isotropic rotator or to special cases of a rigid ellipsoid.[38] However, in an anisotropic environment (e.g., lipid bilayer) the description of the anisotropy, $r(t)$, by one or two exponential relaxations and a constant can only be approximate.[39]

In an experimental environment, the two data sets may be unmatched because of fluctuations in lamp intensity, different durations of measurement of each component, and unequal characteristics of the optics with respect to the polarization directions. The simultaneous analysis of $i_\parallel(t)$ and $i_\perp(t)$ allow one to determine the fluorescence and the anisotropy parameters without the knowledge of this matching factor,[37] even with the linear LAP2 method.[18] In the following $f(t)$ is taken to be monoexponential. Assume that the matching is realized by multiplying $i_\perp(t)$ with the factor γ. The polarized intensities $i_\parallel(t)$ and $i_\perp(t)$ can be rewritten as

[38] M. Ehrenberg and R. Rigler, *Chem. Phys. Lett.* **14**, 539 (1972).
[39] K. Kinosita, S. Kawato, and A. Ikegami, *Biophys. J.* **20**, 289 (1977).
[40] R. E. Dale, L. A. Chen, and L. Brand, *J. Biol. Chem.* **252**, 7500 (1977).

$$i_\parallel(t) = \sum_{i=1}^{m+1} e_i e^{-t/\lambda_i} \tag{38}$$

$$i_\perp(t) = \sum_{i=1}^{m+1} d_i e^{-t/\lambda_i} \tag{39}$$

with

$$\lambda_i = \tau \tag{40}$$

$$\lambda_{i+1} = \left(\frac{1}{\tau} + \frac{1}{\phi_i}\right)^{-1} \tag{41}$$

$$e_1 = \frac{a}{3}(1 + 2r_\infty), \qquad d_1 = \gamma\frac{a}{3}(1 - r_\infty) \tag{42}$$

$$e_{i+1} = \frac{2}{3}a\beta_i, \qquad d_{i+1} = -\gamma\frac{a}{3}\beta_i \tag{43}$$

The two data sets, $i_\parallel(t)$ and $i_\perp(t)$, have the same apparent decay constants λ_i, so that the linking can be performed as indicated for the global analysis of total intensity decays. Note that the fluorescence lifetime corresponds to the largest λ_i. This means that linking with a total intensity decay is possible. If a limiting anisotropy r_∞ is not considered, the factors β_i can be determined from a single polarization component. We found that the data of $i_\parallel(t)$ are to be preferred. In the other situation, the following formulas are suggested in which the factor γ is not present:

$$\beta_i = \frac{3}{2\left(\dfrac{e_1}{e_{i+1}} - \dfrac{d_1}{d_{i+1}}\right)} \tag{44}$$

$$r_\infty = \frac{\dfrac{e_{i+1}}{2e_1} + \dfrac{d_{i+1}}{d_1}}{\dfrac{d_{i+1}}{d_1} - \dfrac{e_{i+1}}{e_1}} \tag{45}$$

Not only $i_\parallel(t)$ and $i_\perp(t)$ can be analyzed simultaneously. Other polarized intensities collected under various orientations of the emission polarizer can be taken into account.[34,37] In general, the polarized fluorescence intensity of a macroscopically isotropic sample, $i(\phi,\theta,t)$, observed by using an excitation polarizer oriented under the angle ϕ and an emission polarizer oriented under the angle θ, both with respect to the normal to the excitation–emission plane, is given by

$$i(\phi,\theta,t) = f(t)/3\ [1 + (3\ \cos^2\phi\ \cos^2\theta - 1)r(t)] \tag{46}$$

In most experimental setups the excitation light is vertically polarized, so that $\phi = 0$ yielding

$$i(0,\theta,t) = f(t)/3[1 + (3\cos^2\theta - 1)r(t)] \tag{47}$$

The expressions for e_i and d_i have then to be changed accordingly.

Deconvolution against Decay of Reference Solution

The instrumental function $l(t)$ is frequently estimated from a measurement at the excitation or emission wavelength of a scatter solution which replaces the sample. However, this procedure ignores the wavelength dependence of the instrument.[41] Several correction schemes have been suggested (for an overview, see, e.g., Ref. 1). When the shape of the recorded excitation profile is only weakly wavelength dependent, the described Q-shift correction can be applied.[16,42] The use of a monoexponential decaying reference solution has been suggested to estimate the function $l(t)$ indirectly.[41] It has also been shown that the deconvolution can be performed directly against the observed decay of the reference, $g_R(t)$,[43,44–49]

$$g_R(t) = l(t) * f_R(t) \tag{48}$$

with

$$f_R(t) = a_R\,e^{-t/\tau_R} \tag{49}$$

The deconvolution against a reference using LAP2 can be achieved as follows. Combination of the Laplace transform of Eqs. (1) and (48) leads to

$$G(s) = G_R(s)\frac{F(s)}{F_R(s)} \tag{50}$$

When $f(t)$ is given by Eq. (2) one obtains

[41] P. Wahl, J. C. Auchet, and B. Donzel, *Rev. Sci. Instrum.* **45,** 28 (1974).

[42] A. Grinvald, *Anal. Biochem.* **75,** 260 (1976).

[43] D. R. James, D. R. M. Demmer, R. E. Verral, and R. P. Steer, *Rev. Sci. Instrum.* **54,** 1121 (1983).

[44] P. Gauduchon and P. Wahl, *Biophys. Chem.* **8,** 87 (1975).

[45] R. W. Wijnaendts van Resandt, R. H. Vogel, and S. W. Provencher, *Rev. Sci. Instrum.* **53,** 1392 (1982).

[46] L. J. Libertini and E. W. Small, *Anal. Biochem.* **75,** 260 (1984).

[47] M. Zuker, A. G. Szabo, L. Bramall, D. T. Krajcarski, and B. K. Selinger, *Rev. Sci. Instrum.* **56,** 14 (1985).

[48] P. Wahl, *Biophys. Chem.* **10,** 91 (1979).

[49] N. Boens, M. Ameloot, I. Yamazaki, and F. C. De Schryver, *Chem. Phys.* **121,** 73 (1988).

$$G(s) = G_R(s) \frac{s + (1/\tau_R)}{a_R} \sum_{i=1}^{n} \frac{a_i}{s + (1/\tau_i)} \tag{51}$$

or equivalently, by using partial fractions,

$$G(s) = G_R(s) \sum_{i=1}^{n} \frac{b_i}{s + (1/\tau_i)} + G_R(s) b_{n+1} \tag{52}$$

with

$$b_i = \frac{a_i}{a_R} \left(\frac{1}{\tau_R} - \frac{1}{\tau_i} \right) \qquad \text{for } i = 1, \ldots, n \tag{53}$$

$$b_{n+1} = \sum_{i=1}^{n} \frac{a_i}{a_R} \tag{54}$$

This means that $g(t)$ can be considered as the sum of a multiexponential decay with the "excitation" $g_R(t)$ and a "scatter" contamination expressed by b_{n+1}. The corresponding expression of Eq. (52) in the time domain is given by

$$g(t) = \sum_{i=1}^{n} b_i \, e^{-t/\tau_i} * g_R(t) + b_{n+1} g_R(t) \tag{55}$$

This result has been used by several authors.[44–49]

The reference deconvolution using the restricted Laplace transforms $\mathcal{L}^T$ can be realized as follows. As $g_R(t)$ plays formally the role of $l(t)$, the reference deconvolution can be performed by straightforward application of Eq. (30), yielding

$$\sum_{i=0}^{n} M_{i,n}[G^T(s_j)]D_i(\tau) = M_{n,n}[G_R^T(s_j)]b_{n+1}D_n(\tau)$$
$$+ \sum_{i=1}^{n} M_{i-1,n}[G_R^T(s_j)]K_i(b,\tau,b_{n+1}) \tag{56}$$

It can be shown from Eq. (56) that[20]

$$\sum_{i=0}^{n} M_{i,n}[G^T(s_j)]D_i(\tau) = \sum_{i=1}^{n} \{M_{i-1,n}[G_R^T(s_j)] + \tau_R M_{i,n}[G_R^T(s_j)]\}E_i(a',\tau) \tag{57}$$

where

$$a_i' = \frac{a_i}{a_R \tau_R} \tag{58}$$

It has to be emphasized that it is not required that $g_R(t)$ vanishes at the end of the time window so that Eqs. (56) and (57) are generally applicable.

Equations (56) and (57) are completely equivalent. In the actual computer implementation, Eq. (57) is selected so that the same program can be used for the deconvolution against $l(t)$ by putting τ_R formally equal to zero. Furthermore, this implementation is also convenient in the global analysis of multiple decay curves as has been described above.

In the case that $g(t)$ is contaminated by scattered excitation light, $S_c l(t)$, the deconvolution against $g_R(t)$ can be shown to be

$$g(t) = \sum_{i=1}^{n} b_i e^{-1/\tau_i} * g_R(t) + \left(b_{n+1} + \frac{S_c'}{\tau_R} \right) g_R(t) + S_c' \frac{d}{dt} g_R(t) \qquad (59)$$

with

$$S_c' = \frac{S_c}{a_R} \qquad (60)$$

In LAP2 no derivative of experimental data is required to deal with scatter contamination in the reference convolution. It can be shown that

$$M_{i-1,n}[L^T(s)] = \frac{1}{a_R \tau_R} \{ M_{i-1,n}[G_R^T(s_j)] + \tau_R M_{i,n}[G_R^T(s_j)] \} \qquad (61)$$

Introducing Eq. (61) in Eq. (30) gives the desired expression. In principle, it is not strictly required that τ_R be known beforehand as its value can be determined from the system of Eqs. (57). However, as has been found,[45,47,49] more accurate results are obtained for multiexponential decays when τ_R is known, especially when there is a component with $\tau_i \cong \tau_R$. This raises the question of the determination of τ_R. Its value may be found from a deconvolution against another monoexponential decay by solving the system formed by Eqs. (57) in an iterative way for both lifetimes. Other procedures have been suggested by Zuker et al.[47] Global analyses with reference convolution in the time domain by using NLLS have also been reported.[30,31]

Implementation of LAP2 Global Analysis

In this section we describe the implementation of the global analysis in which lifetimes and preexponentials can be linked. n different relaxation times are assumed. The linear and nonlinear LAP2 approaches can be used for this purpose. In both LAP2 global analyses, the actual data of the traces in each experiment are used only once. When the transforms $G_k^T(s_j)$ and $L_k^T(s_j)$ (or the transform of the reference decay) and subsequently the values $M_{i,n}[G_k^T(s_j)]$ and $M_{i-1,n}[L_k^T(s_j)]$ are calculated, the collected data in the time domain are no longer needed. At the time that these transforms are determined for each experiment, it takes little effort then to also perform single-curve analysis of each experiment contemporaneously. These two features contribute to the speed of the LAP2 methods as compared to nonlinear least-squares searches.

The linear LAP2 can be implemented in two ways. Both implementations give the same results. The methods differ in the way the $n(q + 1)$ unknowns $E_{k,i}$ and D_i are determined from the system of equations given by Eq. (31). In the first implementation, $E_{k,i}$ and D_i are calculated simultaneously. In most cases, the system of equations will be overdetermined. A solution can then be obtained by applying linear least-squares. The lifetimes are obtained from D_i and the preexponentials thereafter from $E_{k,i}$. For a large number of experiments, memory restrictions may require reprogramming to provide efficient storage and handling of the sparse matrix corresponding to the system of equations. An alternative can be provided as follows. For each experiment k the values $M_{i,n}[G_k^T(s_j)]$ and $M_{i-1,n}[L_k^T(s_j)]$ are saved. The unknown $E_{k,i}$ can be eliminated as follows. It can be shown that[18]

$$\sum_{i=0}^{n} \mathbf{M}_{k,i} D_i = 0 \tag{62}$$

with $\mathbf{M}_{k,i}$ being the determinant defined by

$$\mathbf{M}_{k,i} = \; \mid M_{i,n}[G_k^T(s_j)] \, M_{i-1,n}[L_k^T(s_j)] \ldots M_{0,n}[L_k^T(s_j)] \mid_{\, j=1, \ldots, n} \tag{63}$$

This will finally lead to an overdetermined system in the unknowns D_i. The symmetric matrix $(n \times n)$ of the normal equations in the least-squares is built up while passing through the calculations of the transforms for each experiment. Hence, the elements $\mathbf{M}_{k,i}$ do not have to be saved for each experiment. Estimates for D_i and hence for τ_j are readily obtained. These values and the stored coefficients $M_{i,n}[G_k^T(s_j)]$ and $M_{i-1,n}[L_k^T(s_j)]$ for each experiment are then used to set up a linear system from which $E_{k,i}$ can be determined for each experiment separately.

An outline of this implementation of the linear LAP2 analysis can be given as follows:

Step 1. For each decay experiment k:
Read $g_k(t)$ and $l_k(t)$ [or $g_{R,k}(t)$].
Calculate coefficients in Eq. (57) ($\tau_R = 0$ if lamp deconvolution); save results on disk or keep in memory.
Calculate determinants $\mathbf{M}_{k,i}$ defined in Eq. (63).
Use $\mathbf{M}_{k,i}$ to calculate contribution of experiment k to the matrix

GLOBALD corresponding with the normal equations of the total system. The dimension of GLOBALD is $n \times (n + 1)$ if n relaxation times have to be considered.

Step 2. Determine the n values $D_i(\tau)$ from GLOBALD.

Step 3. Recover τ_j from $D_i(\tau)$ according to Eq. (13).

Step 4. For each decay experiment k:
Read stored coefficients of Eq. (57) from disk or call from memory.
Use recovered values of τ_j to construct linear system for $E_{k,i}$.
Solve linear system for $a_{k,i}$.

For the nonlinear LAP2, it is also recommended that $M_{i,n}[G_k^T(s_j)]$ and $M_{i-1,n}[L_k^T(s_j)]$ be saved because the calculation of the transforms is the most time-consuming step in the analysis. In this way a rapid evaluation of analyses with alternative linkages between the parameters can be performed. To obtain the solution of the overdetermined system, the Newton method can be used, and the derivatives in the linearization can be calculated analytically. The increments for the parameters in each iteration can be obtained by applying the Householder transformation[50] to the overdetermined system. Initial guesses for the nonlinear LAP2 may be estimated from the single curve analyses while the values for $M_{i,n}[G_k^T(s_j)]$ and $M_{i-1,n}[L_k^T(s_j)]$ are calculated.

For the analysis of relaxation curves on the nanosecond time scale, recommended values for the transform parameters are as follows: $s_1 = 0$ and increments Δ of 0.002 and 0.005. The Simpson integration rule was used in the analysis of the examples discussed below.

Examples and Applications

Examples of Single-Curve Analysis

A comparison between the performance of LAP1, LAP2, and nonlinear least-squares (NLLS) in single-curve analysis has been described before.[18,20] Mono-, bi-, and triexponential decay curves were considered. In the following, σ will indicate the standard deviation obtained from NLLS analysis.

When the lamp deconvolution was used and the time window was reduced by considering fewer data channels at a constant channel width, the number of iterations increased significantly for LAP1. The linear LAP2

---

[50] W. Miller and C. Wrathall, Software for roundoff analysis of matrix algorithms. Academic Press, New York, 1980.

gave similar parameter estimates within a single step. These estimates were within the confidence intervals imposed by NLLS.

The performances of the lamp deconvolution methods in the presence of the time-shift artifact were examined on a measurement of the total fluorescence decay of bis[9-(10-phenyl)anthrylmethoxy]methane (DPAA), an intramolecular excimer-forming probe, embedded in unilamellar vesicles of dipalmitoylphosphatidylcholine (DPPC) at 33°. The probe to lipid ratio was 1 : 500. The time-shift Q was determined from a measurement of a degassed solution of anthracene in cyclohexane and appeared to be 0.12 nsec. This value was then used in the Laplace deconvolution. The decay of DPAA in DPPC vesicles was triexponential. The values obtained with LAP2 were as follows: a_1, 0.165; a_2, 0.825; a_3, 0.009; τ_1, 2.5 nsec; τ_2, 8.8 nsec; τ_3, 38 nsec. The results obtained with NLLS were practically identical. The time-shift Q was a freely adjustable parameter in the NLLS analysis, which yielded $Q = 0.13$ ($\sigma = 0.02$). The observed data, the calculated decay curve, and the residuals obtained with LAP2 are shown in Fig. 1. The weighted residuals and the corresponding autocorrelation exhibit the same behavior as expected from a good fit obtained with NLLS.

The reference deconvolution with LAP2 was tested on a simulated triexponential decay (a_1, 0.571; a_2, 0.286; a_3, 0.143; τ_1, 1 nsec; τ_2, 5 nsec; τ_3, 10 nsec). It was found that all decay parameters were well recovered in a single step when the reference lifetime was kept fixed on the correct value, even when the selected reference lifetime was equal to a relaxation time. The determination of reference lifetimes was tested by deconvoluting two monoexponential decays. Although both single lifetimes were freely adjustable in this iterative procedure, the recovery was very good when the lifetimes were sufficiently different. Reasonable values were still obtained for the case where the reference lifetimes were 1 and 2 nsec.

Global Analysis of Multiexponential Decays

An example of global analysis of biexponential decays concerns the experimental data set used by Knutson et al.[28] to demonstrate the performance of the NLLS global analysis. The data consist of the fluorescence relaxations observed at six different emission wavelengths of a mixture of 9-cyanoanthracene and anthracene in methanol. To diminish the difference between the two lifetimes, potassium iodide was added as a quencher. Figure 2 displays the two fluorescence lifetimes recovered by analyzing each experiment separately using the linear LAP2 in lamp deconvolution mode. At first sight, the lifetimes appeared to depend on wavelength. A similar pattern was obtained by performing single-curve analysis using NLLS.[28] However, the lifetimes observed with the individual unmixed

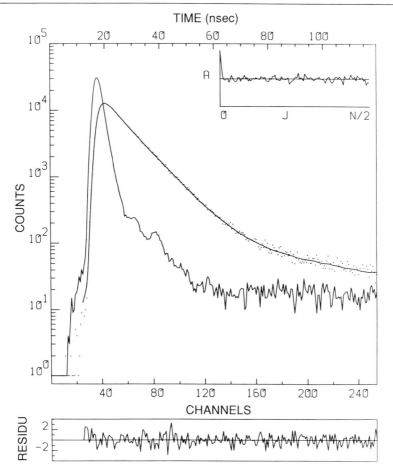

FIG. 1. The decay of DPAA in DPPC vesicles at 33°. λ_{ex}, 358 nm; λ_{em}, 425 nm; channel width, 0.471 nsec. The excitation light was vertically polarized, and the analyzer at the emission side was oriented at 55° with respect to the vertical. The measured excitation profile and the fitting curve are indicated with a solid line. The dots indicate the measured decay. The weighted residuals (RESIDU) and their autocorrelation (A) are also displayed. (Reproduced from the *Biophysical Journal*, 1983, Vol. 44, p. 27, by copyright permission of the Biophysical Society.)

compounds did not show any wavelength dependence and were found to be 4.6 nsec for 9-cyanoanthracene and 3.2 nsec for anthracene. Because of the relatively small difference between the two lifetimes, the results of the single-curve analyses are strongly influenced by the relative contributions of each fluorophore to the observed total intensities. The single-

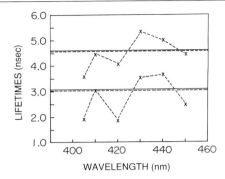

FIG. 2. Fluorescence lifetimes obtained from a mixture of 9-cyanoanthracene and anthracene in methanol quenched by KI: LAP2 single-curve analysis (X), LAP2 global analysis (—), and NLLS global analysis (– – –). The decays were measured over 511 channels; timing calibration, 0.102 nsec/channel; full-width at half-maximum of the excitation profile, ~2 nsec; excitation wavelength, 337 nm. (Reproduced from *Biophysical Chemistry*, 1986, Vol. 23, p. 155, by copyright permission of Elsevier Science Publishers.)

curve analysis by LAP2 and NLLS showed the same variation in the recovered lifetimes over the investigated wavelength region. The global analysis of the six decay curves by the linear LAP2 yields lifetime values of 4.56 and 3.02 nsec. Almost identical values were obtained by NLLS globals, namely, 4.57 nsec ($\sigma = 0.03$) and 3.07 nsec ($\sigma = 0.02$).

The following example illustrates the performance of LAP2 in the global analysis of a fluorescence decay surface in which three relaxation times have to be linked. The fluorescence decay of 2-anilinonaphthalene in cyclohexane in the presence of small amounts of ethanol was measured at seven emission wavelengths at 25°C.[20] The decay curves contained 511 data points. The channel width was 0.05 nsec. The decay surface was analyzed by linking the relaxation times using the linear LAP2 in lamp and in reference mode. The reference solution was 2-anilinonaphthalene in cyclohexane with a lifetime value of 4.35 nsec. The plots of the residuals indicated that a triexponential decay model had to be used. The three relaxation times obtained with LAP2 with reference convolution were 0.70, 1.96, and 5.322 nsec. LAP2 with lamp deconvolution yielded similar parameter values. This implied that for the used apparatus in the considered wavelength region the Q-shift correction was a valid approximation. The relaxation times obtained with NLLS with lamp deconvolution were 0.75 nsec ($\sigma = 0.05$), 1.95 nsec ($\sigma = 0.02$), and 5.378 nsec ($\sigma = 0.006$). All three methods gave comparable values for the preexponential factors.

The LAP2 global analysis of four simulated quadruexponential decays with common relaxation times was also examined. The relaxation times in the simulation were 1, 3, 7, and 10 nsec. The decays contained 511 data channels, and the channel width was 0.102 nsec. The global LAP2 in lamp mode yielded 0.95, 2.9, 7.1, and 10.2 nsec. NLLS results were 1.01 nsec ($\sigma = 0.06$), 2.9 nsec ($\sigma = 0.2$), 7.1 nsec ($\sigma = 0.2$), and 10.1 nsec ($\sigma = 0.2$). It can be concluded that LAP2 global analyses in which relaxation times can be linked over the data surface yield the same results as NLLS in a short computation time.

Resolution of Spectra by Global Analysis

One of the main applications of the global analysis by LAP2 is resolving a steady-state fluorescence spectrum of a heterogeneous emitting sample in its components. Using time-resolved fluorescence decay measurements, the discrimination between the different noninteracting species in the sample can be made on the basis of lifetime[25,26,51] or rotational correlation time.[52,53]

Assume q species of molecules in the excited state after excitation at wavelength λ_{ex}. The fluorescence relaxation at the emission wavelength λ_{em} due to a delta excitation is given by

$$f(\lambda_{ex}, \lambda_{em}, t) = \sum_{j=1}^{q} \alpha_j(\lambda_{ex}, \lambda_{em}) e^{-t/\tau_j} \tag{64}$$

The spectral contours $\alpha_j(\lambda_{ex}, \lambda_{em})$ have been called decay-associated emission spectra (DAS).[51] These spectra reflect the species-associated emission spectra (SAS)[54,55] only when the species are independent. When the species do interact, the SAS can be obtained from a compartmental analysis[54,56] (see also [14], in this volume). For example, the spectrum of the monomer in an excimer reaction can be obtained by adding the preexponentials obtained in a DAS analysis. The determination of DAS can be useful in distinguishing a scatter contribution from a true fluorescence component.

The nature of a DAS experiment is well suited for a global approach: the relaxation times can be linked over the complete decay surface, and variations in the preexponentials with emission wavelength change the relative contribution of each lifetime component. It has been shown that

[51] J. R. Knutson, D. G. Walbridge, and L. Brand, *Biochemistry* **21**, 4671 (1982).
[52] J. R. Knutson, L. Davenport, and L. Brand, *Biochemistry* **25**, 5026 (1986).
[53] L. Davenport, J. R. Knutson, and L. Brand, *Biochemistry* **25**, 1811 (1986).
[54] J. M. Beechem, M. Ameloot, and L. Brand, *Chem. Phys. Lett.* **120**, 466 (1985).
[55] J.-E. Löfroth, *J. Phys. Chem.* **90**, 1160 (1986).
[56] M. Ameloot, N. Boens, R. Andriessen, V. Van den Bergh, and F. C. De Schryver, *J. Phys. Chem.* **95**, 2041 (1991).

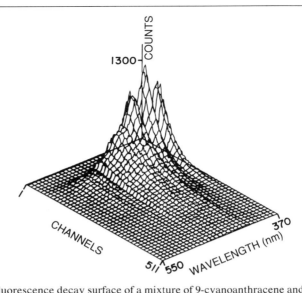

FIG. 3. Fluorescence decay surface of a mixture of 9-cyanoanthracene and anthracene as a function of emission wavelength and time. For clarity, not all decay curves are shown. (Reproduced from *Biophysical Chemistry*, 1986, Vol. 23, p. 155, by copyright permission of Elsevier Science Publishers.)

the curves at the edges of the spectrum contribute the most to a good recovery of the fluorescent lifetimes and the model discrimination.[31] A specific advantage of the global approach concerns the collection dwell times. It appears that the simultaneous analysis of decay curves of only a few hundred counts at the peak will reveal DAS. The preexponentials recovered at the various wavelengths may be not properly scaled owing to fluctuations in excitation intensity and different collection times. The contribution of each species to the steady-state fluorescence at each wavelength can be calculated from the products of the recovered preexponential factors and the lifetimes.

To test the ability to recover DAS, a mixture of anthracene ($\tau = 3.7$ nsec) and 9-cyanoanthracene ($\tau = 11.5$ nsec) in methanol was examined at room temperature.[20] The solution was excited at 357 nm. Fluorescence decay curves were collected every 2 nm in the wavelength region 370–550 nm. The fluorescence decay surface is schematically represented in Fig. 3. The 90 decay curves were analyzed simultaneously by the linear LAP2 in lamp mode by linking the two fluorescence lifetimes over the complete decay surface. The number of unknown parameters for this case is 182. The analysis with LAP2 took about 40 min on an HP 1000 multiuser system. The iterative NLLS global analysis required about 7 hr starting

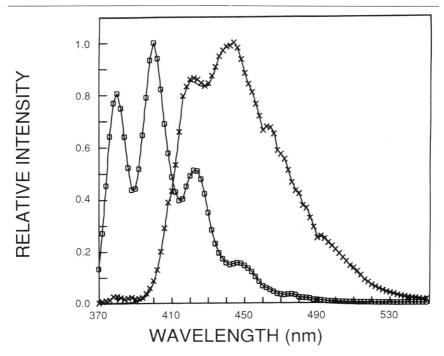

FIG. 4. Recovery of the technical fluorescence emission spectra of 9-cyanoanthracene (X) and anthracene (□) by global analysis of the decay surface shown in Fig. 3. The solid lines represent the spectra of the unmixed samples. (From *Biophysical Chemistry*, 1986, Vol. 23, p. 155, by copyright permission of Elsevier Science Publishers.)

from reasonable initial guesses for the parameters. The recovered preexponentials were well superimposable on the spectra obtained from each component separately (Fig. 4).

Global Analysis of Anisotropy

The analysis of anisotropy experiments by LAP2 has been tested on the measurement of $i_\parallel(t)$ and $i_\perp(t)$ of rhodamine 6G in propylene glycol at 37°.[18] The excitation was at 339 nm, and the emission was monitored at 555 nm. The time shift determined from a separate measurement of the total fluorescence emission was 0.3 nsec. The number of data channels was 254, and the channel width was 0.160 nsec. The simultaneous analysis of $i_\parallel(t)$ and $i_\perp(t)$ by LAP2 yielded $\tau = 4.152$ nsec, $\beta = -0.142$, and $\phi = 3.4$ nsec. The NLLS results were $\tau = 4.151$ nsec ($\sigma = 0.007$), $\beta = -0.140$ ($\sigma = 0.003$), and $\phi = 3.3$ nsec ($\sigma = 0.1$).

Conclusions and Perspectives

It is shown in this chapter that the use of a modified Laplace transform leads to the fast and reliable analysis program LAP2 for fluorescence decay data. There are two versions of the LAP2 analysis program: linear (or noniterative) and nonlinear (or iterative). In both versions, the same program can be used to perform a deconvolution against a monoexponential decay of a reference compound or an excitation profile. In the latter mode the lifetime of the reference compound is formally set equal to zero.

In the linear implementation of LAP2, the functions $D_i(\tau)$ and $E_i(a,\tau)$ of the relaxation times and the amplitudes are determined first. The relaxation times and the preexponentials are determined subsequently. The linear LAP2 analysis is very fast because it is noniterative. No initial guesses for the parameters are required. The observed data are used only once, namely, when the Laplace transforms are determined. The observed data are then replaced by equations corresponding to the various values of the Laplace transform parameter. The functions $D_i(\tau)$ and $E_i(a,\tau)$ are the unknowns in the resulting system of equations.

In the nonlinear LAP2 analysis, the relaxation times and the preexponential factors are determined directly without the determination of their functions $D_i(\tau)$ and $E_i(a,\tau)$. Because the system of equations are nonlinear in the relaxation times and the preexponentials, an iterative procedure has to be used and initial guesses for the parameters have to be provided. The computation time required for this iterative part is much less than for the NLLS procedure because in LAP2 only a few equations have to be considered instead of the numerous data points of the recorded curves. In a single-curve analysis, there is no real need to use the nonlinear version of LAP2, except in the determination of reference lifetimes where two monoexponential decay curves are deconvolved against each other.

The LAP2 programs can be used efficiently in the analysis of fluorescence decay surfaces. LAP2 is the first transform method that can be used in a global analysis, and the linear LAP2 is the only noniterative global analysis described up to now. The larger the number of curves and the number of parameters, the faster is LAP2 as compared to NLLS global analysis. When only relaxation times have to be linked over the data surface, the noniterative LAP2 can be used. A typical application is the recovery of DAS. To achieve the flexibility of NLLS globals in linking parameters, the nonlinear LAP2 has to be used. Again, although this version of LAP2 is iterative, this analysis program is very fast because the numerous data points of the recorded surface are replaced by a system of equations.

LAP2 analyses do not provide the statistics obtained from NLLS.

Only the sequence and the distribution of the weighted residuals can be evaluated. To obtain the remaining statistics it is recommended to run a NLLS analysis using the results of the LAP2 analysis as initial guesses. The NLLS search will then take only a few iteration steps.

It is worthwhile mentioning that the different analyses based on transforms are related to each other. One can show that for the model function $f(t)$ in Eq. (2) the functions $g(t)$ and $l(t)$ are related by the following differential equation[57]:

$$\sum_{i=0}^{n} D_i(\tau) g^{(i)}(t) = \sum_{i=1}^{n} E_i(a,\tau) l^{(i-1)}(t) \tag{65}$$

where $g^{(i)}(t)$ and $l^{(i)}(t)$ denote the ith time derivative of $g(t)$ and $l(t)$. In the different transform methods, Eq. (65) is multiplied by a specific function $\rho(t)$ and an integration over time is performed,[57,58] yielding

$$\sum_{i=0}^{n} D_i(\tau) \int_0^u \rho(t) g^{(i)}(t)\, dt = \sum_{i=1}^{n} E_i(a,\tau) \int_0^u \rho(t) l^{(i-1)}(t)\, dt \tag{66}$$

In LAP2, the integration is performed over the time interval $[0,T]$ and $\rho(t) = \exp(-st)$. In the method of moments the integration is over the same time interval, but the function $\rho(t) = t^m$ is used. The indicated similarity between transform methods provides some evidence that it should be possible to make the method of moments noniterative by circumventing the cutoff problem in a way similar to LAP2.

Two rather recent issues in time-resolved fluorescence are the NLLS analysis in terms of continuous distributions of relaxation times[24,59,60] and the NLLS analysis of excited-state reactions directly in terms of the rate constants and the SAS,[54,56,61] the latter also being denoted as the compartmental analysis. Whether LAP2 can be extended to the analysis of distributions of relaxation times is still under investigation. The method of moments, which is iterative, has already been applied to this type of problem.[62]

[57] J. L. Viovy, J. C. Andre, M. Bouchy, M. Roger, and J. Bordet, in "Deconvolution–Reconvolution" (M. Bouchy, ed.), p. 3. Conference Proceedings, Ecole Nationale Supérieure des Industries Chimique de l'Institut Nationale Polytechnique de Lorraine, Nancy, France, 1982.

[58] A. H. Kalantar, J. Lumin. 28, 411 (1983).

[59] J. R. Alcala, E. Gratton, and F. G. Prendergast, Biophys. J. 51, 587 (1987).

[60] J. R. Lakowicz, H. Cherek, I. Gryzynski, N. Joshi, and M. L. Johnson, Biophys. Chem. 28, 35 (1987).

[61] R. Andriessen, N. Boens, M. Ameloot, and F. C. De Schryver, J. Chem. Phys. 95, 2047 (1991).

[62] L. J. Libertini and E. W. Small, Biophys. Chem. 34, 269 (1989).

The formal description of excited-state reactions and the direct fitting of the data surface for the rate constants and SAS using NLLS is described in [14] in this volume. The theoretical part of that chapter should be consulted to get insight in the extension of LAP2 toward compartmental analysis. The basic idea of compartmental analysis is the following. For species which interact in the excited state the real parameters of interest are not the relaxation times but the rate constants which determine the relaxation times. The rate constants can be arranged in a square matrix **A** whose eigenvalues are the negative of the inverse of the relaxation times. To elucidate excited-state processes, experiments have to be performed under a number of conditions: different concentrations, pH, etc. This implies that, in most cases, the relaxations times can only be linked over the excitation or emission wavelength axis. For example, if the concentrations are changed, the relaxation times will change accordingly. Hence, to apply a global analysis over the complete decay data surface, the rate constants have to be linked instead of the relaxation times. The preexponential factors depend not only on the rate constants but also on the SAS and the absorbances of the species in the ground state. The spectral contribution of the emitting species at each emission wavelength will denoted by the vector **C** whose dimension corresponds to the number of excited-state species. In the same way, the absorbance information will be denoted by the vector **B**.

The extension of LAP2 to compartmental analysis can be implemented as follows. For each decay trace a number of equations of the type of Eq. (31) is obtained corresponding to the values of the transform parameter s. The fitting parameters are the elements of the matrix **A** and of the normalized vectors **B** and **C** (see [14] on compartmental analysis in this volume). A possible approach to determine these parameters is to apply NLLS to the system of equations corresponding to the complete data surface. The linking and mapping scheme can be copied from the NLLS analysis of the observed data. The derivatives required during the search are calculated numerically. For each set of parameter values, the functions $D_{k,i}$ and $E_{k,i}$ can be calculated and their values evaluated by checking the equations of the type of Eq. (31).

Acknowledgments

Part of this work was developed when the author was in the laboratory of Dr. L. Brand (The Johns Hopkins University, Baltimore, MD). The author thanks Drs. J. Beechem and L. Brand for the discussions during this work.

[13] Interpolation Methods

By CAROLE J. SPANGLER

Introduction

The process of interpolation involves deriving a curve, or surface, which passes through a limited number of measured data points in order to predict or determine values of an experimental parameter in between the measured data points. Interpolated data are used in a broad variety of scientific disciplines including molecular dynamics calculations, geological studies (earthquake seismograms and rainfall data), spectroscopy (chromatography and fast Fourier transformation processing), imaging (reconstructing three-dimensional objects from cross sections), speech synthesis, robotics, and automotive engineering. One example of a practical use of interpolation was presented in 1968, when Bézier published a paper describing how Renault used interpolation for car body design and tooling.[1] In this approach, curves were generated between points outlining the perimeter of a car body by serially incrementing vectors emanating out from the initial points which constituted successive sides of a defining polygon.

Serra *et al.*[2] have demonstrated a technique to interpolate amplitude spectra to reproduce short-time spectral variations. Specifically, they have been able to analyze tones from orchestral instruments, and with a compressed recording, resynthesize the original sounds by interpolating between target tones without perceptible loss of realism. Wright and Elliott[3] demonstrate a method for interpolation between target values to generate continuous speech. Noninvasive techniques for temperature measurement utilize interpolation by using reference temperatures to make accurate predictions about the temperature in other locations of the system.[4] Interpolation methods are used in generating three-dimensional objects from serial cross sections, important in medicine and in the study of biological specimens (computerized tomography, magnetic resonance imaging, and ultrasound). In these applications, the spaces between each two-dimen-

---

[1] Bézier, Society of Automotive Engineers Congress in Detroit, Paper No. 6800010 (1968).

[2] M.-H. Serra, D. Rubine, and R. Dannenberg, *J. Audio Eng. Soc.* **38**(3), 111 (1990).

[3] R. D. Wright and S. J. Elliott, *J. Acoust. Soc. Am.* **87**(1), 383 (1990).

[4] M. A. Khan, C. Allemand, and T. W. Eager, *Rev. Sci. Instrum.* **62**(2), 392 (1991).

METHODS IN ENZYMOLOGY, VOL. 210

sional slice are interpolated to generate smooth, continuous surfaces.[5,6] Interpolation has also been used to increase fast Fourier transform resolution without increasing sampling frequency or the number of samples.[7] Interpolation has played a role in molecular dynamics calculations as well, where theoretical protein-folding pathways are generated by interpolating between energy-minimized states.

Since 1670, efforts have been directed toward refining and improving methods for predicting the behavior of experimental systems in regions where data have not been measured. Often, an analytical function describing the behavior of the data points is unknown, and therefore an estimating function must be determined. In many cases, a linear relationship can be assumed between each adjacent set of data points. In this case, the interpolated value is based on the difference between the two determined boundary values and the fractional distance the desired value is between these two points. This approximation can yield reasonable results if the data density is very high and/or if the curvature in the data is low. An example of the successful use of linear interpolation is provided by Wright and Elliott,[3] where they applied this technique to speech synthesis.

However, in biological systems it is more frequent that the relationships are nonlinear and hence, more complex. Interpolation in these cases involves establishing a more complicated function to represent the known experimental data points. To achieve this, one must first decide on the type of function to be used for the approximation. The most commonly used functions include polynomials, rational functions, trigonometric functions, and spline functions. This chapter reviews several methods for interpolating experimental data and includes examples of their application.

Methods of Interpolation

Graphical Techniques

The simplest and least accurate method for interpolation is to sketch a smooth curve which best approximates the data. From this graph, estimates for intermediate values can be readily obtained.[8] This method is useful only if accuracy is not imperative, since drawing of such a curve tends to be arbitrary. However, if the initial observations are not extremely

[5] S.-Y. Chen, W.-C. Lin, C.-C. Liang, and C.-T. Chen, *IEEE Trans. Med. Imaging* **9**(1), 71 (1990).

[6] S. P. Raya and J. K. Udupa, *IEEE Trans. Med. Imaging* **9**(1), 32 (1990).

[7] D. McKee, *EDN* **35**(23), 282 (1990).

[8] B. M. Shchigolev, "Mathematical Analysis of Observations." American Elsevier, New York, 1965.

reliable and a functional relationship is not known, this method of interpolation can be useful.

Polynomial Interpolation

As early as 1670, a method of interpolation based on representing functions by polynomials was introduced by James Gregory.[9] In 1885, Weierstrass proved that every function continuous in a certain range, say between a and b, can be represented in that range by a polynomial, and, more specifically, a set of n data points in a defined range can be fit by a polynomial of degree $n - 1$.[10]

Several algorithms are available for solving the interpolation problem with polynomial functions. The most straightforward, yet computationally demanding, is to define a set of simultaneous equations by requiring that the polynomial necessarily intersect the original data points (where a subset of m data points have been chosen from the set of n total data points):

$$p_{m-1}(x_i) = f(x_i) = y_i \qquad i = 1, 2, \ldots, m \tag{1}$$

Hence, for each data point, one polynomial is generated as follows:

$$
\begin{aligned}
a_0 + a_1 x_1 + a_2 x_1^2 + \ldots + a_{m-1} x_1^{m-1} &= y_1 \\
a_0 + a_1 x_2 + a_2 x_2^2 + \ldots + a_{m-1} x_2^{m-1} &= y_2 \\
&\vdots \\
a_0 + a_1 x_m + a_2 x_m^2 + \ldots + a_{m-1} x_m^{m-1} &= y_m
\end{aligned}
\tag{2}
$$

Simultaneously solving these equations for the values of the coefficients, a_j, would then allow for the prediction of the function at other values of the independent variable x. The obvious drawback of this technique is the requirement of solving a set of simultaneous equations. Further, no error information is generated with this technique.[10,11]

Several other algorithms for solving polynomial interpolations are available. Some require the construction of a difference table of the type shown below. So, for example, $\Delta^3 y_1 = y_4 - 3y_3 + 3y_2 - y_1$. This tabulation is referred to as a diagonal difference table. "Horizontal" tables also exist which contain the same information but in a different format. These

[9] Rigaud, "Correspondence of Scientific Men of the Seventeenth Century," Vol. 2, p. 209 (1670).

[10] A. C. Norris, "Computational Chemistry: An Introduction to Numerical Methods." Wiley, Chichester, 1981.

[11] P. J. Davis, "Interpolation and Approximation." Blaisdell, New York, 1965.

x	y	Δy	$\Delta^2 y$	$\Delta^3 y$	$\Delta^4 y$
x_0	y_0				
		Δy_0			
x_1	y_1		$\Delta^2 y_0$		
		Δy_1		$\Delta^3 y_0$	
x_2	y_2		$\Delta^2 y_1$		$\Delta^4 y_0$
		Δy_2		$\Delta^3 y_1$	
x_3	y_3		$\Delta^2 y_2$		
		Δy_3			
x_4	y_4				

differences are similar to derivatives and are related by the following formulas, which are derived in detail by Vallée-Poussin[12]:

$$\Delta^n f(x) = (\Delta x)^n f^{(n)}(x + \theta n \Delta x) \qquad 0 < \theta < 1 \tag{3}$$

$$\lim_{\Delta x \to 0} \frac{\Delta^n f(x)}{(\Delta x)^n} = f^{(n)}(x) \tag{4}$$

Several different equations exist which use the entries from this table directly to generate interpolated values. For example, in one method for equal-spaced x values, interpolated values from Newton's forward formula are generated as follows:

$$p(x) = y_0 + \frac{\Delta y_0}{\Delta x}(x - x_0) + \frac{\Delta^2 y_0}{2(\Delta x)^2}(x - x_0)(x - x_1) + \ldots +$$

$$\frac{\Delta^n y_0}{n!(\Delta x)^n}(x - x_0)(x - x_1)(x - x_2)\ldots(x - x_{n-1}) \tag{5}$$

This is called the forward formula because it uses only tabular values to the right of the chosen y_0 value. Because of this, this formula is most useful for values near the beginning of a set of tabulated values.

Newton's backward formula [Eq. (6)] is similar in form but contains values of the difference table to the left and is most useful for determining values near the end of a data set. More complicated versions of Newton's

$$p(x) = y_n + \frac{\Delta^1 y_{n-1}}{\Delta x}(x - x_n) + \frac{\Delta^2 y_{n-2}}{2(\Delta x)^2}(x - x_n)(x - x_{n-1}) + \ldots +$$

$$\frac{\Delta^n y_0}{n!(\Delta x)^n}(x - x_n)(x - x_{n-1})\ldots(x - x_1) \tag{6}$$

formulas exist, for example, for data with unequally spaced intervals.

[12] Vallée-Poussin, "Cours d'Analyse Infinitésimale," I, 4th Ed., 1921.

These formulas are fundamental and are usually applicable to most interpolation problems, but they often do not converge rapidly.[13,14]

Other interpolation functions which utilize difference tables are the central difference formulas, which tend to converge easily and are well suited for finding values near the middle of a table of data. Two common formulas of this class are Bessel's interpolation function,[13-15] which is better at interpolation near the middle of the data range, and Stirling's formula,[13-15] which is better at the beginning or the end of the data range. Both of these algorithms are of similar accuracy.

A popular polynomial interpolation method which does not require solving for coefficients, as was necessary in the first example, is the use of Lagrange interpolating polynomials. Again, the goal is to determine the value of the function, expressed as a polynomial, at the desired point, say x:

$$p_{n-1}(x) = a_0 + a_1 x + a_2 x^2 + \cdots + a_{n-1} x^{n-1} \tag{7}$$

This method is easily described for a linear system where two data points are used, (x_1, y_1) and (x_2, y_2), which closely bracket the desired value ($n = 2$). The defining polynomial is then

$$p_1(x) = a_0 + a_1 x \tag{8}$$

The equation defining the line passing through (x_1, y_1) and (x_2, y_2) is

$$p_1(x) = y_1 + \frac{(x - x_1)(y_2 - y_1)}{(x_2 - x_1)} \tag{9}$$

Rearranging,

$$p_1(x) = y_1 \frac{(x_2 - x)}{(x_2 - x_1)} + y_2 \frac{(x - x_1)}{(x_2 - x_1)} \tag{10}$$

$$p_1(x) = y_1 L_1(x) + y_2 L_2(x) \tag{11}$$

where L_1 and L_2 are Lagrange coefficients, and

$$L_1(x) = \frac{(x_2 - x)}{(x_2 - x_1)} \quad \text{and} \quad L_2(x) = \frac{(x - x_1)}{(x_2 - x_1)} \tag{12}$$

Setting x to the desired value, $p_1(x)$ can be easily solved since the Lagrange coefficients can be readily calculated.

[13] J. B. Scarborough, "Numerical Mathematical Analysis," 4th ed., Johns Hopkins Press, Baltimore, Maryland, 1958.

[14] F. B. Hildebrand, "Introduction to Numerical Analysis." McGraw-Hill, New York, 1956.

[15] E. T. Whittaker and G. Robinson, "A Short Course in Interpolation." Van Nostrand, New York, 1923.

Similar in form to Eq. (12), for a case with n data points, it can be shown that the generalized form of the Lagrange coefficient is[10,11,13-16]

$$L_i(x) = \prod_{j=1}^{m} \frac{(x - x_j)}{(x_i - x_j)} \qquad (13)$$

for $i = 1, 2, \ldots, n$ and $i \neq j$. The generalized Lagrange interpolating polynomial is

$$p_{m-1}(x) = \sum_{i=1}^{m} y_i L_i(x) \qquad (14)$$

Again, given (x_i, y_i) data points, the Lagrange coefficients can be calculated in a straightforward manner, and then the value of the function at any intermediate value can be calculated from Eq. (14).

If the degree of polynomial from which the m data points are taken is greater than $m - 1$, truncation errors will occur and must be included in the calculation if an accurate interpolation is to be achieved.[10] Hence, for polynomials of order m and higher, the Lagrange interpolating formula is in the form

$$y = L(x) + \mathcal{R} \qquad (15)$$

and

$$\mathcal{R} = f^{(m)}(\xi) \frac{\prod_{j=1}^{m} (\bar{x} - x_j)}{m!} \qquad (16)$$

where $\mathcal{R}$ is the truncation error and $f^{(m)}(\xi)$ is the mth derivative of function f calculated at some ξ value of x in between x_1 and x_m. Feuillebois[17] has presented an algorithm for numerically calculating these derivatives and has implemented it into a FORTRAN routine for personal computers. One can also directly estimate boundaries for errors using these equations even though ξ is unknown. For example, if the function being interpolated is $y = x^{1/2}$, then the error formula can be expressed as

$$\varepsilon = \frac{(x - x_0)(x - x_1)}{(n + 1)!} y''(x^{1/2}) \qquad (17)$$

$$\varepsilon = \frac{(x - x_0)(x - x_1)}{2!} \frac{(-1)}{[4(x^3)^{1/2}]} \qquad (18)$$

[16] L. G. Kelly, "Handbook of Numerical Methods and Applications." Addison-Wesley, Reading, Massachusetts, 1967.
[17] F. Feuillebois, *Comput. Math. Appl.* **19**(5), 1 (1990).

One can obtain the maximum error by substituting in an extreme value for x in the above formula.

Asadov and Dzhabrailov[18] have demonstrated the use of Lagrange interpolation to compute binodal stratification curves in mercury–dithallium chalcogenide systems. They were able to construct phase equilibrium curves for three of these compounds (Hg–Tl$_2$X^{VI}, where X is S, Se, or Te), systems with limited solubility in the liquid state, and demonstrated that the use of straightforward interpolation avoids more complex solution models. Lowen[19] describes a method of using Lagrange interpolation to approximate "fuzzy" data sets, for example, to approximate a set of $n +$ 1 points when each of these points has a "fuzzy" value rather than a crisp, precise one.

The main drawback to the strict application of this algorithm is the inability to predict the most appropriate degree of polynomial prior to the analysis. Hence, different degree polynomials must be tested separately, each time recalculating the Lagrange coefficients, and then the function value at the desired point until the interpolated value becomes reasonably constant. However, iterative Lagrange formulations do exist, such as the Neville iterated interpolation.[10]

Rational Functions

A rational function is a ratio of polynomials and can be useful if a polynomial is insufficient to approximate the measured data points. For a set of data points (x_i, y_i), one can derive two polynomials, $n(x)$ and $d(x)$, such that

$$y(x) = \frac{n(x)}{d(x)} \qquad n, d \text{ coprime} \tag{19}$$

This rational function interpolates the above points if and only if

$$y(x_i) = y_i \tag{20}$$

The complexity or MacMillan degree of $y(x)$ is defined as

$$deg\ y = \max\ (deg\ n,\ deg\ d) \tag{21}$$

To interpolate with rational functions, one varies the order in both the numerator and denominator of Eq. (19) until the relation in Eq. (20) is satisfied. The interpolation is then described by its complexity, as defined in Eq. (21). A succinct summary of this derivation is given by Antoulas

[18] M. M. Asadov and N. N. Dzhabrailov, *Inorg. Mater.* **24**(11), 1648 (1988).
[19] R. Lowen, *Fuzzy Sets Syst.* **34**(1), 33 (1990).

and Anderson.[20] Rational functions are computationally easy and rapid and are often the simplest function to describe complicated relationships, for example, infinite y values for finite x values.[16]

Rational interpolation can be extended, as shown by Clements,[21] to be implemented in a piecewise cubic interpolation. An algorithm is described which is currently implemented in a computer system for automatic generation of developable ship hulls and other marine applications. This fast, simple algorithm for generating a piecewise rational cubic interpolant is coded in FORTRAN.

Trigonometric Functions

For data that is observed to be periodic, trigonometric functions can be used. The Hermite formula is the periodic equivalent to the nonperiodic Lagrange formula:

$$y = \frac{\sin(x - x_1) \sin(x - x_2) \ldots \sin(x - x_n)}{\sin(x_0 - x_1) \sin(x_0 - x_2) \ldots \sin(x_0 - x_n)} y_0 +$$

$$\frac{\sin(x - x_0) \sin(x - x_2) \ldots \sin(x - x_n)}{\sin(x_1 - x_0) \sin(x_1 - x_2) \ldots \sin(x_1 - x_n)} y_1 + \ldots +$$

$$\frac{\sin(x - x_0) \sin(x - x_1) \ldots \sin(x - x_{n-1})}{\sin(x_n - x_0) \sin(x_n - x_1) \ldots \sin(x_n - x_{n-1})} y_n \quad (22)$$

This function has a periodicity of 2π and is applicable to data of arbitrary spacing.[13]

Spline Functions

Spline functions produce smooth single-variable functions which take on prescribed values at a finite number of points. Excellent reviews are given by de Boor[22] and Gordon and Reisenfeld.[23] Spline functions interpolate by generating piecewise polynomials of low degree, n, between each pair of adjacent data points, joined together such that the function over the entire range of data is continuous and has continuous derivatives up to the order of $n - 1$. To achieve this, coefficients for these piecewise polynomials are calculated globally by adding the constraint that the deriv-

[20] A. C. Antoulas and B. D. O. Anderson, *Linear Alg. Appl.* **122,** 301 (1989).
[21] J. C. Clements, *SIAM J. Numer. Anal.* **27**(4), 1016 (1990).
[22] C. de Boor, "A Practical Guide to Splines." Springer-Verlag, New York, 1978.
[23] W. J. Gordon and R. F. Reisenfeld, *in* "Computer Aided Geometric Design" (R. E. Barnhill and R. F. Reisenfeld, eds.), Academic Press, New York, 1974.

atives (up to $n - 1$) of each function be identical at the points of intersection, also referred to as "knot points."[10]

To demonstrate how this is implemented for a cubic spline interpolation for a set of arbitrary but ordered data points, first polynomials are defined, $p_{3,i}(x)$ between x_i and x_{i+1}, to constrain the system such that the interpolation function is forced through the data points[10]:

$$p_{3,i}(x_i) = y_i \quad i = 1, 2, \ldots, n - 1 \tag{23}$$

$$p_{3,i}(x_{i+1}) = y_{i+1} \quad i = 1, 2, \ldots, n - 1 \tag{24}$$

Note that these constraints imply that for points in between x_i and x_{i+1}

$$p_{3,i-1}(x_i) = p_{3,i}(x_i) \tag{25}$$

For continuity and smoothness, adjacent polynomials must also have the same first-order and second-order (cubic system) derivatives at the knots:

$$p'_{3,i-1}(x_i) = p'_{3,i}(x_i) \tag{26}$$

$$p''_{3,i-1}(x_i) = p''_{3,i}(x_i) \tag{27}$$

Because this is a cubic system, the second derivative at any point within a subinterval will be a linear function of x and can be expressed as a special form of the general Lagrange formula discussed earlier.

$$p''_{3,i}(x) = p''_{3,i}(x_i) \frac{(x_{i+1} - x)}{(x_{i+1} - x_i)} + p''_{3,i}(x_{i+1}) \frac{(x - x_i)}{(x_{i+1} - x_i)} \tag{28}$$

Equation (28) is then integrated and further manipulated to yield the final interpolation equation. Detailed derivations are given elsewhere.[10,24]

Spline functions tend to be more computationally robust than polynomials and can be computed with two or more points (order being equal to the number of points minus one). However, unlike polynomials, higher order splines are often subject to wild oscillations, especially if there are many points distant from the desired point. Three- or four-point spline interpolations are popular, but one must be cautious in using orders higher than six unless careful error analysis is done as well.[24]

There are several advantages of using a spline interpolation scheme. First, the computation is fairly robust. Second, the piecewise approach allows more freedom for the interpolating function to follow the actual data points. Third, various fitting constraints and criteria could be imposed to obtain functions which incorporate customized features.[25]

[24] W. H. Press, B. P. Flannery, S. A. Teukolsky, and W. T. Vetterling, "Numerical Recipes: The Art of Scientific Computing." Cambridge Univ. Press, Cambridge, 1986.
[25] J. M. Chambers, "Computational Methods for Data Analysis." Wiley, New York, 1977.

Conclusions

Interpolation techniques have been successfully applied to a variety of studies for several years owing to a ubiquitous need to derive smooth curves or shapes from a limited number of data points. Several different algorithms are currently available, most rigorously defined mathematically, and many of these algorithms are available on commercial software. For example, graphics packages for personal computers often include an option for spline interpolation between graphed data points, and various interpolation subroutines and functions are usually available in mathematics libraries for all types of computers as well. One of the most powerful uses of interpolation techniques in the biological sciences is probably in the representation of data, for example, to illustrate multidimensional NMR data, or to reconstruct three-dimensional images from electron microscopy data, or simply to produce smooth curves through data points for publication figures. Interpolation functions could also be used for nonlinear calibration curves or other nonlinear predictive-type curves. However, a word of caution must be stated. The methods presented here are valid for obtaining values in between measured data points. Under no circumstances should any of these interpolation methods be used to extrapolate a value outside of the range of measured data. In summary, interpolation techniques are well defined and currently available for application to a myriad of biological problems.

[14] Compartmental Analysis of Fluorescence Decay Surfaces of Excited-State Processes

By MARCEL AMELOOT, NOËL BOENS, RONN ANDRIESSEN, VIVIANE VAN DEN BERGH, and FRANS C. DE SCHRYVER

Introduction

Time-resolved fluorescence measurements unravel the kinetics of excited-state processes. The fluorescence decays are collected under a variety of conditions to elucidate different aspects of the process. The excitation and/or emission wavelength, pH, concentration, temperature, etc., can be varied. The resulting multidimensional data surface is then analyzed toward the parameters of interest, namely, the rate constants and the excitation/emission spectra of the species, the so-called species-associated spectra.

Several approaches have been suggested to extract the required information from the fluorescence data surface. When the relaxation times are not too similar and the ratio of the preexponential factors of the exponential functions is not too large or close to zero, the individual decay curves can be analyzed separately in terms of relaxation times and preexponential factors. However, it is possible to analyze related decay curves simultaneously by linking parameters which are common to the fitting functions. This simultaneous or global analysis allows for a better model discrimination and parameter recovery. For example, by linking relaxation times over decay curves collected at various emission wavelengths, a much better determination of the number of required exponentials and of the values of the fitting parameters can be obtained.[1-6] The combination of the values obtained for the relaxation times and the preexponentials can then lead to the determination of the rate constants of the excited state. In some cases, it may be necessary to assume that at some wavelength the relaxation of just one species can be followed. Once the rate constants of the excited-state process are known, the species-associated spectra can be calculated.[7,8] This is essentially a two-step method to obtain the parameters of interest.

The ideal analysis would be a single-step analysis of all curves collected from the same photophysical system subject to a variety of conditions. This means that the relaxation times can no longer be the type of parameters to be linked. For example, when changing concentration in a system forming intermolecular excimers, the relaxation times change accordingly. This means that one has to fit directly for the parameters which determine the relaxation times, that is, the rate constants. This fitting procedure has been introduced in the field of time-resolved fluorescence spectroscopy by the group of Brand.[4,9,10]

By fitting directly for the underlying parameters, the applicability of the global analysis is extended. In general, once a model can be proposed, the complete data surface should be analyzed directly in terms of the parameters of the model. The general methodology has been reviewed

[1] J. R. Knutson, J. M. Beechem, and L. Brand, *Chem. Phys. Lett.* **102**, 501 (1983).

[2] M. Ameloot, J. M. Beechem, and L. Brand, *Biophys. Chem.* **23**, 155 (1986).

[3] J.-E. Löfroth, *Eur. Biophys. J.* **13**, 45 (1985).

[4] J. M. Beechem, M. Ameloot, and L. Brand, *Anal. Instrum.* **14**, 379 (1985).

[5] N. Boens, L. D. Janssens, and F. C. De Schryver, *Biophys. Chem.* **33**, 77 (1989).

[6] L. D. Janssens, N. Boens, M. Ameloot, and F. C. De Schryver, *J. Phys. Chem.* **94**, 3564 (1990).

[7] J.-E. Löfroth, *Anal. Instrum.* **14**, 403 (1985).

[8] J.-E. Löfroth, *J. Phys. Chem.* **90**, 1160 (1986).

[9] J. M. Beechem, M. Ameloot, and L. Brand, *Chem. Phys. Lett.* **120**, 466 (1985).

[10] M. Ameloot, J. M. Beechem, and L. Brand, *Chem. Phys. Lett.* **129**, 211 (1986).

recently[11] and is also discussed in [2] in this volume. In this chapter, we consider only excited-state processes with time-independent and orientation-independent rate constants. The emphasis is not on numerical parameter estimation but rather on the design of experiments and the kind of information that can be extracted from fluorescence decay surfaces of excited-state processes.

The relaxation of an excited-state process and the exchange between the excited species can be described by a system of coupled linear differential equations of first order. The generation of excited-state state species can be seen as the input to a system of which the fluorescence emission is the output. This means that excited-state systems can be seen as compartmental systems. The fundamental approach in compartmental analysis is to analyze a system by separating it into a finite number of component parts, called compartments or states, which interact through the exchange of material. This type of modeling has application in a variety of areas such as drug kinetics in pharmacology, studies of metabolic systems, analysis of ecosystems, and chemical reaction kinetics (see also [18] on compartmentation analysis and enzyme kinetics in this volume). It is interesting to note that, some time ago, the related systems theory was used by Eisenfeld and co-workers to analyze fluorescence decays of heterogeneous solutions of noninteracting species.[12,13]

One of the goals of compartmental analysis is to deal with the so-called inverse problem, namely, what information can be obtained from the measured data. This involves model specification, structural identifiability, and parameter estimation. Model specification refers to the determination of the number of compartments and the interconnections between compartments.

The study of structural identifiability inquires as to whether the system parameters are uniquely defined under error-free observations, given that the model is completely specified including compartments receiving inputs and those which are observed.[14] The experimental information found and its relation to the model parameters are given by a set of nonlinear algebraic equations, and, as a consequence, the structural identification problem reduces to the question of whether a nonlinear algebraic system of equa-

---

[11] J. M. Beechem, E. Gratton, M. Ameloot, J. R. Knutson, and L. Brand, *in* "Fluorescence Spectroscopy, Volume 1: Principles and Techniques" (J. R. Lakowicz, ed.). Plenum, New York and London, in press.

[12] J. Eisenfeld, *in* "Time-Resolved Fluorescence Spectroscopy in Biochemistry and Biology" (R. B. Cundall and R. E. Dale, eds.), p. 233. Plenum, New York and London, 1983.

[13] J. Eisenfeld and C. C. Ford, *Biophys. J.* **26,** 73 (1979).

[14] D. H. Anderson, "Compartmental Modeling and Tracer Kinetics," Lecture Notes in Biomathematics, Vol. 50. Springer-Verlag, Berlin and New York, 1983.

tions possesses a unique solution.[14] These nonlinear analytical equations allow one to obtain the rate constants and the species-associated emission and/or excitation spectra from the relaxation times and the preexponential factors.

From this identifiability theory in compartmental analysis, it has to be realized that, apart from the numerical aspect of the analysis of excited-state processes, the preexponential factors and the relaxation times are not always related to a unique set of rate parameters and spectral con-tours.[10] When an excited-state system is unidentifiable from a given fluo-rescence data surface, large uncertainties in parameter estimates may result. When performing a least-squares analysis of experimental data, the parameter values may show dependence on initial guesses. This identifi-ability problem is independent of the accuracy of the measured data and is not related to the way in which the decay curves are collected, either with pulse or frequency methods, and analyzed, either individually or globally. Consequently, it is important to know what type and number of experiments have to be performed to obtain a unique solution for the kinetic and spectral parameters in a system undergoing excited-state pro-cesses.

Many papers have dealt with the identifiability problem in compartmen-tal systems.[14–16] However, a separate study of the identifiability of excited-state processes as detected by fluorescence is appropriate. The multidi-mensional nature of the fluorescence phenomenon,[17] specifically the fact that one can use several independent experimental axes to study a given problem, makes it different from the commonly considered compartmental systems. Furthermore, a large number of data points of high quality can be obtained with time-resolved fluorescence measurements. This is usually not possible in other compartmental systems. In addition, the *a priori* information available in the common compartmental systems may also be different in fluorescence studies.

In this chapter the identifiability of two-state excited-state processes with a known concentration dependence in the forward process will be discussed. It is demonstrated that the decay curves in a fluorescence decay surface do not need to be normalized to perform a global analysis in terms of the parameters of interest.[18] Normalization of decay curves is a tedious

[15] J. A. Jacquez, "Compartmental Analysis in Biology and Medicine." Elsevier, Amsterdam, 1972.

[16] K. Godfrey, "Compartmental Models and Their Application." Academic Press, Orlando, Florida, and London, 1983.

[17] I. M. Warner, G. Patonay, and M. P. Thomas, *Anal. Chem.* **57**, 463A (1985).

[18] M. Ameloot, N. Boens, R. Andriessen, V. Van den Bergh, and F. C. De Schryver, *J. Phys. Chem.* **95**, 2041 (1991).

and sometimes even impossible task. The steady-state spectra can be used afterward to obtain the species-associated spectra. This means that, in contrast to the analysis of individual decay traces, no assumptions have to be made about the spectra of the emitting species. No extreme conditions are required to obtain the spectra of the excited species. It is also shown that for the considered system the ratio of the absorbances of the species in the ground state does not have to be known *a priori* but can be determined from the fluorescence decay surface. This has the additional benefit that when this information about the absorbances is combined with the value of the total concentration and of the optical density in the ground state, the equilibrium constant of the ground-state process can be determined together with the respective extinction coefficients of the absorbing species. This means that the equilibrium constants of both the ground- and excited-state process can be determined at once.

The organization of this chapter is as follows. First, there is a theoretical section. We start with a general description of excited-state processes with time-independent rate constants. Then the identifiability equations are given, and the identifiability of a two-state excited-state process is discussed. Consequently, the conclusions with respect to the experimental design are drawn. We then indicate how the species-associated spectra can be calculated using the information obtained from the analysis of the decay surface and steady-state spectra. In the second section some indications are given to extend existing global analysis programs to compartmental analysis. The determination of the ground-state equilibrium constant is then discussed. In the fourth section, examples of the ideas developed in the theoretical section are given.

Theory

General Description of Excited-State Processes

Consider a causal, linear time-invariant system consisting of n different species or compartments in the ground state. The species may interact or not, but they are assumed to be in equilibrium. The concentration of species i will be denoted by x_i, $i = 1, \ldots, n$. Excitation of compartment i leads to the corresponding concentration $x_i^*(t)$ in the excited state.

The relaxation of the system after delta excitation is given by the following system of differential equations,

$$\frac{dx_i^*(t)}{dt} = -\left[k_{0i} + \sum_{j=1, j \neq i}^{n} k_{ji} \right] x_i^*(t) + \sum_{j=1, j \neq i}^{n} k_{ij} x_j^*(t) \tag{1}$$

with the initial conditions

$$x_i^*(0) = b_i \tag{2}$$

The coefficients k_{ij}, $k_{ij} \geq 0$, represent the apparent rate constants of transfer of species j to species i. These coefficients determine the connectivity between the compartments. In the case of a heterogeneous sample of noninteracting species, only the coefficients k_{0i} are different from zero. The subscript 0 denotes the ground state of the considered species. This means that k_{0i} is given by the sum of radiative and nonradiative deactivation rate constants. It is possible that the radiative rate is equal to zero, which means that the corresponding species is a so-called dark species. The coefficient b_i is the fraction of the incident light that is absorbed by species i in the ground state and is given by[18]

$$b_i = \frac{\varepsilon_i x_i}{S}(1 - 10^{-dS}) \tag{3}$$

with

$$S = \sum_{i=1}^{n} \varepsilon_i x_i \tag{4}$$

and where ε_i is the decadic extinction coefficient and d is the path length of the excitation light in the measuring cuvette. Obviously, b_i depends on the considered excitation wavelength and the set of concentrations of the various species in the ground state. Note that not all b_i have to be different from zero.

It is convenient to adopt matrix notation in this context. An introduction to the use of this notation in kinetic rate equations has been described recently.[19] The system of differential equations [Eq. (1)] can be written in matrix notation as

$$\frac{dX^*(t)}{dt} = AX^* \tag{5}$$

where $X^*(t) = [x_i^*(t)]$ is a $(n \times 1)$ vector; $A = [a_{ij}]$ is the $(n \times n)$ system or transfer matrix. The matrix A indicates the connectivity between the compartments in the excited state. In the case of noninteracting species, the matrix A is diagonal. The elements of A are

$$a_{ij} = k_{ij}, \qquad i \neq j \tag{6}$$

[19] M. N. Berberan-Santos and J. M. G. Martinho, *J. Chem. Educ.* **67**, 5 (1990).

$$a_{ii} = -\left[k_{0i} + \sum_{j=1, j \neq i}^{n} k_{ji}\right] \tag{7}$$

The solution of Eq. (5) is given by[14]

$$X^*(t) = \exp(tA)B \tag{8}$$

where the exponential of the matrix A, $\exp(tA)$, is defined by the Taylor series of the function,

$$\exp(tA) = I + \frac{(tA)}{1!} + \frac{(tA)^2}{2!} + \frac{(tA)^3}{3!} + \cdots \tag{9}$$

and where $B = [b_i]$ is the $(n \times 1)$ vector of the initial conditions. The matrix A can be expressed in terms of its eigenvalues γ_i and eigenvectors U_i,

$$A = U\Gamma U^{-1} \tag{10}$$

where U is the matrix with columns U_i; the elements of the matrix Γ are given by $\Gamma_{ii} = \gamma_i$ and $\Gamma_{ij} = 0$ for $i \neq j$.

Equation (8) can then be rewritten as

$$X^*(t) = U \exp(t\Gamma)U^{-1}B \tag{11}$$

The fluorescence delta response, $\phi(t)$, observed within the selected emission band depends on the emission spectrum and the radiative deactivation rate of each species and is given by

$$\phi(t) = CX^*(t) \tag{12}$$

where C is a $(1 \times n)$ vector. The elements c_i are, apart from some instrumental proportionality factors, given by

$$c_i = k_{Fi} \int_{\Delta\lambda^{em}} \rho_i(\lambda^{em}) \, d\lambda^{em} \tag{13}$$

where k_{Fi} is the radiative deactivation rate of species i; $\rho_i(\lambda^{em})$ is the spectral emission density of species i, normalized to the complete emission band; $\Delta\lambda^{em}$ is the emission wavelength interval in which the fluorescence is monitored.

Using Eq. (8) the scalar function $\phi(t)$ can be written as

$$\phi(t) = C \exp(tA)B \tag{14}$$

or equivalently, by using Eq. (11),

$$\phi(t) = CU \exp(t\Gamma)U^{-1}B \tag{15}$$

This means that for each combination of excitation wavelength, concentration, and emission wavelength, $\phi(t)$ will be given by a sum of exponentially decaying functions,

$$\phi(t) = \sum_{i=1}^{n} \alpha_i e^{\gamma_i t} \tag{16}$$

Owing to the properties of the matrix A, the eigenvalues can be shown to be negative.[14] The eigenvalues are the negative of the inverse of the relaxation times usually encountered in the fluorescence literature.

The steady-state emission spectra F_s in terms of A, B, and C are obtained by integration of Eq. (15):

$$F_s = -CA^{-1}B \tag{17}$$

Identifiability Equations

Each combination of experimental conditions such as concentration and emission/excitation wavelength yields a corresponding set $[\alpha_i, \gamma_i]$. The elements of B, A, and C have to be determined from these sets of values. To determine these elements, the following approach has been used.

The characteristic polynomial of the system matrix A is given by

$$\det(\gamma I - A) = \gamma^n + s_1 \gamma^{n-1} + s_2 \gamma^{n-2} + \ldots + s_n \tag{18}$$

where s_m is $(-1)^m$ times the sum of all the m-square principal minors of A, that is, the minors corresponding to the diagonal elements of A. The characteristic polynomial of A is also given by

$$\prod_{i=1}^{n} (\gamma - \gamma_i) = \gamma^n - \sigma_1 \gamma^{n-1} + \sigma_2 \gamma^{n-2} - \ldots + (-1)^n \sigma_n \tag{19}$$

with

$$\sigma_1 \equiv \sum_i \gamma_i, \qquad \sigma_2 \equiv \sum_{i<j} \gamma_i \gamma_j, \qquad \sigma_3 \equiv \sum_{i<j<k} \gamma_i \gamma_j \gamma_k, \qquad \sigma_n \equiv \gamma_i \gamma_j \ldots \gamma_n \tag{20}$$

Comparing the corresponding coefficients in Eqs. (18) and (19) gives

$$s_m = (-1)^m \sigma_m \tag{21}$$

Equation (21) relates the eigenvalues γ_i, or equivalently the relaxation times, recovered from the collected decay curves to the principal minors of the matrix A and hence to the process rates.

The vectors B and C appear in the so-called Markov parameters[14] m_i, defined by

$$m_i \equiv \phi^{(i)}(0) \tag{22}$$

with $\phi(^{(i)}(0)$ being the ith time derivative of the delta response $\phi(t)$ at time zero. This means that the explicit expression for the Markov parameters m_i is given by

$$m_i = CA^iB \tag{23}$$

where A^i denotes the ith power of matrix A. An equivalent expression for the Markov parameter m_i can be found by using Eq. (16):

$$m_i = \sum_{j=1}^{n} \alpha_j \gamma_j^i \tag{24}$$

where γ_j^i denotes the ith power of γ_j.

The Markov parameters can be calculated according to Eq. (24) from the values for α_j and γ_j recovered from each collected decay curve. The calculated values for the Markov parameters m_i are then used in Eq. (23) for the determination of B and C for each value of A. Owing to the Cayley–Hamilton theorem, which states that a matrix is a solution of its characteristic polynomial,[20] the index i can take only n different values for each set of experimental parameters.[14] This means that for an n compartmental system n independent Markov parameters can be constructed from each decay trace.

To summarize, the unknown elements of the system matrix A and the vectors B and C have to be determined from the equations defined by σ_i and m_i. The system is globally identifiable when a unique solution exists. The system is said to be locally identifiable when a finite number of solutions are obtained, and unidentifiable in the case of an infinite number of solutions.

Apart from σ_1, the equations used in the identifiability study are nonlinear in the parameters to be determined. Hence, a completely general treatment for n compartments is very difficult. In the next section only a two-state excited-state process will be considered.

Identifiability of Two-State Excited-State Processes

As an example, only the case where the forward apparent reaction rate constant is concentration dependent will be considered (Scheme I). The forward apparent rate constant at the kth value of the concentration, $k_{21,k}$, is assumed to be of the form

$$k_{21,k} = k_{21}^0 h_{21,k} \tag{25}$$

where k_{21}^0 is a concentration-independent rate constant and $h_{21,k}$ is a given function of the concentration.

[20] S. Lipschutz, "Linear Algebra." McGraw-Hill, New York, 1974.

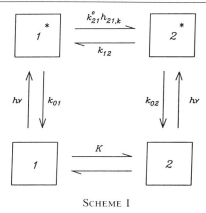

SCHEME I

The rate constants have to be determined from equations of the type indicated by Eqs. (20) and (21). The equations with the Markov parameters are required for the determination of B and C. For the kth value of the concentration, the explicit expressions corresponding to Eqs. (20) and (21) are given by

$$\sigma_{1,k} = \gamma_{1,k} + \gamma_{2,k} = -k_{01} - k_{21}^0 h_{21,k} - k_{02} - k_{12} \tag{26}$$

$$\sigma_{2,k} = \gamma_{1,k}\gamma_{2,k} = k_{01}k_{02} + k_{01}k_{12} + k_{02}k_{21}^0 h_{21,k} \tag{27}$$

The use of two different concentration values leads to the solution for k_{21}^0 and k_{02},

$$k_{21}^0 = \frac{-\sigma_{1,1} + \sigma_{1,2}}{h_{21,1} - h_{21,2}}, \qquad k_{02} = \frac{\sigma_{2,1} - \sigma_{2,2}}{-\sigma_{1,1} + \sigma_{1,2}} \tag{28}$$

If no information about k_{01} is separately available, a quadratic equation will result for either k_{01} or k_{12}. If the quadratic equation yields a positive and a negative value, all the rate parameters can be recovered using the relaxation times only. If two positive solutions result, the system may not be completely identifiable. Including experiments conducted at additional concentrations, different from zero, will not solve the problem. This can be deduced from Eqs. (26) and (27) if the concentration-dependent factors are transferred toward the left-hand side of these equations. The right-hand side of the new expression is then concentration independent. A unique solution may be obtained by use of the preexponential factors in the impulse response function through the Markov parameters.

At the excitation wavelength λ_i^{ex}, the emission wavelength λ_j^{em}, and the kth concentration value, the identifiability equations corresponding to the first two Markov parameters are given by

$$m_{0,ijk} = \alpha_{1,ijk} + \alpha_{2,ijk} = c_{1,j}b_{1,ik} + c_{2,j}b_{2,ik} \tag{29}$$

$$
\begin{aligned}
m_{1,ijk} &= \alpha_{1,ijk}\gamma_{1,k} + \alpha_{2,ijk}\gamma_{2,k} \\
&= -c_{1,j}b_{1,ik}(k_{01} + k_{21}^0 h_{21,k}) + c_{1,j}b_{2,ik}k_{12} + c_{2,j}b_{1,ik}k_{21}^0 h_{21,k} \\
&\quad - c_{2,j}b_{2,ik}(k_{02} + k_{12})
\end{aligned}
\tag{30}
$$

When the various decay curves are normalized as to the number of impinging photons on the sample, Eqs. (29) and (30) can be used directly. However, proper normalization is sometimes difficult to achieve. Therefore, it will be assumed below that the Markov parameters are known up to a proportionality factor d_{ijk}. In the subsequent derivation the vectors B and C are assumed to be normalized to unity, the normalization factor being included in the proportionality factor d_{ijk}. The normalized vectors will be indicated by $\tilde{B}$ and $\tilde{C}$. Equations (22) and (23) can then be rewritten as

$$m_{0,ijk} = d_{ijk}[\tilde{c}_{1,j}\tilde{b}_{1,ik} + \tilde{c}_{2,j}\tilde{b}_{2,ik}] \tag{31}$$

$$
\begin{aligned}
m_{1,ijk} = d_{ijk}[&-\tilde{c}_{1,j}\tilde{b}_{1,ik}(k_{01} + k_{21}^0 h_{21,k}) + \tilde{c}_{1,j}\tilde{b}_{2,ik}k_{12} \\
&+ \tilde{c}_{2,j}\tilde{b}_{1,ik}k_{21}^0 h_{21,k} - \tilde{c}_{2,j}\tilde{b}_{2,ik}(k_{02} + k_{12})]
\end{aligned}
\tag{32}
$$

Elimination of the proportionality factor d_{ijk} leads to

$$\tilde{c}_{1,j}[\tilde{b}_{1,ik}(P_{ijk} - Q_{ijk}) + Q_{ijk}] = \tilde{b}_{1,ik}(R_{ijk} - S_{ijk}) + S_{ijk} \tag{33}$$

where

$$P_{ijk} \equiv m_{1,ijk} + m_{0,ijk}(k_{01} + 2k_{21}^0 h_{21,k}) \tag{34}$$

$$Q_{ijk} \equiv -m_{1,ijk} - m_{0,ijk}(2k_{12} + k_{02}) \tag{35}$$

$$R_{ijk} \equiv m_{0,ijk}k_{21}^0 h_{21,k} \tag{36}$$

$$S_{ijk} \equiv -m_{0,ijk}(k_{02} + k_{12}) - m_{1,ijk} \tag{37}$$

The element $\tilde{c}_{1,j}$ can be eliminated from Eq. (33) by considering the same expression for an additional concentration, say 1. The resulting expression contains the unknown elements $\tilde{b}_{1,ik}$ and $\tilde{b}_{1,il}$ only:

$$T_{ij}(kl)\tilde{b}_{1,ik} + U_{ij}(kl)\tilde{b}_{1,ik}\tilde{b}_{1,il} + V_{ij}(kl)\tilde{b}_{1,il} = W_{ij}(kl) \tag{38}$$

where the coefficients are defined by

$$T_{ij}(kl) \equiv (P_{ijk} - Q_{ijk})S_{ijl} - (R_{ijk} - S_{ijk})Q_{ijl} \tag{39}$$

$$U_{ij}(kl) \equiv (P_{ijk} - Q_{ijk})(R_{ijl} - S_{ijl}) - (P_{ijl} - Q_{ijl})(R_{ijk} - S_{ijk}) \tag{40}$$

$$V_{ij}(kl) \equiv -(P_{ijl} - Q_{ijl})S_{ijk} + (R_{ijl} - S_{ijl})Q_{ijk} \tag{41}$$

$$W_{ij}(kl) \equiv -Q_{ijk}S_{ijl} + Q_{ijl}S_{ijk} \tag{42}$$

Equation (38) forms the basis for the choice of the experiments to be performed. To guarantee a unique solution for $\tilde{b}_{1,ik}$ and $\tilde{b}_{1,il}$, three emission wavelengths have to be selected such that from the six fluorescence decays three independent equations of the type of Eq. (38) are obtained. From the resulting system of equations a unique solution can be obtained. When only two emission wavelengths are used, a quadratic equation in either $\bar{b}_{1,ik}$ or $\bar{b}_{1,il}$ is obtained. The nonnegativity requirement for the unknown elements may yield a unique solution. Once $\bar{b}_{1,ik}$ and $\bar{b}_{1,il}$ are determined, the elements $\bar{c}_{1,j}$ are readily obtained from Eq. (33). When there are two different solutions for the rate constants k_{01} and k_{12}, the determination of the elements $\bar{b}_{1,ik}$ and $\bar{c}_{1,j}$ has to be carried out for each of the two solutions. The nonnegativity requirement for $\bar{b}_{1,ik}$ and $\bar{c}_{1,j}$ can then lead to a unique solution.

Another experimental design can be deduced in the following way. Instead of eliminating $\bar{c}_{1,j}$ from Eq. (33) one can eliminate $\bar{b}_{1,ik}$. Because $\bar{C}$ depends only on the emission wavelength, a different combination of experiments has to be used. $\bar{c}_{1,j}$ can be recovered from experiments conducted at three different concentrations and two emission wavelengths. When only two different concentrations are used, a quadratic equation in $\bar{c}_{1,j}$ will be obtained which might lead to a unique solution.

Identifiability and Experimental Design

Because most of the identifiability equations are nonlinear, the number of equations should preferably exceed the number of parameters to be determined. Because we work with unnormalized decay curves and use the normalized matrices $\bar{B}$ and $\bar{C}$, each decay curve considered in the global analysis requires a scaling factor as an additional fitting parameter. Each set of values for the experimental parameters leads to a set of identifiability equations. However, each decay curve containing n exponentially decaying functions does not necessarily add $2n$ new and independent identifiability equations. For example, for the two-state excited-state process considered in the previous section, four rate parameters had to be determined. Two concentrations different from zero may be sufficient. Possibly, a concentration equal to zero has to be used to obtain a unique value for all the rate constants. However, a third concentration different from zero will not add two independent identifiability equations of the type of Eqs. (26) and (27).

Attention has to be paid in selecting the emission wavelengths in setting up the appropriate set of experiments. Decay curves collected at emission wavelengths where only one and the same species is emitting do not lead to new and independent equations with the Markov parameters. It is recommended to include decay curves collected at both edges of the total emission spectrum and at the region where overlap between the underlying spectra can be expected.

For each solution of the rate constants, the scaling factor for each curve and the different matrices $\tilde{B}$ and $\tilde{C}$ have to be determined from the equations with the Markov parameters. The resulting parameter values may indicate the correct solution for the rate coefficients. To clarify the relation between experimental design and identifiability for these parameters, the two-state excited-state process discussed in the previous section is reconsidered. Assuming that experiments are collected at a single excitation wavelength, the number of experiments is given by the product of the number of concentration values (n_{conc}) times the number of emission wavelengths (n_{em}). The number of parameters $\tilde{b}_{1,ik}$ equals n_{conc}, the number of parameters $\tilde{c}_{1,j}$ is given by n_{em}, and the number of scaling factors equals n_{conc} times n_{em}. For each combination of experimental conditions, two Markov parameters can be obtained. The following necessary condition for identifiability then results:

$$2n_{conc}n_{em} \geq n_{conc} + n_{em} + n_{conc}n_{em} \tag{43}$$

or

$$n_{conc}n_{em} \geq n_{conc} + n_{em} \tag{44}$$

This condition is symmetric in n_{conc} and n_{em}. Based on this condition one can say, for example, that when $n_{conc} = 3$ and $n_{em} = 1$, $\tilde{B}$ and $\tilde{C}$ cannot be determined. When $n_{conc} = 2$ and $n_{em} = 2$ the system might be identifiable.

Species-Associated Spectra

From the values obtained for the normalized vectors $\tilde{C}$ and $\tilde{B}$, species-associated emission spectra (SAEMS) and species-associated excitation spectra (SAEXS) can be calculated using steady-state emission and excitation spectra, respectively. The observed steady-state fluorescence spectrum at emission wavelength λ^{em}, owing to excitation at λ^{ex}, $F_s^0(\lambda^{em} \mid \lambda^{ex})$, can be written as

$$F_s^0(\lambda^{em} \mid \lambda^{ex}) = p \sum_{i=1}^{n} c_i(\lambda^{em})x_{si}^*(\lambda^{ex}) \tag{45}$$

where p is a proportionality constant which also includes instrumental factors and x_{si}^* is the steady-state value of $x_i^*(t)$ given by

$$x_{si}^* = \int_0^\infty x_i^*(t)\, dt = -(A^{-1}B)_i \tag{46}$$

The contribution of species i to the steady-state emission spectrum $F_s^0(\lambda^{em} \mid \lambda^{ex})$ will be called the species-associated emission spectrum of species i, $SAEMS_i(\lambda^{em} \mid \lambda^{ex})$,

$$\text{SAEMS}_i(\lambda^{\text{em}} \mid \lambda^{\text{ex}}) = pc_i(\lambda^{\text{em}})x_{si}^*(\lambda^{\text{ex}}) \tag{47}$$

or equivalently, by using Eq. (17) and the normalized matrices $\tilde{C}$ and $\tilde{B}$,

$$\text{SAEMS}_i(\lambda^{\text{em}} \mid \lambda^{\text{ex}}) = \frac{\tilde{c}_i(\lambda^{\text{em}})[A^{-1}\tilde{B}(\lambda^{\text{ex}})]_i}{\tilde{C}(\lambda^{\text{em}})A^{-1}\tilde{B}(\lambda^{\text{ex}})} F_s^0(\lambda^{\text{em}} \mid \lambda^{\text{ex}}) \tag{48}$$

The spectral emission density $\rho_i(\lambda^{\text{em}} \mid \lambda^{\text{ex}})$ can then be readily calculated,

$$\rho_i(\lambda^{\text{em}} \mid \lambda^{\text{ex}}) = \frac{\text{SAEMS}_i(\lambda^{\text{em}} \mid \lambda^{\text{ex}})}{\displaystyle\int_{\text{full band}} \text{SAEMS}_i(\lambda^{\text{em}} \mid \lambda^{\text{ex}}) \, d\lambda^{\text{em}}} \tag{49}$$

Once the SAEMS are recovered, the radiative rate constant k_{Fi} can be obtained from k_{0i} and the total quantum yield measurement.

The species-associated excitation spectra, SAEXS, can be obtained in a similar way as for the SAEMS. If $E_s^0(\lambda^{\text{ex}} \mid \lambda^{\text{em}})$ denotes the steady-state excitation spectrum of the sample recorded at λ^{em}, $\text{SAEXS}_i(\lambda^{\text{ex}} \mid \lambda^{\text{em}})$ for species i is given by

$$\text{SAEXS}_i(\lambda^{\text{ex}} \mid \lambda^{\text{em}}) = \frac{[\tilde{C}(\lambda^{\text{em}})A^{-1}]_i\tilde{b}_i(\lambda^{\text{ex}})}{\tilde{C}(\lambda^{\text{em}})A^{-1}\tilde{B}(\lambda^{\text{ex}})} E_s^0(\lambda^{\text{ex}} \mid \lambda^{\text{em}}) \tag{50}$$

Program Implementation

As the implementation of a global analysis program for linking relaxation times and preexponentials has been described several times[1,4,11] (see also [2] in this volume), we focus here on the extension for the compartmental analysis. A nonlinear least-squares search routine according to Marquardt[21] is used. First, the program needs the number of compartments to be considered. According to the number of compartments, the program expects initial guesses for the rate constants k_{ij} and the elements of the matrices $\tilde{B}$ and $\tilde{C}$. The rate constants can be linked over decay curves collected under experimental conditions which do not change the frequency factors and the activation energies. To link the elements of $\tilde{B}$, the relative concentrations and the extinction coefficients of the species in the ground state have to be the same for the considered decay curves. To link the elements of $\tilde{C}$, the radiative deactivation rates and the spectral emission density of the species in the excited state must be the same. A freely adjustable scaling factor has to be specified for each decay trace because the normalized matrices

[21] D. W. Marquardt, *J. Soc. Ind. Appl. Math.* **11**, 431 (1963).

$\bar{B}$ and $\bar{C}$ are used and to take instrumental effects into account. This scaling factor is the only local fitting parameter, that is, a parameter which cannot be linked.

When reference convolution[22,23] is used, the lifetime of the reference compound has to be specified. This reference lifetime can be kept fixed or freely adjustable and can be linked. The preexponentials and the relaxation times corresponding to the initial guesses then have to be calculated. This means that eigenvalues and eigenvectors of the matrix A have to be determined. Several scientific computer libraries (e.g., EISPACK, IMSL, NAG) provide subroutines for this purpose. The resulting delta response function of the sample is then multiplied by the local scaling factor and convoluted with either the recorded instrumental response function or the decay of the reference compound. The calculated decay curve can now be compared with the observed trace and the residuals determined.

In the search for the best set of parameter values according to the least-squares criterion, the partial derivatives of the calculated curve with respect to the freely adjustable fitting parameters are required. All the derivatives can be calculated numerically by repeating the procedure for the determination of the calculated curve for various values of the parameter for which the derivative has to be determined, keeping all other fitting parameters fixed to their actual values. The matrices of the normal equations for each individual curve are then mapped in the matrices of the normal equations of the total data surface. This mapping is analogous to the mapping described for the linkage of preexponentials and relaxation times. The increments for the freely adjustable parameters are then calculated by solving the global normal equations.

The new values of the fitting parameters are then verified to check whether their values are physically acceptable. For example, the rate constants, the elements of $\bar{B}$ and $\bar{C}$, the local scaling factor, and the reference lifetime must all be positive. Also, upper bounds can be specified for each fitting parameter. When the new parameter values cross the specified boundaries or when the weighted residuals summed over the total decay surface increase, new increments for the parameters have to be determined by changing the Marquardt parameter on the diagonal of the global matrix of the normal equations. The rest of the procedure is standard to least-squares analysis.

[22] R. W. Wijnaendt van Resandt, R. H. Vogel, and S. W. Provencher, *Rev. Sci. Instrum.* **53**, 1392 (1982).

[23] M. Zucker, A. G. Szabo, L. Bramall, D. T. Krajcarski, and B. Selinger, *Rev. Sci. Instrum.* **56**, 14 (1985).

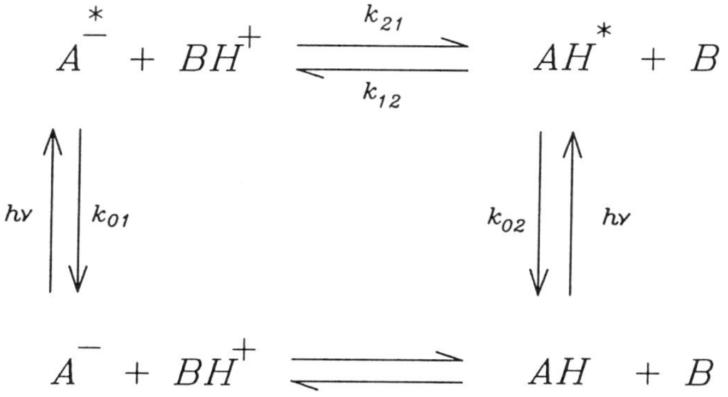

$$\overset{*}{A}^{-} + BH^{+} \underset{k_{12}}{\overset{k_{21}}{\rightleftharpoons}} \overset{*}{AH} + B$$

$$h\nu \Big\updownarrow k_{01} \qquad\qquad k_{02} \Big\updownarrow h\nu$$

$$A^{-} + BH^{+} \rightleftharpoons AH + B$$

SCHEME II

Determination of Ground-State Equilibrium Constant

From the value obtained for $\tilde{b}_{1,ik}$ the ratio of b_{ik} to b_{2k} can be determined

$$\frac{\tilde{b}_{1k}}{\tilde{b}_{2k}} = \frac{\tilde{b}_{1k}}{1 - \tilde{b}_{1k}} = \frac{b_{1k}}{b_{2k}} = \frac{\varepsilon_1 x_{1k}}{\varepsilon_2 x_{2k}} \tag{51}$$

Note that this ratio is equal to the ratio of the concentrations of the excited species at $t = 0$. Additional information about the ground state can be obtained in the following way.

The equilibrium constant of the ground state and the extinction coefficients of the ground-state species, ε_i, can be determined when the total concentration and the optical density are known for each experimental condition. This will be illustrated for the acid–base equilibria (Scheme II).

From the definition of $\tilde{B}$, one has that

$$\tilde{b}_1 = \frac{\varepsilon_1[A^-]}{\varepsilon_1[A^-] + \varepsilon_2[AH]} \tag{52}$$

Using the definition of K_a,

$$K_a = \frac{[BH^+][A^-]}{[AH]} \tag{53}$$

Eq. (52) can be rearranged to

$$\tilde{b}_1 = \frac{\varepsilon_1 K_a}{\varepsilon_1 K_a + \varepsilon_2[BH^+]} \tag{54}$$

Measuring the absorbance at λ^{ex} gives a value for S [See Eq. (4)]:

TABLE I
RATE CONSTANTS FOR PHOTOPHYSICS
OF PYRENE IN CYCLOHEXANE[a]

$k_{01} = 2.25 \times 10^6 \text{ sec}^{-1}$	$k_{02} = 15.5 \times 10^6 \text{ sec}^{-1}$
$k_{21}^0 = 6.7 \times 10^9 \, M^{-1} \text{ sec}^{-1}$	$k_{12} = 6.5 \times 10^6 \text{ sec}^{-1}$

[a] From B. Birks, "Photophysics of Aromatic Molecules." Wiley, London, 1970.

$$S = \varepsilon_1[A^-] + \varepsilon_2[AH] \tag{55}$$

If the total concentration is known, one can calculate

$$q = \frac{[A^-] + [AH]}{\varepsilon_1[A^-] + \varepsilon_2[AH]} \tag{56}$$

which can be rearranged to

$$q = \frac{K_a + [BH^+]}{\varepsilon_1 K_a + \varepsilon_2[BH^+]} \tag{57}$$

For each pH value one has the independent data $\tilde{b}_1$ and q. Simultaneous analysis by nonlinear least-squares of these data at various pH values by linking the fitting parameters K, ε_1, and ε_2 solves the problem.

Alternatively, the parameters K_a and ε_1 can be obtained directly by the global compartmental analysis of the fluorescence decay surface by rewriting $\tilde{b}_1$ as

$$\tilde{b}_1 = \frac{q\varepsilon_1 K_a}{K_a + [BH^+]} \tag{58}$$

Examples

The examples chosen deal with relatively simple systems. The main purpose is to clarify the previous sections.

Intermolecular Excimer Formation of Pyrene

Fluorescence decays of pyrene at several concentrations in hexane were simulated,[24] using the rate constants at room temperature published by Birks *et al.* (Table I). The photophysics of pyrene at concentrations higher than 10^{-5} M in a nonviscous solvent can be pictured in terms of

[24] R. Andriessen, N. Boens, M. Ameloot, and F. C. De Schryver, *J. Phys. Chem.* **95**, 2047 (1991).

compartmental systems as follows. In Scheme I compartments 1 and 1* correspond with the monomer in the ground and in the excited state, respectively. Compartment 2* corresponds with the excimer. Obviously, compartment 2 does not apply as the pyrene dimer in the ground state does not exist in dilute, homogeneous solutions. This means that the data analysis should yield $\bar{b}_2 \cong 0$. The forward reaction rate constant depends on the pyrene concentration as expressed by Eq. (25) with $h_{21,k}$ being the kth value of the pyrene concentration $[M]$. To illustrate the identifiability problem and the usefulness of Eq. (44), data surfaces corresponding to different concentrations and emission wavelengths were generated by the computer. The possible linkages between the fitting parameters in the least-squares analysis are represented in Scheme III.

In the first set of experimental conditions fluorescence decays at two different pyrene concentrations measured at the same emission wavelength are considered. The four rate constants can be determined from Eqs. (26) and (27). The remaining unknown parameters are $\bar{c}_1$, two $\bar{b}_1$ values, and two scaling factors, a total of five. These have to be determined from four Markov parameters [Eqs. (31) and (32)]. Obviously, the system under this set of experimental conditions is not identifiable. The results of a least-squares analysis of this data surface is given in Table II, type A. No accurate parameter recovery is obtained using pyrene concentrations of 0.001 and 0.003 M. Considering the concentrations 0.001 and 0.01 M results in a more accurate but not yet acceptable recovery.

In the second experimental setup, the fluorescence decays of three different pyrene concentrations at a single emission wavelength are used in the global bicompartmental analysis. At first sight, this set of experiments should suffice to calculate all parameters. Indeed, 12 identifiability equations can be constructed to determine 11 parameters: 4 k_{ij}, $\bar{c}_1$, 3 $\bar{b}_1$ values, and 3 scaling factors. However, as has been pointed out above, the decay times of the third concentration do not lead to any new and independent equations. The four rate constants can be calculated using the two decay times at two different concentrations. The Markov parameter equations allow one to construct 6 equations from which 7 parameters have to be determined. Hence, not all the parameters of interest can be determined, as could have been foreseen by applying Eq. (44). Table II, type B displays the results obtained by the least-squares analysis.

In the third experimental setup, the fluorescence decays of two different pyrene concentrations, each at two different emission wavelengths, are combined in one global compartmental analysis. Again, values for the rate constants can be determined from the relaxation times at the two concentrations. The Markov expressions allow one to construct 8 independent equations for the calculation of 8 parameters: 2 $\bar{c}_1$ values, 2 $\bar{b}_1$ values,

TABLE II

RECOVERED PARAMETER VALUES ACCORDING TO SCHEME III FROM SIMULATED PYRENE DATA[a]

Experimental design type	[M] (mol liter^{-1})	$\bar{c}_1$ true	$\bar{c}_1$ est.[b]	$\bar{b}_1$ est.	k_{01} ($\times 10^6$ sec^{-1})	k_{21}^0 ($\times 10^{-9}$ M^{-1} sec^{-1})	k_{02} ($\times 10^{-6}$ sec^{-1})	k_{12} ($\times 10^{-6}$ sec^{-1})
A	0.001	1	0.9 ± 0.9	1 ± 6	0.7 ± 1.7	8.5 ± 1.7	13 ± 1.4	3.9 ± 2
	0.003			1 ± 5				
	0.001	0.75	0.75 ± 4	0.9 ± 1.2	2 ± 6	6.7 ± 8	14.3 ± 1	4.9 ± 9
	0.003			0.9 ± 24				
	0.001	0.75	0.5 ± 5	0.64 ± 0.5	1.6 ± 9	6.9 ± 4	15.1 ± 0.5	4.5 ± 2
	0.01			0.83 ± 70				
B	0.001	0.95	0.95 ± 4	1.0 ± 8	2.2 ± 0.1	6.7 ± 0.1	15.5 ± 0.1	6.3 ± 0.3
	0.003			0.96 ± 13				
	0.01			0.99 ± 34				
	0.001	0.05	0.1 ± 0.8	1.1 ± 1.1	2.27 ± 0.06	6.77 ± 0.09	15.54 ± 0.01	6.8 ± 0.2
	0.003			1.1 ± 1.2				
	0.01			1.1 ± 1.3				
C	0.001	0.95	0.84 ± 0.05	1.5 ± 0.8	0.5 ± 1.3	8.7 ± 1.4	13 ± 1	3.9 ± 1.5
		0.9	0.81 ± 0.05					
	0.003			0.9 ± 0.6				
	0.001	0.95	1.0 ± 0.08	1.1 ± 0.4	2.8 ± 0.2	5.6 ± 0.5	17 ± 1	7.1 ± 0.4
		0.5	0.5 ± 0.03					
	0.003			1.1 ± 0.3				
	0.001	0.95	0.88 ± 0.04	1.3 ± 0.2	2.2 ± 0.1		15.5 ± 0.3	6.6 ± 0.02

Expt	σ	weight	est[b]	$\hat{b}_1$				
	0.003	0.05	0.02 ± 0.08	1.3 ± 0.2		6.8 ± 0.2		
	0.001	0.95	0.931 ± 0.008	1.14 ± 0.06	2.25 ± 0.08	6.7 ± 0.1	15.54 ± 0.08	6.6 ± 0.2
		0.05	0.15 ± 0.03					
	0.01			1.17 ± 0.07				
D	0.001	0.9	0.85 ± 0.04	1.1 ± 0.2	2.4 ± 0.4	6.4 ± 0.8	15.7 ± 1.2	6.4 ± 0.7
		0.75	0.73 ± 0.02					
		0.5	0.54 ± 0.05					
	0.003			1.3 ± 0.6				
	0.001	0.9	0.88 ± 0.03	1.01 ± 0.06	2.3 ± 0.3	6.6 ± 0.2	15.5 ± 0.2	6.5 ± 0.9
		0.75	0.74 ± 0.03					
		0.5	0.50 ± 0.01					
	0.01			1.1 ± 0.3				
	0.001	0.95	0.946 ± 0.004	1.03 ± 0.02	2.29 ± 0.08	6.7 ± 0.1	15.54 ± 0.06	6.7 ± 0.2
		0.5	0.499 ± 0.004					
		0.05	0.077 ± 0.017					
	0.01			1.05 ± 0.02				
E	0.001	0.95	0.89 ± 0.03	1.2 ± 0.12	2.2 ± 0.7	6.7 ± 0.1	15.4 ± 0.14	6.3 ± 0.3
		0.9	0.85 ± 0.03					
	0.003			1.2 ± 0.2				
	0.01			1.5 ± 0.4				
	0.001	0.95	0.941 ± 0.005	1.08 ± 0.04	2.32 ± 0.04	6.67 ± 0.06	15.59 ± 0.06	6.76 ± 0.13
		0.05	0.11 ± 0.03					
	0.003			1.08 ± 0.04				
	0.01			1.10 ± 0.05				

[a] $\hat{b}_1$ true is 1 for all experiments. All errors are two times the standard deviation.
[b] est. = estimated.

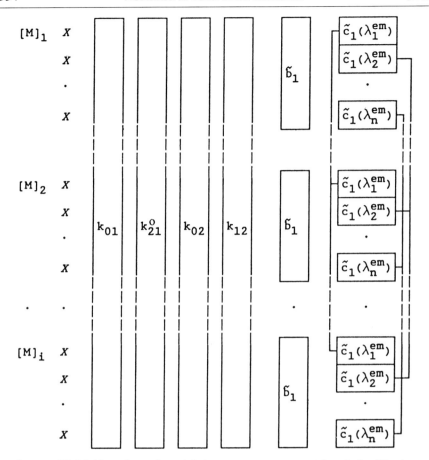

SCHEME III. Linking scheme for a global two-state compartmental analysis of the decays at i different concentrations and n different emission wavelengths λ^{em} collected at a single excitation wavelength λ^{ex}. Boxed parameters are linked, while X denotes the unlinked scaling factors. $\tilde{b}_1$ depends on the concentration $[M]_i$.

and 4 scaling factors. The system can now be identified. As was pointed out above, two different solutions may be obtained. However, the restrictions to positive values for the rate constants, and $\tilde{c}_1$ and $\tilde{b}_1$ being limited to the range between zero and unity, can lead to a unique solution. The least-squares results are shown in Table II, type C. Changing the initial guesses had no influence on the estimated values. As is shown in Table II, type C, the accuracy and the precision of the parameter recovery improve by combining experiments with extreme $\tilde{c}_1$ values. For real exper-

TABLE III
RATE CONSTANTS FOR PYRENE IN ISOOCTANE[a]

$k_{01} = 2.17 \pm 0.04 \times 10^6 \text{ sec}^{-1}$	$k_{02} = 14.8 \pm 0.02 \times 10^6 \text{ sec}^{-1}$
$k_{21} = 7.2 \pm 0.1 \times 10^9 \ M^{-1} \text{ sec}^{-1}$	$k_{12} = 5.7 \pm 0.07 \times 10^6 \text{ sec}^{-1}$

[a] At room temperature estimated with compartmental analysis. All errors are two times the standard deviation.

iments, this implies that including fluorescence decays collected at the edges of the emission spectrum will contribute the most. Increasing the concentration difference enhances the precision of the parameter recovery (Table II, type C).

The fourth experimental setup uses the fluorescence decays of two different concentrations at three different emission wavelengths. Apart from the rate constants, 11 parameters have to be determined: 3 $\bar{c}_1$ values, 2 $\bar{b}_1$ values, and 6 scaling factors. The Markov parameters lead to 12 equations. The least-squares results are shown in Table II, type D. Again, using extreme $\bar{c}_1$ values and increasing the concentration difference improve the accuracy. The results for the fifth set of experimental conditions, three concentrations and two emission wavelengths, are shown in Table II, type E. These results confirm the conclusions drawn in the third and fourth set of experimental conditions.

When the data surface allowed identification of the model parameters, an analysis was also performed with unlinked and unnormalized C matrices. The local scaling factor was then included in the c elements. The resulting accuracy was not as good as for the normalized $\bar{C}$ matrices, and the standard deviations were more than 10 times larger. This demonstrates the advantage of using normalized $\bar{C}$ matrices.

Real data were obtained from measurements of pyrene in isooctane. The pyrene concentrations were 10^{-6}, 1.45×10^{-4}, and $1.53 \times 10^{-3} \ M$. On excitation at 325 nm, the solutions with the two highest concentrations exhibited excimer emission at room temperature. The fluorescence decays of solutions at the three concentrations were collected at various emission wavelengths and time increments. A total of 42 fluorescence decays were combined in one compartmental analysis in which two compartments were considered for each concentration. The values obtained for the rate constants are shown in Table III. The rate constants can also be calculated from the relaxation times.[25,26] The relaxation times were obtained from a global analysis of the decay traces at each concentration. The values for

[25] W. R. Laws and L. Brand, *J. Phys. Chem.* **83,** 795 (1979).
[26] T. C. Cheung and W. R. Ware, *J. Phys. Chem.* **87,** 466 (1983).

TABLE IV
RATE CONSTANTS OF 2-NAPHTHOL AT ROOM TEMPERATURE

Rate constant	Laws et al.[a]	Beechem et al.[b]	This study
k_{01} (nsec^{-1})	0.106	0.106 ± 0.014	0.112 ± 0.001
k_{21}^0 (M^{-1} nsec^{-1})	47	50 ± 5	50 ± 1
k_{02} (nsec^{-1})	0.138	0.135 ± 0.001	0.133 ± 0.001
k_{12} (nsec^{-1})	0.07	0.071 ± 0.03	0.071 ± 0.006

[a] W. R. Laws and L. Brand, *J. Phys. Chem.* **83,** 795 (1979).
[b] J. M. Beechem, M. Ameloot, and L. Brand, *Chem. Phys. Lett.* **120,** 466 (1985).

the rate constants obtained by the two methods are in good agreement (Table III). However, the compartmental analysis yielded smaller confidence intervals. The relative absorbance b_1 was linked at each concentration, and the recovered values and estimated standard deviations were 0.996 ± 0.002 and 0.998 ± 0.002 for the concentrations 1.45×10^{-4} and 1.53×10^{-3} M, respectively. This is in perfect agreement with the fact that, in the studied concentration range, there are no ground-state dimers for homogeneous solutions of pyrene in a nonviscous solvent and that the excitation light is absorbed by the monomeric pyrene molecules only.

Excited-State Proton Transfer Reaction of 2-Naphthol

The excited-state reaction of 2-naphthol is well studied.[9,25,27] 2-Naphthol becomes more acidic in the excited state. The pK_a in the ground state is 9.49, whereas the pK_a^* in the excited state is 2.8.[27] In terms of the compartments of Scheme I, compartments 1 and 1* correspond with 2-naphtholate. The compartments 2 and 2* are associated with 2-naphthol.

The first introduction of the global method in terms of compartmental analysis has been exemplified by measurements on 2-naphthol.[9] In that study pH values were chosen such that the reaction in the excited state was completely reversible. This implies that only one species is to be considered in the ground state. Indeed the experimental data were analyzed by taking $\bar{B} = (1\ 0)$. Unnormalized C matrices were used. It was shown that the rate constants can be obtained by a compartmental analysis of decay curves at pH 2.15 and 3.0 collected at two emission wavelengths. The results are shown in Table IV, where a comparison is also made with the results obtained by single-curve analysis in an earlier report.[25] The

[27] E. Van der Donckt, *Prog. React. Kinet.* **5,** 273 (1970).

species-associated spectra were obtained from decay curves collected every 2 nm from 350 to 500 nm for each of the two pH values. The dwell time for each experiment was only 10 min. This resulted in an experimental decay surface which has a maximum of approximately 2000 counts at the peak, distributed in 511 channels over 51 nsec. The total number of decays was 150 (two pH values, each at 75 different emission wavelengths). The decay curves were analyzed for unnormalized C matrices. The resulting spectra agreed very well with the steady-state spectra obtained at pH 0 and 13.

The difference with the excimer example is that in the case of 2-naphthol for pH values not too much different from the pK_a, two species in the ground state have to be considered. To demonstrate the usefulness of compartmental analysis with normalized $\tilde{B}$ and $\tilde{C}$ matrices in that case, we globally analyzed decay curves collected at pH 8.04, 9.17, and 9.55 at three different emission wavelengths. Under this condition the system is identifiable. However, an inaccurate value was obtained for k_{21}^0 ($k_{21}^0 = 3 \times 10^7 \, M^{-1} \, sec^{-1}$). This can be explained from the fact that at these pH values the equilibrium is strongly shifted toward naptholate. On the other hand, it is very difficult to obtain accurate values for $\tilde{b}_1$ from the global analysis of experiments obtained at pH 1.91, 2.1, and 2.31. At these pH values, practically only 2-naphthol will be present in the ground state. A global compartmental analysis of the experiments at the pH values 2.48, 2.63, 8.75, and 8.95 yielded accurate and precise values for all parameters (Table IV). The calculated pK_a^* is 2.85. The following values and standard deviations were obtained for $\tilde{b}_1$: 0.25 $\pm$ 0.03 (pH 8.75) and 0.27 $\pm$ 0.03 (pH 8.95). The knowledge of the total concentration and the optical density and the use of Eqs. (54) and (57) yielded $pK_a = 9.5 \pm 0.1$, $\varepsilon_1 = 1647 \pm 505 \, M^{-1} \, cm^{-1}$, and $\varepsilon_2 = 1251 \pm 104 \, M^{-1} \, cm^{-1}$.

Conclusion and Perspectives

It is shown in this chapter how fluorescence decay curves collected under various experimental conditions can be analyzed simultaneously. The possibilities of a global analysis of a data surface is extended by fitting for the parameters of interest instead of preexponential factors and relaxation times. Furthermore, the discussed compartmental analysis allows one to use unnormalized decay traces. It is assumed that the rate constants are time independent. However, no assumptions have to be made about the ground-state distribution. On the contrary, this distribution can be obtained from the fluorescence data surface. Spectral contours of the emitting species can be treated in the same way. Even nonemitting species can be detected in this kind of analysis. In this sense, the compartmental analysis can be called a *non a priori* method. In the case where only two species have to be considered in the ground state, knowledge of the concentration and the optical density allows one to calculate the ground-state equilibrium constant and the extinction coefficients.

It is shown that optimal results are obtained when experimental conditions are such that some fitting parameters are substantially influenced while other parameters can be linked. Although the best parameter recovery is obtained from significantly different experimental conditions, it is possible to obtain good parameter determination for less different experimental conditions if the system is identifiable. It is not required that the sample be subjected to extreme conditions in order to determine the parameters.

The importance of a correct number and combination of experiments has been emphasized. A simple condition has been formulated which has to be fulfilled to identify completely the system in terms of the assumed model. When the assumed model is not identifiable from the collected data surface, the parameter estimates may show a dependence on the initial guesses and exhibit large uncertainties.

The explicit identifiability equations described above can be used to determine the parameters of interest from preexponentials and relaxation times when a compartmental analysis program is not available. The relaxation times and the corresponding preexponential factors can in principle be determined from separate single-curve analysis or from a global analysis of curves with the same concentration. The identifiability equations [Eqs. (26), (27), (31), and (32)] can be applied to determine the parameters of interest. It should be noted that this identifiability study is different from the numerical stability problem considered with the analysis of real data.

Objections have been formulated against the use of a global analysis in the sense that the parameters are restricted and the outcome of the analysis may be biased. Indeed, by linking parameters one is restricting the freely adjustable parameters in the fitting procedure. When the linkages between the parameters are not used, one has to test a model in a second step using the values recovered for the preexponentials and the relaxation times. Owing to error propagation the model discrimination power may not be sufficient. It is better to impose the model directly on the data so that the various tests on the residuals can indicate which model is most likely to be correct. Of course, when a new photophysical system is investigated, it is difficult or even impossible to suggest a model directly from the collected data. The following procedure can be suggested.

First, the decay curves of the data surface are analyzed individually as a sum of exponentials. From the resulting preexponentials one can estimate which experiments can be analyzed globally using a linking of relaxation times and/or preexponentials. That global analysis should provide some evidence about the number of exponentials. To start the global compartmental analysis, one has to specify the number of compartments and to provide initial estimates for the rate constants. The number of compartments should at least be equal to the number of exponentials. Assumptions must also be made regarding the connectivity between the compartments, namely, the rate constants different from zero have to be specified. Possibly, estimates for the rate constants can be obtained from the recovered values of the relaxation times by using Eqs. (26) and (27). On the other hand, one can try to analyze the data surface with all elements of the matrix A being different from zero.[9] When very small values are obtained for some rate constants, one can perform an alternative analysis by fixing these rate constants to zero.

The advantage of performing a simultaneous analysis of related experiments is the resulting high model discrimination power. Several competing models should be tested. The appropriateness of a particular model can be evaluated by fitting criteria such as the sum of the weighted squared residuals and the properties of the residuals. It is always recommended to verify the number of solutions for the rate constants. If two acceptable solutions result from Eqs. (26) and (27), the following procedure can be suggested. For each solution one performs a compartmental analysis of the fluorescence decay surface by fixing the rate constants at their respective values. The corresponding initial estimates for $\bar{B}$ and $\bar{C}$ can be obtained from the Markov parameters. The values resulting from the global compartmental analysis for $\bar{B}$, $\bar{C}$, and the scaling factors together with values for the rate constants are then used as initial guesses for a new search in which all parameters are freely adjustable. The final results should then be checked for their acceptability.

In this chapter only the explicit identifiability requirements have been discussed for a two-state excited-state reaction under certain experimental conditions. However, the identifiability equations [Eqs. (20) and (23)] are completely general. For a three-state excited-state reaction the identifiability study becomes very complex. The different connectivity schemes between the compartments require a separate identifiability study. Nevertheless, it is worthwhile considering this problem as many photophysical systems show a triple-exponential behavior.

We discussed only bimolecular reactions for which the concentration axis can be used to unravel the kinetic scheme of the reaction. For unimolecular or intramolecular reactions this experimental axis cannot be used. Other means have to be explored to investigate these systems in terms of compartmental analysis. Recently, Weidner and Georghiou[28] described the combination of global methods for time-resolved and steady-state fluorescence and steady-state absorption spectroscopy in the study of nucleic acid systems for which there is energy transfer and exciplex formation. Until now, global compartmental analysis has mainly been applied to investigate its performance by studying well-characterized systems. Further applications of global compartmental analysis in the life sciences will be found in the near future as more laboratories are acquiring the required software, which is nowadays even commercially available.

Acknowledgments

M.A. acknowledges discussions with J. M. Beechem, R. P. DeToma, and L. Brand. N.B. is a Bevoegdverklaard Navorser of the Belgian Fonds voor Geneeskundig Wetenschappelijk Onderzoek (FGWO). R.A. is an Aspirant Onderzoeker of the Belgian Nationaal Fonds voor het Wetenschappelijk Onderzoek (NFWO). V.V.d.B. is a predoctoral fellow of the Instituut ter Aanmoediging van het Wetenschappelijk Onderzoek in de Nijverheid en de Landbouw, Belgium. The authors of the K. U. Leuven thank the Ministry of Scientific Programming and the NFWO for financial support to the laboratory.

[28] R. Weidner and S. Georghiou, *Proc. SPIE Int. Soc. Opt. Eng.* **1204**, 717 (1990).

[15] Analysis of Discrete, Time-Sampled Data Using Fourier Series Method

By LINDSAY M. FAUNT and MICHAEL L. JOHNSON

Introduction

A key question in the analysis of time-sampled data is whether the data set displays a characteristic periodicity. This was the case with the assays of human luteinizing hormone (LH; see Fig. 1) and melatonin (see Fig. 5) that we are interested in analyzing for circadian and other periodicities. These data consist of measurements of plasma hormone concentration sampled at intervals of between 10 and 30 min over a period of between 24 and 36 hr. In examining these data for periodicities, we explored the application of Fourier series analysis to sets of discrete data with variable sampling intervals and nonconstant variances.

The Fourier series technique is a valuable tool in the analysis of data in the time domain. Typically, the experimenter is presented with a series

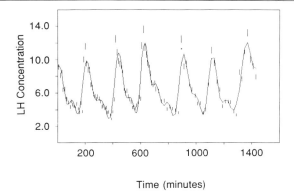

Time (minutes)

F<small>IG</small>. 1. Serum concentration of human luteinizing hormone (LH) sampled at 10- to 20-min intervals over a period of 24 hr (M. J. Sollenberger, E. C. Carlsen, J. D. Veldhuis, and W. S. Evans, unpublished data). The smooth curve is the optimal fit, obtained for truncation order $k_{max} = 16$ in Eq. (1).

of measurements of a parameter $y(x_i)$ (hormone concentration, for instance) as a function of the independent variable x_i (usually time). The uncertainty, or standard deviation, of a particular measurement $y(x_i)$ is denoted σ_i. The task of analysis is to fit the N data points, $y(x_i)$, σ_i, and x_i, to a function, such as a Fourier series, appropriate for the questions one wishes to answer.

The Fourier series is a sum of sine and cosine waves (basis functions):

$$f(x) = A_0 + \sum_{k=1}^{k_{max}} A_k \cos(2\pi kx/T) + B_k \sin(2\pi kx/T) \qquad (1)$$

The fundamental periodicity is denoted T in Eq. (1). Periodicity is a fundamental requirement for Fourier series analysis. The k in Eq. (1) denotes the harmonic of the fundamental frequency to which a specific sine or cosine wave corresponds.

The basis functions, considered as continuous functions, are orthogonal. This means that the integral over one fundamental period of the product of any two distinct basis functions is zero. Under certain conditions, there is an analogous statement for discretely sampled data. If (1) the data points are equally spaced in time, (2) there is an odd number of data points per fundamental period, and (3) the $y(x_i)$ values all have the same experimental uncertainty, then the discrete sum over one fundamental period of the product of any two distinct basis functions is also zero. The basis functions then are said to be orthogonal with respect to the data. Furthermore, if these orthogonality conditions are met then the classic approach to Fourier series analysis provides a simple means of evaluating the Fourier series coefficients, namely, the A_k and B_k values in Eq. (1). However, the experimental data that we are investigating do not meet these conditions.

As generally applied, a Fourier series analysis provides an exact fit of N data points per fundamental period with N equations in terms of M unknowns, where M is equal to $2k_{max} + 1$. The data will normally consist, however, not only of some underlying periodic function, but also of some random experimental uncerrtainty, or noise. Consequently, the low-order terms of the Fourier series correspond to the underlying periodicity of interest, and the high-order terms correspond to a fit of the random experimental noise that is superimposed on the data. We wish to distinguish between these two components. This is done by finding the upper limit k_{max} such that the portion of the data not described by the truncated Fourier series is in fact random noise. We discuss in a later section the means of determining this upper limit.

Often the Fourier basis functions are orthogonal with respect to the data, at least to a good approximation. This is true when there is a high sampling rate, and each data point is equally weighted, that is, all have identical uncertainties. Then, the N equations become decoupled, and the Fourier coefficients can be evaluated independently. This will be shown in the next section. On the other hand, we are interested in interpreting sets of discrete time-series data that do not fulfil such criteria. These hormone concentration data sets consist of unequally spaced data points with substantially differing standard deviations (uncertainties). Furthermore, the nature of the sampling procedure precludes our being able to collect the data in such a way that the requirements of the conventional Fourier analysis are satisfied. In this chapter, we present a method by which the limitations of the standard Fourier analysis are overcome.

Methods

For applications such as ours, where the number N of data points is greater than the number $2k_{max} + 1$ of unknown parameters, the unknown parameters are overdetermined and can be evaluated by maximum likelihood techniques. Put another way, we can obtain values for the Fourier coefficients that have the highest probability, or maximum likelihood, of being correct. This is equivalent to a least-squares method if we assume that (1) the random experimental uncertainties (noise) can be described by a Gaussian distribution, (2) there are no systematic uncertainties in the data, (3) all the uncertainties are in the dependent variable, $y(x_i)$, (4) the experimental observations $y(x_i)$ are independent observations, (5) N is large enough to provide a good sampling of the random uncertainties, and (6) a Fourier series [Eq. (1)] is indeed a correct description of the experimental data. This last assumption is equivalent to requiring that the upper limit of k be approximately correct. We discuss these assumptions in more detail elsewhere in this volume.[1] These assumptions are reasonable for the hormone time-series data of the present example. The least-squares method provides not only a solution for the Fourier coefficients, but also an estimate of the uncertainties in those values.

The least-squares method we used for the estimation of the Fourier coefficients can be formulated by approximating the experimental data $y(x_i)$ by the fitting function [Eq. (1)] evaluated at the maximum likelihood of Fourier coefficients. The Taylor series in Eq.(2) forms the basis of an iterative nonlinear least-squares technique for solving for the Fourier coefficients, represented by the vector **a**. Initial guesses, collectively notated by the vector **g**, for the Fourier coefficients are made, and Eq. (2) is solved for the next set of approximations to **a**. Successive approximations to the coefficients **a** are made by taking the previous iteration's approximation as the guess for the next. This process continues until the error (the difference between successive approximations) is sufficiently small.

$$y(x_i) \cong f(x_i, \mathbf{a}) = f(x_i, \mathbf{g}) + \sum_{j=1}^{M} \frac{\partial f(x_i, \mathbf{g})}{\partial g_j} (a_j - g_j) + \cdots \tag{2}$$

The linear least-squares method that we used is a degenerate case of this approach where the high-order derivatives, represented by the ellipsis in Eq. (2), are all zero. For expansions linear in the coefficients, Eq. (2) can be truncated after the first derivative term and remain an exact equation. Consequently, the initial values **g** for the Fourier coefficients can have nearly any value (we used all zeros), and the maximum likelihood estimates are reached after a single iteration.

Evaluation of Coefficients

The determination of the Fourier coefficients follows from the general Taylor series expansion of a nonlinear function, for each of the N data points. These expansions are of the form of Eq. (2), and are collectively expressed in matrix notation as

---

[1] M. L. Johnson and L. M. Faunt, this volume [1].

$$
\begin{bmatrix}
\dfrac{Z_1(x_1)}{\sigma_1} & \cdots & \dfrac{Z_M(x_1)}{\sigma_1} \\[2ex]
\vdots & \ddots & \vdots \\[2ex]
\dfrac{Z_1(x_N)}{\sigma_N} & \cdots & \dfrac{Z_M(x_N)}{\sigma_N}
\end{bmatrix}
\begin{bmatrix}
a_1 - g_1 \\[2ex]
\vdots \\[2ex]
a_M - g_M
\end{bmatrix}
=
\begin{bmatrix}
\dfrac{y(x_1) - f(x_1,\mathbf{g})}{\sigma_1} \\[2ex]
\vdots \\[2ex]
\dfrac{y(x_N) - f(x_N,\mathbf{g})}{\sigma_N}
\end{bmatrix}
\tag{3}
$$

where $Z_j(x_i)$ is the derivative of the fitting function with respect to each parameter being estimated:

$$
Z_j(x_i) = \frac{\partial f(x_i,\mathbf{g})}{\partial g_j}
$$

Note that for the Fourier series, these derivatives are just the sine and cosine basis functions; therefore, the Z_j values form an orthogonal basis set.

Equation (3) can be expressed in matrix notation as

$$
\mathbf{Ae} = \mathbf{D} \tag{4}
$$

where $\mathbf{A}$ is the matrix of weighted partial derivatives; $\mathbf{D}$ is the vector of weighted differences between the actual data and the fitting function [Eq. (1)] evaluated at the current estimates of the parameters, $\mathbf{g}$; and $\mathbf{e}$ is the vector of differences between the predicted least-squares parameter estimates $\mathbf{a}$ and the current estimates $\mathbf{g}$ of the parameters. If there are more independent observations than parameters to be estimated, then Eq. (4) can be solved for the correction vector $\mathbf{e}$:

$$
\mathbf{e} = \mathbf{a} - \mathbf{g} = (\mathbf{A}^{\mathrm{T}}\mathbf{A})^{-1}(\mathbf{A}^{\mathrm{T}}\mathbf{D}) \tag{5}
$$

For the Fourier series, which is linear in the coefficients $\mathbf{g}$, this process can be simplified. The higher-order derivatives in the Taylor expansion [Eq. (2)] are identically zero. The initial guesses for the Fourier coefficients can without loss of generality be taken to be zero, and the solution set $\mathbf{a}$ is obtained from Eq. (5) in one step:

$$
a_j = \sum_{k=1}^{M} (\mathbf{A}^{\mathrm{T}}\mathbf{A})_{jk}^{-1}(\mathbf{A}^{\mathrm{T}}\mathbf{D})_k \tag{6}
$$

where the subscripts refer to particular vector or matrix elements.

So far little has been said of the differences between the standard formula for a linear expansion, such as the Fourier sum, and the more general derivation that we shall now consider. The distinction arises in the way that the derivatives $Z_j(x)$ are interrelated. Of particular interest is the matrix $\mathbf{A}^{\mathrm{T}}\mathbf{A}$; its elements are

$$(\mathbf{A}^T\mathbf{A})_{jk} = \sum_{i=1}^{N} A_{ji}^T A_{ik}$$

$$= \sum_{i=1}^{N} \frac{Z_j(x_i)Z_k(x_i)}{\sigma_i^2}$$

When the functions $Z_j(x)$ form a basis orthogonal with respect to the data, the sum is nonzero only for j equal to k. This is in fact the definition of orthogonality of a set of functions. A diagonal matrix such as this is inverted simply by taking the scalar inverses of the diagonal elements of the matrix, to obtain

$$C_{jk} \equiv (\mathbf{A}^T\mathbf{A})_{jk}^{-1} = \frac{\delta_{j,k}}{(\mathbf{A}^T\mathbf{A})_{jj}}$$

where the Kronecker delta $\delta_{j,k}$ has value 1 when j equals k, and 0 otherwise.

When the matrix $\mathbf{A}^T\mathbf{A}$ is a diagonal matrix (i.e., all off-diagonal elements are zero), each parameter a_j can be evaluated independently of the others. This means that the parameters are orthogonal, and it implies the decoupling of the N equations of the form of Eq. (2). When applied to real experimental data, the orthogonality of the basis functions, and thus of the expansion coefficients, is assured only for data with high sampling rates and identical spacings and standard deviations σ. As we shall see, these are significant limitations when analyzing some types of experimental data.

When the stringent restrictions on the experimental data are relaxed, the matrix $\mathbf{A}^T\mathbf{A}$ is no longer diagonal. There are numerous methods for solving Eq. (3) that do not require a diagonal matrix. The technique we chose is singular value decomposition, or SVD.[2,3] This method uses the fact that an arbitrary n by m real matrix $\mathbf{A}$ may be expressed as

$$\mathbf{A} = \mathbf{U}\mathbf{\Sigma}\mathbf{V}^T$$

where $\mathbf{\Sigma}$ is a diagonal n by m matrix; and $\mathbf{U}$ and $\mathbf{V}$ are n by n and m by m matrices, respectively, both having the property

$$\mathbf{U}^T = \mathbf{U}^{-1}$$

The matrices $\mathbf{U}$, $\mathbf{V}$, and $\mathbf{\Sigma}$ are calculated in the kernel SVD routine published elsewhere.[4] Equation (4) can be solved to obtain

[2] C. L. Lawson and R. J. Hanson, "Solving Least Squares Problems," p. 18. Prentice-Hall, Englewood Cliffs, New Jersey, 1974.

[3] G. E. Forsythe, M. A. Malcolm, and C. B. Moler, "Computer Methods for Mathematical Computations," p. 192. Prentice-Hall, Englewood Cliffs, New Jersey, 1977.

[4] G. E. Forsythe, M. A. Malcolm, and C. B. Moler, "Computer Methods for Mathematical Computations," p. 227. Prentice-Hall, Englewood Cliffs, New Jersey, 1977.

$$\mathbf{e} = \mathbf{V}\mathbf{\Sigma}^{-1}\mathbf{U}\mathbf{D}$$

where, as before, $\mathbf{g}$ equals 0, hence $f(x_i, \mathbf{g})$ equals 0 [see Eq. (2)] and we have an exact solution for the Fourier series expansion parameters. One may think of the singular value decomposition as providing a transformation of the basis functions to a set actually orthogonal with respect to the data.

Estimation of Uncertainties

The estimation of the uncertainty in the Fourier series expansion coefficients is equally important as the evaluation of the coefficients themselves. For example, to determine if a Fourier coefficient is significant, a comparison of the magnitude of the coefficient and its corresponding uncertainty must be made.

To understand the origin of the uncertainty in the fitted function $f(x_i, \mathbf{a})$, we consider a Fourier series expansion that includes high enough order terms that statistical tests (discussed later) prove the fit to be good. Such a fit $f(x_i, \mathbf{a})$ will not in general coincide with all the data points. This is usually attributed to experimental noise and round-off errors. If we removed the experimental noise and round-off errors from the data before the fitting procedure, we would obtain a function $f'(x_i, \mathbf{a}')$ that would coincide with each data point, and so be somewhat different from the function obtained with the noise. In light of this, we can consider $\delta f = f - f'$ as a measure of the uncertainty in the fitted function attributable to uncertainty in measurements of $y(x_i)$.

The uncertainty in the parameters that follows from the uncertainty in the fitted function is expressed as[5]

$$\delta a_j = \sum_{i=1}^{N} [f(x_i) - f'(x_i)] \frac{\partial a_j}{\partial y(x_i)}$$

$$\equiv \sum_{i=1}^{N} E(x_i) \frac{\partial a_j}{\partial y(x_i)} \tag{7}$$

where possible cross-correlations between the fitting parameters are explicitly excluded. If the experimenter has correctly assessed the standard deviations σ_i in the measurements of $y(x_i)$, the weighted error is approximately equal for all N data points

[5] P. R. Bevington, "Data Reduction and Error Analysis for the Physical Sciences," p. 113. McGraw-Hill, New York, 1969.

$$\frac{E(x_i)}{\sigma_i} \cong E_{\text{rms}} \qquad i = 1, 2, \ldots, N$$

and Eq. (7) can be expressed as

$$\delta a_j = E_{\text{rms}} \sum_{i=1}^{N} \sigma_i \frac{\partial a_j}{\partial y(x_i)} \qquad (8)$$

To estimate the magnitude of this uncertainty, Eqs. (3) and (6) are combined to obtain

$$\frac{\partial a_j}{\partial y(x_i)} = \frac{\partial}{\partial y(x_i)} \left[\sum_{k=1}^{M} C_{jk} \sum_{n=1}^{N} \frac{Z_k(x_n) y(x_n)}{\sigma_n^2} \right]$$

$$= \sum_{k=1}^{M} C_{jk} \frac{Z_k(x_i)}{\sigma_i^2}$$

where $M = 2k_{\text{max}} + 1$ is the number of independent fitting parameters. Substituting this into Eq. (8), and using the assumption that the data points are independent, so that no cross-terms remain, we arrive at

$$\left(\sum_{i=1}^{N} \sigma_i \frac{\partial a_j}{\partial y(x_i)} \right)^2 = \sum_{i=1}^{N} \sigma_i^2 \sum_{k=1}^{M} \frac{C_{jk} Z_k(x_i)}{\sigma_i^2} \sum_{p=1}^{M} \frac{C_{jp} Z_p(x_i)}{\sigma_i^2}$$

$$= \sum_{k=1}^{M} \sum_{p=1}^{M} C_{jk} C_{jp} \sum_{i=1}^{N} \frac{Z_k(x_i) Z_p(x_i)}{\sigma_i^2}$$

$$= \sum_{k=1}^{M} \sum_{p=1}^{M} C_{jk} C_{jp} (\mathbf{A}^{\text{T}} \mathbf{A})_{kp}$$

$$= \sum_{k=1}^{M} \sum_{p=1}^{M} C_{jk} C_{jp} C_{kp}^{-1}$$

$$= \sum_{p=1}^{M} C_{jp} \sum_{k=1}^{M} C_{jk} C_{kp}^{-1}$$

$$= \sum_{p=1}^{M} C_{jp} \delta_{j,p}$$

$$= C_{jj}$$

Therefore the uncertainty in parameter a_j is, by Eq. (8),

$$\delta a_j = E_{\text{rms}} (C_{jj})^{1/2}$$

Note that the derivation of this equation requires the fitting parameters to be uncorrelated; the estimation of uncertainty for correlated parameters is discussed elsewhere in this volume.[1] An expression for the root-mean-square weighted error in terms of the variance of the fit is derived elsewhere[6]

$$E_{rms} \cong \left\{ \frac{\sum\limits_{i=1}^{N} \left[\dfrac{f(x_i, \mathbf{a}) - y(x_i)}{\sigma_i} \right]^2}{N - M} \right\}^{1/2}$$

Note, too, that if the number of parameters equals the number of data points, the uncertainties in the parameters are undefined.

Statistical Tests

In practice, the Fourier sum is evaluated for the set of Fourier coefficients a number of times. Each evaluation, of the form of Eq. (2), is independent in the sense that a different upper limit k_{max} of the sum in Eq. (1) is chosen. All the coefficients must be evaluated for each value of k_{max} because for our data the determined Fourier series coefficients are not orthogonal. Beginning with the zero-order fit [for the Fourier sum this amounts to finding the weighted mean $y(x_i)$ value] and proceeding to integrally higher orders, the data are fit to Eq. (1).

The optimal order of fit is determined by independent statistical tests. The variance of fit is not a good indicator as it is generally a monotonically decreasing function which is equal to zero when $2k_{max} + 1$ is equal to the number N of data points. Two tests in particular are good indicators of goodness-of-fit for the hormone data presented as examples: the run test[7] and the Akaike information content.[8]

The run test measures the randomness of the residuals.[7] The residuals are the differences between the experimental data points and the fitting function evaluated at the fitting parameters (Fourier coefficients) with the highest probability of being correct (the least-squares solution). A run is defined as the occurrence of one or more residuals in a row with the same sign (either positive or negative value). The expected number of runs for a series of residuals is

$$\mu = \frac{2n_1 n_2}{n_1 + n_2} + 1$$

[6] F. B. Hildebrand, "Introduction to Numerical Analysis," p. 268. McGraw-Hill, New York, 1956.
[7] Y. Bard, "Nonlinear Parameter Estimation," p. 201. Academic Press, New York, 1974.
[8] H. Akaike, *IEEE Trans. Autom. Control* **AC-19**, 716 (1974).

where n_1 and n_2 are the number of negative and positive residuals in the series. The expected variance of this quantity is

$$\sigma^2 = \frac{2n_1n_2(2n_1n_2 - n_1 - n_2)}{(n_1 + n_2)^2(n_1 + n_2 - 1)}$$

A Z score can be calculated from the observed number of runs, r, the expected number of runs, and the variance of the expected number of runs as

$$Z = \frac{(r - \mu + 0.5)}{\sigma}$$

According to the run test the residuals will be random (at a 95% confidence) when the run Z score is in the range of ± 2.

The Akaike index measures the "information content" (AIC) of the fit; the fit that conveys the most information corresponds to the order where the first significant minimum (as a function of order) in the Akaike index lies.[8] The Akaike index is given by

$$AIC = N \ln \left\{ \sum_{i=1}^{N} \left[\frac{y(x_i) - f(x_i, \mathbf{g})}{\sigma_i} \right]^2 \right\} + 2M$$

When both tests agree on the order to within 1 or 2 the fit is satisfactory in that evaluations at higher orders do not contribute significantly to the power spectrum. Note that since the run test and the AIC measure different properties of the fitting operation, they will not always agree on the optimal value of k_{max}.

Results

Figure 1 presents a typical example of a human luteinizing hormone (LH) sampled every 10 or 20 min for 24 hr. These data[9] are presented here as an example of the use of the methodology. The results of these tests for the LH data are shown in Figs. 2 and 3. For these data an analysis was performed at an artificially high truncation order k_{max} equal to 25.

The first minimum in the Akaike index occurs at order k equal to 11. This is not a good candidate for the optimal order of fit, as the run Z score there is rather large. Another possibility is k equal to 21. Although the run Z score is marginal at that order, indicating nonrandom residuals, the

[9] LH data courtesy of M. J. Sollenberger, E. C. Carlsen, J. D. Veldhuis, and W. S. Evans, University of Virginia.

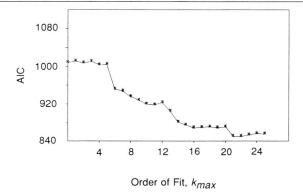

FIG. 2. Akaike information content as a function of order, for the LH data set of Fig. 1 and with $k_{max} = 25$. The minimum ($k = 16$) indicates the fit with maximum information content, which is taken to be the best fit.

situation is complicated by a pronounced peak in the power spectrum there (Fig. 4). There are two feasible explanations of this peak. First, it may be a genuine high-frequency component fitting the sharp nonsinusoidal shape characteristic of the LH pulses. Alternatively, the peak may be the third harmonic of the peak at k equal to 6, an explanation that is reinforced by the presence of a peak near k equal to 14. This would mean it is essentially an artifact of the fitting procedure. The prudent choice of optimal order, however, is k_{max} equal to 16, where there is a minimum in

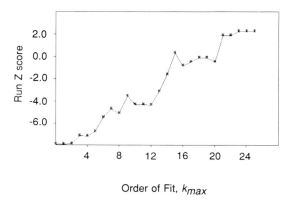

FIG. 3. Run Z score as a function of order, for the LH data set and with $k_{max} = 25$. A value with magnitude less than about 2.0 indicates sufficient randomness of the residuals. Orders between 14 and 20 meet this criterion.

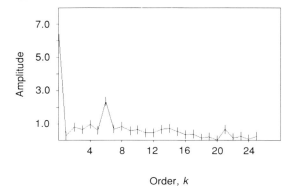

Fig. 4. Power spectrum, or Fourier coefficients as a function of order, for the LH data set and with $k_{max} = 25$. Note the peak at $k = 21$, just above the truncation order given by the Akaike and run Z score tests; this feature is discussed in the text.

the Akaike index. This identification is reinforced by the fact that the run Z score there is close to zero.

Figure 5 presents a typical human serum melatonin concentration time series.[10] Secretion of this hormone is normally circadian, with a broad profile coincident with the sleeping hours for the subject. The statistical tests for these data are shown in Figs. 6 and 7. An additional quantity of interest is the acrophase, or position of maximum hormone concentration.

[10] Melatonin data courtesy of S. S. C. Yen and G. A. Laughlin, University of California at San Diego.

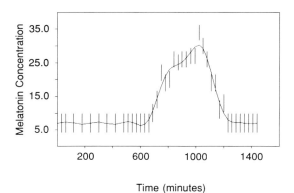

Fig. 5. Serum concentration of melatonin sampled at 30-min intervals over a period of 24 hr. The smooth curve is the optimal fit, obtained for truncation order $k_{max} = 6$ in Eq. (1).

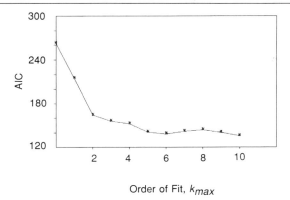

FIG. 6. Akaike information index as a function of order, for the melatonin data set of Fig. 5 and with $k_{max} = 10$. The first minimum (at $k = 6$) indicates the fit with maximum information content, which is taken to be the best fit.

These maxima are tabulated for the melatonin data in Table I. Note that the position and amplitude of the acrophase change with the order of the analysis. Figure 8 shows the power spectrum for the melatonin data set.

Discussion

This application of Fourier analysis affords a practical means of analyzing discrete data without the constraint of orthogonal basis functions. The requirement of basis orthogonality is a significant limitation of the analysis as usually performed. This chapter has outlined a method of Fourier

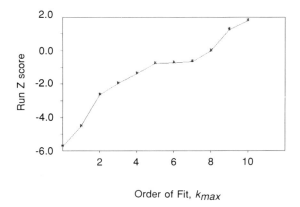

FIG. 7. Run Z score as a function of order, for the melatonin data set and with $k_{max} = 10$. A value with magnitude less than about 2.0 indicates sufficient randomness of the residuals. Orders between 3 and 10 meet this criterion.

TABLE I
Positions and Amplitudes of Acrophases
(Maximum Fitted Values) of
Melatonin Data[a]

Order of fit, k_{max}	Amplitude	Position
0	—	—
1	23.59	960.0
2	28.33	960.0
3	29.75	960.0
4	29.42	990.0
5	29.68	1020.0
6	30.37	1020.0
7	30.59	1020.0
8	30.71	1020.0
9	31.20	1020.0
10	31.85	1020.0
11	32.19	1020.0
12	32.13	1020.0
13	31.96	1020.0
14	32.12	1020.0
15	32.66	1020.0
16	33.13	1020.0
17	33.07	1020.0
18	32.63	1020.0

[a] For varying orders of fit, k_{max}. The optimal fit is at order $k_{max} = 6$.

analysis suitable for a more general class of data. The data points need not be equally spaced. Furthermore, each data point may be accompanied by a uniquely valued uncertainty in the independent variable. An additional advantage to the more general analysis is that the confidence intervals (uncertainties) of the fitting parameters may be independently and confidently determined.

Because the basis functions are not necessarily orthogonal, the Fourier coefficients will usually depend on the particular order at which the series is truncated. A consequence of this is that the amplitudes of the individual Fourier components may not be independent; that is, they cannot be determined without simultaneous knowledge of the other Fourier coefficients. This means that if there is a significant component at a frequency higher than is included in the Fourier sum, the present treatment will attempt to fit that higher component by summing over the allowed orders. This distorts the coefficients away from the "true" values that would otherwise be obtained at the optimum order. This effect is illustrated

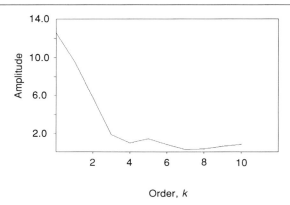

Order, k

FIG. 8. Power spectrum for the melatonin data set. Note the smaller standard deviations of the Fourier coefficients relative to those of Fig. 4 corresponding to the LH data.

in Table II by the variation in the amplitudes of the individual Fourier components and their corresponding uncertainties, as the value of k_{max} increases.

The choice of order of fit is an important consideration. As discussed above, the optimum order of fit is determined to a large extent by the power spectrum in the infinite sum limit. A strong component at a particular frequency may arise from at least two distinct sources. For one, there may be a signal component that actually repeats at that characteristic frequency. This would be the case if the signal were comprised of a series of broad, similarly spaced pulses readily approximated by a sine wave. Second, the nonsinusoidal shape of the pulses influences the shape of the power curve. For example, the series of sharp peaks shown in Fig. 1 (sharp that is with respect to a sine wave at the frequency corresponding to the peak spacing) has a spectrum encompassing a broad range of frequencies required to fit the pulses. This is shown in Fig. 4. An extreme example of this phenomenon is a set of data containing a single sharp peak (delta function) and having a perfectly flat power spectrum.[11]

Another consideration relevant to the power spectra of discrete data sets is introduced by the Nyquist theorem.[12] This theorem states that one must sample at least two points per period to define a sine wave. Thus, the spacing δt of the data places an upper limit on the Fourier sum, above which the Fourier components are ill-defined. This condition is

[11] R. W. Ramirez, "The FFT, Fundamentals and Concepts," p. 30. Prentice-Hall, Englewood Cliffs, New Jersey, 1985.
[12] R. W. Ramirez, "The FFT, Fundamentals and Concepts," p. 129. Prentice-Hall, Englewood Cliffs, New Jersey, 1985.

TABLE II

Amplitudes of First Few Fourier Components (Orders k) for Luteinizing Hormone Data[a]

k_{max}	0	1	2	3	4	5	6
0	5.385 ± 0.165	—	—	—	—	—	—
1	5.436 ± 0.175	0.210 ± 0.342	—	—	—	—	—
2	5.474 ± 0.159	0.115 ± 0.336	0.731 ± 0.329	—	—	—	—
3	5.468 ± 0.159	0.109 ± 0.329	0.667 ± 0.340	0.226 ± 0.339	—	—	—
4	5.569 ± 0.147	0.219 ± 0.301	0.909 ± 0.310	0.349 ± 0.319	0.919 ± 0.313	—	—
5	5.611 ± 0.143	0.373 ± 0.302	0.946 ± 0.296	0.333 ± 0.304	0.934 ± 0.317	0.505 ± 0.309	—
6	6.041 ± 0.085	0.480 ± 0.169	1.000 ± 0.171	0.585 ± 0.174	1.051 ± 0.176	0.623 ± 0.183	1.864 ± 0.177
7	6.063 ± 0.076	0.383 ± 0.161	0.919 ± 0.158	0.553 ± 0.163	0.996 ± 0.164	0.592 ± 0.165	1.867 ± 0.170
8	6.134 ± 0.068	0.415 ± 0.140	0.933 ± 0.144	0.544 ± 0.144	1.058 ± 0.146	0.584 ± 0.147	1.881 ± 0.147
9	6.171 ± 0.062	0.317 ± 0.128	0.861 ± 0.128	0.655 ± 0.134	1.005 ± 0.133	0.601 ± 0.134	1.910 ± 0.133
10	6.220 ± 0.057	0.313 ± 0.117	0.923 ± 0.118	0.674 ± 0.120	1.057 ± 0.125	0.587 ± 0.123	1.953 ± 0.123
11	6.244 ± 0.055	0.348 ± 0.112	0.920 ± 0.113	0.667 ± 0.117	1.064 ± 0.117	0.634 ± 0.121	1.995 ± 0.118
13	6.303 ± 0.045	0.383 ± 0.094	0.841 ± 0.094	0.648 ± 0.096	0.999 ± 0.098	0.634 ± 0.099	2.046 ± 0.098
15	6.393 ± 0.033	0.346 ± 0.068	0.838 ± 0.069	0.616 ± 0.070	1.008 ± 0.071	0.608 ± 0.072	2.130 ± 0.072
17	6.417 ± 0.030	0.317 ± 0.062	0.854 ± 0.062	0.632 ± 0.064	1.017 ± 0.065	0.608 ± 0.065	2.174 ± 0.065
19	6.429 ± 0.029	0.313 ± 0.059	0.821 ± 0.059	0.633 ± 0.061	0.989 ± 0.062	0.604 ± 0.062	2.193 ± 0.062
21	6.459 ± 0.023	0.311 ± 0.048	0.805 ± 0.048	0.628 ± 0.049	0.997 ± 0.050	0.612 ± 0.050	2.213 ± 0.050

[a] With several values of truncation order, k_{max}.

$$k_{max} < \frac{T}{2(\delta t)} \tag{9}$$

This may be considered a guideline estimate of how well-defined are the upper frequencies of a fit. After finding the statistically determined best fit using the run Z score and Akaike tests, one may compare the truncation order with the result of Eq. (9) to assess the validity of the high frequency components of the fit. Conversely, Eq. (9) indicates the minimum frequency at which one must sample to make a statistically viable observation of an event of duration δt. However, if a significant amount of random experimental noise is superimposed on the underlying periodicities, then a significantly higher sampling rate is required to model the data correctly. These factors underscore the caution that must be exercised in the interpretation of the power spectrum, and in the choice of truncation order.

The experimenter need not be limited to the Fourier series expansion. Choice of basis functions should be determined both by the nature of the measured phenomenon as well as by the questions to be addressed. We chose the Fourier series for our data because it lends itself well to the analysis of periodic data. Other expansions (the Legendre polynomials,[13] for instance) are compatible with the methodology outlined here, and may be preferable for their particular functional characteristics. A PC-compatible computer program is available from the authors on written request.

Acknowledgments

The authors wish to express their thanks to S. S. C. Yen and G. A. Laughlin of the University of California at San Diego for allowing us to use some of their melatonin data as an example in this work. We also thank M. J. Sollenberger, E. C. Carlsen, J. D. Veldhuis, and W. S. Evans of the University of Virginia for the use of the LH data.

This work was supported in part by National Institutes of Health Grants GM-28928 and DK-38942, National Science Foundation Grant DIR-8920162, the National Science Foundation Science and Technology Center for Biological Timing, the Diabetes Endocrinology Research Center of the University of Virginia, and the Biodynamics Institute of the University of Virginia.

[13] M. Abramowitz and I. A. Stegun, "Handbook of Mathematical Functions," National Bureau of Standards, Applied Mathematics Series, Vol. 55, p. 332. U.S. Govt. Printing Office, Washington, D.C., 1970.

[16] Alternatives to Consider in Fluorescence Decay Analysis

By JAY R. KNUTSON

Global Analysis of Fluorescence Decay Rate Distributions

The following observations frame a current problem: (1) Pure fluorophores in a single environment are seen to decay as monoexponentials or very nearly so. (2) A superposition of exponential compartments has been *adequate* for most systems analyzed. (3) In some cases[1] the underlying physical process is expected to be nonexponential owing to a distributed interaction parameter (orientation, distance, etc.). Some particular cases (involving transient[2] diffusion) have led to analytic expressions that approximate the real system well. (4) Distributed lifetimes have been considered[3] for some time; only recently, however, has the quantity and quality of data increased to a level supposedly permitting quantitative testing of distributed versus compartmental views. (5) Compartmental (multiexponential) problems have been shown to benefit greatly via association[4] and overdetermination[5] techniques; a key to this approach has been the effort to combine experiments that have *different* mixtures of the decays. (6) Efforts have been made to characterize protein fluorescence decays[6] via distribution functions. Although these models have been difficult to test, they are laudable in the sense that they stimulate debate and provoke efforts to improve the testing. (7) The results presented to date are predominantly experiments at a single wavelength; it seems likely that multiple curve, simultaneous analysis would be of benefit.

In this section it will be shown that a "global" (simultaneous, multicurve) least-squares analysis can be of benefit both in identifying the number of species present and in confining the distributions. In particular, it will be shown that global analysis more correctly recovers the centers and widths of the actual lifetime distributions.

[1] E. Haas, M. Wilchek, E. Katchalski-Katzir, and I. Z. Steinberg, *Proc. Natl. Acad. Sci. U.S.A.* **72**, 1807 (1975).

[2] J. C. Andre, F. Baros, W. Dong, J. Duhamel, and A. T. Reis De Sousa, *Proc. SPIE Int. Soc. Opt. Eng.* **1204**, 415 (1990).

[3] W. R. Ware and T. L. Nemzek, *Chem. Phys. Lett.* **23**, 557 (1976).

[4] J. R. Knutson, D. G. Walbridge, and L. Brand, *Biochemistry* **21**, 4671 (1982).

[5] J. R. Knutson, J. M. Beechem, and L. Brand, *Chem. Phys. Lett.* **102**, 501 (1983).

[6] J. R. Alcala, E. Gratton, and F. G. Prendergast, *Biophys. J.* **51**, 925 (1982).

Much of the theory in this section derives from that provided in Ref. 5. For the demonstrations provided here, however, the fitting parameters for each term are not amplitude and lifetimes; instead, they are the distribution height, width, and center.[7] In particular, although several models can be used, a Gaussian distribution was employed for each term (j) to produce an impulse response f_j:

$$f_j(t) = \sum_{i=1}^{N} \alpha_{ij} \, e^{-t/\tau_i}$$

where $\alpha_{ij} = A_j \, e^{-(\tau_i - \tau_{cj})^2 h_j^2}$. A_j is the height, τ_{cj} is the central lifetime, and h_j is the inverse width, all for the jth term. The τ_i are situated on a discrete mesh: $\tau_i = i\Delta\tau$; $N = \tau_{max}/\Delta\tau$.

These impulse-fitting functions must be convolved with a "lamp" (instrumental response to an impulse) to compare with actual decay curves:

$$F_j(t') = \int_0^{t'} L(t) f_j(t' - t) \, dt$$

or

$$F_j(t') = \int_0^{t'} L(t) \left[\sum_{i=1}^{N} \alpha_{ij} \, e^{-(t'-t)/\tau_i} \right] dt$$

Interchanging the order of summation and integration,

$$F_j(t') = \sum_{i=1}^{N} \alpha_{ij} \int_0^{t'} L(t) \, e^{-(t'-t)/\tau_i} \, dt$$

or

$$F_j(t') = \sum_{i=1}^{N} \alpha_{ij} U_i(t')$$

where $U_i(t')$ is the unit convolution of the ith exponential.

Each of the unit convolutions can be done with the fast recursion of Grinvald and Steinberg.[8] Their algorithm can be truncated slightly, since we do not require the partial derivative with respect to lifetime. The derivatives needed for nonlinear least-squares processing can all be calculated via the chain rule:

$$\frac{dF}{db} = \sum_{i=1}^{N} \frac{\partial F}{\partial \alpha_i} \frac{\partial \alpha_i}{\partial b} = \sum_{i=1}^{N} U_i \frac{\partial \alpha_i}{\partial b}$$

where b is the fitting parameter in question.

[7] J. R. Knutson, *Biophys. J.* **51**, 285a (1987).
[8] A. Grinvald and I. Z. Steinberg, *Anal. Biochem.* **59**, 583 (1974).

In practice, the summation need not extend over the entire mesh; a restricted sum within $\tau_c \pm 3/h$ usually includes enough terms (further truncations create mesh-dependent errors). Nonpositive τ_i values are nonphysical, so the actual summation is the intersection of the Gaussian limits with the positive interval previously defined.

The global mapping procedure in this case is quite similar to that in Ref. 5, except that three parameters are required for the jth local distribution term: A_j, τ_{cj}, and h_j. Thus, the designation while counting parameters proceeds "mod 3" rather than mod 2, but the matrix mapping is unchanged. For example, a system containing *two* distributed decays (mixed differently in two different curves) would use the mapping matrix:

$$P_1, P_2, P_3, P_4, P_5, P_6 \qquad \text{(curve 1)}$$
$$P_7, P_2, P_3, P_8, P_5, P_6 \qquad \text{(curve 2)}$$
$$A_1, \tau_{c1}, h_1; A_2, \tau_{c2}, h_2 \qquad \text{(template)}$$

A nonglobal mapping for the same system would require 12 rather than 8 parameters, since the parameters links between curves would be absent. For a single linked distribution:

$$P_1, P_2, P_3 \qquad \text{(curve 1)}$$
$$P_4, P_2, P_3 \qquad \text{(curve 2)}$$
$$A_1, \tau_{c1}, h_1 \qquad \text{(template)}$$

so 4 (rather than 6) global parameters are sought.

The program TFIT[9] was built to generalize earlier global programs. It was easily modified to conform with the needs outlined above; especially convenient mapping features were incorporated. No effort was made to optimize speed via minimizing recalculation of the u_i functions, but minor coding changes could be made to store these unit convolutions (and convolve new ones on demand). The central amplitudes can be negative or positive; thus, a compartmental (excited-state reaction) system can be explored for distributed rates (data not shown).

Examples

The performance of the fitting program was evaluated by applying it to simulated fluorescence decay curves. Each curve represents a reconvolution of exponential terms with a lamp function, followed by the addition of Gaussian noise appropriate to the count level simulated. These types

---

[9] Programs by M. D. Pritt and J. M. Beechem, laboratory of L. Brand, 1985.

TABLE I
BIEXPONENTIAL DECAY AND ATTEMPTED DISTRIBUTION FITS (NON-GLOBAL)

Amplitude	τ	H parameter	Amplitude	τ	H parameter	χ^2	#	Model type
0.25	2.00	—	0.08	4.00	—	—	a[a]	INPUT "A"
0.08	2.00	—	0.25	4.00	—	—	b	INPUT "B"
0.2534	2.01	—	0.0805	4.02	—	0.97	a	Discrete
0.0732	1.79	—	0.2618	3.95	—	1.05	b	Doubles
	2.68					4.28	a	Discrete
	3.67					1.87	b	Singles
0.0291	2.38	0.735				1.21	a	Gaussian
0.0300	3.51	0.795				1.21	b	Singles
0.0878	2.11	2.80	0.0290	4.37	4.53	1.12	a	Gaussian
0.0613	1.15	10.4	0.0629	3.80	1.91	1.12	b	Doubles
5.097	2.31	1.68				1.15	a	Lorentzian
4.770	3.49	1.97				1.18	b	Singles
2.355	1.64	27.8	2.432	3.26	4.29	1.18	a	Lorentzian
1.783	2.95	6.35	2.691	4.22	8.11	1.17	b	Doubles

[a] a and b refer to the two different mixtures of 2 and 4 nsec used to construct the simulated decays.

of simulations have already proved a reliable tool for predicting the recovery of pulse fluorometric decay parameters.[5,10,11]

Initially, a widely spaced biexponential simulation (different mixtures of discrete 2 and 6 nsec lifetimes) was examined. The ideal parameter recovery for a discrete input case is narrow, properly centered distributions whose integrals yield the right amplitudes. Naturally, discrete lifetime analysis of this "easy" case yields appropriate values, and a discrete global analysis sharpens the confidence in those values.[10]

Analysis of the "easy" biexponential via ordinary (single-curve analysis) lifetime distributions was also judged effective, since one of the two simulated curves could not be fit with a single distribution. Unfortunately, one of the admixtures was easily reconciled with a single Gaussian smear of lifetimes. Conventional distribution analysis also yielded about the correct centers and yields when two distributions were allowed in each curve. A global analysis of the decay distributions was more effective. Not only did it recover the appropriate parameters via narrower distributions, it also signaled the inadequacy of a single distribution with a much larger χ^2 penalty.

[10] J. M. Beechem, E. Gratton, M. Ameloot, J. R. Knutson, and L. Brand, in "Fluorescence Spectroscopy," (J. R. Lakowicz, ed.), Vol. 1, Plenum, New York, in press.
[11] J. B. A. Ross, K. W. Rousslang, and L. Brand, Biochemistry 20, 4361 (1981).

TABLE II
GLOBAL VERSIONS OF DISTRIBUTIONS THAT ARE NARROW (LARGE H) AND CORRECT[a]

Amplitude	τ	H parameter	Amplitude	τ	H parameter	χ^2	#	Model type
0.25	2.00	—	0.08	4.00	—	—	a[b]	INPUT "A"
0.08	2.00	—	0.25	4.00	—	—	b	INPUT "B"
0.149	1.97	5.25	0.039	3.96	4.03	1.12	a	Global, two
0.047	1.97	5.25	0.166	3.96	4.03	1.12	b	Gaussians
0.027	2.94	0.712				14.66		(Global, one″)
3.66	2.07	28.4	1.02	4.09	12.5	1.15	a	Global, two
1.48	2.07	28.4	3.28	4.09	12.5	1.15	b	Lorentzians
4.04	2.88	1.63				14.70		(Global, one″)

[a] When another data dimension is incorporated explicitly in the method of analysis, the recovery of parameters is more accurate. More important, statistical justification for the choice between models can be seen now.

[b] a and b refer to the two different mixtures of 2 and 4 nsec used to construct the simulated decays.

A more stringent test was prepared by selecting mixtures of two closer lifetimes, 2 and 4 nsec. Tables I and II summarize the results of this simulation study. Most notable is the inability of a free (nonglobal) distribution analysis to indicate the need for more than one term, as evinced by the inconclusive ($<10\%$) reduction in χ_R^2 accompanying the addition of a second term. Even if one chose the double for reasons not based on χ_R^2, the free analysis also failed to recover the correct lifetime centers. As expected, the "dominant" lifetime center was properly extracted in each curve. In contrast, as seen in Table II, the global approach (linking distributions between curves) provides accurate recovery of both lifetime centers, and the distribution widths are relatively narrow.

The results of the different analyses for the 2 nsec/4 nsec mixtures are also summarized in Fig. 1, which shows the various distributions on the 200 psec mesh used. The width of globally recovered peaks is clearly narrow compared to the mesh used.

The next simulation was designed to test how much of the recovered width of a distribution may be ascribed to simple counting errors [in this case, the Gaussian (normally) distributed errors are on the order of $N^{1/2}$, where N is the "photon count" in each channel]. A single exponential of 5 nsec was convolved with the lamp, normalized to a series of amplitudes, and subjected to a noise generator that provides the appropriate counting error for those simulated count levels. This set of decay curves contained approximately 4,000, 8,000, 16,000, or 32,000 counts in the peak channel. Analysis with a single distribution on a 50 psec mesh was carried out,

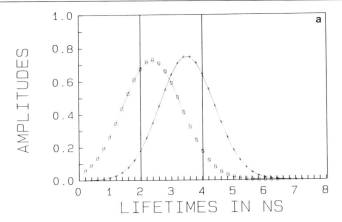

FIG. 1. Single-curve data, when fit with Gaussian distributions, provide a "weighted average" representation of the discrete terms. Mixtures of pure 2 and 4 nsec components were convolved with a lamp function, and noise appropriate to the 10,000 count level was added. Fitting with global (both mixtures) distribution programs more correctly recovers the narrow character and amplitude ratio (see Table II). (a) Conventional analysis via single Gaussians for the two different mixtures (open vs. closed symbols). (b) Conventional analysis via dual Gaussians. (c) Global analysis via Gaussian distributions: dual Gaussians [open and closed as in (a)] or one Gaussian (dashed line).

and a general trend in parameter recovery toward decreasing width with increasing precision was observed.

Amplitude-normalized distributions are shown in Fig. 2, proceeding in order with the exception of the third curve (16K peak). Its width anomalously reversed the trend, lying midway between the first and second recoveries. This erratic behavior suggested that width is not a well-determined parameter in this type of analysis. The ill-determined nature of the width parameter was confirmed by examining the χ_R^2 value on fixing larger widths. Even the precise (32K peak curve) data could be easily reconciled with rather large Gaussian widths; this is shown in Fig. 3. Figure 3 also displays one of the reasons for weak determination: the parameters of central amplitude and width are highly (anti) correlated, such that total integral amplitude (proportional to their product) remains nearly constant. Without multicurve (global) overdetermination techniques, distribution analysis can yield relatively large parameter uncertainty and cross-correlation problems.

The central point of this section is that decay analysis, whether discrete or distributed, should exploit the multidimensional nature of luminescence data. A simple linkage between curves was used herein, but other functional linkages could be as easily accomplished. If, for example, one wishes to postulate models mapping decay rates onto a continuous (e.g., potential

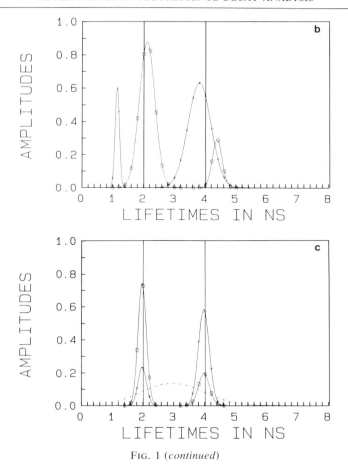

FIG. 1 (*continued*)

energy) surface, it may be desirable to require a concurrent spectral mapping. The Bakhshiev[12] approach to solvent relaxation is an early example of this approach.

An exponential series method[13] has also been demonstrated. In our fitting process, it was observed that the width parameter often oscillated prior to convergence (data not shown). Because the aforementioned method permanently removed critically small amplitude terms on the mesh, oscillations in that system might yield incorrect distributions. Nonlinear least-squares searching is "guaranteed" in a theoretical sense only when the framework of a model is unchanged by the fit process. Their free

[12] Y. T. Mazurenko and N. G. Bakhshiev, *Opt. Spectrosc.* (*Engl. Transl.*) **28,** 490 (1970).
[13] D. R. James and W. R. Ware, *Chem. Phys. Lett.* **126,** 7 (1986).

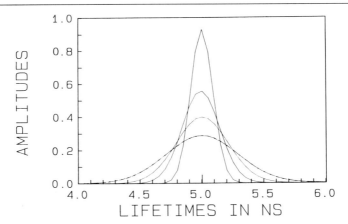

Fig. 2. Single-curve distribution recoveries obtained from data at different levels of precision. They correspond to peak counts of 4,000, 8,000, 32,000, and 16,000, respectively (see text).

histogram approach does, however, carry a significant advantage: the χ_R^2 value in such a model potentially includes all degrees of freedom. Without separate information, one might have difficulty choosing between various distribution models (e.g., Gaussian versus Lorentzian versus polynomial). A least-squares approach is naturally confined to comparisons among the particular functional forms chosen, and these comparisons may be confused by "hidden" degrees of freedom in χ_R^2. Similar statistical considerations hold for the *fixing* of parameters; the values (and especially uncertainties) of free parameters hold true if and only if the fixed values are "true." In analogy, one might imagine each candidate distribution to be a fixed slice through some master distribution surface. It is the position of this slice that serves as the hidden variable in such a picture.

In conclusion, global analysis can be of benefit for decay distribution recovery. More important, the functional linkages (e.g., spectra) directly serve as signatures for states in the system.[4,14] One may reasonably ask what value a decay time has in the absence of *associated* information about spectra or the dependence on other variables (concentration, pH, temperature, etc.). Various spectral association tools (for a review, see Refs. 15 and 16) that currently exploit discrete lifetimes and rates can

[14] S. M. Green, J. R. Knutson, and P. Hensley, *Biochemistry* **29,** 9159 (1990).
[15] L. Brand, J. R. Knutson, L. Davenport, J. M. Beechem, R. E. Dale, D. G. Walbridge, and A. A. Kowalczyk, *in* "Spectroscopy and Dynamics of Molecular Biological Systems" (P. M. Bayley and R. E. Dale, eds.), p. 259. Academic Press, London, 1985.
[16] J. R. Knutson, L. Davenport, J. Beechem, D. Walbridge, M. Ameloot, and L. Brand, *in* "Excited State Probes in Biochemistry and Biology" (L. Masotti and A. Szabo, eds.), in press. Plenum, New York, 1992.

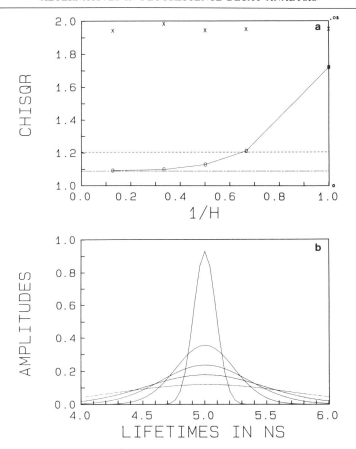

FIG. 3. (a) The χ_R^2 (reduced χ^2 value, 0) is displayed for fits of a discrete (32,000 peak) exponential with fixed values of the width parameter. The product of peak amplitude and width (X) remains fairly constant; this partly explains why the large widths cannot be statistically excluded (width is an ill-determined parameter, see text). (b) Shapes recovered at the fixed widths in (a). Only the widest can be judged "unlikely" via χ_R^2.

readily be adapted to distribution theories. When distributed decay models are reconciled with multidimensional data, they should provide insights unavailable from solitary decay parameters.

Simulated Annealing Methods: A Path to Partial Global Analysis

If, as suggested in the previous section, the choice between discrete and distributed models cannot usually be made by a simple fitting contest, perhaps more protean fitting methods can be of use. Ideally, we would like to have a fitting procedure that would choose a correct distribution shape spontaneously, collapsing to sharp peaks when fitting discrete cases. The maximum-entropy method[17] has been proposed to fill such needs; a wide variety of advantages have been cited, especially the model-free nature of the analysis; the lack of spurious correlations; and the recovery of narrow inputs correctly from high precision data. As disadvantages have not been listed, it is helpful to note two concerns: (1) nontrivial modifications are needed to exploit (rather than suppress) correlation, for example, it is not easily made "global," and (2) the regularization function $\alpha \log \alpha$ or ($S \log S$) makes physically important *negative* α's (preexponentials) available only in restricted (fixed α ratio) cases. A more cynical view of "all-poles" methods can be found in Ref. 18. In fluorescence decay analysis, the maximum entropy method (MEM) has made a significant contribution already, and it is likely to remain a useful tool for visualizing the difficult Laplace inversion that maps $I(t)$ into $\alpha(\tau)$.

More important, MEM is potentially customizable and represents a prototype for the class (one of many ways) to do constrained optimization of $\alpha(\tau)$. In general, one would like to obtain $\alpha(\tau)$ by combining traditional optimization (χ^2, maximum likelihood, or other norms) with some other modeling information that appears as a constraint. With MEM, this constraint (for choice among feasible set) appears in the "regularizer" (see above); in the exponential series method[13] it is the vanishing of certain segments (an effect difficult to assign physical meaning to, but apparently reproducible).

Recently, we have become interested in the method of simulated annealing (SA), a proven performer in the area of structural optimization.[19] As elegantly demonstrated by Press *et al.*,[18] SA schemes can be formulated around many definitions for system energy; in particular, the energy metric can be augmented with user-defined "penalty" functions. Before going on to a particular approach for harnessing this versatility, it is useful to review the SA approach briefly (the more thorough[18] discussion is highly recommended for further exploration).

In synopsis, a system having very many possible configurations and an energy metric is subjected to a random series of reconfiguration attempts. The success or failure of each event is determined probabilistically by comparing the resulting change in energy with kT, the temperature index for the system. Typically, a Boltzmannlike ($e^{-\Delta E/kT}$) probability is assigned to the event. Initially (at high kT) the factor is large (likely) for

[17] A. K. Livesey and J. C. Brochon, *Biophys. J.* **52,** 693 (1987).
[18] W. H. Press, B. P. Flannery, S. A. Teukolsky, and W. T. Vetterling, "Numerical Recipes—The Art of Scientific Computing." Cambridge Univ. Press, London and New York, 1989.
[19] J. D. Forman-Kay, G. M. Clore, P. T. Wingfield, and A. M. Gronenborn, *Biochemistry* **30,** 2685 (1991); and references therein.

a wide range of events, permitting large ΔE (energy increases). The system is "cooled" slowly and becomes more selective; that is, as kT is reduced, only *small* positive ΔE are allowed (negative ΔE improvements are of course always accepted, with probabilities truncated above one). The term simulated annealing refers to the metallurgical analogy for seeking optimal domain structures; slow cooling avoids trapping in local minima, i.e., avoids the consequences of fast quenching. Materials analogies aside, the method has wide applicability to otherwise difficult problems such as the "traveling salesman,"[18] a classic route optimization task faced by transport and communications companies.

Key requirements of the SA approach include a set of distinguishable configuration changes that can be tested individually and an energy metric meaningful to the problem. A general scheme includes the following steps: (1) Start trials at some "high" temperature. (2) Generate a large set of small random reconfigurations. (3) Accept or deny each according to the Boltzmann term. (4) Cool (reduce kT) and start over at step 2 until all attempts at reconfiguration are denied.

In molecular structure refinement, the application is obvious. Individual bond angles and distances yield energies through a series of potential functions, and the so-called computer energies have direct ties to measurable quantities. The traveling salesman problem explained in Ref. 18 is less intuitive; "energy" is replaced by a "cost function" there.

Let us return to our problem in decay analysis, namely, the recovery of $\alpha(\tau)$. If we break the continuous representation up into a τ_i mesh at smoothly spaced intervals, we can recognize that configuration is a set of $\alpha_i(\tau_i)$. Reconfiguration can be achieved by a small increment or decrement in a randomly chosen α_i, and "energy" can easily be constructed from the fitting error (e.g., χ^2) plus penalty functions chosen to achieve other modeling goals. For example, if energy includes a multiple of the sum of $(\alpha_i \log \alpha_i)$, one can expect that MEM-like solutions will result.

A wide range of choices is available to the programmer, both in penalty and trial structures. A random generation of new α_i can be done many ways. For example, a trial can be based on accepting pairs of random α_i changes instead of single α_i events. Further, the size (more correctly, the distribution) of α_i changes is a critical choice. One may encounter accidental bias or traps; for example, generating $\alpha_i = \alpha_i f(\text{rand})$ leads to a trap when $\alpha_i = 0$ (akin to behavior in the "fast quenched" exponential series approach) unless $|\alpha_i|$ is coded to stay above a "floor" (minimum) value.

Results

A series of SA fit programs have been constructed and tested on fluorescence decay data. The use of χ^2 alone is problematic, since the number of effective degrees of freedom is so large. Accidental degeneracy occurs on fine τ_i mesh, owing to the ability of offsetting positive and (adjacent) negative α_i's to cause only small overall fit errors. A penalty function for (frequent) zero crossing was the simplest solution.

Some typical SA distribution recoveries are shown in Fig. 4. The jagged appearance can be suppressed by added terms for smoothness, if desired. At count levels below 100,000, the resolution of sharp versus broad features is difficult, but no method can be expected to resolve discrete versus distributed from a single curve, unless that single curve is precise beyond current limits for either phase or pulse methods (see section on resolution comparison).

Clearly, a great deal more work will be needed to make SA fit programs a predictable tool in decay analysis. It is likely, however, that these highly customizable search strategies will be profitable. Most important, one can construct a partial global analysis in this framework. By partial I mean lacking the rigid functional linkage between curves mentioned before. An example of partial global linkage is the adjustment of an α (amplitude) versus τ (lifetime) and λ (wavelength) surface using a penalty function that regulates the tilt of any ridgelike features (using ratios of the partial derivatives of α with respect to τ and λ). Tilted ridges that incur large penalties will break up during annealing into smaller, more parallel features. The heterogeneous mix of pure decay-associated spectra (DAS) that characterizes many interesting fluorescence systems will probably break up with little penalty, while "truly" distributed[12] model systems should be refractory to this term.

The long-standing controversy of discrete versus distributed decay in biophysical systems will likely benefit from such adjustable "partial global" approaches, although ancillary information (quenchers, energy transfer acceptors, excitation spectra, etc.) may be the best avenue to resolve most cases. Simulated annealing methods are certainly computationally demanding, but as powerful computers become available to more users, their customizable features will grow in attractiveness.

Examining the Fit: Power Spectral Density and Jackknifing

The examination of residual trends is an important part of model testing in fluorescence. Direct viewing of weighted residuals has been augmented by runs tests and trend magnifiers such as autocorrelation.[20] In addition to magnification, one might hope for diagnostic features, and the power spectral density (PSD) of residuals holds some promise. The Fourier transform of autocorrelation, PSD has a shape characteristic of the different

---

[20] A. Grinvald and I. Z. Steinberg, *Anal. Biochem.* **59**, 583 (1974).

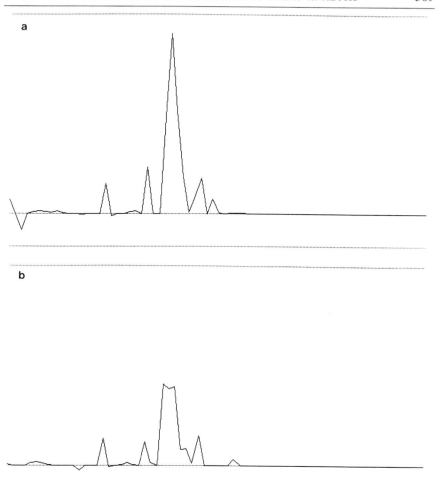

Fig. 4. Results of the simulated annealing program. The $\alpha(\tau)$ map is shown in the range 0–14 nsec, -0.1 to $+0.6$, for two different tests. In (a) a "pure" 5.4 nsec exponential was synthesized with approximately 18,000 peak counts. In (b) the input was an approximately 1 nsec wide distribution, centered on 5.4 nsec. Both cases used random α increments (with $\sigma_\alpha \cong 0.02$) and a penalty function (see text), multiplying deviations from an "ideal" sum of second derivatives ($\Sigma\ d^2\alpha/d\tau^2$). A smoothed version would provide similar fit quality.

types of noise sources in the experiment. For example, the "exponential cosine" noise signature is indicative of Brownian motion. The theory of PSD in random noise systems is thoroughly discussed in Refs. 21 and 22, and the reader is urged to look there for better examples.

Instead of passive examination, one might wish to prod the residuals and test their relative influence on recovered parameters. A reemphasis of certain subsets of data falls under the auspices of methods called "jackknifing," a term explained in Ref. (23). The advantage of jackknifing over passive perusal is clear: one may examine the influence that a group of points has on key parameters. The principles of jackknifing have already been seen in "lambda invariance," a transform reweighting approach unique to the method of moments.[24] Interestingly, it was noted that exponential reweighting should be useful in other methods. Although the criteria for invariance are often subjective, one could establish benchmarks with appropriate simulation.[25]

More direct reweighting schemes are available. In Fig. 5, an example is given where a trace of an unresolved component is detected via jackknifing. This simple program continues where TFIT ends, reweighting a group (section) of residuals, finding the new minimum, moving to the next section, etc. The variation of parameters that results not only demonstrates the existence of an unexplained term, it also gives some clue about its width. With practice, this diagnostic may become more useful, especially when coupled with simulation.

As an aside, all data synthesis done herein used a Poisson noise generator; more common Gaussian generators fail to predict important artifacts that result when one fits with $(data)^{-1}$ weights when many channels contain less than 200 counts.[7,26] This will become very important in evaluating "KINDK"[26,27] data as collection times shrink. Sparse data surfaces may be correctly reweighted by $(fit)^{-1}$ near the χ^2 minimum.[28,29] Recently,

[21] J. S. Bendat, "Principles and Applications of Random Noise Theory." Wiley, New York, 1958.

[22] R. E. Uhrig, "Random Noise Techniques in Nuclear Reactor Systems." Ronald Press, New York, 1970.

[23] P. Diaconis and E. Bradley, *Sci. Am.* **248**, 116 (1983).

[24] I. Isenberg and E. W. Small, *J. Chem. Phys.* **77**, 2799 (1977).

[25] M. D. Barkley and F. Chaudhuri, personal communication, 1988.

[26] D. G. Walbridge, J. R. Knutson, and L. Brand, *Anal. Biochem.* **161**, 467 (1987).

[27] S. M. Green, P. Hensley, and J. R. Knutson, *Proc. SPIE Int. Soc. Opt. Eng.* **1204**, 54 (1990).

[28] M. Ameloot, personal communication, 1984.

[29] B. K. Selinger and C. M. Harris, *in* "Time-Resolved Fluorescence Spectroscopy in Biochemistry and Biology" (R. B. Cundall and R. E. Dale, eds.), p. 155. Plenum, New York, 1983.

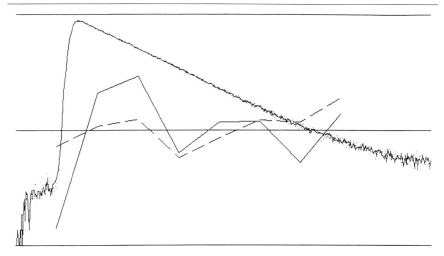

FIG. 5. Examples of jackknifing applied to a mixture of exponentials. Two different convolved mixtures of exponentials, differing only by a "trace" component, are shown plotted on a 512-channel span (~18,000 peak, 88 psec/channel; $0.1\ e^{-t/0.5} + 0.1\ e^{-t/5}$; trace $= 0.01\ e^{-t/2}$). Tenfold weighting was applied to 40-channel intervals, centered at every fiftieth channel. The change in the short lifetime caused by reweighting is plotted (± 150 psec full scale) for a case with (—) and without (---) the unresolved trace.

another method for very large decay surfaces has been developed[30] that is based on momentlike norms. The noise sensitivity of this method is not yet clear, but early tests were promising.

Resolution Comparison: Ideal Pulse and Phase

Prior to the 1980s, phase fluorometry was mired in the technology of Debye–Sears tanks and algebraic data analysis. The mean lifetime was reported with great precision, frustrating users who knew the technique held more power in reserve. The 1980s brought two large improvements: (1) Global analysis[5] quickly moved from pulse to phase[31] and quenching (QDAS)[32] domains. The method was further modified by those seeking to fit their modulation and phase directly. (2) Multifrequency fluorometers extended the key developments by Gratton and Limkeman.[33]

[30] K. Sasaki and H. Masuhara, *Appl. Opt.* **30**(8), 977 (1991).

[31] J. M. Beechem, J. R. Knutson, J. B. A. Ross, B. W. Turner, and L. Brand, *Biochemistry* **22**, 6054 (1983).

[32] J. R. Knutson, S. H. Baker, A. G. Cappuccino, A. G. Walbridge, and L. Brand, *Photochem. Photobiol.* **37**, 521 (1983).

[33] E. Gratton and M. Limkeman, *Biophys. J.* **44**, 315 (1983).

TABLE III
CONFIDENCE INTERVALS FOR SIMULATED DATA[a]

Method	Precision	Lower bound	Center	Upper	F-statistic
P/M	0.2°P/0.005M	4.93	5.03	5.06	1.52
TCSPC	18,000 peak	4.95	4.99	5.03	1.15
TCSPC	9000 peak	4.94	5.00	5.06	1.15

[a] A 5-nsec monoexponential, at 90% confidence.

The improvements were exhilarating, and a number of demonstrations were made that suggested unprecedented resolution. These heady times were met with a combination of skepticism and wonder in the pulse community. For one thing, it seemed like fanfare was accompanying resolution capabilities that time-correlated single photon counting (TCSPC) had taken for granted for decades. Perhaps unaware of this history, phase partisans were quick to claim superiority. The benefits of this friendly competition are many, but adversarial stances were often based on (mutual) unfamiliarity with each others' techniques.

In this brief section, simulation will be used to demonstrate a general equivalence. I shall set aside some issues that eventually need answers, namely, (1) (pulse) At what count levels does Gaussian noise fall below instrument noise from other sources, and what is the character of that noise floor? (2) (phase) Are phase/modulation residuals random? Are they normally distributed? If not, how will one reweight and obtain confidence limits without losing independent fit quality assessment?

By postponing these important issues, we can move to "ideal" comparisons: synthetic data, noise, and analysis. For this task, the aforementioned TFIT program (and accompanying TCON) were modified to provide phase and modulation fits (now PFIT, PCON). Pure Gaussian noise was added to each according to "typical" measurements (~9000 or 18,000 peak counts in pulse, 0.2 degrees phase, and 0.5% modulation). The results for an ordinary monoexponential are compared in Table III. The most obvious result is that the 90% confidence limits on each lifetime are very similar, with the phase result slightly closer to the 9000-peak-count simulation.

This may seem surprising, since, in phase, a 10% reduced χ^2 change is considerably more narrow. This quandary arises because that "rule of thumb" from pulse history corresponds to over 200 degrees of freedom. Rigorous joint confidence interval testing methods developed by Johnson, Halvorson, and Ackers[34,35] use the F-statistic (see Press et al.[18] and/or

[34] M. L. Johnson, H. R. Halvorson, and G. K. Ackers, *Biochemistry* **15**, 5363 (1976).

[35] M. L. Johnson and S. G. Frasier, this series, Vol. 117, p. 301.

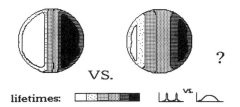

FIG. 6. Even if aromatic side chains in proteins uniformly occupy a wide range in conformation space, they need not exhibit smooth lifetime distributions. If the local environment is dominated by short-range interactions, a highly degenerate map can arise, that is, a multitude of positions can share similar lifetimes. More discriminatory methods will be needed to test these issues with rigorous statistical criteria.

Bevington[36], and the ratio of reduced χ^2's needed for a typical phase/modulation run (with about 40 degrees of freedom) is consequently higher (see Table III). Although the 10% rule was previously misapplied, more recent works (see elsewhere in this volume) correctly examine the region near the χ^2 minimum.

Virtuoso efforts in each method (100,000 peak counts or 0.1°/0.1% mod) have been reported, but neither method can claim a clear advantage in general purpose fluorometry. Phase partisans will note very accurate recovery of few picosecond lifetime terms by multigigahertz instruments, but carefully optimized TCSPC reconvolutions achieve similar results [down to about 20% of full width half-maximum (FWHM)].[37] Pulse partisans note that analog mixing requires about an order of magnitude more photocurrent than TCSPC, but the availability of several milliwatts of UV light from cavity-dumped dye laser systems makes that issue unimportant, except in very low-yield systems. Comparably equipped systems of either type should recover tremendous detail, assuming both take care to use the same global (especially multiwavelength) approach.

Discussion and Comments

The last few years have seen a resurgence of topics in biochemical fluorescence that tug at the edge of our resolution. As pointed out at the beginning of this chapter, many attempts to recover distributions can be equally explained by discrete terms (and vice versa). For proteins in particular, the proponents of distributions remind us of side-chain mobility

[36] P. R. Bevington, in "Data Reduction and Error Analysis for the Physical Sciences." McGraw-Hill, New York, 1969.
[37] G. R. Holtom, in "Time-Resolved Laser Spectroscopy in Biochemistry II" (J. R. Lakowicz, ed.), p. 2. SPIE Proc., Vol. 1204. Society of Photo-Optical Instrumentation Engineers, Bellingham, Washington, 1990.

and the wide dynamic range of conformers. With this one cannot disagree, but it is not the central difference in philosophy. Figure 6 illustrates the unknown link: does position smoothly map lifetime, or do several positions share nearly common lifetimes?

The answer is not present in any data seen to date, although certain proteins seem refractory to distributed analyses. Certainly one can expect cases where either picture holds. This controversy awaits precision improvements in both techniques (and/or new tests with "partial" linkage). For now, it is important that we continue current efforts to distinguish popularity from statistical proof.

Acknowledgments

Many thanks go to the inspirational mentor and many co-workers in the Brand laboratory during the 1980s. Special thanks go to Joe Beechem and Ben Turner for discussions about F-tests, to Marcel Ameloot for information about weighting, and to Carol Kosh for patience with the manuscript.

[17] Practical Aspects of Kinetic Analysis

By JULIEN S. DAVIS

Introduction

Kinetic analysis is the art of effectively linking experiment design and algebraic methods together in order to determine the dynamic behavior of a system. This might at first appear to be rather an odd statement since, by definition, it is feasible to fit any reaction profile generated by a system to the correct mechanism. However, it is rare for each individual step of a reaction to make a resolvable contribution to an experimental record. For example, transient intermediates can form and decay below the limits of detection; steps of similar rate can overlap and remain unresolved. Reaction mechanisms are, for these and related reasons, seldom fully determined by data collected under one set of experimental conditions.

Bearing the above in mind, it is clear that kinetic reactions are best determined by the simplest possible kinetic scheme that relates the available data. This places a considerable burden on the researcher, since appropriate experiments have to be designed to critically test the proposed mechanism. Correct reaction schemes are not obtained by devising overly complex mechanisms that fit the experimental data, but rather by proposing a simple mechanism and elaborating on it if inconsistencies between it and the data appear. For example, the simplest mechanism that can be attributed to an exponential reaction profile is a single step, irreversible first-order reaction. A small perturbation of a reversible, higher-order reaction would generate a similar reaction profile. The possibilities are endless. The preferred approach then, is to develop strategies that bring to light hidden kinetic features that will add to the mosaic of kinetic data which ultimately determine the reaction sequence. Deficiencies in the experimental data have to be first recognized and then overcome—a rich and appropriate area in which to apply numerical analysis, the subject of this volume.

Few question the validity of applying kinetic analysis to the study of biological systems. Dynamics are inseparable from life itself. It is useful, however, to ask whether the kinetics of nucleic acid, protein, organelle, and cellular function warrant a different approach to that used for kinetic studies in a field like chemistry. Biological reactions differ from those of chemistry, insofar as they are mediated by macromolecules selected by design from the offerings of chance—unexpected solutions to biological requirements through natural selection. This means that the rules governing the selection of a particular mechanism for some biological function will at times appear somewhat obscure. The net consequence is that one has to rely more heavily on experiment when developing ideas, and has to be open to the mechanistically unexpected.

This chapter serves to introduce the reader to various strategies that can be used to link experiment with data analysis. It is written as an introductory text from the perspective of the research worker who has limited kinetic expertise, many questions, and few answers and wants to create sense out of apparent kinetic chaos. The focus is exclusively on computer-based methods, since equipment for the digital recording of kinetic data is widely available and the advantages offered by computer-based numerical analysis are considerable. An absolute minimum of the algebraic formalism necessary to understand kinetics is presented first with an emphasis on basic principles and their application to analysis. This is followed by a section on the strategies used to simplify complex kinetic mechanisms for analysis. The important concept of the observed (apparent, composite) rate constant is introduced. Ways in which different methods of initiating reactions (relaxation kinetics, steady-state kinetics, stopped-flow kinetics, and the like) affect analysis are considered. A section follows on commonly encountered reaction profiles and their analysis. The ultimate goal of the chapter is to present sufficient information for the reader to devise an approach that will lead ultimately, through cycles of experiment and analysis, to the complete kinetic description of a system. Most physical chemistry textbooks have useful introductory material on kinetics; monographs devoted to kinetics,[1-3] relaxation kinetics,[4] and biological kinetics[5] will be referenced where appropriate.

Rate Equations and Application to Kinetic Analysis

Rate equations can be written in either differential or integrated form. Both define the relationship between the experimental record and the rate constants and reactant, intermediate, and product concentrations that govern the kinetics. Integrated rate equations are used for the analysis of concentration versus time data, whereas the much less familiar rate versus time data are analyzed using the differential form. Examples of each are given below for the commonly encountered first-order reaction mechanism of Eq. (1):

$$A \xrightarrow{k} B \tag{1}$$

In the differential equation [Eq. (2)], the rate of the reaction, $d[A]/dt$ at any time is given by the product of the first-order rate constant k and the concentration of reactant A, $[A]$, at that instant

$$d[A]/dt = -k[A] \tag{2}$$

Integration yields

$$[A] = [A_0] \exp(-kt) \tag{3}$$

For the integrated rate equation [Eq. (3)], $[A]$ is the concentration of reactant at time t, $[A_0]$ is the initial concentration of reactant at time zero, and k is the rate constant governing the conversion of A to B.

Two Different Methods for Kinetic Analysis

Before proceeding with our discussion of differential and integrated rate equations, we consider two quite distinct methods used for the analysis of kinetic data. Their selection is independent of the form (differential or integrated) of the rate equations.

In the direct method, experimental data are substituted into the designated rate equation, which is then solved either graphically or algebrai-

[1] J. W. Moore and R. G. Pearson, "Kinetics and Mechanism," 3rd Ed. Wiley, New York, 1981.
[2] G. G. Hammes, "Principles of Chemical Kinetics." Academic Press, New York, 1978.
[3] C. Capellos and B. H. J. Bielski, "Kinetic Systems," Reprint Ed. Robert E. Krieger Publ., Huntington, New York, 1980.
[4] C. F. Bernasconi, "Relaxation Kinetics." Academic Press, New York, 1976.
[5] H. Gutfreund, "Enzymes: Physical Principles." Wiley, London, 1975.

cally. A familiar example would be the graphical solution of the integrated rate equation [Eq. (4)], for the mechanism of Eq. (1), using a semilog plot of $(\ln[A] - \ln[A_0])$ versus t, the slope of which yields the rate constant k:

$$\ln[A] - \ln[A_0] = -kt \tag{4}$$

The indirect method, on the other hand, is an iterative procedure in which the reaction profile is first simulated and then compared with the experimental record. Estimation of the kinetic parameters is initiated by providing approximate values (best guesses) of rate constants [e.g., k of Eq. (3)] and overall reaction amplitudes, {e.g., $[A_0]$ of Eq. (3)} for the simulation. These chosen values are then optimized using a least-squares fitting routine (detailed elsewhere in this volume[6]) to minimize the deviation between the simulated curve (calculated values of [A] at various time intervals) and the experimental record. An advantage of the method is that statistical data are readily generated, allowing a quantitative assessment to be made of the correspondence between the proposed mechanism and the experimental data. It also allows analysis of hitherto unanalyzable, complex mechanisms by fitting experimental data to differential equations by numerical integration. The technique is computer based, since it involves a large number of repetitive calculations. Characteristics of differential and integrated rate equations, and their application to kinetic analysis, are now considered in further detail.

Differential Rate Equations

Differential rate equations are considered first, since their relation to kinetic mechanisms are direct and generally easy to establish by observing a few simple rules.[1,2,4] They also offer a suitable point of departure for those who wish to familiarize themselves with the relationship between kinetic mechanisms and rate equations. Their prime use is to formulate rate equations for later algebraic or numerical integration. They are seldom used directly for analysis.

By way of illustration, Eq. (5) represents a frequently encountered biological reaction mechanism in which a ligand A reacts with macromolecule M in a reversible bimolecular step to form the collision complex AM, which in turn undergoes an isomerization to the AM* complex—generally considered the ligand recognition step:

[6] M. L. Johnson and L. M. Faunt, this volume [1].

$$A + M \underset{k_{-1}}{\overset{k_1}{\rightleftharpoons}} AM \underset{k_{-2}}{\overset{k_2}{\rightleftharpoons}} AM^* \tag{5}$$

A set of four simultaneous differential equations [Eqs. (6)–(8)] can be written for the changes in concentration of each species with time:

$$d[A]/dt = d[M]/dt = k_{-1}[AM] - k_1[A][M] \tag{6}$$

$$d[AM]/dt = k_1[A][M] - (k_{-1} + k_2)[AM] + k_{-2}[AM^*] \tag{7}$$

$$d[AM^*]/dt = k_2[AM] - k_{-2}[AM^*] \tag{8}$$

The mechanistic implication of these four simultaneous equations is that *all* the individual kinetic constants and reactant concentrations collectively determine the kinetics of the system. Frequently stoichiometric and mass conservation relationships are used to simplify these equations. The relationships selected generally depend on the conditions under which the experiments are performed, and which reactant is being measured.

As mentioned, differential equations are seldom directly used for the analysis of experimental data, since they specify the rate of change of reactants over an infinitely small interval of time (a point in time). Commonly encountered concentration versus time profiles can, however, be converted to rate versus time profiles by simply determining the slope (tangent) of the record at selected time intervals. Previously used mechanical procedures for drawing of tangents to reaction profiles have been replaced by faster and more accurate computer-based procedures. The Savitzky–Golay[7] method for the computer-based evaluation of the first derivative of reaction profiles is one such method. Differential equations similar in form to Eqs. (2) and (6)–(8) would be solved in order to determine the kinetic parameters. However, this is not an extensively used method and is included mainly for didactic purposes.

Integrated Rate Equations

The limitations inherent in the application of differential rate equations to the analysis of concentration versus time data can be overcome by integration. We shall consider the more traditional approach first in which the differential rate equations are algebraically integrated, a procedure that has the advantage of providing us with analytic (exact) kinetic equations for use in analysis. Numerical integration, on the other hand, has greater flexibility and can readily be applied to the analysis of multistep kinetic schemes that defy algebraic integration [the mechanism of Eq. (5), for example]. The main disadvantages are that the technique is approximate and relatively slow. Either method can be adapted to the computer-

---

[7] A. Savitzky and M. J. E. Golay, *Anal. Chem.* **36,** 1627 (1964).

ized analysis of experimental data using nonlinear least-squares fitting routines.[6]

As noted above, algebraic integration suffers from the limitation that not all differential rate equations can be integrated, and, even for those that can, it is often a time-consuming process. First, the differential equations for the mechanism have to be written in a suitable form for integration. An important requirement is that the rate equations be formulated in terms of the reactants (dependent variables) that can be measured experimentally. Tables of integrals of various differential equations can help; the requirement here is that the differential equation of interest be written in an equivalent form. An extensive literature exists of various kinetic reaction schemes that have been integrated by others. Minor changes to the algebra can frequently adapt these equations for use, thus saving time. Kinetics textbooks[1-5] provide a useful first source in such a search. There is, therefore, no need for the newcomer to kinetics to reinvent the wheel, rather simply ensure that the equations used are correct. This is not an idle comment. Kinetic equations are notorious for the number of typographical and algebraic errors that appear in published material. To summarize, if a suitable integrated rate equation is available for the reaction scheme being investigated, use it. If not, numerical integration, discussed below, offers a more than satisfactory alternative.

Application of numerical integration to kinetic analysis has increased significantly with the widespread availability of computers and appropriate software. There are a number of algorithms (mathematical procedures) available for numerical integration. As a class, these mathematically sophisticated routines are designed to use the minimum of computer time and cope with stiff differential equations (equations with reaction steps differing markedly in rate).

A simple procedure for numerical integration (Euler's method) is presented to illustrate the underlying principles. In the procedure, kinetic constants of the reaction mechanism and the initial concentrations of reactants are used to generate a concentration versus time profile by summing the computed concentration changes of all reactants over a large number of sequential, small intervals of time. The reversible reaction of Eq. (9) illustrates the method:

$$A + B \underset{k_{-1}}{\overset{k_1}{\rightleftharpoons}} C \tag{9}$$

$$d[A]/dt = k_{-1}[C] - k_1[A][B] \tag{10}$$

For a small, finite interval in time (Δt), the following approximation holds

$$d[A]dt \cong \Delta[A]\Delta t \tag{11}$$

This allows the differential rate equation [Eq. (10)] to be rewritten in the form of Eq. (12):

$$\Delta[A]/\Delta t = k_{-1}[C] - k_1[A][B] \tag{12}$$
$$\Delta[A] = k_{-1}[C]\Delta t - k_1[A][B]\Delta t \tag{13}$$

The products of the various rate constants, concentrations, and the chosen small time interval are summed according to Eq. (14); the concentration of reactant A at $t = 0$ is $[A_0]$:

$$[A] = [A_0] + \sum_{t=0}^{t} \Delta[A] = [A_0] + \sum_{t=0}^{t} (k_{-1}[C]\Delta t - k_1[A][B]\Delta t) \tag{14}$$

A reaction profile of concentration of A versus time is thereby simulated. The time-dependent changes in concentrations of B [Eq. (15)] and C [Eq. (16)] can be obtained from mass conservation relations; $[B_0]$ and $[C_0]$ are the initial reactant concentrations in each case:

$$[B] = [B_0] + \sum_{t=0}^{t} \Delta[A] \tag{15}$$

$$[C] = [C_0] - \sum_{t=0}^{t} \Delta[A] \tag{16}$$

The technique is approximate insofar as the smaller the chosen time interval, the greater the accuracy of the computed reaction profile. For the analysis of experimental data, selected values of the rate constants and concentrations are optimized in repetitive simulations by using a least-squares fitting routine[6] to minimize the difference between the experimental and calculated profiles.

Making Complex Reactions Tractable for Analysis

Analysis requires that one force a kinetically analyzable response from the system under study. In the introduction, we noted that complex reactions frequently give rise to deceptively simple experimental profiles. It is evident that the complexity of a reaction profile will decline as fewer and fewer of the participating species make a measurable contribution. At times one is fortunate, and initial experiments yield data that can be readily analyzed: the task is then to probe the reaction further and to discover whether it is as straightforward as appearances suggest. On other occasions the reaction profile will have a complex form that disallows analysis and provides no clue as to mechanism: conditions then have to be altered with the aim of eliciting a kinetically simpler, and therefore analyzable, response. There are two ways to achieve this. The first and most commonly applied method is to manipulate the concentrations of individual reactants and observe the effect this has on the form of the reaction profile. The second approach is to alter rate constants by changing a parameter like temperature or the pressure under which the reaction takes place. Global analysis, in which both reactant concentrations and rate constants are varied and analyzed as a single data set, presents another option.[8] Details of these approaches are presented below.

Manipulating Concentrations

Manipulation of reactant concentrations is a very powerful and extensively used method for resolving kinetic data and testing reaction mechanisms. The method is, of course, only applicable to reactions that are concentration dependent (nothing is gained by altering the concentrations of a participating species in a mechanism consisting of first-order reactions). Two distinct avenues of approach are open to the investigator. Individual reactant concentrations can either be varied or held constant. In practice, a series of experiments using different reactant concentrations are performed, and the concentration dependence of the kinetics is duly noted. Alternatively, the concentration of one or more of the reactants may be held constant, thereby simplifying the kinetics by converting a dependent variable (a changing concentration) to a constant [see, e.g., Eqs. (19), (26), and (37)]. The pseudo-order reactions that result from this procedure are less complex, and generally easier to analyze, than the full reaction. In practice, the reagent has to be present in large excess or buffered to prevent a change in concentration during the time course of measurement—for example, a reaction that utilizes a proton carried out in the presence of a pH buffer. Similarly, a reactant can be present in such marked excess that its concentration remains essentially constant over the time course of measurement. Classic examples are reactions in which solvent water participates. Hydrolysis reactions [Eq. (17)], typically considered first-order, are in fact second-order [Eq. (18)] reactions operating under pseudo-first-order conditions with one reagent in excess. The reaction profiles are first-order in form with a composite rate constant (k_{obs}) that contains a concentration term [Eqs. (19), (26), and (37)]:

$$AB + H_2O \overset{k}{\rightleftharpoons} A + B \tag{17}$$

$$-d[AB]/dt = k[AB][H_2O] \tag{18}$$

[8] J. B. Beechem, this volume [2].

In aqueous media

$$k_{obs} = k[H_2O] \tag{19}$$

Yet another example is the common use of the standard (differential) form of Michaelis–Menten steady-state kinetic equations for enzyme-catalyzed reactions in which initial velocity data are collected under conditions where the concentration of substrate changes little and the reverse reaction is negligible.[5] It is always important to check algebraically whether the actual, albeit small, changes in concentration encountered have a negligible effect on the values assumed by individual kinetic parameters.[9]

Manipulating Rate Constants

Rate constants can be altered by changing temperature, pressure, electric field, etc. The effect seen is mediated through the influence of these intensive thermodynamic variables on the transition state of the reactions under consideration. Changing these parameters is useful for two reasons. First, they can serve the practical purpose of differentially altering the values of two similar rate constants so they are no longer indistinguishable from one another. Second, a differential effect of the parameter on the forward and reverse rates of a reaction will alter its equilibrium constant. This effect can be exploited in various ways. If, for example, a reaction consists of a temperature-sensitive, rapid preequilibrium prior to a rate-limiting step, then the concentration of the intermediate prior to the rate-limiting step can be altered at will by changing the temperature. The rate-limiting step itself could change with the result that hitherto undetected intermediates appear. All this information aids the process of kinetic analysis. As we shall see later, perturbations of this sort are frequently used to initiate reactions.

Global Analysis

Global analysis[8] of kinetic data has great potential, since it allows one to vary both reactant concentrations and rate constants and analyze the various experimental records obtained as a composite set of data. To do this, conventional rate equations are extended to include the temperature dependencies (Arrhenius activation energies) or pressure dependencies (activation volumes) of the individual rate constants. Skill and judgment are required in setting up the equations in a form suitable for analysis. This is an advanced topic beyond the scope of this chapter[8] but worth bearing in mind for solving that apparently intractable mechanism.

---

[9] M. Straume and M. L. Johnson, this volume [5].

Analysis and Methods of Initiating Reactions

Reactions being dynamic processes have to be initiated by some means in order to be studied. This applies even at equilibrium, where the redistribution of a fluorescent, radioactive, or other label between reactants is used to follow the kinetics: a nonequilibrium distribution of label is required at the start. Altering reactant concentrations and changing kinetic constants are the two extensively used methods for initiating reactions. Changing concentrations is generally brought about by mixing; rates are altered using perturbation techniques.

Mixing Reactants and Related Techniques

Unlike most of the perturbation techniques described in the next section, mixing can be applied to the study of both reversible and virtually irreversible reactions. There are, however, certain limitations in the use of mixing to initiate reactions. For one, it takes a finite time to fully mix two solutions. Stopped-flow equipment used in the study of rapid reaction kinetics can achieve this in a minimum of 1 msec. This probably sounds fast to those familiar with studying reactions by mixing reagents in a conventional spectrophotometer. As we shall show, it does limit the types of reactions that can be studied and can be a serious limitation. First-order rate constants have an upper limit of 10^{13} sec^{-1} for reactions with low activation energies. A much slower first-order reaction with a rate constant of 300 sec^{-1} and a half-time ($t_{1/2} = 0.69/k$), of the order of 2 msec is probably the fastest rate constant that can be directly determined in rapid mixing experiments. Second-order rate constants for ligand–macromolecule and protein–protein interactions characteristically have rates of the order of 5×10^7 and 1×10^5 M^{-1} sec^{-1} respectively. If we consider a reaction between a ligand and a macromolecule occurring at 5×10^7 M^{-1} sec^{-1} with the two reactants present in equal concentrations, the first half-time ($t_{1/2} = 1/k[A_0]$, where $[A_0]$ is the initial concentration of one or other reaction partner) would be 20 msec at 1×10^{-6} M and 20 μsec at 1×10^{-3} M. This once again highlights the constraints placed on kinetic analysis by the practicalities of experimentation. As we shall see below, perturbation techniques can, under certain circumstances, be used to overcome these speed limitations.

A variant on the mixing theme is to generate a reactant from an inert precursor already present in the reaction mixture by, for example, photolysis. The use of "caged" compounds (e.g., caged ATP, caged calcium) fall into this category. In these reactions an inert precursor is converted to the active compound by photolysis of the protecting group. In some instances, photorelease of reactant factor is faster than the mixing time in a stopped-flow experiment.

Perturbation Techniques

This is where potentially faster techniques that change rate constants by altering temperature, pressure, or other parameters enter. They are collectively known as relaxation kinetic techniques[4] and can, depending on the sensitivity of the reaction to the perturbant, produce either large or small changes in the position of equilibria. As we shall see later, small perturbations allow the rate equations to be linearized (elimination of products of concentrations); the kinetic consequence of this is that experimental records appear as the sum of one or more exponentials.[10] This simplification, as can be imagined, has a profound effect on the way in which the kinetics are analyzed. Application of the technique is limited to processes at equilibrium, or in a steady state, in which significant populations of the various reactants are present. Perturbation of an equilibrium in which one set of equilibrium partners are present in low concentrations (the reaction far over to one side) will result in a negligible perturbation of reactant concentrations.

The major advantages over mixing are the speed with which changes in rate constants can be imposed and the nondisruptive nature of the technique, allowing it to be applied to complex systems. For example, muscle fibers, cells, and delicate supramolecular aggregates either cannot be mixed or would suffer physical damage from the shear forces encountered in the mixing chamber of a stopped-flow apparatus.

The method chosen will depend primarily on the nature of the reaction being studied. In many cases it will be found that mixing and perturbation techniques usefully complement each other.

The Response: Reaction Profiles

Analysis proper starts with the reaction profile. A time-dependent signal proportional to concentration is the prime requirement. Ideally, each reactant, intermediate, and product should have its own unique signal so that the concentration of each can be measured independently of the others. This would solve many of the problems encountered in kinetic analysis; it is unfortunately a rare occurrence. The signal can be an optical property proportional to reactant concentration, or a quite different measure of concentration such as the force produced by specific intermediates

[10] H. R. Halvorson, this volume [3].

in a mechanochemical transduction system—for example, a record of tension versus time during the hydrolysis of ATP by the actomyosin ATPase in muscle. For the purposes of analysis, concentration changes are best monitored on a near continuous basis and the reaction profile digitized for computer processing. Apart from aiding analysis, weak signals of low signal-to-noise ratio can be enhanced by averaging and/or digital filtering to provide data of improved quality. Signal-to-noise ratios improve with the square root of the number of reaction profiles averaged. Care should be taken to avoid averaging cumulative (in phase) artifacts. If this type of error is suspected, the minimum number of data sets needed for reliable analysis should be averaged.

Observed or Apparent Rate Constant in Analysis

At the outset, there is no way of telling whether one is dealing with a single-step reaction governed by an elementary (true) rate constant or the degenerate response of a more complex mechanism governed by a composite rate constant. This ambiguity is coped with by using an operationally defined rate constant, variously known as the observed (k_{obs}) or apparent rate constant (k_{app}) for unknown mechanisms. The term k_{obs} is usually used to describe an experimentally determined composite rate constant, whereas k_{app} is more frequently used for theoretically developed mechanisms. This distinction need not be considered absolute but is worth noting. As shown in Eqs. (26) and (37), apparent rate constants may include a combination of rate and equilibrium constants and constant concentrations of reactant. The examples here [Eqs. (20)–(26)], and elsewhere [Eqs. (19) and (37)], illustrate these points.

$$A + B \underset{k_{-1}}{\overset{k_1}{\rightleftharpoons}} C \overset{k_2}{\rightarrow} D \qquad (20)$$

$$d[D]/dt = k_2[C] \qquad (21)$$

Assume a rapid preequilibration (k_1, $k_{-1} \gg k_2$) of the first step

$$K_1 = [A][B]/[C] \qquad (22)$$

The following concentration relations hold:

$$[A_0] = [A] + [C] + [D] \qquad (23)$$
$$[B_0] \cong [B] \qquad (24)$$

Thus

$$d[D]/dt = \{k_2/(1 + K_1/[B_0])\}([A_0] - [D]) \qquad (25)$$

which has the form of a first-order rate equation in which the apparent first-order rate constant k_{obs} has the following form

$$k_{obs} = k_2/(1 + K_1/[B_0])$$ (26)

k_{obs} is the rate constant for the formation of D at a specific concentration of $[B_0]$. At very high concentrations of $[B]$, it is apparent that k_{obs} will asymptotically approach k_2.

Clearly, all pseudo-order rate constants are composite rate constants. Specialized nomenclatures, different to the above, are encountered on occasion. Small perturbation relaxation experiments in which the response is always exponential (either single or the sum of many) employ the term relaxation time or tau (τ). The reciprocal relaxation time $1/\tau$ is more commonly used in practice, since it has the same units ($\sec^{-1}$) as a first-order rate constant.

Identifiable Reaction Profiles

The procedures outlined below are used to obtain one or more rate constants (either composite or elementary) from experimental records of recognizable form. These analyses are made independently of their associated reaction mechanisms, since different mechanisms can give rise to reaction profiles of similar form. Discussion is limited to the more commonly encountered forms assumed by reaction profiles. In practice this is not a real limitation since most biological reactions involve photochemical or catalytic reactions (zero- and pseudo-zero-order reactions), shape changes in macromolecules (first-order isomerizations), and reactions between ligands and macromolecules (second-order reactions). The majority of kinetic problems in biology can be solved using this limited repertoire. These different types of reactions are frequently combined into multistep mechanisms that may on occasion branch or run in parallel. The strategy with reactions of this type is, as we have discussed, to search for ways of simplifying the kinetic response to enable the system to be analyzed. The criterion, established in our earlier discussions, that it is preferable to use algebraically integrated rate equations whenever available will be adhered to here. Those wishing to obtain information on other forms, or to extend their knowledge beyond the information given here, are advised to refer to the several excellent monographs on the subject.[1-5]

Zero-Order Reactions

Zero- or pseudo-zero-order reactions have units of molar per second (M $\sec^{-1}$). Reaction profiles are linear in form and are therefore independent of both reactant concentration and time. Analysis of this type of reaction is straightforward, in that rate constants are obtained directly from the slope of concentration versus time profile. In Eq. (27) we have a mechanism in which the rate of interconversion of A to B remains unchanged. Equation (28) reveals that the rate of the reaction is equal to k. In the case of zero-order reactions we could use the differential relation directly to obtain k, since the rate of the reaction is invariant and therefore does not require the drawing of tangents to curved rate profiles as discussed in the section on differential equations. In Eq. (29), [A] is the concentration of reactant at time t, with $[A_0]$ the initial concentration of the reactant at time zero.

$$A \xrightarrow{k} B \tag{27}$$
$$-d[A]/dt = k \tag{28}$$
$$[A] = [A_0] - kt \tag{29}$$

A familiar application of zero-order kinetics to analysis is found in initial rate (v) measurements of enzyme-catalyzed reactions in order to determine Michaelis–Menten parameters. The elementary kinetic constants that comprise K_m and V_{max} depend on, and are unique to, the enzyme mechanism under consideration.[5]

First-Order Reactions

Reaction profiles with the appearance of a first-order response (exponential) arise from mechanisms in which rate is proportional to the first power of the concentration of one reactant. Their rate constants, whether they be composite or elementary, have the units of per second (sec^{-1}). A diagnostic feature is that sucessive half-times measured during the course of the reaction are invariant.

Because reactions exponential in form are frequently encountered, their analysis is worth considering in some detail. There are two reasons for their widespread occurrence. Biological macromolecules frequently isomerize while they function, and all quadratic and higher-order rate equations reduce to an exponential response in small-perturbation relaxation kinetic experiments where the reactants are perturbed marginally from equilibrium. Similar kinetics hold for label exchange experiments carried out on systems at equilibrium. Of course, first-order rate equations, being linear in form, exhibit exponential responses independent of the size of perturbation. The reader is referred to the excellent volume on relaxation kinetics by Bernasconi[4] for further details. Single and multiexponential responses of systems and the complexities of their analysis will now be addressed.

Let us consider a reaction of unknown mechanism that generates a reaction profile with the form of a single exponential. The integrated rate equation for such a process is

$$[A] = [A_0] + [A_1] \exp(-k_1 t) \tag{30}$$

At zero time, the reactant being measured will have a total concentration of $([A_0] + [A_1])$, of which an amount, $[A_1]$, will either be consumed or be generated during the time course of the reaction (this depends on whether reactant or product is measured experimentally). The exponential term, $\exp(-k_1 t)$, governs the rate at which $[A_1]$ (the preexponential term) declines. At zero time this proportionality term assumes a value of 1; as time increases and the reaction proceeds its value declines rapidly at first, then more slowly to asymptotically approach zero at the end of the reaction. At this point the concentration $[A_0]$ of reactant will remain. A nonlinear least-squares fitting routine[6] based on Eq. (30) would be used to determine $[A_0]$, $[A_1]$, and k_1 from the experimental record.

Reactions comprised of more than one exponential (termed a relaxation spectrum in small-perturbation relaxation experiments) requires the addition of terms to Eq. (30). A triexponential reaction would be analyzed by fitting the data to

$$[A] = [A_0] + [A_1] \exp(-k_1 t) + [A_2] \exp(-k_2 t) + [A_3] \exp(-k_3 t) \tag{31}$$

In this case a nonlinear least-squares fitting routine would be used to determine the final value of $[A_0]$ at equilibrium, the preexponential amplitude terms $[A_1]$, $[A_2]$, and $[A_3]$, and rate constants k_1, k_2, and k_3. In relaxation kinetics,[4] the rate constants k_1, k_2, and k_3 would be replaced by the reciprocal relaxation times $1/\tau_1$, $1/\tau_2$, and $1/\tau_3$.

Unfortunately all is not so simple as appearances suggest (see Halvorson[10]). Difficulties arise when the rates of the individual exponential processes assume similar values, in which case they become impossible to separate, and their individual contributions to the overall reaction cannot be resolved. Generally speaking, exponential phases with rate constants less than a factor of 3 apart are difficult to resolve from one another. This can be particularly troublesome if the amplitudes of the reactions differ significantly; phases of large amplitude will tend to dominate the analysis. This can be circumvented if each of the individual components of the process possesses a unique signal and, by dint of the nature of the mechanism, are kinetically uncoupled (independent). Under these conditions each exponential phase can be determined in isolation. Changing temperature, pressure, or some other parameter can be used to alter rates and separate the component reactions on the time axis. Complex relaxation spectra with a large number of exponentials contributing to the overall reaction can be usefully analyzed by a special technique that enables a mean relaxation time and the width of the relaxation spectrum to be determined.[4]

Second-Order Reactions

Second-order rate constants have units of per molar per second (M^{-1} sec^{-1}) and have less easily recognizable reaction profiles compared to zero- and first-order reactions. Reaction half-times only serve as a useful diagnostic for second-order reactions in which two like-species react ($t_{1/2} = 1/2k[A_0]$) or two dissimilar reaction partners are present at the same concentration ($t_{1/2} = 1/k[A_0]$). The general integrated rate equation for a second-order reaction [Eq. (38)] can be simplified under certain experimental conditions: (1) if two like species interact

$$A + A \xrightarrow{k} C \tag{32}$$
$$1/[A] - 1/[A_0] = 2kt \tag{33}$$

(2) if both reactants have the same initial concentrations ($[A_0] = [B_0]$)

$$A + B \xrightarrow{k} C \tag{34}$$
$$1/[A] - 1/[A_0] = kt \tag{35}$$

(3) if pseudo-first-order conditions hold in which the concentration of one reactant is in large excess ($[A_0] \gg [B_0]$) over the other. Under these conditions, the second-order rate constant is obtained from a secondary plot of the pseudo-first-order rate constant (k_{obs}) versus $[A_0]$:

$$[B] = [B_0] \exp(-k[A_0]t) \tag{36}$$
$$k_{obs} = k[A_0] \tag{37}$$

The full integrated rate equation for a second-order without imposed constraints has the following form:

$$\{1/([B_0] - [A_0])\} \ln\{([A_0]/[B])/([B_0]/[A])\} = kt \tag{38}$$

Complex Reaction Profiles

Our kinetic analysis of reaction profiles has been limited to those with a recognizable form. The rate constants obtained can be either composite or elementary. The question arises as to how one proceeds from here. It is useful to appreciate the point that we have been viewing the reaction through the equivalent of a kinetic window, a simplified view that allowed us to initiate analysis. From a practical perspective, there are various strategies that one can adopt to expand the view of this window to further develop the mechanism. This presupposes that the primary information (apparent rate constants obtained under different experimental conditions) has been extracted from these experiments.

Secondary plots of the data feature prominently here. For example, in pseudo-order experiments, k_{obs} values would be obtained at a series of different concentrations of the quasi-constant (fixed concentration) reactant. Equation (37) provides a simple example in which a secondary plot of k_{obs} versus $[A_0]$ provides the elementary rate constant, k. Similarly, for Eq. (26), values for K_1 and k_2 are obtained from the slope and intercept, respectively, of a double-reciprocal plot of $1/k_{obs}$ versus $1/[A_0]$. As wide a range of concentrations as allowable (under which pseudo-order conditions hold) should be used in these analyses; deviation from the relationship would indicate that the mechanism is inappropriate.

Testing Complex Reaction Mechanisms

At this point the investigator will have a series of rate constants (preferably elementary, obtained from primary and possibly secondary plots of the data) that can be substituted into a general mechanism. The wider applicability of the mechanism, beyond the limits of the kinetic window used to collect data, is now open for exploration.

The overall equilibrium constant (determined independently of the kinetics) for the reaction should equal the product of the forward rate constants divided by the product of the reverse rate constants. A discrepancy in the values indicates an unidentified step or steps in the reaction mechanism; at best this will present an opportunity to determine independently the missing parameter in the kinetic scheme and at worst alert one to deficiencies in the mechanism.

Numerical integration and nonlinear least-squares analysis offers a powerful method for testing the proposed general (full) mechanism to see if it holds up under different experimental conditions. Complex reaction profiles of unrecognizable form (to our limited vision) can be readily tested in this way. The mechanism should hold over a wide range of reactant concentrations. If a discrepancy between the experimental record and mechanism appears, steps must be taken to ensure that it is not due to a change in the mechanism itself. Macromolecules are particularly sensitive to their environment; ionic strength effects, phase changes, and the like can cause new reaction pathways to function.

Acknowledgments

I would like to thank Dr. Michael E. Rodgers for comments on the manuscript. This is contribution No. 1474 from the Department of Biology, The Johns Hopkins University.

[18] Compartmental Analysis of Enzyme-Catalyzed Reactions

By PRESTON HENSLEY, GLENN NARDONE, and MERYL E. WASTNEY

Introduction

An understanding of the detailed nature of the chemistry of enzyme-catalyzed reactions has been the goal of physical and biological chemists since these processes were first observed. Berzelius, in 1835, invented the term catalysis for the transformation process of either animate or inanimate origin, and in 1877 Kuhne coined the term enzyme (in yeast) for catalytic agents of biological origin.[1,2] These processes were first given a theoretical foundation in 1902 by Brown, who proposed the enzyme–substrate complex, and in 1903 by Henri, who derived the first hyperbolic rate equation for a single-substrate enzyme-catalyzed reaction.[3,4] In his formulation, Henri proposed that the enzyme and substrate were in rapid equilibrium. In 1925 Briggs and J. B. S. Haldane proposed a similar formulation but, in contrast to Brown and Henri, proposed that the enzyme–substrate complex was in steady state, that is, that its rate of formation and breakdown were equal.[5]

Subsequently, these approaches were generalized and now are able to describe quantitatively the catalytic processes of most of the 3000 or so enzymes now cataloged.[6] The character of these kinetic approaches is due, not surprisingly, to the catalytic nature of the majority of the enzymes which have been studied. The most important aspect in this regard is their very high turnover rate, with k_{cat} in the range of 10^2 to 10^7 sec^{-1}, and k_{cat}/K_m approaching the diffusion-controlled limit of 10^8 M^{-1} sec^{-1}.[7] For the rates of these enzyme-catalyzed processes to be measured by standard technologies, the enzyme concentrations employed are often in the range of 10^{-8} to 10^{-10} M, with substrate concentrations in the range of 10^{-6} to 10^{-3} M. When the enzyme concentrations are vastly lower that the sub-

[1] J. J. Berzelius, Translated from "Lehrbuch der Chemie" (F. Wohler, transl.), 3rd Ed., Vol. 6, p. 22. Arnoldische Buchlandlung, Dresden and Leipzig, 1837.

[2] W. Kuhne, *Physiol. Inst. Univ. Heidelberg Unters.* **1**, 293 (1877).

[3] A. J. Brown, *Chem. Soc. J. (Trans.)* **81**, 373 (1902).

[4] V. Henri, "Lois Generales de l'Action des Diastases" (A. Hermann, ed.), p. 85. Librairie Scientifique, Paris, 1903.

[5] G. E. Briggs and J. B. S. Haldane, *Biochem. J.* **19**, 338 (1925).

[6] I. H. Segel, "Enzyme Kinetics." Wiley (Interscience), New York, 1975.

[7] A. Fersht, "Enzyme Structure and Mechanism," p. 152. Freeman, New York, 1984.

strate concentrations, all enzyme species can usually be assumed to be in a steady state (and in very low concentration).

High turnover and low substrate affinity are the signatures of the enzymes of intermediary metabolism. It is important for these processes to be catalyzed with high efficiency. In contrast, for the replication and transformation of DNA and RNA, it is important that this class of reactions be catalyzed with high fidelity. This means that the substrates must bind with very high specificity (affinity), often with K_d values of 10^{-9} to 10^{-12} M. However, the price which is paid for high affinity is slow turnover. For the restriction enzyme-catalyzed reactions discussed below, k_{cat} values of the order of 10^{-2} sec^{-1} are found.[8,9]

The fact that this class of enzymes has extremely slow substrate turnover rates means that the enzyme concentrations may be raised to near or equal that of the substrate concentrations and still have measurably slow catalytic rates. Unfortunately, under these conditions, the steady-state and rapid equilibrium approaches discussed above are not applicable as the concentrations of the various substrate and enzyme forms may vary significantly during the time course of the reaction. However, this experimental circumstance has a distinct computational advantage, namely, that the intermediates in the catalytic process, discussed at length in the steady-state literature but rarely observed, may be significantly populated and, hence, quantifiable.

In the analysis of enzyme kinetics by the steady-state approach, it is the differential (rate) equation which is directly analyzed. For a simple case, the equation has the form

$$-d[S]/dt = v = V_{max}[S]/(K_m + [S])$$

where V_{max} is a phenomenological constant representing the maximal velocity and K_m is a phenomenological constant representing the substrate concentration at half-maximal velocity. Nonlinear least-squares analysis of data of the form v versus $[S]$ directly yields the value of K_m and V_{max}. However, under the slow substrate turnover, high enzyme conditions discussed above, many intermediates in the enzyme-catalyzed reaction are populated, and the time-resolved kinetic processes are more appropriately analyzed by the solution of a series of differential equations, one for each quantifiable enzyme or substrate form.

[8] P. Hensley, G. Nardone, J. G. Chirikjian, and M. E. Wastney, *J. Biol. Chem.* **265**, 15300 (1990).

[9] G. Nardone, M. E. Wastney, and P. Hensley, *J. Biol. Chem.* **265**, 15308 (1991).

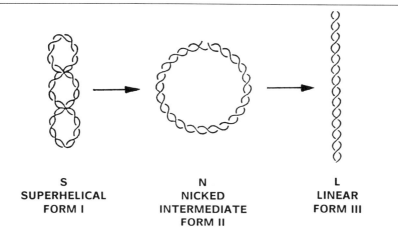

S
SUPERHELICAL
FORM I

N
NICKED
INTERMEDIATE
FORM II

L
LINEAR
FORM III

FIG. 1. Chemical steps in the endonuclease-catalyzed cleavage of superhelical plasmid DNA. (From Ref. 8.)

In the example presented here, the analysis of the time-resolved kinetics of a restriction endonuclease-catalyzed reaction was accomplished by compartmental analysis using methodology developed by Berman and colleagues.[10–14] Traditionally, in the analysis of complex kinetic data, a model is proposed and the corresponding differential equations are written and solved numerically. This sequence must be repeated for succeeding models. The process is tedious, computationally intensive, and, therefore, not often employed except for simple systems.

With the approach of Berman, the kinetic data are analyzed first to determine the relations between compartments using an interactive program for model simulation and data analysis. Subsequently, the compartmental model is interpreted in terms of chemical processes. These processes are represented as transfers of species from one compartment to another. A compartment may be a unique chemical species, or it may be

[10] M. Berman, in "Advances in Medical Physics" (J. S. Laughlin and E. W. Webster, eds.), p. 279. Second International Conference on Medical Physics, Inc., Boston, Massachusetts, 1971.

[11] M. Berman and M. F. Weiss, SAAM-27 Manual, DHEW Publication No. (NIH) 78–180 (1978).

[12] M. Berman, W. F. Beltz, R. Chabay, and R. C. Boston, "CONSAM User's Guide." U.S. Govt. Printing Office, Washington, D.C., 1983.

[13] R. C. Boston, P. C. Greif, and M. Berman, Comput. Programs Biomed. 13, 111 (1981).

[14] D. M. Foster and R. C. Boston, in "Compartmental Distribution of Radiotracers" (J. D. Robertson, ed.), p. 73. CRC Press, Cleveland, Ohio, 1983.

Minutes   0   0.5   1   3   5   10   20

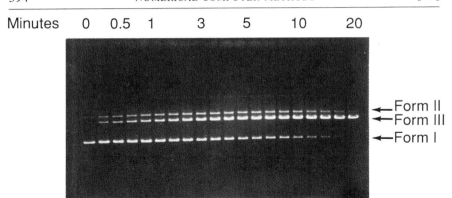

FIG. 2. Agarose gel electrophoresis of endonuclease cleavage products. The upper bands are the nicked intermediate (Form II); the middle bands, linear product (Form III); and the lower bands (Form I), the superhelical substrate. The DNA concentration was 10 nM, and the protein concentration was 6 nM. The bands (0.4–0.6 μg DNA per lane) were visualized by ethidium bromide staining, excised, and then quantified by scintillation counting of the radioactive DNA. (From Ref. 8.)

several chemical species. The program (SAAM/CONSAM,[15] an interactive program for simulation, analysis, and modeling) has the advantage that models may be easily developed and modified. Here, we present an application of this type of kinetic analysis to a restriction endonuclease using *Bam*HI as a model system and show that the cleavage process may be easily resolved into separate binding and hydrolytic steps.

*Bam*HI Hydrolysis of Superhelical Plasmid DNA

The general restriction endonuclease cleavage pattern for a circular superhelical DNA molecule containing a single recognition site is shown in Fig. 1. The enzyme binds to superhelical DNA (S, or Form I) and cleaves one strand. The DNA then relaxes to form a circular, nicked (N, Form II) intermediate with no superhelical turns. Finally, the opposite strand is cleaved, yielding a linear (L, Form III) species. The restriction endonucleases are of special interest, as mentioned above, because the substrate turnover rates are slow, and as a result an intermediate in the enzyme-catalyzed reaction (the singly nicked circle) may be visualized,

---

[15] SAAM/CONSAM software (including a 386 version) and manuals are available free by calling 301-496-8914, by FAX at 301-480-2871, or by writing SAAM, Laboratory of Mathematical Biology, Rm. 4B56, Bldg. 10, National Cancer Institute, National Institutes of Health, Bethesda, MD 20892.

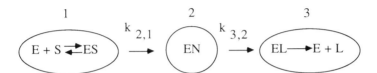

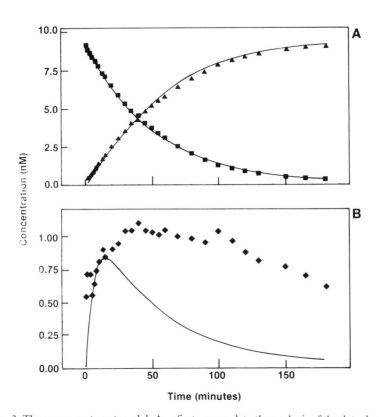

FIG. 3. Three-compartment model. As a first approach to the analysis of the data, because only three DNA species are visible in the primary data [superhelical (S), nicked intermediate (N), and the linear product (L), see Fig. 2], the simplest model that could be proposed to fit the data is a three-compartment model, with one compartment for each of the three DNA species. A binding step is included in the compartment containing the superhelical DNA substrate, and a release step is included in the compartment containing the linear product, as these are required chemically. In this model, $k_{2,1}$ is the rate constant describing the overall conversion of superhelical DNA to the nicked intermediate, and $k_{3,2}$ is the rate constant describing the overall conversion of the nicked intermediate to the linear DNA product. (A) Time course of the decrease in the concentration of the superhelical substrate (squares) and the time course of the increase in the concentration of the linear product (triangles). (B) Time course of the increase and decrease of the concentration of the nicked intermediate (diamonds). As can be seen in (B), the maximal concentration of the nicked intermediate ($\sim$1 nM) exceeds the total concentration of BamHI endonuclease (0.75 nM). (From Ref. 8.)

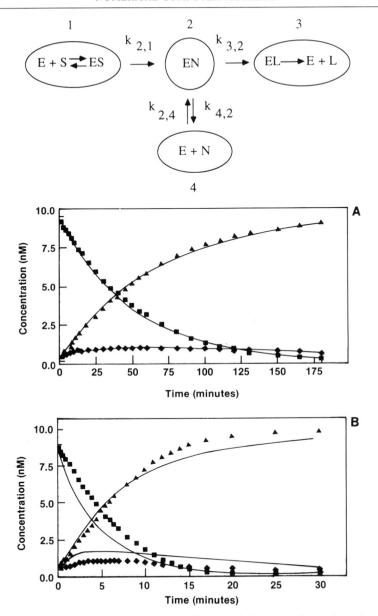

FIG. 4. Four-compartment model. A second approach for the analysis of the data is to invoke a model wherein the complex of the nicked DNA and endonuclease (EN) dissociates to free enzyme (E) and nicked DNA (N). This requires a new (fourth) compartment for the dissociated species. This extension in the model is justified as the concentration of the nicked

isolated, and quantified. The purification, structural characteristics, and overall catalytic properties of the enzyme have been established.[16]

The hydrolysis of superhelical pBR322 DNA by *Bam*HI can be followed by sequentially quenching aliquots of the reaction mixture and resolution of the DNA species by agarose gel electrophoresis. The electrophoretogram in Fig. 2 shows a typical time course. The bands were identified, and the radioactive DNA was quantified. The usefulness of the compartmental method for the analysis of enzyme kinetic data of this sort will be illustrated through a sequence of modeling approaches.

First Approach: Three-Compartment Model

As is clear from Fig. 2, the DNA can be resolved into only three species, a superhelical substrate (S), a nicked intermediate (N), and a linear product (L). The simplest model which could be used to fit data of this type would be a three-compartment model such as that shown in Fig. 3. Here, the only rate processes which can be observed are the overall conversion of the superhelical DNA to the nicked form, described by the rate constant $k_{2,1}$, and the overall conversion of the nicked intermediate to the linear product, described by the rate constant $k_{3,2}$. The model assumes that the association of S and N to the enzyme is rapid relative to the rate of covalent bond cleavage.

As can be seen in Fig. 3A, this model describes well the time courses of the decrease of the superhelical substrate and the increase of the linear product. However, there is a systematic misfit in the analysis of the nicked intermediate seen in Fig. 3B. The data in Fig. 3B suggest the reason why the three-compartment model is inappropriate. Here the peak concentration of the nicked intermediate is roughly 1 nM. However, the total enzyme concentration in this analysis was 0.75 nM. Therefore, the concentration of the nicked intermediate is greater than the total enzyme concentration and suggests that some nicked intermediate is released from the enzyme-

[16] G. Nardone and J. G. Chirikjian, "Gene Amplification and Analysis, Volume 5: Restriction Endonucleases and Methylases" (J. G. Chirikjian, ed.), p. 147. Elsevier, New York, 1987.

---

intermediate exceeds the total concentration of endonuclease (see the legend to Fig. 3B). (A) Data for the reaction at 0.75 nM endonuclease. Using a four-compartment model, the data at low enzyme concentration are well described. As a result of the extension of the model to explicitly include a dissociation step for the EN complex, the systematic misfit seen in Fig. 3B is removed. (B) Data for the cleavage at 6.0 nM endonuclease. At the high enzyme concentration using the four-compartment model there is a systematic misfit for time courses of the concentration changes of all three DNA species. For symbols, see Fig. 3. (From Ref. 8.)

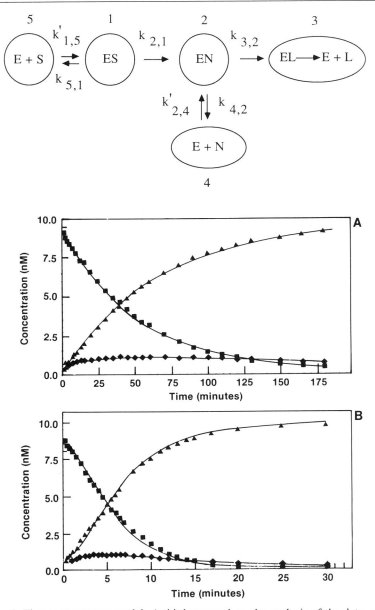

FIG. 5. Five-compartment model. A third approach to the analysis of the data, as the observed disappearance of S is slower than calculated (Fig. 4B), is to extend the model to include separate compartments for the dissociated (S) and enzyme-bound (ES) superhelical DNA species. (A) The kinetics are well described at low (0.75 nM) enzyme concentration.

bound form and reassociates at a rate slower than the rate of cleavage of EN to EL.

Second Approach: Four-Compartment Model

The release of the nicked intermediate from its enzyme-bound form may be treated explicitly by adding a fourth compartment to the model as shown in Fig. 4. This model was first proposed for the cleavage of super-helical DNA by the restriction endonuclease *Eco*RI by Rubin and Mo-drich.[17] When data are analyzed by the four-compartment model (Fig. 4), it is clear that the time courses of the concentration changes of the three DNA forms as the lower enzyme concentration (0.75 nM, Fig. 4A) are well described. However, Fig. 4B shows the time courses of the concentration changes of the three DNA species at high enzyme concentration (6.0 nM). When analyzed in terms of this four-compartment model, there is a systematic deviation for all three DNA species.

Third Approach: Five-Compartment Model

The data in Fig. 4B show that the cleavage of superhelical DNA is slower than would be predicted by a model wherein free enzyme and free superhelical DNA rapidly equilibrate in a single compartment. This lag in the decrease in the concentration of the superhelical DNA substrate with time might be explained if the association rate for enzyme and superhelical substrate, under the conditions of the assay, were slow compared to the first cleavage step. This suggests a third approach to the analysis of the time course of the cleavage of the DNA by *Bam*HI, namely, to separate the superhelical DNA into two compartments, one (compartment 5) containing free superhelical DNA substrate (S) and another (compartment 1) containing bound superhelical DNA (ES). This new model is shown in Fig. 5. Here we see that, using the five-compartment model, the time courses of the concentration changes of the three DNA species for both enzyme concentrations (0.75 nM, Fig. 5A, and 6.0 nM, Fig. 5B) are well described. The systematic misfit has been eliminated.

In this model, $k_{5,1}$ is denoted although it could not be defined by the data

---

[17] R. A. Rubin and P. Modrich, *Nucleic Acids Res.* **5,** 2991 (1978).

(B) The kinetics at high (6.0 nM) enzyme concentration are also well described. The solid lines represent the results of analysis of the data in terms of the five-compartment model. With the five-compartment model, the systematic misfit at high and low enzyme concentration (Fig. 4B) is removed. For symbols, see Fig. 3. (From Ref. 8.)

here. The ratio $k_{5,1}/k_{1,5}$ was fixed at 10^{-11} M as was found for EcoRI.[18,19] It was found that $k_{5,1}$ could be changed by a factor of 10 without affecting the results, suggesting that the dissociation rate process for the superhelical DNA from the enzyme is too slow to be quantified in these studies.

A logical extension of the current line of reasoning is to separate the linear DNA species (EL and L) into separate compartments. When the data at high and low enzyme concentrations were analyzed in terms of this model, the fit was good; however, the parameter for the release of the linear product was not well determined. Hence, the data are not sufficient to resolve this last process, and release of L is assumed to be a rapid process relative to the rate of cleavage of EN.

Test of the Model

A particularly stringent test of the five-compartment model would be to analyze it against a series of data sets wherein the enzyme concentration varied. Simultaneous fitting of data from six enzyme concentrations was therefore done, and it allowed the enzyme concentration-dependent rate constants to be determined. The values of the rate constants, determined by fitting the five-compartment model (Fig. 5) to the data at each enzyme concentration, were plotted against enzyme concentration (Fig. 6). As can be seen, $k'_{1,5}$ (the apparent rate constant describing the association of the DNA substrate and the free enzyme, Fig. 5) and $k'_{2,4}$ (the apparent rate constant describing the association of the nicked DNA and the free enzyme, Fig. 5) are linear with respect to enzyme concentration (data shown in Fig. 6A,C). These rate constants are well determined and are both linear functions of enzyme concentration over a 15-fold change in the enzyme concentration. In contrast, $k_{2,1}$ (the rate constant describing the first strand cleavage event, Fig. 5) and $k_{3,2}$ (the rate constant describing the second strand cleavage event, Fig. 5) are independent of enzyme concentration (data shown in Fig. 6B,D). These results are consistent with the model.

The enzyme concentration dependence of $k'_{1,5}$ and $k'_{2,4}$ was then included into the five-compartment model (Fig. 5), resulting in the model described here (Fig. 7) with $k_{1,5}$ and $k_{2,4}$ denoting enzyme concentration-independent constants. The differential equations describing this model are given below:

$$d[S]/dt = -k_{1,5}[E][S] + k_{5,1}[ES]$$
$$d[E]/dt = -k_{1,5}[E][S] + k_{5,1}[ES] - k_{2,4}[E][N] + k_{4,2}[EN] + k_{3,2}[EN]$$

[18] B. J. Terry, W. E. Jack, R. A. Rubin, and P. Modrich, *J. Biol. Chem.* **258**, 9820 (1983).
[19] L. Jen-Jacobson, M. Kurpiewsky, D. Lesser, J. Grable, H. Boyer, J. M. Rosenberg, and P. J. Greene, *J. Biol. Chem.* **258**, 14638 (1983).

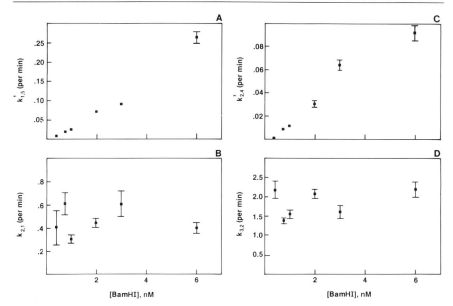

FIG. 6. Individual rate constants [obtained by fitting the data to the five-compartment model (Fig. 5)] as functions of enzyme concentration. The relationships between enzyme concentration and the values of $k'_{1,5}$ (A), $k_{2,1}$ (B), $k'_{2,4}$ (C), and $k_{3,2}$ (D) are shown. The error bars (shown for all values) represent 1 standard deviation. $k'_{1,5}$ and $k'_{2,4}$ are here denoted with primes as they are not true second-order rate constants (with units M^{-1} min^{-1}) as suggested in the models, but are apparent first-order rate constants (with units of min^{-1}). When the enzyme concentration is factored out the primes are removed. (From Ref. 8.)

$$d[ES]/dt = k_{1,5}[E][S] - k_{5,1}[ES] - k_{2,1}[ES]$$
$$d[EN]/dt = -k_{4,2}[EN] - k_{3,2}[EN] + k_{2,1}[ES] + k_{2,4}[E][N]$$
$$d[N]/dt = -k_{2,4}[E][N] + k_{4,2}[EN]$$
$$d[L]/dt = k_{3,2}[EN]$$

Mass was conserved throughout using the following equations:

$$E_{total} = [E] + [ES] + [EN]$$
$$DNA_{total} = [S] + [ES] + [EN] + [N] + [L]$$

The fit of the data from three DNA species at six enzyme concentrations by this model can be seen in Fig. 8.

The values of the rate constants from this analysis are given in Table I. All four rate constants were well determined at the individual enzyme concentrations (the standard errors were all less than 10%). The mean values for the rate constants are given in the last line of Table I. The

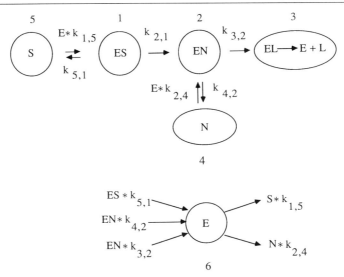

FIG. 7. Proposed model with the enzyme concentration dependence of the two rate constants explicitly included. This is a modification of the five-compartment model (Fig. 5), with the rate constants $k_{1,5}$ and $k_{2,4}$ defined as linear functions of enzyme concentration and a new compartment (compartment 6) added which contains the free enzyme. The rate constants $k_{2,1}$ and $k_{3,2}$ remain enzyme concentration independent. (From Ref. 8.)

standard errors of the means are all about 20%. The first three columns in Table I are the calculated initial concentrations of free superhelical DNA, nicked DNA, and linear DNA. The concentrations of substrate, nicked intermediate, and linear product were measured at the start of the experiment. Although some was measured as nicked (<7% of total), linear product was never observed. The calculated values may represent experimental artifact or a small fraction (<10%) cleaved by a different pathway.

Two parameters of the model, $k_{4,2}$ and $k_{3,2}$, were found to be highly correlated (0.94). As a result, $k_{4,2}$ was fixed a value of 0.22 min^{-1} (the value determined by fitting all data from all experiments simultaneously was 0.26 ± 0.03 min^{-1}). The resulting parameter correlations are given in Table II. The highest correlation of parameters is now 0.86, which is still high. The effect of this correlation is tested by varying the (fixed) value of $k_{4,2}$ by a factor of 2 and fitting the data. It was found that (1) no changes occurred in parameter values other than $k_{3,2}$ and that (2) the ratio of $k_{4,2}/k_{3,2}$ remained constant. With the values of $k_{4,2}$ and $k_{3,2}$ determined, the fraction of nicked intermediate bound to enzyme (EN) which is cleaved without being first released may be determined from the ratio $k_{3,2}/(k_{3,2} +$

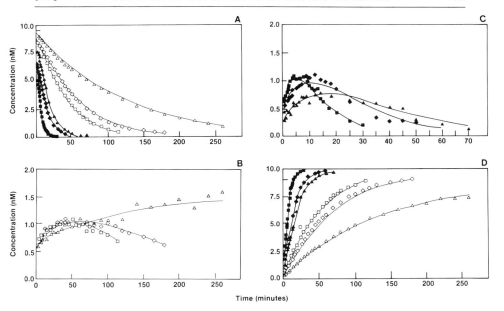

FIG. 8. Simultaneous analysis of data representing three DNA species and six enzyme concentrations in terms of the model (Fig. 7). (A) Cleavage of superhelical DNA. (B) Nicked intermediate at three lower enzyme concentrations. (C) Nicked intermediate at three higher concentrations (note the change in time scale). (D) Appearance of linear DNA. In all cases, the filled squares are the results at 6 nM enzyme, the filled diamonds are at 3 nM enzyme, the filled triangles are at 2 nM enzyme, the open squares are at 1 nM enzyme, the open diamonds are at 0.75 nM enzyme, and the open triangles are at 0.4 nM enzyme. All curves are the calculated fits according to the model (Fig. 7). (From Ref. 8.)

$k_{4,2}$) and is 90%. Hence, most of the DNA substrate is cleaved as a result of a single enzyme–DNA recognition event.

In these studies we analyzed the kinetics of the cleavage of the superhelical DNA into a nicked circular intermediate and finally into a linear product. Our approach was to use compartmental analysis of the data to describe the kinetics of the process. The kinetics were subsequently interpreted in terms of discrete chemical models. In this *a posteriori* approach, conclusions about the nature of the processes were arrived at based on observation and experiment rather than beginning analysis by requiring that the data fit discrete chemical models and introducing simplifying assumptions (often not testable) so that the data can be fitted. The rate constants for the processes were simultaneously determined over a 15-fold range of enzyme concentration.

Using the compartmental approach developed by Berman,[10–14] we

TABLE I

ANALYSIS OF pBR322 DNA CLEAVAGE AS A FUNCTION OF *Bam*HI CONCENTRATION USING SIX-COMPARTMENT MODEL[a,b]

Enzyme concentration (nM)	$IC_5{}^c$ (nM)	$IC_4{}^c$ (nM)	$IC_3{}^c$ (nM)	$k_{1,5}$ (nM^{-1} min^{-1})	$k_{2,1}$ (min^{-1})	$k_{2,4}$ (nM^{-1} min^{-1})	$k_{3,2}$ (min^{-1})
0.40	8.93	0.66	0.41	0.024	0.85	0.0^e	2.18
	$(0.18)^d$	(0.03)	(0.07)	(0.0007)	(0.029)		(0.22)
0.75	9.00	0.59	0.34	0.030	0.88	0.012	0.162
	(0.09)	(0.021)	(0.017)	(0.0005)	(0.08)	(0.001)	(0.13)
1.00	9.09	0.52	0.36	0.033	0.67	0.014	1.55
	(0.04)	(0.010)	(0.008)	(0.0004)	(0.05)	(0.001)	(0.068)
2.00	9.13	0.18	0.65	0.047	0.68	0.023	1.83
	(0.14)	(0.015)	(0.039)	(0.001)	(0.031)	(0.002)	(0.118)
3.00	9.11	0.39	0.50	0.038	0.76	0.022	1.59
	(0.11)	(0.025)	(0.024)	(0.001)	(0.034)	(0.002)	(0.11)
6.00	8.70	0.49	0.82	0.047	0.62	0.016	2.28
	(0.19)	(0.045)	(0.056)	(0.001)	(0.06)	(0.001)	(0.19)
Mean^f	8.99	0.47	0.51	0.036	0.74	0.017	1.84
	(0.16)	(0.169)	(0.188)	(0.009)	(0.104)	(0.004)	(0.32)

[a] As in Fig. 7. From Ref. 8.

[b] In all calculations reported here, $k_{4,2}$ was fixed at a value of 0.22 min^{-1}.

[c] The calculated initial concentration (IC) of DNA in compartments 5 (free S), 4 (N), and 3 (EL and L). The data were fitted by assuming that the initial concentration of EN (compartment 2) was 0.0 nM.

[d] Standard deviation of the parameter value.

[e] At low enzyme concentration, this parameter did not differ significantly from 0.

[f] The mean and standard deviation of the mean for the appropriate parameter for the six enzyme concentrations. The values convert to $k_{1,5} = 6 \times 10^5$ M^{-1} sec^{-1}, $k_{2,1} = 1.2 \times 10^{-2}$ sec^{-1}, $k_{2,4} = 2.83 \times 10^5$ M^{-1} sec^{-1}, $k_{3,2} = 3.1 \times 10^{-2}$ sec^{-1}, and $k_{4,2} = 3.7 \times 10^{-3}$ sec^{-1}.

TABLE II

PARAMETER CORRELATION MATRIX FOR SEVEN-PARAMETER MODEL[a,b]

	$k_{1,5}$	$k_{2,4}$	IC_5	IC_4	IC_3	$k_{2,1}$	$k_{3,2}$
$k_{1,5}$	1.00	—	—	—	—	—	—
$k_{2,4}$	0.23	1.00	—	—	—	—	—
IC_5	0.37	0.09	1.00	—	—	—	—
IC_4	−0.22	−0.46	−0.26	1.00	—	—	—
IC_3	−0.08	0.22	−0.32	−0.10	1.00	—	—
$k_{2,1}$	−0.57	−0.01	0.04	−0.25	−0.15	1.00	—
$k_{3,2}$	−0.12	−0.86	0.04	0.60	−0.32	−0.13	1.00

[a] From Ref. 8.

[b] $k_{4,2}$ is fixed at a value of 0.22 min^{-1}.

show that the minimum model required to describe the kinetics of cleavage of plasmid DNA consisted of five compartments for DNA, one for free enzyme, and contained two enzyme concentration-dependent steps and two enzyme concentration-independent steps. The association of enzyme with superhelical and nicked (relaxed circular) DNA is described by rate constants of similar magnitude. Similarly, the rate constants describing the first and second phosphodiester bond cleavages appear to be similar within a factor of 2. These observation suggest that (1) the topology of the DNA (based on the current data) does not affect the rate of association of the enzyme for the DNA and (2) the cleavage of one strand does not affect the rate of bond cleavage in the adjoining strand. These conclusions are, of course, tentative and await kinetic studies where the topology is systematically varied.

Finally, the mechanism of cleavage, represented by the model, appears to be similar under a variety of substrate conditions, namely, heating the DNA, changing the topology of the DNA, and assaying in the presence of varying concentrations of salt.[9] However, an additional mechanism was required to explain the data for highly supercoiled DNA, where some of the substrate was cleaved by an additional, more rapid mechanism. Details of this more rapid mechanism have yet to be clarified from additional experimental data.

The application of compartmental modeling to the analysis of the steps of the restriction endonuclease-catalyzed reaction demonstrates the facility with which new chemical hypotheses may be formulated and tested using this approach. Here we have shown the sequence which ultimately led to a model which described the data. Along the way there were many models which were tried but which failed in one aspect or another. Much was also learned in these exercises. Knowledge is not necessarily gained when models fit data, but it is when they can be rejected.

[19] Analysis of Site-Specific Interaction Parameters in Protein–DNA Complexes

By Kenneth S. Koblan, David L. Bain, Dorothy Beckett, Madeline A. Shea, and Gary K. Ackers

Introduction

Quantitative determination of the energetics in molecular biological systems is prerequisite to elucidation of the physicochemical forces of functional significance. Many systems, for example, those involving transcriptional control, operate as "combinatorial switches" in which regulatory proteins bind with specific configurations at multiple DNA sites to control function. Experimental data provide a set of "overall" or composite interaction parameters (e.g., a fraction of DNA molecules reacted with protein). These equilibrium parameters measured as a function of thermodynamic variables (e.g., protein concentration, temperature, pH, salt) characterize the system of interest. The measured interaction properties frequently represent averages over site-specific subsystem properties such as local binding and cooperative interactions. Consequently the approach used to analyze these data is often based on some conceptual mechanism or model, thought to represent the system in question. The goal is to estimate a set of values for the "model parameters" which best describes the experimental data. A related issue is the demonstration that the conceptual model is the correct mathematical description of the system under investigation. In this chapter we present examples to illustrate nonlinear least-squares analyses for estimation of such interaction parameters. The examples are from studies that have been carried out in our laboratory during the past few years. Most of these studies have been carried out on the λ cI repressor–operator system.

Interactions between many proteins and specific DNA sequences occur with high affinity, characterized by Gibbs energies in the range -10 to -15 kcal/mol. A number of experimental techniques developed in recent years are capable of monitoring the high-affinity binding typical of proteins which ligate DNA. Included in this group are the filter-binding technique,[1] the gel mobility shift method,[2,3] the gel chromatographic technique,[4] the footprint titration method,[5,6] and the fluorescence technique.[7] Each experimental study provides a different set of relationships between the dependent and independent variables, defined by a particular mathematical function in which the constants of interest are the parameters to be estimated. Resolution of these constants from an experimental data base involves three processes: (i) estimation of the parameter values; (ii) determination of the respective confidence limits for errors associated with the estimated parameter values; and (iii) tests for validity of the analysis and assessment of physical significance.

[1] A. Riggs, H. Suzuki, and S. Bourgeois, *J. Mol. Biol.* **48,** 67 (1970).
[2] M. M. Garner and A. Revzin, *Nucleic Acids Res.* **9,** 3047 (1981).
[3] M. Fried and D. M. Crothers, *Nucleic Acids Res.* **9,** 6505 (1981).
[4] A. D. Frankel, G. K. Ackers, and H. O. Smith, *Biochemistry* **24,** 3049 (1985).
[5] D. J. Galas and A. Schmitz, *Nucleic Acids Res.* **5,** 3157 (1978).
[6] M. Brenowitz, D. F. Senear, M. A. Shea, and G. K. Ackers, this series, Vol. 130, p. 132.
[7] D. E. Draper and L. Gold, *Biochemistry* **19,** 1774 (1980).

Least-Squares Parameter Estimation

The least-squares approach to parameter estimation makes three assumptions: (1) all error in the data can be ascribed to the dependent variable; (2) there are no systematic errors in the data; and (3) random errors of the data conform to a Gaussian distribution. Least-squares analysis[8,9] is a numerical technique for choosing an optimal set of parameters, α, for a function, G, that describes a set of data points, X_i and Y_i:

$$Y_i = G(\alpha, X_i) + \text{experimental noise} \tag{1}$$

The method depends on an initial guess for the search vector α. An algorithm is then applied which iteratively finds better guesses for α until vector α does not change within some predetermined criterion. Least-squares analysis dictates that the precision of the determination of the independent variable (X_i) is significantly greater than, and uncorrelated with, the precision of determining dependent values (Y_i). In an equilibrium ligand-binding experiment, the extent to which ligand binds to the macromolecule of interest (saturation) is some function of the free ligand concentration in solution. The data are normally presented as saturation versus ligand concentration or saturation versus the logarithm of ligand concentration. The parameters to be estimated would be, in their final form, a set of free energies. The error statistics depend on randomness of errors in determination of the dependent variable.

A large number of algorithms for nonlinear least-squares parameter estimation have been published. The program of choice in this laboratory[8,9] employs a variation of the Gauss–Newton procedure[10] in order to determine the best-fit model-dependent parameters which yield a minimum in the variance. The resolved variance ratio is predicted by an F-statistic[11] in order to determine the worst case joint confidence intervals for the estimated parameters. Reported confidence intervals (67%) correspond to approximately one standard deviation.

Simultaneous Analysis of Protein–DNA Interactions

To improve the accuracy of a set of resolved parameters it is advantageous to analyze data from different experiments simultaneously. In such an analysis, information from independent data sets are analyzed as a composite in order to resolve the model parameters that simultaneously

[8] M. Johnson, H. Halvorson, and G. K. Ackers, *Biochemistry* **15**, 5363 (1976).
[9] M. L. Johnson and S. G. Frasier, this series, Vol. 117, p. 301.
[10] F. B. Hildebrand, "Introduction to Numerical Analysis." McGraw-Hill, New York, 1956.
[11] G. D. P. Box, *Ann. N.Y. Acad. Sci.* **86**, 792 (1960).

minimize the least-squares value for the entire composite. Because it is often difficult to assess the magnitude of systematic error that may exist in a given data set, the analysis of multiple data sets, preferably from different experimental techniques, yields greater confidence in the accuracy, or "physical validity," of resolved parameters as well as confidence in the tests employed for a given model. In addition the different data sets may, in composite, provide lower correlation between parameters, thus enhancing their resolvability.

Simultaneous analysis is often crucial to unique resolution of parameters for protein–nucleic acid interactions. These systems frequently involve multiple sites at which ligands bind, often with cooperative interactions. Consequently, the appropriate model equations contain terms that are often highly correlated. A solution to this problem is frequently provided by including binding data not only from the complete system but also from mutant DNA templates in which the number of binding sites has been reduced. In this manner the mathematical expressions containing some or all of the parameters are altered in the simultaneous analysis, leading to better resolution.

This chapter describes the application of nonlinear least-squares techniques to data obtained using footprint titration, filter binding, gel chromatography, and mobility shift techniques for analyzing site-specific binding of proteins to multiple sites on DNA. Most of these applications will be discussed using a common model system, namely, the binding of bacteriophage λ cI repressor to the right operator region (O_R). The λ right operator consists of three specific sites to which cI dimers bind cooperatively.[12,13] Interaction of cI dimers with the three operator sites yields nine configurations (Table I).[14] Species 8 and 9 are isomeric forms of the same triligated operator complex but have different configurations of pairwise cooperativity. Each intrinsic free energy, ΔG_1, ΔG_2, or ΔG_3, reflects binding to the respective site in the absence of binding at the others. ΔG_{ij} is the difference between the total free energy to fill two adjacent sites i and j simultaneously and the sum of the intrinsic binding energies ($\Delta G_i + \Delta G_j$). These cooperativity terms are depicted in Table I as ΔG_{12} and ΔG_{23}.

Individual-Site Binding Expressions for Footprint Titrations

Most classic binding techniques cannot distinguish between interactions at the individual, specific sites on DNA. It is therefore impossible to

[12] A. D. Johnson, B. J. Meyer, and M. Ptashne, *Proc. Natl. Acad. Sci. U.S.A.* **76,** 5061 (1979).

[13] G. K. Ackers, A. D. Johnson, and M. A. Shea, *Proc. Natl. Acad. Sci. U.S.A.* **79,** 1129 (1982).

[14] M. A. Shea, Ph.D. Dissertation, Johns Hopkins University, Baltimore, Maryland (1983).

TABLE I

MICROSCOPIC CONFIGURATIONS AND ASSOCIATED FREE ENERGY CONTRIBUTIONS FOR
THE λ cI REPRESSOR–OPERATOR SYSTEM, $O_R{}^a$

Species	Operator configuration			Free energy contributions
	O_R1	O_R2	O_R3	
1	0	0	0	Reference
2	R_2	0	0	ΔG_1
3	0	R_2	0	ΔG_2
4	0	0	R_2	ΔG_3
5	R_2 ↔	R_2	0	$\Delta G_1 + \Delta G_2 + \Delta G_{12}$
6	R_2	0	R_2	$\Delta G_1 + \Delta G_3$
7	0	R_2 ↔	R_2	$\Delta G_2 + \Delta G_3 + \Delta G_{23}$
8	R_2 ↔	R_2	R_2	$\Delta G_1 + \Delta G_2 + \Delta G_3 + \Delta G_{12}$
9	R_2	R_2 ↔	R_2	$\Delta G_1 + \Delta G_2 + \Delta G_3 + \Delta G_{23}$

[a] Individual operator sites are denoted by 0 if vacant or R_2 if occupied by cI dimers. Cooperative interactions are denoted by (↔). ΔG_i (i = 1, 2, or 3) are the intrinsic Gibbs free energies for binding to each of the three operator sites. ΔG_{ij} are free energies of cooperative interaction between liganded sites. Free energies are related to the corresponding microscopic equilibrium constants, k_i, by the relationship $\Delta G_i = -RT \ln k_i$.

resolve the intrinsic free energies of interaction.[15] However quantitative DNase footprint titration is a rigorous method capable of distinguishing interactions at the individual sites. For a cooperative multisite system such as O_R, even individual-site binding data for the wild-type operator is incapable of uniquely resolving all interaction energies of interest. Combinations of free energies are numerically correlated with one another owing to effects of coupling between the three sites.[15] It is possible, however, to resolve the experimental parameters by simultaneously considering binding data for the wild-type and mutant operators in which the number and types of interaction are altered.

Gibbs energies for both site-specific repressor binding and pairwise interactions are resolved by analyzing data using expressions that separately consider the individual sites. Binding expressions are constructed by considering the relative probability, f_s, of each operator configuration of Table I:

$$f_s = \frac{\exp(-\Delta G_s/RT)[R_2]^j}{\displaystyle\sum_s \exp(-\Delta G_s/RT)[R_2]^j} \tag{2}$$

ΔG_s is the sum of free energy contributions for configuration s, R is the

[15] D. F. Senear, M. Brenowitz, M. A. Shea, and G. K. Ackers, *Biochemistry* **25,** 7344 (1986).

gas constant, T the absolute temperature, $[R_2]$ the concentration of free dimer, and j the repressor stoichiometry in operator configuration s. The DNase I footprint titration technique determines the fractional occupancies of each operator site, $\bar{Y}_i$, as a function of free $[R_2]$. Mathematical expressions for each of these individual-site isotherms can be constructed by summation of the probabilities f_s for the operator configurations of Table I that include bound repressor at the respective site, for example, for the wild-type operator:

$$\bar{Y}_{O_R1} = f_2 + f_5 + f_6 + f_8 + f_9 \qquad (3a)$$
$$\bar{Y}_{O_R2} = f_3 + f_5 + f_7 + f_8 + f_9 \qquad (3b)$$
$$\bar{Y}_{O_R3} = f_4 + f_6 + f_7 + f_8 + f_9 \qquad (3c)$$

Relationships are obtained in an analogous manner for mutant operators where specific binding sites have been eliminated by single base pair substitutions. For example, only configurations 1, 3, 4, and 7 (Table I) can exist for an O_R1^- mutant operator:

$$\bar{Y}_{O_R1} = 0 \qquad (4a)$$
$$\bar{Y}_{O_R2} = f_3 + f_7 \qquad (4b)$$
$$\bar{Y}_{O_R3} = f_4 + f_7 \qquad (4c)$$

For these relationships to be quantitatively compatible with Eqs. (3a)–(3c) the interactions at any competent binding site must be quantitatively unperturbed by the mutation.[12]

The signal in a footprint titration experiment is not the fractional saturation at a given site but rather the degree of protection from DNase I cleavage afforded that site by a given concentration of ligand. Protection at a given site is linearly related to fractional saturation. A saturating concentration of ligand does not offer complete protection at any site, and therefore apparent fractional saturation at each site does not span the range from 0 to 1. For this reason, in our model-dependent fits we treat the data as a transition curve and include the end points as adjustable parameters to be resolved by least-squares analysis. This treatment dictates that each experiment be conducted over a wide concentration range (e.g., 10^{-12} to 10^{-6} M) in order to obtain reliable transition end points and energetics from the complex curve shapes.

Analysis of Experimental Data

Representative data for binding to the individual sites of wild-type O_R and three mutant operators are shown in Fig. 1. The simultaneous analysis of these data sets resolves the three intrinsic free energies, ΔG_1, ΔG_2, ΔG_3, and the two cooperativity terms, ΔG_{12} and ΔG_{23}. The problem of

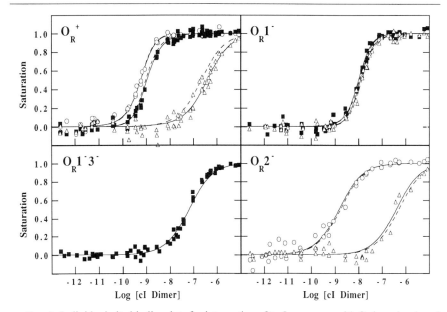

FIG. 1. Individual-site binding data for interaction of λ cI repressor with O_R^+, and reduced valency O_R mutants, at pH 7.00, 200 mM KCl, and 37°. The data shown represent eleven separate experiments. Graphs show O_R^+, O_R1^-, $O_R1^-3^-$, and O_R2^- operators. (○) Site O_R1; (■) site O_R2; (△) site O_R3. Solid curves represent simultaneous analysis of the data in all graphs according to the alternate pairwise coupling model. Dashed curves represent analysis according to the extended pairwise model (see text).

variations in experimental noise between data sets (typically 4–6% in footprint experiments) can be addressed by careful selection of weighting factors. It is possible to determine a reasonable estimate for the amount of experimental noise by analyzing each experiment to obtain the best "phenomenological" fit. Combinations of ΔG_{ij} terms are input as fixed parameters in these separate analyses for cases in which cooperative interactions occur, while ΔG_1, ΔG_2, ΔG_3, and the titration end points are fitted parameters. The shapes of the resolved isotherms are sensitive to changes in ΔG_{ij} over a small range, allowing a relatively rapid approximation of the best combination.[15] The resolved variances for the best combination of ΔG_{ij} terms are assumed to indicate the precision of that set of individual data points. Normalized weighting factors[16] are calculated by

[16] P. R. Bevington, "Data Reduction and Error Analysis for the Physical Sciences." McGraw-Hill, New York, 1969.

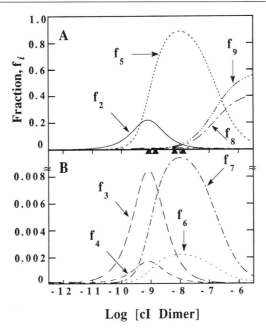

Log [cI Dimer]

FIG. 2. Species populations for the different configurations of bound sites (Table I) at 37°, 200 mM KCl, pH 7.00. Singly ligated species are f_2, f_3, and f_4; doubly ligated species are f_5, f_6, and f_7; triply ligated species are f_8 and f_9. (A) Distributions for all nine species (species 1 reference state) from resolved equilibrium constants and the statistical thermodynamic model of Table I. Symbols (▲) indicate maxima for individual ligation species which are not significantly populated. (B) Distribution of species which are not highly populated in (A). Note that the ordinate scale is 1/1000 that of (A).

$$\sigma_{ij} = N_t(1/\sigma_j^2)/\sum_j (N_j/\sigma_j^2) \tag{5}$$

where σ_{ij} is the weight assigned to data point i of experiment j, σ_j^2 and N_j are the variance and number of data points for experiment j, respectively, and $N_t = \Sigma_j N_j$.

The experimental individual-site isotherms (Fig. 1) are uniquely defined by the concentrations of various liganded species and their dependence on ligand activity [Eq. (3a)–(3c)]. The probability f_s [Eq. (2)], of each of the nine microscopic configurations of Table I represents the fraction of operators which exist in that configuration at a given concentration of repressor dimer. The distribution of ligation species according to the model described in Table I is shown in Fig. 2.

When considering all potential single ligation events, species 2, 3, and 4 of Table I, it is clear that species 2 (cI bound to O_R1) dominates this ligation state. For all doubly liganded species only species 5 (cI bound and cooperatively interacting at sites O_R1 and O_R2) is significantly populated. It is important to note that the doubly liganded species 7 (cI bound and cooperatively interacting at sites O_R2 and O_R3) is never highly populated (Fig. 2B). In this system species 7 and in particular the ΔG_{23} term contribute virtually no "information" to the individual-site isotherms. However, experiments conducted on the reduced valency template O_R1^- do allow resolution of the ΔG_{23} term.

The nature of free energy coupling in this and similar systems[17] dictates that the intermediate ligation states ($j = 1$ or 2) are highly unpopulated or dominated by a single ligation species. The consequence is that it is extremely difficult to discern between mechanistic models which embody subtle differences between these unpopulated ligation species. Such is the case in the λ O_R system. Only by considering the three-site problem as a subset of two-site problems and employing reduced valency templates (O_R1^-, O_R2^-, $O_R1^-3^-$) are we able to resolve all of the relevant interaction parameters.

The individual-site binding isotherms (for wild-type and mutant operators) resolved in each experiment provide a unique measure of the total chemical work to bind a mole of ligand at the respective site, while stoichiometric amounts of ligand are also bound to other sites. This quantity is the individual-site "loading" free energy $(\Delta G_{L,i})$[18] and includes all contributions that affect binding of a ligand at site i. The loading free energies can be evaluated directly from the individual-site isotherms, in a model-independent manner by numerical integration:

$$\Delta G_{L,i} = RT \ln \bar{X}_i = RT \int_0^1 \ln X \, d\bar{Y}_i \tag{6}$$

where $\bar{X}_i$ is the median ligand activity[19] and $\bar{Y}_i$ is the fractional saturation at ligand activity X.

It is possible to compare the resolved $\Delta G_{L,i}$ values between experiments in a completely model-independent test of reproducibility (Table II). As illustrated in Table II there is close agreement between separate experiments and with the isotherms resolved from the model-dependent nonlinear least-squares analysis. The sum of the individual-site loading free energies equals the total free energy of interaction within the system: $\Delta G_T = \Delta G_{L,1} + \Delta G_{L,2} + \Delta G_{L,3}$. This total must reflect the product of the microscopic interaction constants resolved by any model-dependent

[17] G. K. Ackers and F. R. Smith, *Annu. Rev. Biophys. Biophys. Chem.* **16**, 583 (1987).
[18] G. K. Ackers, M. A. Shea, and F. N. Smith, *J. Mol. Biol.* **170**, 223 (1983).
[19] J. Wyman, Jr., *Adv. Protein Chem.* **19**, 224 (1964).

TABLE II

INDIVIDUAL-SITE LOADING FREE ENERGIES ($\Delta G_{L,i}$) FROM
SEPARATE EXPERIMENTS FOR EACH OPERATOR[a]

Operator	Experiment[b]	$\Delta G_{L,1}$	$\Delta G_{L,2}$	$\Delta G_{L,3}$
$O_R{}^+$	Experiment	-13.05	-12.60	-9.61
	Model[c]	-13.08	-12.66	-9.69
O_R1^-	Experiment	—	-11.58	-11.39
	Model[c]	—	-11.53	-11.38
O_R2^-	Experiment	-12.46	—	-9.47
	Model[c]	-12.54	—	-9.66
$O_R1^-3^-$	Experiment	—	-10.62	—
	Model[c]	—	-10.54	—

[a] $\Delta G_{L,i}$ values (in kilocalories per mole) were calculated by
numerical integration using Eq. (6).
[b] The fractional saturation data for each isotherm were used
in the calculation.
[c] Calculated from isotherms resolved from model-dependent
simultaneous analysis of data from all operators.

analysis {e.g., $\Delta G_T = -RT \ln[k_1 k_2 k_3(k_{12} + k_{23})]$}. The sum of $\Delta G_{L,i}$ values
are found to agree quite well with the sum of model-dependent ΔG_i and
ΔG_{ij} values (e.g., at 37° $\Sigma \Delta G_{L,i} = -35.4$ kcal/mol versus $\Sigma \Delta G_{T,model-}$
$_{dependent} = -35.3$ kcal/mol). This close agreement provides an important
verification that the model employed is physically correct.

To conduct a simultaneous analysis it is critical to define which parameters are common to each isotherm and which form of the binding equations
[Eqs. (3a)–(3c) and (4a)–(4c)] are to be applied. As an example, the appropriate sets of simultaneous equations [e.g., Eqs. (3a)–(3c) and (4a)–(4c)]
were solved for various combinations of the wild-type and mutant operator
data in Fig. 1. The combined data set for all four operators (or for the
subset including $O_R{}^+$, O_R1^-, and $O_R1^-3^-$) uniquely resolved the values
of all five interaction energies, which are given in Table III. Resolution of
all the interaction parameters was not possible with any other subset of
the four templates. For these cases the cooperative energies (ΔG_{12}, ΔG_{23})
were fixed, equal to the values determined from all the data, and the
intrinsic free energies were estimated. Agreement between results from
all combinations of the operators was excellent. Confidence intervals for
the intrinsic free energies are small, 0.2–0.3 kcal/mol. Confidence intervals
reflect the cumulative systematic errors between both different experiments and different operators. The fact that confidence intervals for the
ΔG_{ij} terms are only slightly less precise (0.3–0.5 kcal/mol) than the intrin-

TABLE III

MICROSCOPIC GIBBS ENERGIES OF REPRESSOR–O_R INTERACTIONS RESOLVED BY FOOTPRINT TITRATION OF O_R^+ AND REDUCED VALENCY MUTANT OPERATORS[a]

Operators	ΔG_1	ΔG_2	ΔG_3	ΔG_{12}	ΔG_{23}	s[b]
O_R^+, O_R1^-, O_R2^-, $O_R1^-3^-$[c]	-12.5 ± 0.2	-10.5 ± 0.3	-9.4 ± 0.3	-2.9 ± 0.4	-2.9 ± 0.5	0.049
O_R^+, O_R1^-, O_R2^-, $O_R1^-3^-$	-12.5 ± 0.3	-10.5 ± 0.2	-9.5 ± 0.2	-2.7 ± 0.3	-2.9 ± 0.5	0.047
O_R^+, O_R1^-, $O_R1^-3^-$	-12.6 ± 0.3	-10.5 ± 0.2	-9.5 ± 0.3	-2.6 ± 0.4	-2.8 ± 0.5	0.041
O_R^+, O_R1^-, O_R2^-	-12.5 ± 0.3	-10.6 ± 1.0	-9.5 ± 0.3	-2.6 ± 1.1	-2.8 ± 1.1	0.047
O_R^+, O_R2^-, $O_R1^-3^-$	-12.6 ± 0.2	-10.5 ± 0.2	-9.5 ± 0.3	(-2.7)	(-2.9)	0.046
O_R^+, O_R1^-	-12.6 ± 0.3	-10.4 ± 0.3	-9.6 ± 0.2	(-2.7)	(-2.9)	0.042
O_R^+, O_R2^-	-12.5 ± 0.3	-10.5 ± 0.3	-9.5 ± 0.2	(-2.7)	(-2.9)	0.049
O_R^+, $O_R1^-3^-$	-12.6 ± 0.3	-10.5 ± 0.2	-9.6 ± 0.3	(-2.7)	(-2.9)	0.042
O_R^{++}	-12.6 ± 0.3	-10.4 ± 0.3	-9.6 ± 0.6	(-2.7)	(-2.9)	0.043
O_R1^-	—	-10.5 ± 0.4	-9.5 ± 0.4	—	(-2.9)	0.041

[a] Free energies (in kilocalories per mole) are presented with 67% confidence intervals according to the alternate pairwise cooperativity model (see text) except as noted. Values in parentheses were fixed for that analysis.

[b] Square root of the variance of the fitted curves.

[c] Free energies were resolved according to the extended pairwise coupling model.

sic parameters is evidence of the reproducibility and internal consistency obtainable with this technique.

The earliest quantitative analyses of the cI–O_R interactions employed a statistical mechanical model[13] which contained eight operator species. That model embodies features of coupling at O_R proposed by Johnson *et al.*[12] One of the assumptions of the earliest model was the "selection rule" which states that cooperative interactions between adjacently bound repressors at O_R2 and O_R3 occurs only when O_R1 is vacant. In our analysis we have employed two different coupling methods: (1) the alternate pairwise model in which a "selection rule" restricts cooperative interactions to sites O_R1 and O_R2 in the triply liganded operator (lines 1–8 of Table I); and (2) the extended pairwise model in which there exists no selection rule. This gives rise to species 9 in Table I.[14] For this data set (Fig. 1) the isotherms pertaining to the alternate pairwise model (solid lines) give a better description of the experimental data than the extended pairwise model (dashed lines), with the major difference found in the O_R3 isotherms. The significance of this comparison is difficult to assess given the noise of the data. Results obtained in this laboratory at different experimental conditions[20] (Fig. 3) indicate that the extended pairwise model fits the experimental data better than does the alternate pairwise model. A combination of additional experimental data and a correct coupling model will yield internally consistent results over a wide range of experimental conditions.

To verify that the parameters resolved and the model employed are physically meaningful, it is important to consider several lines of evidence. (1) The parameters resolved from the simultaneous analysis describe each of the individual data sets (wild type and mutant). This indicates that our model describes the subset of interactions which occur in mutant templates very well. (2) The individual-site loading free energies calculated for each isotherm agree quite well with the parameters resolved in the model-dependent least-squares analysis. (3) The random distribution in the residuals for each of the different operators indicates that the individual data points are independent of one another.

Analytical Gel-Permeation Chromatography

An important feature among regulatory proteins studied to date is the observation that many bind as oligomers. In the λ system, as in many others, there are coupled equilibria between oligomer association and DNA binding. Physicochemical dissection of coupled protein–protein and

---

[20] D. F. Senear and G. K. Ackers, *Biochemistry* **29**, 6568 (1990).

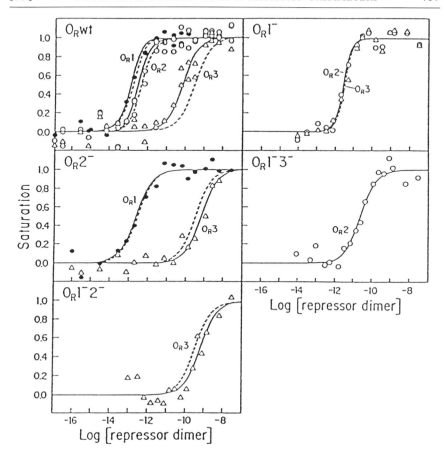

FIG. 3. Individual-site binding of cI repressor to O_R^+, and to reduced valency O_R mutants, at pH 5.00, 200 mM KCl, and 20°. Graphs show O_R^+, O_R1^-, O_R2^-, $O_R1^-3^-$, and $O_R1^-2^-$ operators. (●) Site O_R1; (○) site O_R2; (△) site O_R3. The solid curves represent the simultaneous analysis of the data in all graphs using the extended pairwise model; dashed curves represent analysis according to the alternate pairwise model.

protein–DNA interactions must include quantitative measurements of all relevant protein polymerization reactions.[21–23] Lack of independent information concerning the state and energetics of polymerization would introduce a systematic error into the independent variable (active ligand concentration). This systematic error can result in highly inaccurate (even

[21] K. S. Koblan and G. K. Ackers, *Biochemistry* **30**, 7817 (1991).
[22] K. S. Koblan and G. K. Ackers, *Biochemistry* **30**, 7822 (1991).
[23] K. S. Koblan and G. K. Ackers, *Biochemistry* in press (1992).

though precise) parameter estimates, or incorrect mechanistic interpretation of results.

Analytical gel chromatography[24] provides an especially sensitive and accurate method for studying the assembly reaction of cI repressor. A version of this technique recently developed in this laboratory permits analysis from the picomolar to micromolar range of protein concentration.[25] In a "large zone" experiment a sample of protein solution is applied to a sieving column to establish a plateau in the elution profile of concentration versus volume (Fig. 4A). For an associating system, as the plateau concentration C_T is varied, a shift occurs in the relative populations of each species according to the law of mass action. Each species has a characteristic partition coefficient σ, which is a measure of the extent of solute penetration within the interior solvent region of the gel. For systems containing multiple species,

$$\bar{\sigma}_w = \sum_i f_i \sigma_i \tag{7}$$

where $\bar{\sigma}_w$ is the weight average partition coefficient for all species and f_i is the fraction of each species. Elution volumes of large zones are determined as the equivalent sharp boundaries (centroids) of the leading or trailing edges of the solute profile. The centroid elution volume V_e, provides a measure of $\bar{\sigma}_w$:

$$V_e = V_0 + \bar{\sigma}_w V_i \tag{8}$$

where V_0 and V_i are the void and internal volumes of the column, respectively. The resulting dissociation curve of $\bar{\sigma}_w$ versus concentration may then be analyzed to resolve the stoichiometry and equilibrium constants for the interactions between species. For the λ cI repressor a monomer–dimer stoichiometry was found to provide the best fit to the data (Fig. 4b).[21]

Binding Expressions for Gel Mobility Shift Method

The gel mobility shift assay[2,3] experimentally measures the fraction of DNA molecules with i ligands bound (θ_i, $i = 0$ to n for a macromolecule with n sites). In the absence of ligand-induced bending or kinking of the DNA at specific sites, the technique resolves each stoichiometric complex independent of the configuration of the complex. The expressions for repressor binding to the three-site right operator determined by the gel mobility shift method are

[24] R. Valdes, Jr., and G. K. Ackers, this series, Vol. 61, p. 125.
[25] D. Beckett, K. S. Koblan, and G. K. Ackers, *Anal. Biochem.* **196,** 69 (1991).

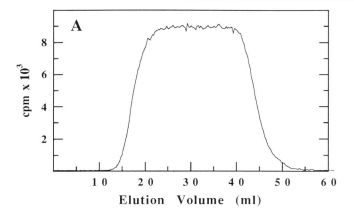

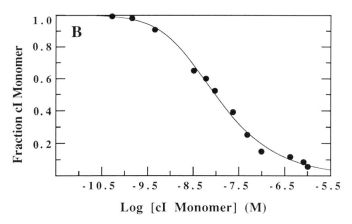

FIG. 4. (A) Characteristic large zone profile for λ cI repressor, plateau concentration, $C_T = 5.1 \times 10^{-11}\ M$, pH 7.00, 20°. (B) Representative dissociation curve at pH 7.0, 200 mM KCl, and 37°. Circles (●) correspond to fraction monomer values determined by chromatography. The solid curve represents parameters resolved by least-squares analysis according to a monomer–dimer model.

$$\theta_0 = 1/(1 + K_1[R_2] + K_2[R_2]^2 + K_3[R_2]^3) \tag{9a}$$

$$\theta_1 = K_1[R_2]/(1 + K_1[R_2] + K_2[R_2]^2 + K_3[R_2]^3) \tag{9b}$$

$$\theta_2 = K_2[R_2]^2/(1 + K_1[R_2] + K_2[R_2]^2 + K_3[R_2]^3) \tag{9c}$$

$$\theta_3 = K_3[R_2]^3/(1 + K_1[R_2] + K_2[R_2]^2 + K_3[R_2]^3) \tag{9d}$$

where the K_i terms are macroscopic equilibrium constants. Each of these parameters is a sum over the terms for each species with i ligands bound.

The following relationships between the K_i values and microscopic equilibrium constants k_1, k_2, k_3, k_{12}, and k_{23} (corresponding to the Gibbs energies of Table I) can be written:

$$K_1 = k_1 + k_2 + k_3 \tag{10a}$$
$$K_2 = k_1k_2k_{12} + k_1k_3 + k_2k_3k_{23} \tag{10b}$$
$$K_3 = k_1k_2k_3(k_{12} + k_{23}) \tag{10c}$$

It is important to note that this technique samples a different set of operator configurations than does a footprint titration ($\theta_0 = f_1$; $\theta_1 = f_2 + f_3 + f_4$; $\theta_2 = f_5 + f_6 + f_7$; $\theta_3 = f_8 + f_9$ of Table I). Therefore, the energetic parameters to be resolved by the least-squares procedure will have different correlation properties and altered resolvabilities. The wild-type isotherms define three composite macroscopic parameters which contain combinations of the microscopic interaction constants. The macroscopic constants are incapable of uniquely resolving the five microscopic interaction free energies. Simultaneous analysis of data obtained using a mutant operator in which the number and type of interactions has been reduced allows resolution of the microscopic interaction constants. The binding expressions for an O_R1^- operator define two additional parameters, $K_1' = k_2 + k_3$ and $K_2' = k_2k_3k_{23}$

$$\theta_0' = 1/(1 + K_1'[R_2] + K_2'[R_2]^2) \tag{11a}$$
$$\theta_1' = K_1'[R_2]/(1 + K_1'[R_2] + K_2'[R_2]^2) \tag{11b}$$
$$\theta_2' = K_2'[R_2]^2/(1 + K_1'[R_2] + K_2'[R_2]^2) \tag{11c}$$

which are in principle sufficient to resolve all five microscopic constants.

It should be noted, however, that although the theory for analysis of gel mobility shift data is easily applicable, the accuracy of the resolved parameters rests on the assumption that the sampled distribution represents the equilibrium state. Failure to demonstrate that the equilibrium state is not perturbed during electrophoresis can result in grievous underestimations of parameters. This issue of generating valid and accurate data is crucial to dissecting the physicochemical forces responsible for the strength and specificity of protein–DNA interactions. Our laboratory is currently developing a novel approach in using the gel mobility shift assay as a technique for rigorous biophysical studies of protein–DNA systems.[26]

Binding Expressions for Filter-Binding Method

The filter-binding technique is based on the observation that protein ligands bound to double-stranded DNA are retained by nitrocellulose filters while unliganded DNA is not.[1] The fraction of DNA retained by the

---

[26] K. S. Koblan, D. L. Bain, and G. K. Ackers, manuscript in preparation.

filter as a function of free ligand concentration is described by Clore *et al.*[27]:

$$\theta = [\text{DNA}_{\text{Bound}}]/[\text{DNA}_{\text{Total}}] = (Z - 1)/Z \tag{12}$$

where Z represents the binding polynomial[18] given by

$$Z = \sum_{i=0}^{n} K_i[\text{L}]^i \tag{13}$$

K_i is again the macroscopic association constant for the binding of i ligands, [L] is the free ligand concentration, and n is the number of binding sites on the macromolecule. The binding expression for the three site λ operator is

$$\theta = \frac{K_1[\text{R}_2] + K_2[\text{R}_2]^2 + K_3[\text{R}_2]^3}{1 + K_1[\text{R}_2] + K_2[\text{R}_2]^2 + K_3[\text{R}_2]^3} \tag{14}$$

where K_1, K_2, and K_3 are defined by Eqs. (10a)–(10c). The three macroscopic equilibrium constants of Eq. (14) are incapable in principle of uniquely defining the five microscopic interaction free energies of Table I.

A complication in analysis of experimental filter-binding data is the observation that protein–DNA complexes are retained by the filters with less than 100% efficiency. The experimental signal is the fraction of DNA (F_R) retained by the filters. To compare the experimental data to the macroscopic constants K_1, K_2, and K_3, a term which accounts for the probability of retention, r_i, for each state of ligation must be included:

$$F_R = \frac{r_1 K_1[\text{R}_2] + r_2 K_2[\text{R}_2]^2 + r_3 K_3[\text{R}_2]^3}{1 + K_1[\text{R}_2] + K_2[\text{R}_2]^2 + K_3[\text{R}_2]^3} \tag{15}$$

Woodbury and von Hippel[28] have defined a retention efficiency parameter based on the assumption that each liganded site on the DNA exhibits a fixed and independent probability of retention. An alternative model assumes the retention efficiency for any liganded DNA as a constant, that is, r_i is independent of the stoichiometry of the complex.[15]

Analysis of the relative filter-binding retention efficiencies of ligation states for the λ system has been discussed elsewhere.[15] The macroscopic equilibrium constants (K_i) and the retention factors (r_i) of Eq. (15) are highly correlated with one another. Variations in any one of these parameters is compensated by a corresponding change in one or more of the others. Least-squares analysis of the experimental data reveals that there

[27] G. M. Clore, A. M. Gronenborn, and R. W. Davies, *J. Mol. Biol.* **155**, 447 (1982).
[28] C. P. Woodbury and P. H. von Hippel, *Biochemistry* **22**, 4730 (1983).

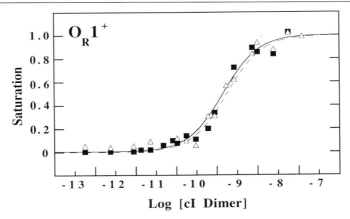

FIG. 5. Comparison of footprint titration and filter-binding titration of repressor binding to O_R1^+. The triangles are footprint titration measurements, and the solid curve is the fit of Eq. (16), which yields $\Delta G_1 = -12.5 \pm 0.1$ kcal/mol ($\text{Var}^{1/2} = 0.06$). The squares are the means of replicate filter-binding experiments, scaled by the maximum retention efficiency to yield fractional saturation. Analysis yields $\Delta G_1 = -12.4 \pm 0.2$ kcal/mol ($\text{Var}^{1/2} = 0.07$; dashed curve).

is no simple relationship between the number of ligands bound and the retention efficiency of the complex. The retention factors, r_i, are themselves complex functions of ligand concentration. Hence, the filter-binding technique does not permit valid determination of even macroscopic interaction constants in multisite systems.

Application of the filter-binding technique to a single-site system is valid and unambiguous. The fractional saturation ($\bar{Y}_1$) and the fraction of liganded DNA (θ) are both given by the familiar Langmuir isotherm:

$$\bar{Y}_1 = \theta = \frac{k_1[R_2]}{1 + k_1[R_2]} \tag{16}$$

In this case the retention efficiency r_1 is simply a normalization factor which relates F_R to θ and can be resolved by the least-squares analysis. Comparison of footprint titration and filter-binding titrations of cI to a single O_R1 site is shown in Fig. 5. The isotherms resolved by the two techniques and the experimental precision are identical, attesting to the accuracy of both techniques and the validity of the filter-binding method when applied to single-site systems.

Gel Chromatographic Technique for Protein–DNA Binding

Analytical gel-permeation chromatography also affords a simple and convenient approach to the accurate determination of equilibrium binding constants of protein–DNA complexes.[4] In this technique a small zone of protein is chromatographed on a gel permeation column that has been preequilibrated with a solution of DNA. Conditions can be arranged so that the DNA molecules and their complexes with protein will occupy only the void space V_0, while unbound protein will readily partition into the interior solvent spaces of the gel. Hence, when a "small zone" sample containing the protein is passed through the preequilibrated column a competition is set up for the fraction of time that the protein spends in the interior stationary phase of the column versus that spent in the mobile void phase. This partitioning of the protein between DNA sites and interior gel spaces is reflected in the experimental elution volumes according to Eq. (17), written for a nonassociating protein binding noncooperatively to DNA:

$$f = \frac{V_p - V_c}{V_c - V_d} \tag{17}$$

where f is the fraction of protein bound to the DNA, V_p is the elution volume of protein in the absence of DNA (usually determined by separate experiment), V_d is the elution volume of DNA by itself (usually the column void volume, V_0), and V_c is the elution volume of protein in the presence of DNA. The experiment is arranged such that the concentrations of free (i.e., unbound) DNA and total DNA are effectively equal. Then the position of the equilibrium in this experiment is determined solely by the concentration [D] of DNA with which the column has been equilibrated. The fraction f of protein bound to the DNA is given by

$$f = \frac{K[D]}{1 + K[D]} \tag{18}$$

where [D] is the concentration of DNA sites and K is the equilibrium association constant per site. Note the contrast between this binding expression and the more conventional fraction of binding sites occupied, as given by Eq. (16). In the gel chromatography technique [D] is varied in a series of experiments, and the corresponding values of f are determined as the right-hand side of Eq. (17). With fixed V_p and V_d, Eqs. (17) and (18) are parametric relations that determine a best value of K from simultaneous fitting of all the experimentally derived pairs of V_c and [D]. Figure 6 shows such a binding isotherm for the interaction of *Hin*fI restriction

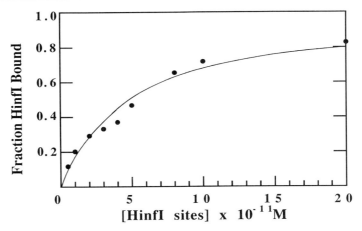

FIG. 6. Binding isotherm for the interaction of *Hin*fI restriction endonuclease with pBR322 supercoiled DNA. DNA concentration is expressed as the concentration of *Hin*fI sites (10 per molecule).

endonuclease with pBR322 supercoiled DNA. Table IV shows fitted values obtained by this technique for binding of the *Hin*fI protein to a series of different DNA substrates, exhibiting a wide range of equilibrium constants while using only catalytic amounts of protein. The method is best suited to systems such as enzymes that have an assay independent of the binding itself.

Delineation of Thermodynamic Driving Forces

The methods described in this chapter provide tools for resolving energetic contributions of the various types of interactions characteristic

TABLE IV
BINDING CONSTANTS RESOLVED
FOR *Hin*fI–DNA INTERACTIONS

Substrate	K_{obsd} (M^{-1})
pBR322 supercoiled DNA	$(1.98 \pm 0.36) \times 10^{10}$
pBR322 *Taq*I fragments	$(3.88 \pm 1.45) \times 10^{10}$
pBR322 *Hin*fI fragments	$(5.20 \pm 0.48) \times 10^{4}$
pDI10 supercoiled DNA	$<1.43 \times 10^{5a}$

[a] This value is an upper limit due to the high substrate concentrations required to observe binding.

TABLE V
THERMODYNAMIC LINKAGE OF O_R AT 20°

System[a]	$\Delta H°$ (kcal/mol)	$T\Delta S°$ (kcal/mol)	Δv_{H^+} "absorbed"	Δv_{KCl} "released"
O_R1	− 23.3	− 10.1	0.4	3.7
O_R2	− 14.7	− 4.1	1.1	5.2
O_R3	− 22.7	− 12.5	0.6	3.8
Cooperativity	~0.0	+ 3.0	0.0	0.0
Dimerization	− 15.9	+ 4.9	0.3	− 1.5

[a] $\Delta v_{H^+} = (d \ln k)/(d \ln a_{H^+})$; $\Delta v_{KCl} = (d \ln k)/(d \ln [KCl])$. "Cooperativity" denotes values derived from the cooperative interaction parameters ΔG_{12} and ΔG_{23}. "Dimerization" indicates cI dimerization equilibrium parameters.

of these systems: ligand binding, protein assembly, and cooperativity (conformational changes are also implicit). These "primary" energetic terms in turn contain contributions from interactions with ions (salt, protons), plus any other small molecules including solvent. Therefore, it is often desirable to study processes outlined in this review as a function of thermodynamic variables including temperature and the chemical potentials of the interacting small species. An example of such a dissection into "linked functions" is shown in Table V. Experimental results from footprint titration and analytical gel chromatography studies conducted on λ cI repressor and O_R were carried out as a function of pH, temperature, and [KCl].[20–23] Experimental delineation of these linkages provides a crucial basis for interpreting the roles of the various structural elements in the functional mechanism.

Acknowledgments

We thank Michael D. Brenowitz, Alan Frankel, Bertrand Garcia-Moreno, and Donald F. Senear for contributions to various aspects of research conducted in our laboratory on the problem reviewed in this chapter. This work was supported by National Institutes of Health Grants GM39343 and R37-GM24486.

[20] Analysis of Circular Dichroism Spectra

By W. Curtis Johnson, Jr.

Introduction

In general, molecules absorb light when they undergo a transition from one state to a higher energy state. Here we are concerned with electronic absorption, that is, the spectra that can be measured for a molecule undergoing a transition from its ground electronic state to some higher energy electronic state. Most biological molecules contain a number of electronic units that absorb light nearly independently, called chromophores, which are asymmetrically disposed in space. Such asymmetric molecules will absorb left circularly polarized light differently from right circularly polarized light. Circular dichroism (CD) is the difference in absorption between left and right circularly polarized light, so that CD spectral bands occur wherever there are normal electronic absorption bands in an asymmetric molecule.

Normal absorption spectroscopy obeys Beer's law:

$$A(\lambda) = \varepsilon(\lambda)lc \tag{1}$$

where A is the measured absorption that is unitless and varies with wavelength λ, l is the path length of the sample cell in centimeters, c is the concentration of the sample in moles liter^{-1}, and ε is the characteristic of the molecule in liters mole^{-1} cm^{-1} called the extinction coefficient. A graph of ε versus λ gives bands of a finite width, and the integrated intensity under these bands is measured by the dipole strength,

$$D = \frac{3 \times 10^3 hc \ln 10}{8\pi^3 N_0} \int \frac{\varepsilon(\lambda)}{\lambda} d\lambda \tag{2}$$

where h is Planck's constant, c is the speed of light, and N_0 is Avogadro's number.

In CD spectroscopy, both left (L) and right (R) circularly polarized light obey Beer's law, so the difference obeys the following equation:

$$A_L(\lambda) - A_R(\lambda) = \Delta A(\lambda) = [\varepsilon_L(\lambda) - \varepsilon_R(\lambda)]lc = \Delta\varepsilon(\lambda)lc \tag{3}$$

The integrated intensity under CD bands is measured by the rotational strength,

$$R = \frac{3 \times 10^3 hc \ln 10}{32\pi^3 N_0} \int \frac{\Delta\varepsilon(\lambda)}{\lambda} d\lambda \tag{4}$$

Theoretically, the rotational strength can be calculated quantum mechanically from interactions among the transitions in the various chromophores. However, as yet such calculations are not sufficiently accurate for interpretation of CD spectra. Thus CD spectra are usually interpreted empirically, and this chapter discusses two mathematical methods that will help in empirical analysis. Taylor series fitting separates the two distinct types of interactions between chromophores that give rise to CD spectra. For oligomers and polymers, this separation gives insight into the type of secondary structure, facilitates comparison among different molecules, or provides an accurate measure of the changes in a specific molecule when some parameter is altered. Taylor series fitting has been applied primarily to oligo- and polynucleotides as a function of solution conditions. Singular value decomposition (SVD) is an extremely powerful method that can be applied to any series of any type of spectra. It will give the number of independent components, can be used to average among the spectra, and will provide an integrated parameter that varies with changes in experimental conditions, such as temperature and ligand concentration. It has even been used to analyze the CD of proteins for secondary structure.

This chapter is restricted to mathematical methods for the analysis of CD spectra. Many other reviews of CD spectroscopy are available, including a very basic and detailed treatise on CD[1] and articles covering applications to nucleic acids,[2,3] proteins,[4-7] and carbohydrates.[8]

Fitting with Taylor Series

Nearly all asymmetric molecules can be divided into symmetric chromophores; then CD bands may be considered to arise from interactions among the transitions of these symmetric chromophores, which are asymmetrically disposed. The interactions are coulombic in nature and are the interactions among the charge distributions that result from the transitions. Interacting transitions may occur at similar energies, in which case they

[1] W. C. Johnson, Jr., in "Methods of Biochemical Analysis" (D. Glick, ed.), Vol. 31, p. 61. Wiley, New York, 1985.

[2] I. Tinoco, Jr., C. Bustamante, and M. F. Maestre, Annu. Rev. Biophys. Bioeng. 9, 107 (1980).

[3] I. Tinoco, Jr., and A. L. Williams, Annu. Rev. Phys. Chem. 35, 329 (1984).

[4] R. W. Woody, in "The Peptides" (E. R. Blount, F. A. Bovey, M. Goodman, and N. Lotan, eds.), Vol. 7, p. 15. Academic Press, New York, 1985.

[5] W. C. Johnson, Jr., Annu. Rev. Biophys. Biophys. Chem. 17, 145 (1988).

[6] W. C. Johnson, Jr., Proteins 7, 205 (1990).

[7] J. T. Yang, C.-S. C. Wu, and H. M. Martinez, this series, Vol. 130, p. 208.

[8] W. C. Johnson, Jr., Adv. Carbohydr. Chem. Biochem. 45, 73 (1987).

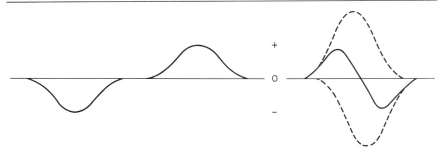

FIG. 1. Nondegenerate interactions (left) and degenerate interactions (right) between two chromophores yield two CD bands of equal magnitude and opposite sign.

are called degenerate, or at different energies, in which case they are called nondegenerate.

The interaction between two nondegenerate transitions produces a negligible increase in their energy difference and, for an asymmetric molecule, CD intensity in both of the transitions. The rotational strength induced in one transition will be equal in magnitude and opposite in sign to the rotational strength produced in the other transition, as illustrated in Fig. 1. The shape of the CD bands will be similar to the corresponding absorption bands.

The interaction between two degenerate transitions produces two new transitions that are split in energy. The splitting will be twice the coulombic interaction between the transition charge distributions, which is at most a few thousand cm^{-1}. Again, for an asymmetric molecule the rotational strength corresponding to each of the new transitions will be equal in magnitude and opposite in sign, and will have a shape similar to the corresponding absorption bands. Because most molecular absorption bands are broad compared to the splitting, there is a great deal of overlap and the CD bands largely cancel (Fig. 1). Thus, the observed CD from the degenerate interaction between two transitions resembles the derivative of the absorption band. As a practical matter, the observation of a derivative-shaped CD band immediately tells the CD spectroscopist that there are degenerate interactions. In proteins these are the amide–amide interactions one might expect from the formation of an α helix or a β sheet. For nucleic acids these are the interactions one might expect in a helical structure with stacked bases.

Of course, a measured CD spectrum is the net result of all the interactions among all of the transitions. The degenerate interactions may be

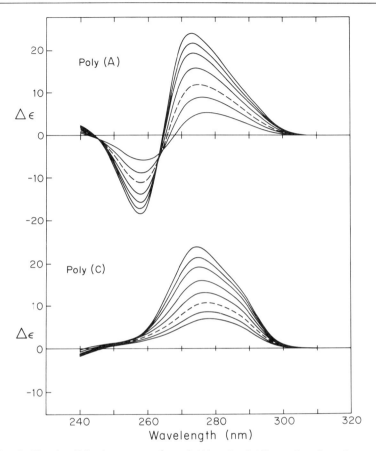

Fig. 2. Circular dichroism spectra for poly(A) and poly(C) as a function of temperature. [Redrawn from G. C. Causley, P. W. Staskus, and W. C. Johnson, Jr., *Biopolymers* **22**, 945 (1983).]

obvious as we see in Fig. 2 for poly(A), or they may be completely obscured by the nondegenerate interactions as we see for poly(C).[9]

The fact that a CD spectrum is made up of a nondegenerate part with the shape of the absorption spectrum and a degenerate part with the shape of the derivative of the absorption spectrum led Tinoco[10] to expand the CD spectrum of an oligomer as the first two terms of a Taylor series. The

[9] G. C. Causley, P. W. Staskus, and W. C. Johnson, Jr., *Biopolymers* **22**, 945 (1983).
[10] I. Tinoco, *J. Chem. Phys.* **65**, 715 (1968).

first term will contain the CD bands with the shape of the monomer absorption, while the second term will contain the CD bands with the shape of the derivative of the monomer absorption. It is not necessary to understand the derivation to make use of the Taylor series expansion. Nevertheless, a brief derivation is given below for completeness. A more detailed derivation, which is somewhat different from the original treatment by Tinoco,[10] can be found in Ref. 9.

Derivation of Taylor Series Expansion

The absorption band for a monomer with the maximum at ν_a can be written in the form

$$\varepsilon_m(\nu) = D_a \nu f(\nu - \nu_a) \tag{5}$$

where D_a is the dipole strength of the band, as in Eq. (2). Because we assume the corresponding CD band has the same shape,

$$\Delta\varepsilon_m(\nu) = 4R_a \nu f(\nu - \nu_a) \tag{6}$$

where the definition for the rotational strength, R_a, in Eq. (4) gives rise to the factor of 4.0.

An oligomer with N monomeric units will have N oligomer transitions corresponding to the monomer transition at ν_a. Summed together through index k, they give the net oligomer CD:

$$\Delta\varepsilon_0(\nu) = \frac{4\nu}{N} \sum_{k=1}^{N} R_k f(\nu - \nu_k) \tag{7}$$

The Taylor series expansion for the function $f(\nu - \nu_k)$ is carried out in incremental form about $\nu - \nu_a$. Keeping only the first two terms, we have

$$f(\nu - \nu_k) = f(\nu - \nu_a) + (\nu_a - \nu_k)\frac{\partial f(\nu - \nu_a)}{\partial \nu} \tag{8}$$

Equation (5) gives $f(\nu - \nu_a)$ in terms of the monomer absorption band. Differentiating Eq. (5) yields

$$\frac{1}{D_a}\left(\frac{\partial\varepsilon_m(\nu)}{\partial\nu}\right) = \frac{\partial[\nu f(\nu - \nu_a)]}{\partial\nu} = \nu\frac{\partial f(\nu - \nu_a)}{\partial\nu} + f(\nu - \nu_a) \tag{9}$$

which gives $\partial f(\nu - \nu_a)/\partial\nu$ in terms of the monomer absorption band and its derivative. Thus Eq. (8) for $f(\nu - \nu_k)$ can be written in terms of the monomer absorption band and its derivative, so that Eq. (7) for the CD of the oligomer now becomes

$$\Delta\varepsilon_0(\nu) = \frac{4}{ND_a} \sum_{k=1}^{N} \left\{ R_k \varepsilon_m(\nu) + [(\nu_a - \nu_k)R_k] \left[\frac{\partial \varepsilon_m(\nu)}{\nu} - \frac{\varepsilon_m(\nu)}{\nu} \right] \right\} \quad (10)$$

The nondegenerate interactions are measured by the coefficient of the first term,

$$A = \frac{4}{ND_a} \sum_{k=1}^{N} R_k \quad (11)$$

The degenerate interactions are measured by the coefficient of the second term,

$$B = \frac{4}{ND_a} \sum_{k=1}^{N} (\nu_a - \nu_k)R_k \quad (12)$$

Thus the Taylor series expansion for the CD of an oligomer may be written in the following final form

$$\int \Delta\varepsilon_0(\nu)\, d\nu = A \varepsilon_m(\nu)\, d\nu + B \left[\frac{\partial \varepsilon_m(\nu)}{\partial \nu} - \frac{\varepsilon_m(\nu)}{\nu} \right] \quad (13)$$

Here the unknowns are A and B. They can be found by doing a least-squares fit of the known monomer absorption spectrum, $\varepsilon_m(\nu)$, and the related shape, $\partial \varepsilon_m(\nu)/\partial \nu - \varepsilon_m(\nu)/\nu$, to the measured CD spectrum. Because the Taylor series will only be accurate over a limited range, this is best done over the width of a single absorption band.

Application of Taylor Series Fitting

As an example, we apply the Taylor series fitting to each CD spectrum measured by Causley et al.[9] for poly(A) and poly(C). Some of these spectra are shown in Fig. 2. Least-squares fitting of the two shapes corresponding to the first two terms in the Taylor series gives the values for the coefficients A and B found in Fig. 3. The CD of poly(A) in Fig. 2 shows a small amount of nondegenerate interaction, and thus the value of A for this polymer is never large. Even the small nondegenerate interaction measured by A decreases with increasing temperature as the interactions are melted out. The CD of poly(A) has a clear derivative shape, indicating a strong stacking interaction. This is mirrored in the Taylor series coefficient, B. At 0°, the stacking interaction is exceedingly large, giving rise to a concomitantly large value for B. As the temperature is increased the stacking is decreased, and the value of B drops rapidly above 30°.

The CD of poly(C) as seen in Fig. 2 has no obvious derivative shape; however, Taylor series fitting shows that our initial impressions are wrong,

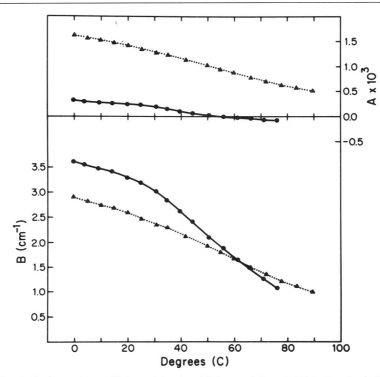

FIG. 3. Taylor series coefficients A (top) and B (bottom) for poly(A) (—) and poly(C) ($\cdot\cdot\cdot$) as a function of temperature. [Redrawn from G. C. Causley, P. W. Staskus, and W. C. Johnson, Jr., *Biopolymers* **22**, 945 (1983).]

and the value of B is quite large. As expected, the value of B decreases with increasing temperature since the stacking is melted out. The CD due to the degenerate stacking interactions in poly(C) are overwhelmed by the nondegenerate interactions. Thus the Taylor series coefficient A for poly(C) has about five times the value of A for poly(A) at $0°$. It decreases with increasing temperature but still has a substantial value at $90°$.

Taylor series fitting of a CD spectrum yields decomposition into degenerate and nondegenerate interactions. This may reveal far more degenerate interaction due to stacking than is obvious in the undecomposed spectrum, as we have just seen for poly(C). Furthermore, the Taylor series coefficient B as a measure of stacking can be used as a melting curve to determine thermodynamic parameters. Although the completely stacked and completely unstacked end points are not contained in these data, Powell *et*

TABLE I

TAYLOR SERIES COEFFICIENTS AND THERMODYNAMIC PARAMETERS FOR POLYMER
MELTING USING VAN'T HOFF MODEL[a]

Compound	B_{st} (cm^{-1})	B_{mp} (cm^{-1})	A^b ($\times 10^4$)	A_{mp} ($\times 10^4$)	$-\Delta H$ (kcal)	$-\Delta S$ (e.u.)
Poly(A)	3.7	1.9	3.2	0.6	11.7	36.0
Poly(C)	3.2	1.4	16.2	7.5	6.5	19.0

[a] Data from G. C. Causley, P. W. Staskus, and W. C. Johnson, Jr., *Biopolymers* **22**, 945 (1983).

[b] The value of A at the lowest temperature measured.

$al.$[11] have shown that it is still possible to fit models for polymer melting. The two-state model using the van't Hoff equation can be written in terms of the stacked (B_{st}) and unstacked (B_{un}) conformations:

$$B(T) = B_{un} + (B_{st} - B_{un})/[1 + \exp(\Delta H/RT - \Delta S/R)] \qquad (14)$$

This will yield ΔH and ΔS in terms of an unknown cooperative unit. Similarly, the Ising model can be written in terms of these parameters as well as a cooperativity factor, σ:

$$B(T) = B_{un} + (B_{st} - B_{un})\alpha$$
$$\alpha = 0.5 + 0.5(s - 1)[(1 - s)^2 + 4\sigma s]^{-1/2} \qquad (15)$$
$$s = \exp\left(\frac{-\Delta H}{RT} + \frac{\Delta S}{R}\right)$$

Without the end points, the van't Hoff equation has four unknowns: B_{un}, B_{st}, ΔH, and ΔS. The Ising model has a fifth unknown, σ. The models can be fit to the data in a standard least-squares calculation, using the simplex algorithm.[12] Results of fitting the van't Hoff equation to the data for poly(A) and poly(C) are given in Table I. If the melting point (mp) is defined to be the temperature at which there is 50% stacking, then the Taylor series coefficient B_{mp} for the melting point will be one-half $B_{st} - B_{un}$. Comparing Table I and Fig. 3, we deduce melting temperatures of 55° for poly(A) and 65° for poly(C). In addition, the Taylor series coefficients A for the melting temperature can be found on Fig. 3 and are given in Table I. Using A_{mp} and B_{mp}, it is possible to deduce the CD spectrum for the half-melted polymer (K_{eq} of 1), which are given in Fig. 2 as the dashed curves.

When comparing CD spectra among different kinds of polymers or

[11] J. T. Powell, E. G. Richards, and W. B. Gratzer, *Biopolymers* **11**, 235 (1972).
[12] S. N. Deming and S. L. Morgan, *Anal. Chem.* **45**, 278A (1973).

TABLE II
STACKING PARAMETERS AND THERMODYNAMIC CONSTANTS
FOR POLYMER MELTING USING ISING MODEL[a]

Compound	σ	B_{st} (cm^{-1})	B_{mp} (cm^{-1})	$-\Delta H$ (kcal)	$-\Delta S$ (e.u.)
Poly(A)	0.4	3.8	2.0	7.2	22
	0.5	3.8	2.0	8.1	25
	0.6	3.7	2.0	8.9	27
Poly(C)	0.8	3.2	1.5	5.8	17
	0.9	3.2	1.4	6.2	18
	1.0	3.2	1.4	6.5	19

[a] Data from G. C. Causley, P. W. Staskus, and W. C. Johnson, Jr., *Biopolymers* **22,** 945 (1983).

among related oligomers, it is not reasonable to compare data at the same temperature. Workers should really compare data for the fully stacked samples, but of course these are generally unavailable. Using Taylor series fitting, it is possible to obtain CD spectra for 50% stacking, so that data can be compared under conditions where K_{eq} is 1 for all samples.

The five unknowns in the Ising model were too much for these data, but it was possible to fit the data using the Ising model with fixed values of σ. Good fits of poly(A) were found for cooperativities of 0.4–0.6, and of poly(C) for cooperativities of 0.8–1.0. We see in Table II that the value of B_{st} is the same as for the van't Hoff model. The lack of cooperativity in the melting of poly(C) gives about the same values for ΔH and ΔS as the van't Hoff model. Cooperativity in the melting of poly(A) yields smaller values for ΔH and ΔS, because they are per monomer rather than cooperative unit, the undefined parameter of the van't Hoff model.

In summary, Taylor series fitting of CD spectra allows the separation of degenerate and nondegenerate contributions. The coefficient of the derivative term in the Taylor series, which is related only to base stacking in the case of poly- and oligonucleotides, can be used to study melting. The magnitude of this term versus temperature can be fit with melting models, such as the van't Hoff and Ising equations. Even when the melting curves are not complete, analyses can give the melting temperature and the corresponding CD spectrum. Various samples can then be compared under the common condition of K_{eq} equal to 1, rather than at the same temperature where their conditions are really unrelated. Decomposition gives the amount of stacking, which may not be obvious from inspection of the data.

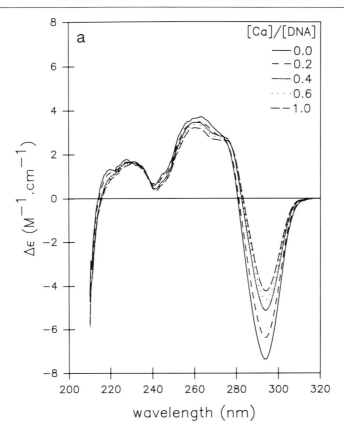

FIG. 4. Circular dichroism spectra for the Ca^{2+} titration of poly[d(Gm^5C)·d(Gm^5C)] on a nucleotide basis that are averaged by SVD and smoothed by the Savitzky–Golay method. [From L. Zhong and W. C. Johnson, Jr., *Biopolymers* **30**, 821 (1990).]

Singular Value Decomposition

We work best with continuous functions that can be expressed in closed form, and we like to visualize functions as graphs. Thus CD spectra are presented in journal articles in graphical form, as we see in Fig. 4 for poly[d(Gm^5C)·d(Gm^5C)] in 30% ethanol as a function of added calcium.[13] In contrast, digital computers deal in specific numbers, so that CD spectra must be digitized to be entered as data. This is generally done by picking a wavelength interval, say 1 nm, and entering the value of the CD at each wavelength. In Cartesian coordinates a vector is a list of three numbers

[13] L. Zhong and W. C. Johnson, Jr., *Biopolymers* **30**, 821 (1990).

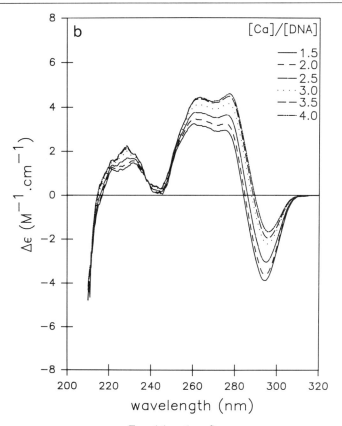

FIG. 4 (*continued*)

that represents its value as projected on the x, y, and z axes. In analogy, a digitized CD spectrum is a vector where the axes correspond to each wavelength. The CD spectrum as a vector is a list of numbers, each of which corresponds to the value of $\Delta\varepsilon$ at a particular wavelength. Thus, for the raw measured CD spectrum of poly[d(dGm^{5}C · d(Gm^{5}C)] with no added calcium, shown smoothed in Fig. 4, we would have 111 digitized entries, one at each wavelength beginning at 210 and ending at 320 nm: $(-4.68, -2.97, -2.24, -1.33, -0.32, \ldots, 0.03)$.

When there are a series of spectra, such as the calcium titration of Fig. 4, the corresponding vectors can be taken together to form a matrix. CD spectra were measured for 15 different calcium titrations, and they form a matrix **A** with 15 columns representing each of the spectra and 111 rows representing the 1-nm increments between 210 and 320 nm, as shown in

nm	r	0.0	0.1	0.2		4.0
210		-4.68	-4.59	-4.20	···	-4.48
211		-2.97	-3.55	-3.21	···	-3.47
212		-2.24	-1.80	-1.77	···	-1.51
213		-1.33	-0.29	-1.43	···	-1.86
214		-0.32	0.28	-0.05	···	-0.34
⋮		⋮	⋮	⋮	⋮	⋮
320		0.03	-0.02	-0.11	···	-0.04

FIG. 5. Matrix **A** of the raw measured CD spectra for the Ca^{2+} titration of poly[d(Gm^5C)·d(Gm^5C)]. The ratio r is calcium ions to nucleotides.

Fig. 5. Representing a series of CD spectra as a matrix allows us to make use of matrix mathematics. Indeed, because matrix mathematics is completely general, the methods described here can be applied to any series of any type of digitized data, not just CD spectra.

An extremely powerful theorem to emerge from matrix mathematics is the singular value decomposition (SVD) theorem. It has been discussed in detail by Noble and Daniel[14] and in [8] in this volume. The SVD theorem says that any matrix can be decomposed into a product of three matrices,

$$\mathbf{A} = \mathbf{USV}^{T} \tag{16}$$

which have properties that are particularly valuable to the spectroscopist. The matrix **U** is a unitary column matrix of orthogonal basis vectors, which in practice has the same dimensions as matrix **A**. The matrix **S** has entries called singular values on the main diagonal, and zeros elsewhere. Thus each singular value in **S** corresponds to and multiplies a particular column basis vector in **U**. The product matrix **US** has orthogonal columns that can be considered a new basis for the CD spectra, which are given in terms of the wavelength basis. The matrix **V**^{T} is a unitary row matrix of coefficients whose entries fit the basis CD spectra of **US** to the original data of **A** in the least-squares sense.

SVD became useful with the advent of computers with the power to carry out the decomposition, and today even personal computers are powerful enough to decompose any reasonable-sized matrix. A subroutine to carry out the singular value decomposition, written in FORTRAN, can

[14] B. Noble and J. W. Daniel, "Applied Linear Algebra," 2nd Ed., p. 323. Prentice-Hall, Englewood Cliffs, New Jersey, 1977.

	U					S						V^T				
	1	2	3		15											
210	-0.145	-0.018	0.330	···	-0.056											
211	-0.106	-0.041	0.230	···	0.062		1	2	3		15		0.0	0.1	0.2		4.0
212	-0.067	-0.008	0.274	···	-0.064	1	113.1	0	0	···	0	1	0.32	0.31	0.29	···	0.23
213	-0.036	-0.013	0.085	···	-0.116	2	0	40.58	0	···	0	2	-0.36	-0.30	-0.25	···	0.51
214	-0.001	-0.018	0.021	···	0.097	X 3	0	0	8.73	···	0	X 3	0.29	0.21	0.44	···	0.18
⋮	⋮	⋮	⋮	⋮	⋮	⋮	⋮	⋮	⋮	⋮	⋮	⋮	⋮	⋮	⋮	⋮	⋮
320	-0.000	-0.000	-0.002	···	0.015	15	0	0	0	···	4.83	15	0.07	-0.19	0.10	···	0.36

FIG. 6. Singular value decomposition of matrix **A** into the product of three matrices, **U**, **S**, and **V**ᵀ.

be found in Forsythe *et al.*[15] The program requires the matrix to have more rows than columns, one reason for expressing the original spectra as column vectors. The other reason is that the basis vectors end up in **U**, rather than **V**ᵀ. SVD of raw data for the calcium titration of poly [d(Gm⁵C) · d(Gm⁵C)] in 30% ethanol is illustrated in Fig. 6.

Information Content

The singular values in the **S** matrix weigh the importance of each orthogonal basis vector in **U**. The number of nonzero singular values is the number of independent components in a perfect set of data. Because real data are noisy, there will be no singular values that are identically zero. However, the number of singular values that are significant for real data will give the information content above the noise.

In our example of the calcium titration of poly[d(Gm⁵C) · d(Gm⁵C)] in 30% ethanol, the three largest singular values are 133.1, 40.6, and 8.73. There are only 2 significant singular values, and the remaining 13 correspond to noise in the data. An information content of 2 means that there are only two independent species with CD spectra in matrix **A** that are significant above the noise. This in turn means that we are dealing with a two-component system, even though the set of data in Fig. 4 do not show an isosbestic point. The **S** matrix obtained from SVD is valuable to spectroscopists because it gives the number of independent variables above the noise in a series of spectra.

Averaging among Spectra in a Series

It is not necessary to repeat each spectrum in a series in order to obtain an average spectrum. SVD allows averaging *among* spectra in a series even though they look different. The singular values weigh the importance

[15] G. E. Forsythe, M. A. Malcolm, and C. B. Moler, "Computer Methods for Mathematical Computations," p. 229. Prentice-Hall, Englewood Cliffs, New Jersey, 1977.

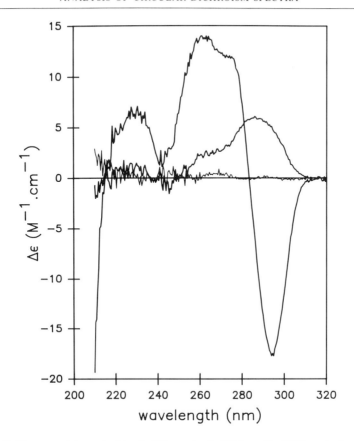

Fig. 7. Three most important basis CD spectra in **US** from the SVD analysis of 18 CD spectra for the Ca^{2+} titrations. [From L. Zhong and W. C. Johnson, Jr., *Biopolymers* **30**, 821 (1990).]

of each basis vector in **U**. Thus the significant singular values determine the significant basis CD spectra in **US**. Insignificant singular values are attributed to noise and can be set equal to zero to eliminate basis CD spectra that are below the noise level of the measurement. The original data can then be regenerated using only the significant basis CD spectra and the appropriate least-squares coefficient in $\mathbf{V}^T$ to yield data that are effectively averaged among the series of spectra.

The three most important basis CD spectra from the **US** matrix are shown in Fig. 7. Clearly, only the first two are significant above the noise. Other basis CD spectra are even more noiselike than the least important basis CD spectrum shown here. When the CD spectra from the titrations

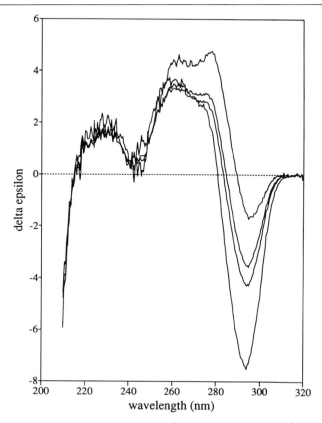

FIG. 8. Four averaged CD spectra for the Ca^{2+} titration of $poly[d(Gm^5C) \cdot d(Gm^5C)]$ that result from an SVD reconstruction using the two most important basis CD spectra.

are regenerated using only the two significant basis CD spectra and their corresponding coefficients from $\mathbf{V}^T$, we obtain average spectra, some of which are shown in Fig. 8.

SVD averaging is interspectral smoothing, which removes noise at each wavelength that is uncorrelated among the experimental spectra. Each spectrum can be further smoothed by a weighted averaging that combines data over a range of wavelengths. This is the intraspectral smoothing that is commonly used to remove noise that is uncorrelated within each spectrum. Typical methods are Fourier transform smoothing,[16] which eliminates high-frequency sinusoidal components in the data, and

[16] E. E. Aubanel and K. B. Oldham, *Byte* **10,** 207 (1985).

the Savitzky and Golay method,[17] which fits a wavelength range to a polynomial. Another advantage of the SVD procedure is that multiple experiments can be combined without explicit averaging of the original spectra. Spectra for the calcium titration of poly[d(Gm⁵C) · d(Gm⁵C)] in 30% ethanol that are both averaged by reconstruction through SVD and smoothed by the Savitzky–Golay method are shown in Fig. 4.

There are an infinite number of sets of basis vectors that could be used to describe a series of spectra. One basis set are the sine and cosine functions of the Fourier series. This set is popular because we have an intuitive feeling for what they mean, and they can be described in closed form; however, they do not give any particularly incisive information about the data. The basis set given in the U matrix from SVD does not have simple intuitive shapes, nor can it be expressed in closed form. Furthermore, without additional information the basis set cannot be related to the CD spectra of the individual species that produced the series of spectra. However, the basis vectors are quite useful to the spectroscopist because they show what is common to the spectra in the series. Thus the most significant basis vector is the single shape that best describes each of the spectra in the series. A linear combination of the two most significant basis vectors is the best least-squares description of each of the spectra in the series that can be found with only two component shapes. Similar statements can be made about including further basis vectors in a description of the spectra. Thus the U matrix is important to spectroscopists because it shows the shapes that are common among the spectra in a series. The significant basis vectors in the U matrix can then be used to average among the diverse spectra. SVD averaging of a spectral series with an information content of 2 will often reveal isosbestic points that are obscured by the noise.

Following the Variable in a Series of Spectra

The V^T matrix contains the least-squares coefficients that fit the basis CD spectra in US to the original data. The most significant basis CD spectrum represents the common features in the series; the least important basis CD spectrum that is still significant represents the most changeable features in the series above the noise. Thus the coefficients in V^T for the least important but significant basis CD spectrum monitor the changes that are occurring in the series as integrated over the entire wavelength range.

Figure 9 plots the coefficients in V^T as a function of added calcium to poly[d(Gm⁵C) · d(Gm⁵C)] in 30% ethanol for the three most important basis

[17] A. Savitzky and A. E. Golay, *Anal. Chem.* **36,** 1627 (1985).

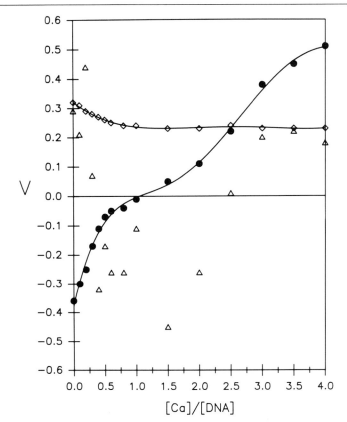

FIG. 9. The three row vectors ($\diamond$, $\bullet$, $\triangle$) in $\mathbf{V}^T$ corresponding to the basis CD spectra in Fig. 7. [From L. Zhong and W. C. Johnson, Jr., *Biopolymers* **30**, 821 (1990).]

CD spectra. The coefficient for the most important basis CD spectrum decreases somewhat at the beginning of the titration and then levels out, giving a particularly simple curve. The coefficients of the third most important basis CD spectrum are scattered over the graph, another indication that this basis is not significant above the noise. This can be investigated analytically by using the first-order autocorrelation function, C, as a measure of smoothness[18]:

$$C = \sum_{i=1}^{n-1} X_i X_{i+1} \tag{17}$$

[18] T. I. Shrager and R. W. Hendler, *Anal. Chem.* **54**, 1147 (1982).

where the X_i are the elements of the vector. For the data in Fig. 9, the significant basic vectors gave values of C greater than 0.3, whereas the meaningless vectors gave values less than 0.1.

The coefficients for the second most important basis CD spectrum are a titration curve that represents the variation in CD with added calcium. We see at once why there is no isosbestic point in the original data; the curve has two distinct straight-line portions, which we interpret as the binding of calcium to two distinct binding sites. It is this type of curve that could be fit with a binding model to determine thermodynamic parameters, such as binding constant, cooperativity, binding stoichiometry, enthalpy, or entropy, as appropriate to the titration. The van't Hoff and Ising models were given as Eqs. (14) and (15) above for a series of CD spectra that follow melting. Cooperative binding of a ligand, such as calcium, could be modeled by the McGhee–von Hippel equation.[19]

We expect the least-squares coefficients that fit the significant basis CD spectra to the original series to fall on a smooth curve, similar to the curves fitted to the data in Fig. 9. The measured coefficients do not fall precisely on the smooth curves, presumably because of experimental error in parameters such as the concentration of ethanol, the concentration of poly[d(Gm^{5}C) · d(Gm^{5}C)], and the concentration of added calcium. These experimental errors can be eliminated from the data by choosing least-squares coefficients from the smooth curves rather than the experimentally measured coefficients when reconstructing the data using SVD. Thus SVD can be used to average among the curves for magnitude, as well as for shape.

The $\mathbf{V}^\mathrm{T}$ matrix is valuable to spectroscopists because it provides points for melting or titration curves that are integrated over entire spectra. These curves may be fit with models to deduce thermodynamic parameters. In addition, smooth curves for the coefficients can be used to regenerate the original data to eliminate experimental errors.

Predicting Protein Secondary Structure from Circular Dichroism Spectra

The CD of a protein is due to the rotational strength induced in the transitions of the amide chromophore by the superasymmetry of the various secondary structures, and to the rotational strength induced in the transitions of the chromophoric side chains by the asymmetric surroundings. In favorable situations, CD is due only to the amide groups, and variations in the secondary structures such as α-helix chain length or twists

[19] J. D. McGhee and P. H. von Hippel, *J. Mol. Biol.* **86,** 469 (1974).

in β sheet, as well as the transitions in the side chains, are not significant. How many independent variables are there when only the secondary structures contribute to the CD? How many equations does the CD spectrum of a protein represent? One way to analyze the CD spectrum of a protein with unknown structure is to fit it with the CD spectra of polypeptides that represent pure secondary structures. However, this basis ignores the effect of variations in secondary structure, as well as the contributions of chromophoric side chains, such as the aromatics. How can we include these contributions if we do not know what they are? The CD spectra of proteins include all these features, so another method is to use CD spectra of proteins with known structure as the basis. However, when the CD spectra for a large number of proteins are used as a basis in the analysis, the problem is overdetermined and solutions are unstable. How can we avoid this problem? SVD can provide answers to all of these questions.

Variables cannot be eliminated by summing them together. If parallel and antiparallel β sheets have different CD spectra, then there is no gain in considering these two contributions in a single category designated β sheet. Because the two types of β sheet will contribute in different proportions to each protein CD spectrum, it does not make any more sense to sum them than it does to sum two variables in a set of equations. We can see how many independent variables are represented in the classic secondary structures by considering the structures for a set of proteins with known structure.[20] The secondary structures α helix (H), antiparallel β sheet (A), parallel β sheet (P), β turns types I, II, and III, remaining types of β turn (T), and other structures not included in the foregoing categories (O) are given in Table III for 16 proteins (including one polypeptide) that have been investigated by X-ray diffraction. The remaining turns (T) are summed because they are rarely represented in the proteins, while the other category (O) represents amides that are not in classic secondary structures.

Table III is a structure matrix, $\mathbf{F}$, and as such can be explored by SVD. The five most important basis vectors and their corresponding singular values are given in Table IV. Three structure basis vectors reproduce the original matrix with 29 entries out of 128 in error by more than 3% (two more than 10%). Four structure basis vectors give 14 entries in error by more than 3%, and none more than 5%. Five structure basis vectors give 14 entries in error by more than 2%. Thus all the classic secondary structures for these 16 proteins are well represented by only five variables. Variations in protein secondary structure and the chromophoric side chains also contribute to CD, but it not clear how to include these features

[20] J. P. Hennessey, Jr., and W. C. Johnson, Jr., *Biochemistry* **20**, 1085 (1981).

TABLE III
STRUCTURE MATRIX, F, FOR 16 PROTEINS IN BASIS SET

Protein	H	A	P	I	II	III	T	O
α-Chymotrypsin	0.10	0.34	0.00	0.11	0.03	0.02	0.04	0.36
Cytochrome c	0.38	0.00	0.00	0.03	0.06	0.01	0.07	0.45
Elastase	0.10	0.37	0.00	0.09	0.07	0.03	0.03	0.31
Hemoglobin	0.75	0.00	0.00	0.08	0.01	0.04	0.01	0.11
Lactate dehydrogenase	0.41	0.06	0.11	0.01	0.02	0.05	0.03	0.31
Lysozyme	0.36	0.09	0.00	0.14	0.06	0.07	0.05	0.23
Myoglobin	0.78	0.00	0.00	0.02	0.00	0.08	0.02	0.10
Papain	0.28	0.09	0.00	0.05	0.02	0.03	0.04	0.49
Ribonuclease A	0.24	0.33	0.00	0.02	0.02	0.02	0.08	0.29
Flavodoxin	0.38	0.00	0.24	0.06	0.02	0.03	0.05	0.22
Glyceraldehyde-3-phosphate dehydrogenase	0.30	0.09	0.13	0.01	0.02	0.04	0.07	0.34
Prealbumin	0.07	0.38	0.07	0.04	0.03	0.00	0.07	0.34
Subtilisin BPN	0.30	0.02	0.07	0.12	0.01	0.03	0.05	0.40
Subtilisin novo	0.31	0.02	0.08	0.07	0.00	0.01	0.03	0.48
Triose-phosphate isomerase	0.52	0.00	0.14	0.01	0.00	0.00	0.10	0.23
Polyglutamic acid	1.00	0.00	0.00	0.00	0.00	0.00	0.00	0.00

in the analysis for the number of variables. Nevertheless, because the fifth most important structure basis vector is not particularly important, there is hope that there are only five or six independent variables that contribute to the CD spectrum of a protein.

To solve for five unknowns, we need five equations. To how many equations does the CD spectrum of a protein correspond? The CD spectra of proteins are quite diverse, as shown for three representative proteins

TABLE IV
FIVE MOST IMPORTANT BASIS VECTORS AND CORRESPONDING SINGULAR VALUES
FROM PROTEIN STRUCTURE MATRIX, F

Basis vector	H	A	P	I	II	III	T	O	Singular value
1	0.845	0.146	0.091	0.089	0.036	0.054	0.078	0.488	2.105
2	−0.511	0.485	0.055	0.106	0.063	0.010	0.090	0.691	1.060
3	0.147	0.835	−0.258	0.015	0.041	0.008	−0.027	−0.459	0.498
4	0.009	−0.188	−0.929	0.158	0.058	0.034	−0.175	0.205	0.274
5	−0.035	−0.052	0.172	0.907	0.163	0.290	−0.102	−0.148	0.173

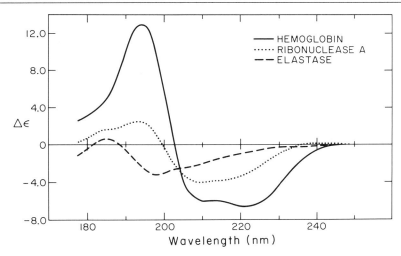

FIG. 10. Circular dichroism spectra of three proteins with different amounts of the various secondary structures. [Redrawn from J. P. Hennessey, Jr., and W. C. Johnson, Jr., *Biochemistry* **20**, 1085 (1981).]

in Fig. 10. If we consider the 16 proteins from Table III as a series, we can make a matrix of their 16 CD spectra and analyze it for information content with SVD. The singular values do not drop off rapidly for this diverse series. However, with CD spectra measured to 178 nm, it takes five basis vectors to reproduce the shape and magnitude of the original CD spectrum within experimental error.[20] Thus the matrix of CD spectra measured to 178 nm for the 16 proteins has an information content of 5; each CD spectrum represents five independent equations that can be used to solve for only five unknowns. Similarly, CD spectra truncated at 190 nm represent three to four equations, whereas spectra truncated at 200 nm represent only two equations.[21]

Research into protein secondary structure as represented in a CD spectrum has shown conclusively that the spectra must be measured to at least 184 nm to provide the five equations that only minimally solve for secondary structure.[21] CD spectra of proteins that have been truncated at longer wavelengths do not have the information content to solve for secondary structure, and workers are simply trying to solve for too many unknowns with too few equations. A particular method of analysis may appear to work on a particular truncated CD spectrum. However, this is

[21] P. Manavalan and W. C. Johnson, Jr., *Suppl. J. Biosci.* **8,** 141 (1985).

fortuitous. The favored method will vary from one protein CD spectrum to another.

The sum of secondary structures for a particular protein should be 100%. This constraint provides another equation, but it has been shown that the use of this constraint does not improve analysis of truncated data.[21] Forcing the sum of structures to be 100% makes an analysis look believable. Without the constraint, a poor sum shows that the analysis has simply failed.[22]

The fractions of secondary structure should also be positive, and this, too, can be added as a constraint. When this constraint is included with the constraint for the sum of structures, analyses become increasingly unreliable.[21] Even the amount of α helix, which is usually correct regardless of the method and wavelength span used because the α-helix CD dominates the spectrum, becomes totally unreliable.

Because the effect of variations in secondary structure and of chromophoric side chains is included in the CD of proteins, these factors are implicitly included in an analysis that uses the CD of proteins with known secondary structure as a basis, even though only classic secondary structures may be explicitly predicted in the analysis. However, when the CD spectra for a large number of proteins are used as the basis, the problem is overdetermined. This problem can be avoided by using CD basis spectra from the **US** matrix derived by SVD from the matrix of protein CD spectra. When only the significant basis CD spectra are used in the analysis (five for spectra measured to 178 nm), then the noise in the CD spectra that generates instability is eliminated.

Thus SVD can be used to determine the number of independent variables in the secondary structure of a protein, to find the number of equations that the CD spectrum of a protein represents, and to provide a method of analysis that is stable and implicitly includes all the factors that determine the CD spectrum of a protein. Even when the CD spectra of proteins are extended to 178 nm, the information content of the data is marginal. Analyses can be improved significantly by using variable selection,[23] which helps make up for the fact that there are often still not enough equations in the CD spectrum to solve for all of the unknowns.

Acknowledgments

This work was supported by National Science Foundation Grant DMB-8803281 from the Biophysics Program and National Institutes of Health Grant GM-21479 from General Medical Sciences.

[22] C. C. Baker and I. Isenberg, *Biochemistry* **15**, 629 (1976).
[23] P. Manavalan and W. C. Johnson, Jr., *Anal. Biochem.* **167**, 76 (1987).

[21] Fluorescence Quenching Studies: Analysis of Nonlinear Stern–Volmer Data

By William R. Laws and Paul Brian Contino

Introduction

Many parameters can be obtained by fluorescence spectroscopic methods to provide insights into the environment, structure, and dynamics of a fluorescent probe that is either covalently bound or liganded to a biological molecule. One important, commonly used method is the addition of a quenching agent to reduce the fluorescence quantum yield of the probe. By comparing the quenching efficiency of different types of quenching agents under various conditions, the environment of the probe, and thus a specific region of the biomolecule, can be characterized in terms of neighboring ionic groups and solvent accessibility. Often, however, steady-state fluorescence quenching data do not follow the standard linear Stern–Volmer expression.[1] These deviations mean that other processes are occurring besides dynamic (collisional) quenching. If the nonlinear data are analyzed in a systematic way, it is possible that even more information may be obtained about the system.

Many situations can cause steady-state quenching data to be nonlinear when plotted by the standard Stern–Volmer relationship; these fluorescence quenching mechanisms have been reviewed by Eftink and Ghiron.[2] In this chapter, we show how two of these situations, static quenching and multiple species, cause the nonlinear deviations. We then evaluate the ability of the commonly used Marquardt nonlinear least-squares algorithm[3] to recover known parameters from synthetic data. The results of this study point out the experimental and analysis criteria that must be met to obtain optimal information from fluorescence quenching studies.[4] In particular, we demonstrate the need to perform time-resolved fluorescence quenching studies. In this way, the dynamic quenching parameters are either con-

[1] O. Stern and M. Volmer, *Phys. Z.* **20,** 183 (1919).
[2] M. R. Eftink and C. A. Ghiron, *Anal. Biochem.* **114,** 199 (1981).
[3] P. R. Bevington, "Data Reduction and Error Analysis for the Physical Sciences." McGraw-Hill, New York, 1969.
[4] Inherent in this discussion is the assumption that all experimental corrections to the steady-state intensities have been made and are not contributing to the nonlinear behavior of the data. This includes any dilution corrections as a result of the titration, as well as corrections for both primary and secondary inner filter effects due to the absorption of light at the excitation and emission wavelengths, respectively, by the quenching agent (see Ref. 8).

firmed or evaluated, the distinction between single and multiple species is verified, and the existence of static quenching, or other process, is established.

Theory

Single Species with Dynamic Quenching

The fluorescence intensity, F_0, of a chromophore, A, in the absence of added quenching agents depends on the initial concentration of the excited state, $[A_0^*]$, and the rates of the depopulation processes. Assuming that a quencher, Q, does not absorb light or affect the extinction of A, then the fluorescence intensity, F, of A in the presence of added quenching agents also depends on $[A_0^*]$. In this case, F will be less than F_0 since the decay kinetics now include the collisional quenching process. The ratio of these two intensities is given in Eq. (1):

$$F_0/F = ([A_0^*]\tau_0/\tau')/([A_0^*]\tau/\tau') = \tau_0/\tau \tag{1}$$

where τ' is the natural (in vacuo) lifetime of A^*, τ_0 is the lifetime in an interacting system without added quencher, and τ is the lifetime at a particular [Q]. Both τ_0 and τ are defined in terms of sums of rate constants: $\tau_0 = (k_{nr} + k_f)^{-1}$ and $\tau = (k_{nr} + k_f + k_q[Q])^{-1}$, respectively, where k_f is the rate of fluorescence in the absence of any interactions, k_{nr} is the sum for all the nonradiative rates in interactive systems lacking the quencher, and k_q is the rate of collisions with the quencher that result in the radiationless deactivation of A^*. The intensity ratio can therefore be expressed [Eq. (2)] in terms of the rate constants:

$$F_0/F = (k_f + k_{nr})^{-1}/(k_f + k_{nr} + k_q[Q])^{-1} \tag{2}$$

By rearranging and defining $K_{sv} = k_q\tau_0 = k_q/(k_f + n_{nr})$, Eq. (3) is obtained:

$$F_0/F = 1 + K_{sv}[Q] \tag{3}$$

which is the well-known, linear Stern–Volmer relationship[1] with K_{sv} the Stern–Volmer constant. An example of this dynamic quenching of a single species is shown by curve A in Fig. 1 where $K_{sv} = 8\ M^{-1}$.

Single Species with Dynamic and Static Quenching

Static quenching mechanisms are directly implicated when the quantum yield of the probe as measured by steady-state methods (number of photons emitted divided by the number of photons absorbed) is less than the quantum yield as measured by time-resolved methods (kinetics of the

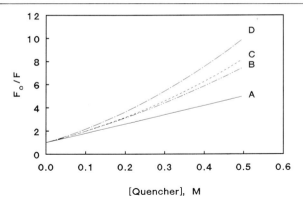

FIG. 1. Curves representing quenching functions for a single chromophore with $K_{sv} = 8$ M^{-1}. (A) Equation (3); (B) and (D) Eq. (12) with K_a values of 1 and 2 M^{-1}, respectively; (C) Eq. (6) with $V = 1 \ M^{-1}$.

decay of the excited state). A static quenching mechanism is a very rapid process that removes a fraction of the A* population before it can be dynamically quenched. Therefore, time-resolved fluorescence measurements are not dependent on $[A_0^*]$, and static quenching processes are only observed by steady-state intensity measurements. Consequently, the F_0/F ratio will (1) increase at a rate greater than that due to dynamic quenching alone and (2) have upward curvature since the static quenching mechanism is also a function of [Q]. The sphere of action and the ground-state complex static quenching models are discussed here.

Sphere of Action Static Quenching Model. Any quenching interaction requires that the chromophore and the quencher be within a certain distance of one another. In solution, this critical distance defines an interaction sphere of volume V. On excitation of the chromophore, a quencher molecule may already be within this volume and thus be able to quench without the need for a diffusion-controlled collisional interaction.[2] The probability of the quencher being within this volume at the time of excitation depends on the volume and on the quencher concentration. Assuming that the quencher is randomly distributed in solution, the probability of static quenching is given by a Poisson distribution, $e^{-V[Q]}$; the corrected fluorescence intensity ratio is expressed in Eq. (4):

$$F_0/F = ([A_0^*]\tau_0/\tau')/([A_0^*] \ e^{-V[Q]} \ \tau/\tau') \tag{4}$$

On substitutions and rearrangements, Eq. (5) can be obtained:

$$(F_0/F) \ e^{-V[Q]} = \tau_0/\tau \tag{5}$$

Combination of this static quenching mechanism with the dynamic quenching described by Eq. (3) results in Eq. (6):

$$F_0/F = (1 + K_{sv}[Q]) \, e^{V[Q]} \qquad (6)$$

An example of the curvature resulting from a single chromophore undergoing both dynamic quenching and sphere of action static quenching [Eq. (6)] is shown by curve C in Fig. 1 where $V = 1 \, M^{-1}$. For comparison, it has been generated with the same K_{sv} value (8 M^{-1}) used for the linear Stern–Volmer plot (curve A). Both K_{sv} and V must be evaluated when nonlinear steady-state fluorescence quenching data with upward curvature are analyzed by Eq. (6). The dynamic aspects of the decay of the excited state of the chromophore are still represented by K_{sv}. Thus, the value of K_{sv} should be consistent with the known lifetime in the absence of quencher, τ_0, and a reasonable value for the bimolecular quenching constant, k_q. The V term must be greater than zero since it represents a volume in which quenching occurs without the need for diffusion.[5] Furthermore, V must be within a range of values that denotes a radius that is not much larger than van der Waals radii; a range of $1–3 \, M^{-1}$ yields acceptable radii (<10 Å) for diffusionless interactions.[2] It has been previously noted that V tends to be about 10% of K_{sv} for acrylamide quenching of indole compounds[2] and a fluorescent estrogen analog.[6] However, this is most likely just a function of the individual lifetimes and quenching rates for these two systems. In general, V is independent of the Stern–Volmer constant and should be within a range for an effective radius of interaction.

Ground-State Complex Static Quenching Model. A quencher can interact with the chromophore to form a ground-state complex, $A \cdot Q$, as detailed in Eq. (7):

$$A + Q \rightleftharpoons A \cdot Q \qquad (7)$$

This complex is able to reach the initial Franck–Condon state; however, because the quencher is part of the complex, quenching occurs "instantaneously" and efficiently without the need for a diffusion-controlled interaction.

In the absence of quencher, F_0 will be proportional to the entire chromophore concentration, $[A_{total}]$. F, however, will be proportional to the free chromophore concentration, $[A_f]$, and will be a function of $[Q]$. The equilibrium expression for this reaction is given in Eq. (8), where K_a is the association constant:

$$K_a = [A \cdot Q]/[A_f][Q] \qquad (8)$$

---

[5] To account for units, V represents the volume per mole of the chromophore and the quencher at the critical interaction distance; division by Avogadro's number yields the volume for one chromophore and one quencher.

[6] E. Casali, P. H. Petra, and J. B. A. Ross, *Biochemistry* **29**, 9334 (1990).

By conservation of mass, $[A_{total}] = [A_f] + [A \cdot Q]$, and, with rearrangement of Eq. (8), an expression for $[A_f]$ can be derived [Eq. (9)]:

$$[A_f] = [A_{total}]/(1 + K_a[Q]) \tag{9}$$

Thus, the excited-state concentration of chromophore in the presence of quencher will be less than in its absence, and the ratio of fluorescence intensities becomes

$$F_0/F = ([A_0^*]\tau_0/\tau')/([A_0^*](1 + K_a[Q])^{-1}\tau/\tau') \tag{10}$$

Equation (10) can be rearranged in a manner similar to the previous case:

$$(F_0/F)(1 + K_a[Q])^{-1} = \tau_0/\tau \tag{11}$$

Therefore, by combining this type of static quenching mechanism with the dynamic quenching described in Eq. (3), the expression in Eq. (12) is derived:

$$F_0/F = (1 + K_{sv}[Q])(1 + K_a[Q]) \tag{12}$$

Equation (12), which represents a single species undergoing both dynamic quenching and ground-state complex static quenching, is a quadratic expression with upward curvature. As shown in Fig. 1, Eq. (12) (curve B) has a similar amount of curvature with a K_a of 1 M^{-1} as that generated by Eq. (6) (curve C).

Both K_{sv} and K_a need to be determined in an analysis of nonlinear steady-state fluorescence quenching data by the ground-state complex static quenching model detailed in Eq. (12). The previous criteria still apply to the value for K_{sv}, which represents the dynamic quenching portion of the interaction. As shown by curves B and D in Fig. 1, K_a does not have to be very large to induce significant curvature. The K_a of 1 M^{-1} in curve B and the K_a of 2 M^{-1} in curve D are of a magnitude expected for very weak or nonspecific binding. This range of values for K_a is obviously too small to be easily measured by most experimental techniques and is in the range suggested for the affinity of acrylamide to proteins.[7] Note that the additional term in Eq. (12) is actually linear with respect to [Q] in contrast to the experimental term in Eq. (6); under the special circumstance of no dynamic quenching, this may allow the two static quenching mechanisms to be distinguished.[6]

Multiple Species with Dynamic Quenching

Multiple chromophores, each with its own distinct K_{sv} term, can also lead to nonlinear F_0/F plots. Multiple chromophores can arise in several ways even in a chemically pure system. If a protein has more than one

[7] M. R. Eftink and C. A. Ghiron, *Biochim. Biophys. Acta* **916**, 343 (1987).

tryptophan residue, for example, each one could have a different K_{sv} term due to either a different k_q, τ_0, or both. If a macromolecule has been labeled with an extrinsic probe, either covalently or by adsorption, there could be more than one site of interaction. Finally, there is the possibility that a single chromophore, either intrinsic or extrinsic, located at a single site might have several different environments due to different structural conformations in the ground state. For example, based on our previous tyrosine studies,[8-10] the three rotamers for the tyrosine side chain about its C^α–C^β bond are a source of ground-state heterogeneity. Therefore, multiple species in the ground state lead to multiple excited states, and these unique excited-state species could have different environments resulting in distinct quenching constants.

If each species can be spectrally isolated, then the normal Stern–Volmer expression, the sphere of action model, or the ground-state complex model [Eqs. (3), (6), or (12)] would be appropriate for an analysis depending on the linearity of the F_0/F versus [Q] plot. However, often the excitation and/or emission energies of each species overlap. The measured F_0 and F values, therefore, consist of weighted sums of the F_0 and F values for each species. This means that the fluorescence intensity ratio will be a function of $\Sigma(1 + K_{sv}(i)[Q])$, which is an extension of Eq. (3) where i denotes a specific quenching constant/species. Assuming that the individual species do not interact with one another, the complete expression for the dynamic quenching of multiple species must weight each element of this summation by the fractional intensity, $f(i)$, of each emitting species. These $f(i)$ are a function of the concentration, extinction, quantum yield, and emission spectrum of each species; by definition, $\Sigma f(i) = 1$. The complete expression for multiple (n) species undergoing only dynamic quenching is given in Eq. (13):

$$F_0/F = \left[\sum_{i=1}^{n} \frac{f(i)}{\{1 + K_{sv}(i)[Q]\}} \right]^{-1} \tag{13}$$

Depending on the values of the $K_{sv}(i)$ and $f(i)$ terms, the F_0/F expression may be linear or curved downward.[2] Generally, downward curvature will be observed since the initial slope of the intensity ratio is influenced by the largest K_{sv}. As [Q] increases, the effect of the smaller K_{sv} terms will begin to be seen and the "average" K_{sv} will be less, resulting in the

[8] P. B. Contino and W. R. Laws, *J. Fluoresc.* **1**, 5 (1991).

[9] W. R. Laws, J. B. A. Ross, H. R. Wyssbrod, J. M. Beechem, L. Brand, and J. C. Sutherland, *Biochemistry* **25**, 599 (1986).

[10] J. B. A. Ross, W. R. Laws, A. Buku, J. C. Sutherland, and H. R. Wyssbrod, *Biochemistry* **25**, 607 (1986).

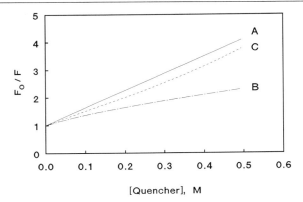

FIG. 2. Effect of multiple species on the F_0/F ratio. In all cases, $f(1) = f(2) = 0.5$ and $K_{sv}(2) = 8\ M^{-1}$. (A) Equation (13) with $K_{sv}(1) = 5\ M^{-1}$; (B) Eq. (13) with $K_{sv}(1) = 1\ M^{-1}$; (C) Eq. (14) with $K_{sv}(1) = 1\ M^{-1}$ and $V(1) = V(2) = 1\ M^{-1}$.

downward curvature. This dependence of the degree of curvature on the extent of the differences in the $K_{sv}(i)$ terms is shown in Fig. 2. For two similar K_{sv} values, the expression appears to be linear (curve A). The downward curvature in curve B is a result of the K_{sv} values being significantly different.

Multiple Species with Dynamic and Static Quenching

Each species may also undergo static quenching by either the sphere of action or the ground-state complex model.

Sphere of Action Static Quenching Model. If we include an $e^{V[Q]}$ term for each species, the development will be the same as for one chromophore. The resulting expression for the multiple species/sphere of action model is Eq. (14):

$$F_0/F = \left[\sum_{i=1}^{n} \frac{f(i)}{\{1 + K_{sv}(i)[Q]\}\, e^{V(i)[Q]}} \right]^{-1} \tag{14}$$

The trend in any data following the quenching mechanism detailed by Eq. (14) will of course depend on the specific values of $f(i)$, $K_{sv}(i)$, and $V(i)$. It is important to note that the addition of static quenching can override any downward curvature due to different K_{sv} values and linearize the data or even cause plots to have positive curvature. This is shown by curve C in Fig. 2. The same $f(i)$ and $K_{sv}(i)$ values used in curve B were used in Eq. (14), but a V term of $1\ M^{-1}$ for each species is able to induce upward curvature. In fact, curve C is actually a complex function since the upward curvature occurs after a small amount of downward curvature.

Ground-State Complex Static Quenching Model. As shown in Eq. (15), nonlinear quenching data could be due to the multiple species/ground-state complex model:

$$F_0/F = \left[\sum_{i=1}^{n} \frac{f(i)}{\{1 + K_{sv}(i)[Q]\}\{1 + K_a(i)[Q]\}} \right]^{-1} \tag{15}$$

This expression is also able to cause upward curvature even when the K_{sv} terms are different enough to have downward curvature.

Analysis

In this section, we show that unless steady-state quenching experiments are performed to sufficiently high quencher concentrations, an incorrect analysis can occur because data that appear linear are actually nonlinear. We also show that it is difficult to distinguish between the two static quenching models for both single species and multiple species systems. We further demonstrate that it is necessary to obtain the dynamic quenching parameters as well as the fractional intensities to enable an analysis of multiple species quenching data.

Concentration Range and Linearity versus Nonlinearity

A typical upper limit for quencher concentration used for many fluorescence quenching studies in the literature is approximately 0.2 M; this concentration range usually requires minimal experimental corrections,[4] and the signal-to-noise ratio does not become a problem for the quenched sample. Considering the nonlinear examples displayed in Figs. 1 and 2, most could be interpreted as linear over this limited concentration range, particularly if experimental data with the inherent noise are considered. To demonstrate this point, we synthesized four data sets with noise (see Appendix at the end of this chapter) consisting of 11 concentrations equally distributed between 0 and 0.2 M. These data sets were generated using the respective equations and parameters for the curves shown in Fig. 1. Each data set was then analyzed by linear least-squares regression assuming a fixed intercept of 1.0 since, by defintion, $F_0/F = 1.0$ at $[Q] = 0$. In each case, the correlation coefficients for the regressions were around 0.99, which could be interpreted as an acceptable fit to a straight line. This was true even for the situation represented by curve D in Fig. 1, which is clearly nonlinear by 0.2 M.

The apparent linearity of these expressions over this small concentra-

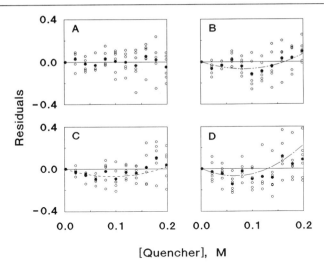

FIG. 3. (A) through (D) represent the residuals (data − fit) for curves A through D of Fig.
1, respectively, resulting from fitting data generated by that function to a straight line over
a [Q] range of 0 to 0.2 M. The smooth line denotes the residuals for noiseless data. Open
circles are the residuals for the entire data set. Solid circles are the averages of the residuals
at each [Q].

tion range can also be demonstrated by the residuals of the synthetic data
and the linear fit. For data without added noise (represented by curves B,
C, and D in Fig. 1), the residuals calculated from a linear fit, as expected,
exhibit systematic deviations as shown by the smooth lines in the corresponding panels of Fig. 3. For the residuals of the data with noise, these
systematic trends are lost; this is true even for the averages of the residuals
at each [Q]. It could be argued that there is a trend in the average residuals
for plots B, C, and D (Fig. 3) when contrasting them against the residuals for noiseless data or by comparing them to the randomness of the residuals for the normal Stern–Volmer situation (panel A), but this is an
artificial comparison resulting from synthetic data. Consequently, the nonlinear examples shown in Fig. 1 cannot be distinguished from a straight
line over this quencher concentration range; a similar situation exists for
the examples shown in Fig. 2 which represent the quenching of multiple
species.

   If the quenching of a single species involves more than collisional
interactions, a consequence of such a linear analysis over this small concentration range is the incorrect estimation of K_{sv}. For the synthetic data
representing curves B and C in Fig. 1, the K_{sv} obtained by fitting to a
straight line was 10.5 M^{-1} instead of the value of 8 M^{-1} used to generate

the data. An analysis of the data representing curve D in Fig. 1 gave a K_{sv} value of 12.1 M^{-1}. Because the characterization of the environment of a chromophore requires a comparison of the k_q terms for different quenchers under different conditions, and because $k_q = K_{sv}/\tau_0$, an incorrect K_{sv} value could lead to an incorrect hypothesis concerning the nature of the environment.

Quenching studies carried out to higher quencher concentrations and with a greater data density may permit the linear/nonlinear behavior of the data to be resolved. These studies could require large corrections owing to dilution and the extinction coefficient of the quencher, but careful experimental procedures and evaluation of these corrections permit highly reproducible data to be obtained. To further test the linearity/nonlinearity of the data, or to examine the effectiveness of any quenching model to fit the data, residuals should always be examined to ensure against systematic behavior.

Single Species with Dynamic and Static Quenching

Sphere of Action Static Quenching Model. To test the ability of non-liner least-squares to recover K_{sv} and V in the sphere of action model, we synthesized data sets with noise (see Appendix) consisting of 17 quencher concentrations equally spaced over 0 to 0.5 M. These data sets were generated by Eq. (6) with V equal to 1.0 M^{-1} and K_{sv} equal to either 2, 4, 8, or 16 M^{-1}. In all cases, both recovered parameters were consistently within 2 to 3% of the correct value, and the random distribution of the residuals for all points and for the averaged residuals indicated excellent fits. The same convergence point (same values for the parameters and the same minimum on the χ^2 surface[3]) was obtained for iterations starting with different guesses spanning physically relevant ranges for both parameters.

Ground-State Complex Static Quenching Model. Data sets with noise (see the sphere of action model above) were generated by Eq. (12) with K_a equal to 1 M^{-1} and K_{sv} equal to either 2, 4, 8, or 16 M^{-1} and then analyzed by the ground-state complex model using nonlinear least-squares. Although excellent fits were obtained for each data set, and the convergence point was not particularly sensitive to the initial guesses for the parameters, there were problems in recovering the parameters. These problems are a direct result of the symmetry of Eq. (12) since there is no mathematical difference between K_{sv} and K_a. Recovery of the parameters depended on the relative magnitudes of the parameters used to generate the data. If K_{sv} and K_a were similar, as for the case of K_{sv} equal to 2 M^{-1} and K_a equal to 1 M^{-1}, then K_{sv} equaled K_a. If K_{sv} and K_a were different, as for the case of K_{sv} equal to 4 M^{-1} and K_a equal to 1 M^{-1}, then the correct values were recovered. However, the symmetry of Eq. (12) made the assignment of the iterated parameters to a specific variable (K_{sv} or K_a) ambiguous.

Cross Analyses. To test whether the data sets generated by either of the two static quenching models uniquely describe that particular model, we analyzed each data set by the other model. For data generated by either model, equivalent fits were achieved by the other model, and the values for the parameters were physically reasonable. Therefore, these two static quenching models applied to a single species cannot be resolved with this data density over this concentration range by these statistical criteria. If this approach is experimentally feasible, data collected to even higher quencher concentrations might be able to resolve the different curvatures of the two models.

Multiple Species with Dynamic Quenching

If multiple species are known to exist, or if the F_0/F data have downward curvature, then the steady-state fluorescence quenching data must be analyzed by Eq. (13) or some form of Eq. (13) that includes static quenching and/or other processes. Analysis by the multiple species model requires that $K_{sv}(i)$ be determined for n distinct species and $f(i)$ be determined for $n - 1$ species.[11] Therefore, the minimum number of parameters that must be iterated for two species is three.

To evaluate the ability of nonlinear least squares to recover parameters from a multiple species quenching situation, we simulated data as above using the same parameters used to generate curve A in Fig. 2 [$f(1) = f(2) = 0.5$, $K_{sv}(1) = 5$ M^{-1}, and $K_{sv}(2) = 8$ M^{-1} in Eq. (13)]. Using various starting guesses, different convergence points were reached if the algorithm converged. These convergence points yielded equivalent χ^2 minima and adequate residuals but gave different values for the parameters. At no time, however, were the original values for the parameters recovered; this was true even when the known values were given as the guesses. It is often difficult to obtain a unique solution in a multiple parameter fitting problem if two or more parameters are similar in magnitude. To check whether the near equivalence in the two K_{sv} values was a factor, we generated data using the parameters for curve B of Fig. 2 [$f(1) = f(2) = 0.5$, $K_{sv}(1) = 1$ M^{-1}, and $K_{sv}(2) = 8$ M^{-1}]. This near-equivalence was not a factor; similar analysis problems were observed. As would be expected, data synthesized for the quenching of three species had similar

---

[11] Only $n - 1$ $f(i)$ terms must be evaluated since $f(n) = 1 - \sum_{i=1}^{n-1} f(i)$.

analysis problems. These problems are symptomatic of trying to fit for too many cross-correlated parameters.

Multiple Species with Dynamic and Static Quenching

Each species could also be quenched by a static mechanism. For both the sphere of action and the ground-state complex models discussed here, $3n - 1$ parameters must be iterated for n species. Fitting for too many fitting parameters was already shown to hinder the analysis of multiple species undergoing only dynamic quenching. Consequently, as expected, our attempts to analyze data generated by either Eqs. (14) and (15) were futile.

Independently Determined Parameters

If steady-state quenching data are not collected to high enough concentrations of quencher, then a linear regression could adequately fit the data and an incorrect value would be obtained for the Stern–Volmer constant. If the trend in the F_0/F data for a single species is nonlinear, there is a problem deciding which static quenching process is causing the effect since both the sphere of action and the ground-state models can fit the data equivalently. If multiple species are involved, then it is difficult to find a unique solution since too many parameters have to be evaluated.

One possible solution to these problems is to obtain additional information about the system by independent means. With this information, the fitting algorithm may be aided by reducing the number of dependent parameters and/or restricting the search to a specific set of conditions. This additional information can be easily determined by time-resolved fluorescence intensity decay measurements which can provide K_{sv} for a single species system and, depending on the number of species, may be able to obtain $K_{sv}(i)$ and $f(i)$ for a multiple species system.

From Eq. (1), $F_0/F = \tau_0/\tau$; therefore, τ_0/τ can be substituted into Eq. (3), and the slope of the lifetime ratio versus [Q] provides K_{sv}. This plot is linear since static quenching does not affect the lifetime of the excited state. This analysis requires that the fluorescence intensity decay of a single species obeys a single exponential decay law, and that only the lifetime is affected by the addition of quencher.

Additional information about a presumed single species system can be obtained from time-resolved fluorescence quenching studies. If the system consists of a single species with only dynamic quenching, then K_{sv} as determined by F_0/F must be the same as K_{sv} determined by τ_0/τ. However, if the K_{sv} term obtained from steady-state quenching data is larger than the K_{sv} term obtained from time-resolved data, then a static quenching mechanism is implicated even though the F_0/F data appear to be linear. Finally, if on the addition of quencher the intensity decay law is no longer a single exponential, then the presumed single species system is actually a multiple species system. This situation will occur when the τ_0 values for the multiple species are similar (unresolvable) and the $K_{sv}(i)$ values are different.[12] As a result, the fluorescence lifetime of one species (assuming only two for this example) will be affected to a greater extent than the other, and the lifetimes become resolvable.

If a single species has been verified by time-resolved quenching studies but static quenching is occurring, the K_{sv} determined from the τ_0/τ plot represents the dynamic aspect of the quenching interaction. This term can now be used as a constant in the fitting process. This approach was applied to the data generated by either Eq. (6) or Eq. (12) which were previously shown to be equally analyzed by both static quenching models. For both sets of data, inclusion of the known value for K_{sv} as a constant in the algorithm did not differentiate the static quenching mechanisms; both models still fit data generated by either model equally well.

We have shown above that the number of iterated parameters hinders the analysis of multiple species sysetms. To estimate how many $K_{sv}(i)$ and $f(i)$ values need to be independently known to allow the fitting algorithm to recover the remaining dependent parameters, we reduced the number of dependent parameters one at a time by making that variable a constant in the analysis of the above data sets. All possibilities of known and iterated parameters were considered; from these analyses, two basic conclusions emerged. If the $f(i)$ values were provided, all but one of the $K_{sv}(i)$ values also had to be known to recover the other $K_{sv}(i)$. If all the $K_{sv}(i)$ values were used as constants, then all but two of the $f(i)$ were required: one iterated and the other a difference.[11] This was demonstrated for the multiple species model where Eq. (13) was used to generate data for both two and three species. If data simulating multiple species with either static quenching model were evaluated, all the $K_{sv}(i)$ and $f(i)$ terms had to be constants to recover the static quenching parameters.

Time-resolved fluorescence intensity decay studies can provide many of the parameters necessary for the analysis of quenching data from multiple species. If each species has a unique, resolvable lifetime and if the species do not interact, then the fractional intensities, $f(i)$, for each species can be determined by obtaining the intensity decay parameters for the sample in the absence of quencher. From this experiment, the fractional intensities are calculated according to Eq. (16):

---

[12] It is assumed that the quencher does not alter the basic environment of each species.

$$f(i) = \alpha(i)\tau_0(i) \bigg/ \sum_{i=1}^{n} \alpha(i)\,\tau_0(i) \qquad (16)$$

where $\tau_0(i)$ represents the unquenched lifetime of species i and $\alpha(i)$ is the preexponential amplitude term for each species when the intensity decay is analyzed by a sum of exponentials.[13,14]

If the species have unique, resolvable lifetimes, then time-resolved quenching studies will provide the $K_{sv}(i)$ terms. These values are obtained by extending Eq. (3) (for lifetime ratios) to multiple species as given in Eq. (17):

$$\tau_0(i)/\tau(i) = 1 + K_{sv}(i)[Q] \qquad (17)$$

Time-resolved experiments can provide all the dependent parameters in Eq. (13); if the steady-state data deviate from a fit generated by Eq. (13) with these parameters, then static quenching is implicated. These same parameters may also be used in an analysis of multiple species with both dynamic and static quenching as represented by Eqs. (14) and (15). This approach has been used to explain the nonlinear F_0/F acrylamide quenching data for tyrosinamide.[8] By obtaining the $K_{sv}(i)$ and $f(i)$ values for the three tyrosine rotamers about the C^α–C^β bond, it was shown that each rotamer experienced both dynamic and static quenching by acrylamide. Unfortunately, differentiation between the two static quenching models could not be achieved. The same types of analysis problems which existed for a single species with static quenching were also found to exist for multiple species. We have verified this analysis behavior with synthetic data. Both the sphere of action and the ground-state complex models [Eqs. (14) and (15)] fit the data generated by either model equally well once the $K_{sv}(i)$ and $f(i)$ terms are made constants in the analysis.

Conclusions

We have shown that it is essential to have time-resolved fluorescence quenching parameters to analyze steady-state fluorescence quenching studies properly. In the case of a single chromophore in a single environment, the time-resolved studies can verify that in fact there is a single species. Furthermore, if the K_{sv} values determined from both the time-resolved and the steady-state quenching studies do not agree, then other quenching processes such as static interactions have to be included in the overall mechanism. In the case of multiple species, the only way an

[13] A. E. W. Knight and B. K. Selinger, *Chem. Phys. Lett.* **10**, 43 (1971).
[14] A. Grinvald and I. Z. Steinberg, *Anal. Biochem.* **59**, 583 (1974).

analysis of steady-state data can be attempted is to first determine the $K_{sv}(i)$ and the $f(i)$ values from time-resolved experiments. Otherwise, there are too many iterated parameters, and a unique solution cannot be found.

Without time-resolved quenching experiments, incorrect conclusions about a system are easily made. For example, the environment of the steroid binding site for the serum sex steroid-binding protein was examined using just steady-state quenching of a fluorescent steroid analog, equilenin. A quenching mechanism that was consistent with the data was then used to develop a hypothesis about the hormone–protein interaction and the accessibility of the steroid binding site.[15] Subsequent studies on this system, which included time-resolved quenching experiments, demonstrated that an entirely different quenching mechanism was operating. As a result, the steroid–protein binding interaction had to be reevaluated, and different conclusions were drawn regarding the nature of the steroid binding site.[6]

The quenching parameters that are obtained by time-resolved studies can be included in the F_0/F expression either for a single species or for multiple species to detect the existence of additional quenching processes. We have shown, however, that if static quenching is involved it is difficult to determine the particular static quenching mechanism even under optimal experimental conditions. This was found for both single and multiple species systems. It should be possible to resolve the different curvatures induced by the sphere of action and ground-state complex static quenching models by going to quencher concentrations above 0.5 M. This will be experimentally feasible, however, only for chromophores that have excitation and emission bands well removed from the absorption bands of the quencher itself. This situation diminishes the magnitude of the inner filter corrections and leaves only the problems associated with measuring the weak fluorescence intensity of the sample at high quencher concentrations.

In this chapter, we have examined several quenching situations with synthetic data sets. Before drawing conclusions about the environment of a chromophore based on fluorescence quenching studies, similar simulation studies should always be performed to ensure (1) that the parameters can be recovered from steady-state quenching data and (2) that such conclusions are reasonable. If problems occur in the analysis of quenching data, such as multiple solutions, nonconvergence, or recovery of parameters with no physical relevance, it is possible that other nonlinear least-squares algorithms[16] could be more sensitive or that a different formalization of the quenching expression other than the F_0/F ratio may be a

[15] A. Örstan, M. F. Lulka, B. Eide, P. H. Petra, and J. B. A. Ross, *Biochemistry* **25**, 2686 (1986).

[16] M. L. Johnson and S. G. Frasier, this series, Vol. 117, p. 301.

better way to represent the data. But regardless of which chromophores, quenchers, conditions, algorithms, and formulations are employed, to examine a system and obtain the most information possible it is imperative that both steady-state and time-resolved fluorescence quenching studies be performed.

Appendix

To simulate experimental data for a single titration, F_0/F values were calculated for the stated number of concentrations evenly distributed over the given [Q] range. Each data point for [Q] > 0 then had Gaussian-distributed noise added with a standard deviation of 0.05, a representative error based on our experimental experience. Finally, this process was done a total of six times, representing six separate titrations, to give the entire data set used for the analysis of a particular quenching model. This number of data points and concentration range are based on our ability to recover the quenching parameters from nonlinear synthetic data without noise [Eq. (6)] for a single titration to within 10% of the known values. The overall data density provided by six titrations is our present standard experimental protocol for steady-state quenching studies.

Acknowledgments

This work has been supported by National Institutes of Health Grants DK-39548 and GM-39750. Our discussions with Drs. Carol A. Hasselbacher and J. B. Alexander Ross are greatly appreciated. We also wish to thank Evan Waxman for programming assistance and helpful discussions.

[22] Simultaneous Analysis for Testing of Models and Parameter Estimation

By DONALD F. SENEAR and DAVID WAYNE BOLEN

Introduction

Physical measurements on biological systems are designed to extract some thermodynamic quantity or molecular property which describes the molecule or system of interest. The property or quantity in question generally originates from some theoretical framework or model imagined to emulate the real system. The experimentalist is then faced with the problem of obtaining realistic estimates of the parameters of interest while, at the same time, assessing how well the model or theory conforms to the experimental data. These two goals are intimately related. Several issues and techniques for data analysis which address this problem are detailed throughout this volume. In this chapter, we present two examples to illustrate common principles used in nonlinear least-squares analyses for parameter estimation and assessment of how well the model applied fits the experimental data.

Nonlinear least-squares statistical packages are readily available for analyses of the type we discuss. The validity of the nonlinear least-squares approach is based on three important assumptions: (1) the experimental errors are random and distributed in a Gaussian manner, (2) all of the experimental error is in the dependent variable (Y axis), and (3) no systematic error exists in the data. The last two assumptions dictate that the actual experimental measurement be plotted in Y, rather than any transformation of it. Many nonlinear transformations of equations have the effect of propagating the random experimental error into both the ordinate and abscissa of the data plot. This should be avoided. In addition, all known sources of systematic error, zero point calibration and baseline correction, should be included in the formulation of the model and explicitly fit to the data. The more usual practice is to measure the zero point or saturation value independently, or to take the first and/or last data point as a perfect measurement of that value, and apply it as a constraint to the fitting. Because these are experimental measurements subject to the same uncertainty as any other, this turns any random error in measurement of the zero point or saturation value into a systematic error for the rest of the data.

The effect and magnitude of systematic error, for example, introduced by fixing the zero point as an errorless quantity, are generally difficult to assess. One way to accommodate certain kinds of systematic error is to analyze data from different experiments simultaneously, preferably from different methods of observing the same system behavior. Simultaneous analysis means that two or more independent sets of data are analyzed at the same time such that the error contributed by all sets of data are figured into the overall least-squares fit. In this way, the confidence limits for the parameters obtained are more likely to encompass the actual physical values, and so reflect a more meaningful estimate of the accuracy of the results, rather than just the precision of a potentially inaccurate experiment. In addition, the extent to which the individual experiments agree with one another, and with the simultaneous fit, provides a measure of the accuracy of the experiments and a test of the correctness and completeness of the model.

Simultaneous analysis is also extremely useful in situations where it may not be possible to resolve uniquely all of the experimental parameters of the model from any single experiment. Used in this context, resolve means to obtain reasonably bounded parameter estimates. Often, parameters which are resolvable in principle are not in practice, owing to numerical correlations. Common examples are in studies of ligand interactions with multiple site receptors, where it is often difficult to resolve uniqely the parameters that describe the intermediate ligation states. One strategy to obtain the parameters in these cases is to find ways of reducing the number of states (e.g., by modification of the receptor to eliminate some sites), and therefore of parameters in the model, in order to better resolve those remaining. Proper combination of the results of experiments on the simplified and complete systems leads to better resolution for all of the parameters.

In this chapter, we describe the application of the least-squares method to two examples of current problems which benefit from simultaneous analyses. The two examples are the linear extrapolation method for determination of protein unfolding free energy changes and the individual site binding approach to determination of protein–DNA interaction free energy changes. In the former case, we find that by properly formulating the model to apply the least-squares criterion to baseline extensions and extrapolation of free energy changes, more realistic limits to the estimate of the unfolding free energy change are obtained. Simultaneous analysis of the effects of different denaturants increases the precision of the estimates and provides confidence in some of the assumptions of the model. In the later case, proper combination of data for binding of a repressor ligand to a multisite operator DNA and to reduced valency mutants is necessary to resolve all of the interaction free energy changes.

Testing Linear Extrapolation Model for Protein Unfolding

One of the most fundamental quantities of interest in protein folding is the free energy change associated with conversion of the native state of the protein to the fully unfolded species (ΔG°_{N-U}). Only urea-based and guanidinium-based denaturants are known to cause complete unfolding of proteins. Because (ΔG°_{N-U}) is supposed to be the unfolding free energy change in the absence of denaturant, methods have been devised to extract this quantity from urea- and guanidinium salt-induced data. The most commonly used method, which is known as the linear extrapolation method or model (LEM), makes several assumptions about

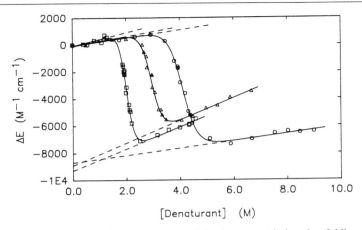

[Denaturant]   (M)

Fig. 1. Difference spectral measurements of the denaturant-induced unfolding of PMS-Ct at pH 4.0. Data represented are as follows: circles, urea; triangles, 1,3-dimethylurea; squares, guanidinium chloride. Solid symbols indicate reversibility points. The dashed lines for pre- and postdenaturational concentration regions are calculated by using the nonlinear best-fitted estimates of $\Delta\varepsilon_N^\circ$, $\Delta\varepsilon_U^\circ$, m_N, and m_U, as defined in Eq. (1). The solid lines are the results of nonlinear least-squares best fits for each of the denaturants from Eq. (4). Reversibility points were not included in the nonlinear least-squares analysis. The nonlinear least-squares program (M. L. Johnson and S. G. Frasier, this series, Vol. 117, p. 301) uses a variation of the Gauss–Newton procedure (F. B. Hildebrand, "Introduction to Numerical Analysis." McGraw-Hill, New York, 1956) to determine the model-dependent parameter values corresponding to a minimum in the variance. [Reprinted with permission from M. M. Santoro and D. W. Bolen, *Biochemistry* **27**, 8063 (1988). Copyright 1988 by the American Chemical Society.]

the functional behavior of denaturant-induced unfolding transitions.[1,2] Historically, the method is largely based on empiricism, and testing and validation of the empirical model have been piecemeal over the years.[3,4]

Typical plots of denaturant-induced unfolding of phenylmethanesulfo-nyl-α-chymotrypsin (PMS-Ct) are presented in Fig. 1, in which sharp cooperative changes in the extinction coefficient of the protein are observed as a function of denaturant concentration.[3] For analysis, the pre- and postunfolding baselines are extended into the transition region as shown in Fig. 1. These extensions are assumed to represent the spectral characteristics of native and fully unfolded species within the transition region. Mathematically, these extensions are

[1] C. N. Pace, *Crit. Rev. Biochem.* **3**, 1 (1975).
[2] C. N. Pace, this series, Vol. 131, p. 266.
[3] M. M. Santoro and D. W. Bolen, *Biochemistry* **27**, 8063 (1988).
[4] D. W. Bolen and M. M. Santoro, *Biochemistry* **27**, 8069 (1988).

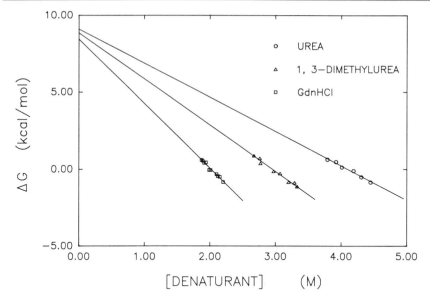

FIG. 2. Dependence of free energy for unfolding of PMS-Ct as a function of each of the three denaturants given in Fig. 1. Solid lines are the consequence of using the ΔG°_{N-U} and m_G values obtained from the individual nonlinear least-squares best fitting of the data of Fig. 1. [Reprinted with permission from M. M. Santoro and D. W. Bolen (1988). *Biochemistry* **27**, 8063 (1988). Copyright 1988 by the American Chemical Society.]

$$\Delta \varepsilon_N = \Delta \varepsilon^{\circ}_N + m_N[D]$$
$$\Delta \varepsilon_U = \Delta \varepsilon^{\circ}_U + m_U[D] \tag{1}$$

where $\Delta \varepsilon_N$ and $\Delta \varepsilon_U$ represent the denaturant concentration-dependent difference extinction coefficients for native and unfolded protein, respectively, while m_N, m_U and $\Delta \varepsilon^{\circ}_N$, $\Delta \varepsilon^{\circ}_U$ are the corresponding slopes and intercepts for native and unfolded species. Equation (2)

$$K_{\text{obsd}} = \frac{\Delta \varepsilon_N + \Delta \varepsilon}{\Delta \varepsilon + \Delta \varepsilon_U} \tag{2}$$

describes how these baselines are used in constructing equilibrium constants at fixed denaturant concentrations assuming two-state behavior to describe the transition. K_{obsd} represents the observed equilibrium ratio of unfolded/native protein species at the fixed denaturant concentration in the transition zone, with $\Delta \varepsilon$ being the measured difference extinction at that denaturant concentration. The K_{obsd} values determined in the transition zone are converted to free energy quantities ($\Delta G_{\text{obsd}} = -RT \ln K_{\text{obsd}}$) and presented as a function of denaturant concentration in Fig. 2. The linear extrapolation method consists of extrapolating the ΔG_{obsd} quantities to zero denaturant to give an intercept ($\Delta G^\circ_{\text{N}-\text{U}}$) which is interpreted to represent the free energy difference between folded and unfolded protein in the absence of denaturant.[1,2] Equation (3) expresses the relationship assumed by the LEM:

$$\Delta G_{\text{obsd}} = \Delta G^\circ_{\text{N}-\text{U}} + m_{\text{G}}[\text{D}] \tag{3}$$

where m_{G} is the slope of the plot, $\Delta G^\circ_{\text{N}-\text{U}}$ the intercept, and [D] the denaturant concentration.

Until recently, the usual method of evaluation of $\Delta G^\circ_{\text{N}-\text{U}}$ by the linear extrapolation model involved construction of two plots (Figs. 1 and 2), and estimates of the error in $\Delta G^\circ_{\text{N}-\text{U}}$ were obtained from linear least-squares analysis of Fig. 2 [see Eq. (3)]. However, the evaluation of errors by this procedure is compromised by the fact that the errors in baselines [Eqs. (1)] are not at all considered in the analysis, since the baselines are taken as errorless quantities for the purpose of obtaining a value for $\Delta G^\circ_{\text{N}-\text{U}}$. Although this two-plot procedure has enjoyed widespread use, the LEM is much better served by nonlinear least-squares treatment of the data, since such treatment overcomes several deficiencies in the conventional analysis involving two plots.[3,4] Nonlinear least-squares treatment can be accomplished by combining Eqs. (1)–(3) with the relationship $\Delta G_{\text{obsd}} = -RT \ln K_{\text{obsd}}$ to give Eq. (4), in which $\Delta \varepsilon^\circ_{\text{N}}$, $\Delta \varepsilon^\circ_{\text{U}}$, m_{G}, m_{N}, m_{U}, and $\Delta G^\circ_{\text{N}-\text{U}}$ are fitting parameters:

$$\Delta \varepsilon = \frac{(\Delta \varepsilon^\circ_{\text{N}} + m_{\text{N}}[\text{D}]) + (\Delta \varepsilon^\circ_{\text{U}} + m_{\text{U}}[\text{D}]) \, e^{-(\Delta G^\circ_{\text{N}-\text{U}}/RT + m_{\text{G}}[\text{D}]/RT)}}{1 + e^{-(\Delta G^\circ_{\text{N}-\text{U}}/RT + m_{\text{G}}[\text{D}]/RT)}} \tag{4}$$

The solid curves in Fig. 1 are the nonlinear least-squares best fits to the unfolding of phenylmethanesulfonyl-α-chymotrypsin by three separate denaturants, namely, urea, 1,3-dimethylurea, and guanidinium chloride.

Table I gives results obtained using the usual procedures involving two plots in comparison with the results obtained by nonlinear least-squares fit to the original data. It is seen that the nonlinear least-squares method provides errors which are not symmetrical and at least twice as large as those from the usual method of analysis. The error using the usual method of analysis is much smaller in part because no accounting is taken of the errors in baseline values of $\Delta \varepsilon_{\text{N}}$ and $\Delta \varepsilon_{\text{U}}$. By contrast, the nonlinear least-squares analysis fully accounts for errors arising from $\Delta \varepsilon_{\text{N}}$ and $\Delta \varepsilon_{\text{U}}$ and provides additional information on correlation of parameters, overall variance of fit, etc.

Because the LEM is largely empirical, nonlinear least-squares methods have been useful in exploring the characteristics of the model. It is noted,

TABLE I

FREE ENERGY CHANGES FOR UNFOLDING OF
PHENYLMETHANESULFONYL-α-CHYMOTRYPSIN[a]

Denaturant	ΔG°_{N-U} (kcal/mol)	$-m_G$ (kcal/mol M)
Guanidinium chloride	8.45 (7.41, 9.54)	4.19 (3.66, 4.74)
	8.69 (± 0.44)[b]	
Urea	9.10 (8.03, 10.21)	2.22 (1.96, 2.49)
	9.19 (± 0.45)[b]	
1,3-Dimethylurea	8.85 (7.64, 10.18)	3.00 (2.59, 3.44)
	8.83 (± 0.56)[b]	
Simultaneous fit	8.78 (8.07, 9.54)	

[a] Results of nonlinear least-squares fitting of data to Eq. (4). The 67% confidence limits (in parentheses) are obtained by searching the N-dimensional parameter space for the variance ratio predicted by an F-statistic [M. L. Johnson, *Biophys. J.* **44,** 101 (1983)]. This method does not assume the confidence limits to be symmetrical about the optimal values. The values reported are worst case joint confidence intervals.

[b] Denotes ΔG°_{N-U} and m_G values with 67% confidence limits calculated in the conventional manner by (1) performing linear least-squares fits on the pre- and postunfolding baselines and extending them into the transition zone, (2) evaluating apparent equilibrium constants for unfolding in the transition zone by assuming the baselines are errorless quantities and then calculating apparent free energy changes for unfolding, and (3) performing a linear least-squares fit of the ΔG (apparent) versus denaturant plot and extrapolating to zero denaturant concentration to get ΔG°_{N-U}.

for example, in Fig. 1 that the extensions of the pre- and postunfolding baselines cluster around common intercepts for $\Delta\varepsilon$ values of 0 and -9000 M^{-1} cm^{-1}, respectively. Common values for these intercepts provide some assurance that the functional dependence of the linear extensions for pre- and postunfolding baselines is reasonably linear and that each denaturant is producing the same unfolded species in the limit of zero denaturant. Also, in Fig. 2, the intercept values of all three denaturants cluster around a common value for ΔG°_{N-U}. In fact, this is expected. If ΔG°_{N-U} is an intrinsic property of the protein it should be totally independent of denaturant.

Whether it is reasonable to claim that $\Delta\varepsilon^{\circ}_{N}$, $\Delta\varepsilon^{\circ}_{U}$, and ΔG°_{N-U} are intrinsic properties of the protein may be tested statistically by performing a simultaneous fit on the three sets of data in Fig. 1.[3] Since there are parameters which are dependent on the nature of denaturant, we need to construct an expression with common parameters of $\Delta\varepsilon^{\circ}_{N}$, $\Delta\varepsilon^{\circ}_{U}$, and ΔG°_{N-U} involving the observable, $\Delta\varepsilon$, as a function of the three different

TABLE II
DATA FILE FOR SIMULTANEOUS FIT TO LINEAR EXTRAPOLATION MODEL[a]

Rows	Columns				
	$\Delta\varepsilon$	Weight	x_1	x_2	x_3
1	0.0	1.0	0.0001	0.0	0.0
2	335	1.0	0.81	0.0	0.0
.	.	.	.	.	.
.	.	.	.	.	.
23	− 5500	1.0	4.58	0.0	0.0
24	0.0	1.0	0.0	0.0001	0.0
25	0.01	1.0	0.0	0.56	0.0
.	.	.	.	.	.
.	.	.	.	.	.
43	− 6420	1.0	0.0	8.99	0.0
44	0.0	1.0	0.0	0.0	0.0001
45	0.01	1.0	0.0	0.0	0.50
.	.	.	.	.	.
.	.	.	.	.	.
64	− 3447	1.0	0.0	0.0	6.66

[a] Data entries are provided for guanidinium chloride (rows 1–23), urea (rows 24–43), and 1,3-dimethylurea (rows 44–64); the meaning of each of the five columns of the data entries is given in the text. To save space, many of the rows of data have been omitted.

independent variables, [guanidinium chloride], [1,3-dimethylurea], and [urea]. An appropriate expression is given by Eq. (5):

$$A_1 = \Delta\varepsilon_N^\circ + m_{N1}x_1 + m_{N2}x_2 + m_{N3}x_3$$

$$A_2 = \Delta\varepsilon_U^\circ + m_{U1}x_1 + m_{U2}x_2 + m_{U3}x_3$$

$$B_1 = \exp\left(\frac{-\Delta G_{N-U}^\circ - m_{G1}x_1 - m_{G2}x_2 - m_{G3}x_3}{RT}\right) \qquad (5)$$

$$\Delta\varepsilon = \frac{B_1 * A_2 + A_1}{1 + B_1}$$

Table II gives the data file to be fitted, in conjunction with Eq. (5). Column 1 is the dependent variable ($\Delta\varepsilon$ data in this case), and column 2 is the weighting of each data point (all data equally weighted in this case). Columns 3, 4, and 5 are the independent variables including concentrations of guanidinium chloride (x_1), 1,3-dimethylurea (x_2), and urea (x_3), respectively. This data file represents one way of setting up a simultaneous analysis in which three separate functions [of the form of Eq. (4)] and three independent variables are combined into one composite function

[Eq. (5)], with some common parameters (here, $\Delta\varepsilon_N^\circ$, $\Delta\varepsilon_U^\circ$, and ΔG_{N-U}°) along with some previously defined parameters (m_{N1}, m_{N2}, m_{N3}, m_{U1}, m_{U2}, m_{U3}, m_{G1}, m_{G2}, and m_{G3}) which are uniquely dependent on each of the three independent variables.

The best fitted value for ΔG_{N-U}° by simultaneous analysis falls between those given by the individual fits. The confidence interval of ΔG_{N-U}° from simultaneous analysis is smaller than that of any of the individual fits, and the variance of the simultaneous fit is smaller than the variances of the fit for two of the three individual analysis. The increased precision of ΔG_{N-U}° and general improvement of variance in the simultaneous fit gives justification for the expectation that ΔG_{N-U}° is a property of the protein, independent of the nature of denaturant.

Given that urea and guanidinium chloride solutions are markedly different in character, the improved variance of fit and smaller confidence intervals provided by simultaneous analysis demonstrate that the linear extrapolation model gives a consistent quantity believed to be the free energy change for complete protein unfolding in the absence of denaturant. Whereas the analysis gives no clue to the origin of the linear relationship between ΔG_{obsd} and denaturant concentration, it does verify the empirical relationship as applicable to strong denaturants and the parameter of interest (ΔG_{N-U}°) as having the property of independence of denaturant, as expected of a thermodynamic parameter. Further testing of the authenticity of ΔG_{N-U}° as a free energy term has focused on whether it satisfies criteria of being a thermodynamic function of state.[4]

Free Energy Changes in Cooperative, Site-Specific
Protein–DNA Interactions

The following example, involving multiple cooperative site-specific protein–DNA interactions, illustrates a more complex problem in data analysis, in which no single experiment provides sufficient information to resolve all of the theoretical parameters uniquely. In this case, parameter estimation relies on incorporation of different types of data derived from different experiments. That is, experiments are designed which result in different mathematical expressions, each involving some or all of the parameters to be estimated. Such different constructs provide a means for resolving highly correlated parameters as well as assessing whether the model conforms to the data.

The system of interest involves a molecular description of the lysogenic/lytic switch of the bacteriophage λ. Central to the regulation of this switch are the interactions of the λ cI repressor protein, with three operator sites at the λ right operator (O_R). The system features cooperative interac-

TABLE III

OPERATOR CONFIGURATIONS AND ASSOCIATED FREE ENERGY STATES
FOR λ cI REPRESSOR AND O_R[a]

Species	Operator configurations			Free energy contributions
	O_R1	O_R2	O_R3	
1	O	O	O	(Reference state)
2	R_2	O	O	ΔG_1
3	O	R_2	O	ΔG_2
4	O	O	R_2	ΔG_3
5	$R_2 \leftrightarrow R_2$		O	$\Delta G_1 + \Delta G_2 + \Delta G_{12}$
6	R_2	O	R_2	$\Delta G_1 + \Delta G_3$
7	O	$R_2 \leftrightarrow R_2$		$\Delta G_2 + \Delta G_3 + \Delta G_{23}$
8	$R_2 \leftrightarrow R_2 \leftrightarrow R_2$			$\Delta G_1 + \Delta G_2 + \Delta G_3 + \Delta G_{123}$

[a] Operator sites are denoted by O if vacant or by R_2 if liganded by cI repressor dimers. Cooperative interactions are denoted by $\leftrightarrow$. The total Gibbs free energy of each configuration (ΔG_s) is given as the sum of contributions from six free energy terms. ΔG_i values (i = 1, 2, or 3) are the intrinsic free energies for binding to the operator sites; $\Delta G_{ij(k)}$ are the free energies of cooperative interaction between liganded sites, defined as the difference in free energy to fill the sites simultaneously (ΔG_T) and the free energy to fill them individually ($\Sigma_i \Delta G_i$). The fractional probabilities of the operator configurations are given by $f_s = e^{-\Delta G_s/RT} [L]^j /$ $\Sigma_{sj} e^{-\Delta G_s/RT} [L]^j$, where ΔG_s is the sum of free energy contributions for configuration s, R is the gas constant, T is the absolute temperature, [L] is the concentration of free protein ligand, and j is the number of ligands bound in to configuration s. The free energies are related to the corresponding microscopic equilibrium constants, k_i, by the standard relationship $\Delta G_i = -RT \ln k_i$.

tions between repressor molecules bound to the operator sites in different configurations[5,6] as indicated in Table III. The goal of the analysis is to evaluate the six free energy changes for the intrinsic binding and cooperative macromolecular interactions. This is possible by utilizing an experimental method (quantitative DNase footprinting[7,8]) that distinguishes the interactions of the repressor with each of the operator sites, to yield separate, individual-site binding curves. However, for highly cooperative

[5] A. D. Johnson, B. J. Meyer, and M. Ptashne, *Proc. Natl. Acad. Sci. U.S.A.* **76**, 5061 (1979).

[6] D. F. Senear and G. K. Ackers, *Biochemistry* **29**, 6568 (1990).

[7] M. Brenowitz, D. F. Senear, M. A. Shea, and G. K. Ackers, this series, Vol. 130, p. 132.

[8] M. Brenowitz, D. F. Senear, M. A. Shea, and G. K. Ackers, *Proc. Natl. Acad. Sci. U.S.A.* **83**, 8462 (1986).

systems such as O_R, even data on individual site binding does not provide unique resolution of all of the interaction free energy changes occurring with the wild-type operator. Some of the free energy changes are nearly infinitely numerically correlated in the binding expressions, as a direct consequence of the thermodynamic coupling between the sites.[9] Unique resolution is possible only by proper consideration of additional binding data for mutant operators, in which the number of operator sites has been reduced in various combinations.

Typical data for binding to the individual sites of O_R and of three mutants in which specific binding to one or two sites has been eliminated (reduced valency) by introduction of single base pair substitutions are presented in Fig. 3. Binding expressions for the individual-site isotherms are obtained by summation of the fractional probabilities for the appropriate operator configurations (Table III); for example, for the wild-type operator,

$$\bar{Y}_{O_R1} = f_2 + f_5 + f_6 + f_8 \tag{6a}$$

$$\bar{Y}_{O_R2} = f_3 + f_5 + f_7 + f_8 \tag{6b}$$

$$\bar{Y}_{O_R3} = f_4 + f_6 + f_7 + f_8 \tag{6c}$$

where $\bar{Y}_{O_Ri}$ denotes the fractional occupancy of site O_Ri. The model is easily extended to the mutants, by assuming that the mutations (single base pair substitutions) leave the interactions with the remaining sites quantitatively unaltered.[10] This leads to binding expressions; for example, for an O_R2^- operator in which only configurations 1, 2, 4, and 6 (Table III) exist,

$$\bar{Y}_{O_R1} = f_2 + f_6 = e^{-\Delta G_1/RT}[L]/(1 + e^{-\Delta G_1/RT}[L]) \tag{7a}$$

$$\bar{Y}_{O_R2} = 0 \tag{7b}$$

$$\bar{Y}_{O_R3} = f_4 + f_6 = e^{-\Delta G_3/RT}[L]/(1 + e^{-\Delta G_3/RT}[L]) \tag{7c}$$

As is evident, it is possible first to analyze separately the data for the noncooperative operators (O_R2^- and single-site) to resolve estimates of the ΔG_i's uniquely, and then to treat these as known quantities in the separate analysis of the cooperatively interacting, multisite operators, to obtain the $\Delta G_{ij(k)}$'s. However, more realistic estimates of the free energy changes are obtained by including the uncertainty in the estimation of the ΔG_i's, in a simultaneous analysis of all of the data.

[9] D. F. Senear, M. Brenowitz, M. A. Shea, and G. K. Ackers, *Biochemistry* **25,** 7344 (1986).

[10] Evaluation of the validity of this assumption for O_R has been extensively described [D. F. Senear and G. K. Ackers, *Biochemistry* **29,** 6568 (1990); D. F. Senear, M. Brenowitz, M. A. Shea, and G. K. Ackers, *Biochemistry* **25,** 7344 (1986)].

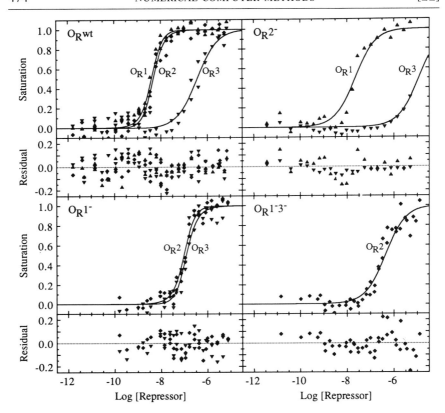

FIG. 3. Individual-site titration data for the interaction of λ cI repressor with wild-type and mutant λ right operators at pH 7.0, 200 mM KCl, and 20°. The data comprise 14 individual-site binding isotherms and represent 7 separate titration experiments. The panels showing the wild type, O_R1, and O_R2^- data each represent two independent experiments. Triangles, O_R1; diamonds, O_R2; downward triangles, O_R3. The solid lines are the results of simultaneous nonlinear least-squares fits according to the model expressed in Table II, and Eqs. (6) and (7), as described in the text and in Table III. The residuals calculated from these lines are plotted in the graphs below each of the data plots.

The basis of the footprint titration method is that a bound ligand (repressor) protects the DNA from an external cleavage reagent (e.g., DNase I). The quantity that is obtained in the experiment is not $\bar{Y}_{O_Ri}$, but rather the fractional protection of the site, denoted by P_{O_Ri}. Because even saturating protein ligand does not confer absolute protection, the P_{O_Ri} terms are linearly related to but not equal to $\bar{Y}_{O_Ri}$. Thus, the final form of the equations used in the nonlinear least-squares fitting is $P_{O_Ri,obs} = P_{O_Ri,o} + (P_{O_Ri,f} - P_{O_Ri,o})\bar{Y}_{O_Ri}$, where $P_{O_Ri,o}$ and $P_{O_Ri,f}$ are fitted lower and upper end points to the observed transition curves. The need to obtain both these transition end points from the data, as well as the detailed shapes of the transitions (see below), dictates a very wide range of concentrations in the experiments and the choice of a log concentration scale for the independent variable.

Separate Analysis of Individual Experiments

Simultaneous analysis of different experiments to resolve parameters that no one experiment can resolve places severe demands on the accuracy of the different experiments. Therefore, preliminary separate analysis of the individual experiments is essential. Systematic errors in the data for any one operator can cause highly inaccurate (though possibly still quite precise) parameter estimates. Although it may never be possible to eliminate this possibility, some insight into the potential for inaccurate experiments for any one operator is gained by assessing the reproducibility of the experiments for each of the different operators. This analysis is complicated by the fact that it is not possible to resolve all of the free energy changes from any one of the experiments. A particularly useful tool for this assessment is the individual site loading energy, $\Delta G_{1,i}$, derived by Ackers *et al.*[11] This quantity is the free energy change associated with saturating site i, including the effect of cooperative interactions with other sites that are also binding ligand. Based on Wyman's concept of the median ligand concentration,[12] it is calculated by numerical integration of the observed isotherm according to

$$\Delta G_1 = RT \int_{Y=0}^{1} \ln(X) \, d\bar{Y} \tag{8}$$

where X is the ligand activity and $\bar{Y}$ is the fractional saturation. Comparison of ΔG_1's between experiments provides a particularly sensitive and model-independent measure of reproducibility. Table IV illustrates this comparison. The data indicate reproducibility in line with the uncertainty from the separate experiments.

A second reason for preliminary analysis of the separate experiments is to calculate weighting factors for the simultaneous analysis. Weights account both for the different magnitudes of the fractional protection data for the different sites, and in the different experiments, and for different levels of random experimental noise (inherently high in footprint experiments). Both derive from the fact that the fractional protection afforded

[11] G. K. Ackers, A. D. Johnson, and M. A. Shea, *Proc. Natl. Acad. Sci. U.S.A.* **79**, 1129 (1982).
[12] J. Wyman, *Adv. Protein Chem.* **19**, 223 (1964).

TABLE IV
INDIVIDUAL-SITE LOADING FREE ENERGIES $(\Delta G_{1,i})^a$

Operator	Experiment[b]	$\Delta G_{1,1}$	$\Delta G_{1,2}$	$\Delta G_{1,3}$
$O_R$wt	A	-11.30 ± 0.26	-11.19 ± 0.27	-8.80 ± 0.40
	B	-11.40 ± 0.21	-11.09 ± 0.22	-8.33 ± 0.34
	Fitted[c]	-11.29	-11.14	-8.65
O_R1^-	A		-9.28 ± 0.23	-9.47 ± 0.23
	B		-9.30 ± 0.38	-9.29 ± 0.36
	Fitted[c]		-9.44	-9.24
O_R2^-	A	-10.15 ± 0.21		n.d.
	Fitted[c]	-10.26		-7.08
$O_R1^-3^-$	A		-8.91 ± 0.27	
	B		-8.53 ± 0.18	
	Fitted[c]		-8.57	

[a] From separate experiments for each operator and from the fitted curves obtained by simultaneous analysis. $\Delta G_{1,i}$'s were calculated by trapezoidal integration using Eq. (8).

[b] Different experiments are denoted A or B. The (unsmoothed) fractional protection data, transformed to $\overline{Y}_{O_R i}$ using the transition end points from the best phenomenological fit to each experiment (see text), were used in the calculation. Confidence limits for the calculated values, representing approximately one standard deviation, are estimated as described [D. F. Senear and G. K. Ackers, *Biochemistry* **29,** 6568 (1990)].

[c] Calculated from the fitted curves from the simultaneous analysis of the binding data from the four operators.

by bound ligand varies significantly from site to site. This dictates separate weighting of the data from each isotherm. Normalized weighting factors[13] are calculated according to

$$\omega_{ij} = N_t(1/\sigma_j^2)/\sum_j (N_j/\sigma_j^2) \tag{9}$$

where ω_{ij} is the weight for data point i of experiment (or isotherm) j, σ_j^2 and N_j are the variance and number of data points for experiment j, and $N_t = \Sigma_j N_j$.

A reasonable way to estimate σ_j^2 for a given isotherm is to analyze separately the experiment in a model-dependent but "phenomenological" sense. The goal is to find the minimum in the variance for each isotherm, independent of the other isotherms and experiments. Individual-site isotherms conform to no simple phenomenological model, for example, the Hill model, and, in fact, can be quite asymmetric with respect to their midpoints. Consequently, it is not possible to shortcut their analysis ac-

[13] P. R. Bevington, "Data Reduction and Error Analysis for the Physical Sciences." McGraw-Hill, New York, 1969.

cording to the complete model for the interactions (Table III). Because the $\Delta G_{ij(k)}$ terms are not resolved by any single experiment, different combinations are fixed as invariant input parameters, and the ΔG_i's (and transition end points) are fitted to the data. For two-site operators, ΔG_{ij} is varied monotonically. The shapes of the isotherms, hence the fits to the data, are only sensitive to ΔG_{ij} over a relatively modest range,[9] making this simple in practice. The wild-type operator features an infinite variety of combinations of $\Delta G_{ij(k)}$. In practice, a few carefully chosen combinations usually suffice to approximate the minimum σ_j^2.

Simultaneous Analysis of Wild-Type and Mutant Operator Titrations

Setting up the simultaneous analysis poses a slightly different problem than did the LEM example. Here, repressor concentration is the only experimental variable (x_1), but the different operators (and sites) have different intrinsic properties, expressed by different functional forms of the binding equation. In addition, the transition end points represent parameters unique to individual experiments. The mapping of parameters and functional forms is accomplished by introduction of two control variables, which are included in the data files as if they were additional independent variables, x_2 and x_3. x_2 is used to select a unique pair of parameters for transition end points for each isotherm. It also identifies the operator site (i.e., 1, 4, 7, . . . indicates O_R1; 2, 5, 8, . . . indicates O_R2; 3, 6, 9, . . . indicates O_R3) to select the correct form (a, b, or c) of the binding equation. x_3 identifies the binding competency of the operator. Sites are scored 1 for wild type (competent to bind) and 0 for mutant, and the binary digits encoded as the base 10 values 1–7. Thus, 5 (101) indicates O_R2^- and eliminates configurations 3, 5, 7, and 8 of Table III (i.e., those with ligand bound to O_R2) to give Eq. (7). Table V presents the data to be fitted, illustrating the application of x_2 and x_3.

The interaction free energy changes determined by simultaneous analysis of the data in Fig. 3 are as follows: $\Delta G_1 = -10.25 \pm 0.29$, $\Delta G_2 = -8.53 \pm 0.18$, $\Delta G_3 = -6.62 \pm 0.46$, $\Delta G_{12} = -3.62 \, (-3.98, -3.24)$, $\Delta G_{23} = -3.53 \, (-3.98, -3.05)$, and $\Delta G_{123} = -5.63 \, (-6.09, -5.13)$. The confidence intervals for the intrinsic free energies are similar to limits for individual isotherms, approximately 0.2–0.3 kcal/mol. These parameters are strongly influenced by the data for the noncooperative operators and tend to reflect the experimental noise and reproducibility of those experiments. The wide interval for ΔG_3 reflects the technical difficulty in saturating O_R3 in the O_R2^- operator, owing to very weak binding at this experimental condition. The $\Delta G_{ij(k)}$'s reflect the differences between wild-type and mutant operator isotherms. Consequently these suffer from all system-

TABLE V
DATA FILE FOR SIMULTANEOUS FIT OF INDIVIDUAL-SITE BINDING DATA[a]

| | | Columns | | |
Rows	$P_{O_R i, obs}$	Weight	x_1	x_2[b]	x_3[c]
1	-0.0306	0.8385	1.60×10^{-12}	1	7
.	.	.	.	.	.
.	.	.	.	.	.
22	0.9624	0.8385	8.35×10^{-6}	1	7
23	-0.0676	0.8026	1.60×10^{-12}	2	7
.	.	.	.	.	.
44	0.9751	0.8026	8.35×10^{-6}	2	7
45	0.0407	0.7401	1.60×10^{-12}	3	7
.	.	.	.	.	.
.	.	.	.	.	.
66	0.8805	0.7401	8.35×10^{-6}	3	7
67	-0.0438	1.585	4.84×10^{-12}	4	7
.	.	.	.	.	.
88	1.081	1.585	6.82×10^{-6}	4	7
89	-0.0976	1.457	4.84×10^{-12}	5	7
.	.	.	.	.	.
110	1.087	1.457	6.82×10^{-6}	5	7
111	-0.1252	0.5842	4.84×10^{-12}	6	7
.	.	.	.	.	.
132	0.9104	0.5842	6.82×10^{-6}	6	7
133	0.0264	1.237	3.67×10^{-12}	7	5
.	.	.	.	.	.
.	.	.	.	.	.
154	1.035	1.237	4.91×10^{-6}	7	5
155	0.0991	0.7637	3.67×10^{-12}	9	5
.	.	.	.	.	.
176	0.7543	0.7637	4.91×10^{-6}	9	5

[a] Representative data entries for two $O_R$wt and one $O_R 2^-$ operator experiments are in lines 1–66, 67–132, and 133–176, respectively.

[b] Isotherm and site identifier. A different value for each titration fits a unique pair of transition endpoints ($P_{O_R i, o}$ and $P_{O_R i, f}$) to the protection data ($P_{O_R i, obs}$). By iteratively subtracting 3 until a value, 1–3, is obtained, the titration is identified as $O_R 1$, $O_R 2$, or $O_R 3$.

[c] Operator binding competency. A 7 (111) indicates $O_R$wt and fits Eq. (6) to the data. A 5 (101) indicates $O_R 2^-$ and fits Eq. (7).

TABLE VI

VARIANCES FROM SEPARATE AND SIMULTANEOUS ANALYSIS OF WILD-TYPE AND
MUTANT OPERATORS[a]

Operator	σ from separate analysis[b]	σ from simultaneous analysis[c]	Variance ratio[d]	F[e]
$O_R$wt	0.0827	0.0853	1.063	1.074
O_R1^-	0.0829	0.0888	1.147	1.106
O_R2^-	0.0728	0.0745	1.049	1.135
$O_R1^-2^-$	0.0939	0.0972	1.069	1.134

[a] Variances for the data shown in Fig. 3.
[b] Square roots of the variances (σ) from separate analysis of the different operators.
These reflect the best phenomenological fit to those data, as described in the text.
[c] Square roots of the variances (σ) calculated separately for the data for each operator,
using the free energy changes and transition end points from the simultaneous analysis.
[d] Ratio of variances (σ^2) from simultaneous analysis to those from separate analysis.
[e] Variance ratio corresponding to the 65% confidence level predicted by the F statistic
(Table I).

atic errors and deviations from the model, as reflected in wider and less symmetrical confidence intervals.

The issue of how well the data for each of the different operators are described by the simultaneous analysis is crucial here, because of the assumption made in the application of the molecular model to the reduced valency operator mutants. Three criteria are useful in judging the goodness of fit and completeness of the model. First, the individual site loading free energy changes calculated for the fitted curves (Table IV) fall nicely within the limits imposed by the individual experiments in every case. Second is the apparently random distribution of residuals for each of the different operators (Fig. 3). The third criterion is the favorable comparison between the variances calculated from the simultaneous analysis for each of the separate operators and those from the best phenomenological fits to each of the operators (Table VI).

The later two criteria are particularly important in considering the consistency of the model with the data. These provide the only means to assess correspondence between the observed shapes of the individual-site isotherms for the different operators and the fitted curves. The shapes are indicative of cooperative interactions; for example, O_R1–O_R2 cooperativity, which greatly affects the binding to wild-type O_R, is evident in the steepness of the transitions for O_R1 and O_R2 (despite the appreciable noise in the data). The mathematical model and parameter values constrain the shapes of the isotherms, as well as their midpoints. Of course, by allowing any combination of $\Delta G_{ij(k)}$'s that best fit the data, the best phenomenological fits to the individual operators consider virtually all possible curve shapes. In this case, the favorable comparison between the variances from the simultaneous and best phenomenological fits (Table VI) provides confidence that the model correctly describes the macromolecular interactions between repressor and both wild-type and mutant operators.

The traditional method for analyzing mutant and wild-type binding or rate of transcription data for gene regulatory systems is to analyze each experiment separately and to take ratios of apparent macroscopic equilibrium constants or titration midpoints to estimate intrinsic and cooperative microscopic equilibrium constants. Often the macroscopic constants are taken as errorless quantities in the comparisons. Simultaneous nonlinear least-squares analysis offers significant advantages over this approach, which more than justify the extra effort. First, by simultaneously applying the least-squares criterion to all of the data more realistic estimates of the constants and their precision are obtained. At the usual expense of some loss of apparent precision, a greater chance of accuracy is achieved. For example, reports of 2-fold cooperative effects from comparisons of ratios of apparent constants are common. However, our analysis points to the difficulty in measuring apparent constants to a precision better than about a factor of 2 and indicates that cooperative effects of less than 3- to 4-fold are probably meaningless. Because cooperativity provides the means for control of the transcriptional activity of promoters by regulatory factors that bind to DNA, proper characterization of cooperative effects is crucial to developing an understanding of the molecular mechanism of the regulation. Second, simultaneous analysis provides the only means to evaluate critically whether a proposed model is consistent with all of the data, namely, by taking into account the detailed shapes of the isotherms for the wild-type and mutant operators.

Concluding Remarks

Our purpose in this chapter was 2-fold: (1) to introduce nonlinear least-squares techniques which provide direct means of analysis for complex models and (2) to illustrate by example the powerful technique of simultaneous analysis for resolving parameters and testing the adequacy of the model used by the investigator. Transforming the mathematics of a complex model to some linearized form in order to evaluate parameters of interest has marked a long and important phase in the development of science. But the complexities of error propagation in these linearized equations makes it extremely difficult to provide adequate confidence intervals for the parameters of interest. The ready availability of nonlinear least-squares analysis programs which can operate on nearly any type of computer obviates the need for linearization transforms and allows the investigator to fit the primary data in a direct manner without having to produce and analyze primary and secondary plots.

Rigor in analysis of error and objectivity in model-dependent analysis can and should be an essential part of all published work. Simultaneous analysis provides a means to achieve these ends by offering a way to resolve highly correlated parameters as well as a way of testing the adequacy of the models being used. For most investigators, such information provides both guidance and enlightenment.

Acknowledgments

Research was supported in part by National Institutes of Health Grant GM41465 to D.F.S. and by NIH Grant GM22300 to D.W.B.

[23] Numerical Analysis of Binding Data: Advantages, Practical Aspects, and Implications

By CATHERINE A. ROYER and JOSEPH M. BEECHEM

Introduction

Progress in recent years in the development of sensitive and accurate instrumentation for the study of macromolecular complexes has had the general result of providing biophysical chemists with much more information concerning the important role of energetically coupled interactions in regulating biological activity. The need therefore arises for binding data analysis algorithms which can easily incorporate this increasing complexity. Utilization of more complicated binding models often requires abandoning the simple closed form analytical expression derived for simple systems. When faced with such complicated schemes, investigators in many areas of physics and chemistry choose to consider their particular problem in terms of systems of equations to be solved using a numerical methodology. In addition to expanding the scope of the models which can be examined, moving to a numerical method of calculating binding profiles presents the advantage of generalizing the problem under consideration.

In the field of biological chemistry, investigators are increasingly confronted with binding problems of relatively high complexity. For example, many systems of importance in the field of transcriptional regulation involve simultaneous equilibria between monomers, oligomers, and heterooligomers, all interacting with small ligands and other effectors, as well as with nucleic acids. Although most biological binding data analysis methodologies still rely on the derivation of analytical forms for the binding profile, a few investigators, owing to the complexity of their systems, have developed numerical methods for binding equilibrium calculations. Rodbard and Munson[1] developed a program, LIGAND, for ligand receptor binding that uses an algorithm which solves for the root of a polynomial describing their ligand-binding partition function. While quite useful in treating the receptor binding problems, this approach has not been adapted to cases such as those mentioned above in which the polypeptide as well as multiple ligand binding equilibria must be simultaneously considered. Avery,[2] in the context of investigations of complex biological kinetic phenomena, developed an algorithm for calculating the equilibrium concentration of the components in the system to be used as a starting point in perturbation kinetics experiments. The algorithm was based on a projection method for simultaneously solving the multiple free energy equations, and it uses a conceptual framework quite similar to that of the algorithm described in this chapter. It was, however, never integated into a general-purpose binding data analysis program.

We have recently developed a computer program (BIOEQS) for the analysis of ligand binding and subunit assembly data for macromolecular complexes.[3] It consists of a nonlinear least-squares data analysis program, in which the calculated binding surface is obtained using a numerical constrained optimization chemical equilibrium solver first implemented in the fields of atmospheric and combustion chemistry and refined by Smith and Missen.[4,5] In fact, Brinkley[6] in 1946 first developed a numerical algorithm for the calculation of chemical equilibrium in complex chemical systems. His algorithm was a very early version of that used in the program presented here, which is based on the Brinkley–NASA–RAND algorithm (BNR).[4] The major feature of the numerical solver is that its direct output corresponds to the concentrations of all of the possible species in a proposed model. These species concentrations can be combined to generate any possible binding profile related to some observable. In macromolecu-

[1] P. J. Munson, this series, Vol. 92, p. 543.
[2] L. Avery, *J. Chem. Phys.* **76**, 3242 (1982).
[3] C. A. Royer, W. R. Smith, and J. M. Beechem, *Anal. Biochem.* **191**, 287 (1990).
[4] W. R. Smith and R. W. Missen, "Chemical Reaction Equilibrium Analysis." Wiley, New York, 1982.
[5] W. R. Smith and R. W. Missen, *Can. J. Chem. Eng.* **66**, 591 (1988); The simulations program, EQS, is available from Technical Database Systems, 10 Columbus Circle, New York, NY 10019.
[6] S. R. Brinkley, *J. Chem. Phys.* **14**, 563 (1946).

lar systems, this observable quantity is often, though not necessarily, the fractional ligand saturation or degree of protein oligomerization.

The data analysis program, BIOEQS, has been designed for the simultaneous analysis of multiple binding data sets. Thus, for example, data sets indicative of the degree of ligation can be simultaneously analyzed with data sets indicative of the degree of oligomerization for a consistent set of free energy parameters defining the macromolecular system. It has incorporated the general features of the global analysis software which has been recently developed for the analysis of fluorescence spectroscopic data.[7-9] These features include general ASCII input/output, simulations capabilities, and calculation of rigorous recovered parameter error confidence intervals.

Because the numerical details of the program can be found elsewhere,[3-5] this chapter will concentrate on other important aspects of the application of a numerical methodology to the binding question. In the first section, the advantages of the numerical method over analytical methods will be discussed. The second section will provide an overview of the numerical methodology utilized within BIOEQS for solving complicated binding equilibria. The third section will serve to illustrate how an example system is treated using a numerical solver. Typical binding data will be simulated for the model presented, and a demonstration of the fit and output will be given. Finally, in the last section we will explore the information content of multidimensional binding data and discuss the unique recovery of the free energy values in complex systems.

Advantages of Numerical Approach

A large number of investigators have, over the last 70 years, employed various analytical forms to describe ligand binding to biological macromolecules. The extent of this literature is such that no attempt is made here to review these analytical forms. By analytical form, it is meant any analysis procedure which utilizes an explicit binding equation of the form

$$\bar{v} = f(x) \tag{1}$$

where $\bar{v}$ corresponds to some binding observable and $f(x)$ is an explicit function of the independent variables of the system.

[7] J. M. Beechem, J. R. Knutson, J. B. A. Ross, B. W. Turner, and L. Brand, *Biochemistry* **22,** 6054 (1983).

[8] J. M. Beechem and E. Gratton, *Proc. SPIE Int. Soc. Opt. Eng.* **909,** 70 (1988).

[9] J. M. Beechem, E. Gratton, M. A. Ameloot, J. R. Knutson, and L. Brand, *in* "Fluorescence Spectroscopy: Principles" (J. R. Lakowicz, ed.), Vol. II, Ch. V. Plenum, New York, in press.

When using a data fitting program based on a closed form analytical equation for the calculated binding profile, the investigator is confronted with two types of constraints. The first constraint is experimental. Even if one is to consider relatively simple systems (e.g., a nondissociating protein molecule which binds multiple ligands of the same kind), the concentration of free ligand must be supplied in order to calculate a binding profile. From an experimental point of view, this obliges investigators to carry out their binding studies under conditions in which the ligand concentration is much higher than the protein concentration. In this manner the free ligand concentration can be considered to be approximately equal to the total ligand concentration. Unfortunately, such an approach severely limits the concentration range over which the system can be studied. In fact, the regulatory events of interest to biophysical chemists often occur under conditions where the protein and ligand are at similar concentrations in the cell. Another alternative is to work with systems in which the free ligand concentration can be very accurately measured. However, not all systems are amenable to this requirement. In addition, a certain degree of experimental error is necessarily associated with such measurements. Using a numerical method for calculating binding profiles allows for the system to be explored under any concentration conditions which are within the detection capabilities of the observation apparatus. In fact, the ability to treat data under any concentration conditions may well prompt investigators to employ additional observation techniques in order to expand the concentration range. Experimentally, one is no longer limited by the ability to analyze the data but, more appropriately, by the ability to collect them.

Another advantage of the numerical methodology is theoretical in nature. Many macromolecular systems of interest in biochemistry are much more complex than the previous example and involve many simultaneous equilibria, including protein–protein, protein–ligand, and/or protein–nucleic acid interactions. Often the ligand or nucleic acid binding equilibria are energetically coupled to the subunit association equilibria. From a mechanistic point of view, this means that interactions at the interface between subunits are used to regulate the binding of ligand or nucleic acid. Such information is crucial to the understanding of the underlying physical mechanisms of regulation. To determine if such interactions play a role in the regulation of binding for a particular protein system, the study of the subunit association equilibria as well as the study of the ligand binding equilibria should be undertaken. Using analytical expressions for the binding isotherms limits the number of species which can be considered, and thus, only a small subset of the possible models for binding can be treated.

This limitation can be easily understood by considering a very simple oligomeric system in absence of any ligand. Let us take as an example a tetrameric protein molecule which can dissociate into its subunits. The ability to consider the existence of the dimeric form of the protein under any particular conditions of total protein concentration necessitates the use of a numerical algorithm. The analytical form of the equilibrium may only be used to express direct dissociation of tetramer to monomer as follows:

$$K_d = 256P_o^3\alpha^4/(1 - \alpha) \tag{2}$$

where α is the degree of tetramer dissociation to monomer and P_o is the total protein concentration expressed as tetramer. The probability of such a fourth order collisional event is highly unlikely. Where the subunit interactions of allosteric oligomeric proteins have been investigated (i.e., hemoglobin, *trp* and *lac* repressors), it has been demonstrated that the dissociation occurs in a stepwise fashion. The relative stability of the dimer with respect to the monomer and tetramer determines the protein concentration range over which oligomer dissociation occurs. Thus, to express a stepwise model for tetramer (T) dissociation to dimer (D) followed by dimer dissociation to monomer (M), a numerical algorithm for the calculation of the concentration of all of the three possible species must be employed.

Possible free energy diagrams for a T–D–M system are shown in Fig. 1. In Fig. 1a, the dimer is shown to be relatively stable with respect to the monomer and tetramer. The relative chemical potential for the dimer is -7.5 kcal/mol, whereas that of the tetramer is -18.0 kcal/mol. The monomer level is taken to be the zero level. In Fig. 2A are plotted the fraction of tetramer, dimer, and monomer as well as the degree of tetramer dissociation obtained from simulations using the above free chemical potentials. It can be seen that as the concentration of protein decreases (right to left on the x axis) monomer formation is separated from tetramer dissociation by a large concentration range in which dimer attains nearly 90% of the total protein. In the free energy diagram in Fig. 1b, the stability of the dimer with respect to tetramer has been decreased by 1.5 kcal/mol. As can be seen in the plots in Fig. 2B, the decrease in fraction of monomer is now overlapping the increase in fraction of tetramer, and the fraction of dimer only reaches 40%. The profile of the degree of oligomer dissociation, α, now decreases over approximately 2.5 log units of protein concentration compared to the 5 log units in Fig. 2A. In Fig. 1c, the stability of the dimer has been further decreased to only -3.0 kcal/mol. In this case (Fig. 2C) there is no appreciable dimer in solution, although the mechanism necessarily includes this step. The degree of dissociation now decreases from 90 to 10% with increasing protein concentration over approximately

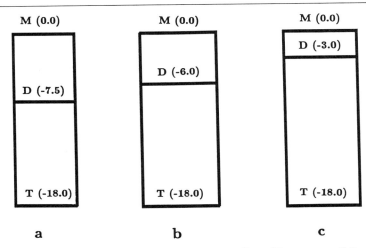

FIG. 1. Free energy diagrams for three tetramer (T)–dimer (D)–monomer (M) systems. Free energy values correspond to the chemical potentials of the relative species and are in units of kcal/mol. (a) Relatively stable dimer, chemical potential of -7.5 kcal/mol; (b) slightly less stable dimer, chemical potential of -6.0 kcal/mol; (c) highly unstable dimer, chemical potential of -3.0 kcal/mol.

1.6 log units, approaching the log span of the analytical expression in Eq. (1).

Clearly, the ability to consider the role of protein–protein interactions in regulating ligand affinity, and thus function, is necessary for the elucidation of cooperative and allosteric phenomena. Of course, in certain instances, the affinities are such that some of the possible species may be ignored. Such is the case for human hemoglobin, where the affinities between the monomers in the $\alpha\beta$ dimer are so high that monomer is, for all practical purposes, never observed. However, this is not a general rule. For a number of interesting genetic regulatory proteins (repressors, transcription factors, and receptors), the subunit affinity is linked with nucleic acid and ligand binding. In fact, the analytical forms generally utilized in binding analysis constitute only a small subset of the possibilities available with a numerical methodology. Thus, the application of a numerical methodology to the binding question can be viewed as more than a technical advance. Its scientific value lies in its ability to free the imagination of the investigator to consider a number of different binding models. In many cases, more than one model may be found to be consistent with the data. To discriminate between models, the investigator is prompted to increase the quantity, quality, and dimensionality of the data. In this manner, the data analysis environment serves both as interpreter of and impetus to additional experimentation.

Inner workings of BIOEQS Program

Any macromolecular assembly can be defined by a system of N species made up of M elements which are linked by a system of R independent equilibrium equations. For these macromolecular assemblies, we may think of the elements of the system as corresponding to the smallest nondissociable protein unit (monomer, for example), or nucleic acid, or small effector molecules. The (R) independent free energy equations of these systems can be expressed in terms of macroscopic equilibrium constants,

$$\Delta G_j = -RT \ln K_j; \qquad j = 1, 2, \ldots, R \qquad (3)$$

where ΔG is the standard free energy change for the jth chemical reaction and

$$K_j = \prod_{i=1}^{N} x_i^{\nu_{i,j}}; \qquad j = 1, 2, \ldots, R \qquad (4)$$

where N is the number of species, ν_{ij} is the stoichiometric coefficient of the ith species in the jth equation, and x_i is the concentration of species i. The stoichiometric coefficients, ν_{ij}, are the elements of the stoichiometric matrix for the system, which satisfies Eq. (5):

$$\sum_{i=1}^{N} a_{k,i} \nu_{i,j} = 0; \qquad k = 1, 2, \ldots, M; \qquad j = 1, 2, \ldots, R \qquad (5)$$

where $a_{k,i}$ or (A) is the formula matrix. In our case, the matrix (ν_{ij}) is, in general, of rank R, where $R = N\text{-Rank}(A)$. In most situations $M = \text{Rank}(A)$. A more detailed treatment of the formation of these matrices may be found elsewhere.[3,4]

Given Eq. (1), one can find a set of standard chemical potentials, μ_i°, one for each species, which satisfy Eq. (6):

$$\sum_{i=1}^{N} \nu_{ji} \mu_i^\circ = \Delta G_j^\circ; \qquad j = 1, 2, \ldots, R \qquad (6)$$

Thus, EQS, the equilibrium solver used in BIOEQS, solves for the concentrations of all of the possible species defined in the system given their relative chemical potentials. From the species concentrations all possible binding profiles can be generated. In biochemistry, we are more familiar with equilibrium constants defining particular binding reactions. However,

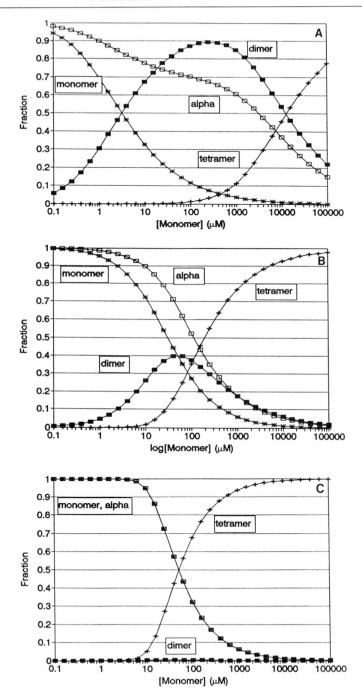

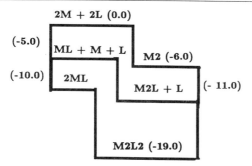

FIG. 3. Free energy diagram for a dimeric protein which binds two ligand molecules in a cooperative manner. Ligand binding favors dimerization, and thus the system is of second order; that is, it is the binding of the second ligand which brings about the most significant increase in subunit affinity. Free energy values in kcal/mol correspond to relative species chemical potentials.

it is possible to convert these equilibrium constants, and thus the free energies of the system, into a set of relative species chemical potentials. In fact, this transformation simply consists in finding another set of R independent free energy relations corresponding to the reactions in which the species are formed directly from the elements. To obtain such a solution to Eq. (4) one must set the chemical potentials of the elements to zero and equate the chemical potentials of the remaining species to their free energies of the formation. When working with macromolecular assemblies, it is always possible to set up the systems using this scheme, because the elements in such systems also exist as species. It is thus useful, when defining models for data analysis or simulation, to draw free energy diagrams such as those shown in Figs. 1 and 3. The relative species chemical potentials are obtained by summing the appropriate binding energies. For example, in Fig. 1a, the chemical potential of the tetramer (-18.0 kcal/mol) is actually the sum of the free energy corresponding to dimerization (-7.5 kcal/mol) and that for tetramerization (-11.5 kcal/mol).

FIG. 2. Simulations of subunit interactions for the three systems represented in Fig. 1. The fractional population of the three species, monomer (*), dimer (■), and tetramer (+), are plotted as a function of the concentration of protein expressed as monomer. In addition, the degree of tetramer dissociation (alpha, □) is also plotted. The dimer was considered to be 67% dissociated and the monomer, 100%. The plots in (A) correspond to simulations from the stable dimer system (Fig. 1a), those in (B) to the slightly destabilized dimer (Fig. 1b), and those in (C) to the highly unstable dimer (Fig. 1c).

The species concentration vector satisfies the law of conservation of mass equations:

$$\sum_{i=1}^{N} a_{ki} x_i = b_k; \qquad k = 1, 2, \ldots, M \tag{7}$$

where b_k is the elemental abundance (or total concentration) of element k. EQS, the equilibrium algorithm at the heart of the BIOEQS program, proceeds by minimizing G, the Gibbs free energy of the system, with respect to x, subject to the constraints of Eq. (7), where

$$G = \sum_{i=1}^{N} x_i \mu_i = \sum_{i=1}^{N} x_i (\mu_i^\circ + RT \ln x_i) \tag{8}$$

We illustrate here, using a very simple case found in the text by Smith and Missen,[4] the use of Lagrange multipliers in solving chemical equilibrium problems. Consider a system with one element (protein monomer, M) and two species (protein monomer M and protein dimer M_2). In such a system as state above, R (the number of independent equations) equals N (the number of species) minus M (the number of elements). This single equation corresponds to the association of two monomeric subunits to form dimeric protein. Calculating the concentrations of monomer and dimer at any given total protein concentration is essentially a problem of minimizing the total free energy of the system

$$G(x_1, x_2) = x_1 \mu_1 + x_2 \mu_2 \tag{9}$$

subject to the mass balance constraint

$$a_1 x_1 + a_2 x_2 = b \tag{10}$$

as per Eq. (7), where a_1 and a_2 are the stoichiometric coefficients of the element M in monomer ($a_1 = 1$) and dimer ($a_2 = 2$) and where x_1 is the concentration of monomer (M) and x_2 is that of the dimer (M_2). When G is at a minimum,

$$dG = \mu_1 \, dx_1 + \mu_2 \, dx_2 = 0 \tag{11}$$

Thus,

$$dx_2/dx_1 = -\mu_1/\mu_2 \tag{12}$$

and from Eq. (10),

$$dx_2/dx_1 = -a_1/a_2 \tag{13}$$

It follows that

$$a_1/a_2 = \mu_1/\mu_2 = \lambda \tag{14}$$

where λ is the single Lagrange multiplier for the single element in this system. The solution for the concentrations of x_1 and x_2 (monomer and dimer in this case) can be envisioned as the point where a line representing x_1 versus $x_2 = b$ (the constraint line) becomes tangent to a contour of values of x_1 and x_2 describing a constant free energy.

For the more general case of more than one element ($M > 1$) and more than two species ($N > 2$), the Gibbs function is minimized using an adjustment vector, δx, for the species concentrations given on iteration m by

$$\delta x_i^{(m)} = x_i^{(m)} \left(\sum_{k=1}^{M} \lambda_k^{(m)} a_{ki} - \mu_i^{(m)} \right); \qquad i = 1, 2, \ldots, N \qquad (15)$$

where λ_k is the Lagrange multiplier for the kth element of the system. The concentrations at the end of each iteration are given by:

$$x_i^{(m+1)} = x_i^{(m)} + \omega^{(m)} \delta x_i^{(m)}; \qquad i = 1, 2, \ldots, N \qquad (16)$$

where ω is a step length parameter that varies between zero and one. To satisfy the mass balance constraints when $\omega = 1$;

$$\sum_{i=1}^{N} a_{ki} \delta x_i^{(m)} = b_k - b_k^{(m)} \equiv \delta b_k^{(m)}; \qquad k = 1, 2, \ldots, M \qquad (17)$$

The main equation used in the constrained minimization algorithm is obtained from Eqs. (15)–(17):

$$\sum_{k=1}^{M} \lambda_k^{(m)} \sum_{i=1}^{N} a_{ki} a_{li} x_i^{(m)} = \delta b_l^{(m)} + \sum_{i=1}^{N} a_{li} x_i^{(m)} \mu_i^{(m)}; \qquad l = 1, 2, \ldots, M \qquad (18)$$

Thus, only M linear equations are in fact solved iteratively for the values of λ_k. These values of λ_k are substituted into Eq. (15) for the calculation of dx^m, which is then used in Eq. (16) to calculate the species concentration vector, x^{m+1}, for iteration $m + 1$. The algorithm iterates until the ratio dx^m/x^m is less than a small number (e.g., 0.000005), at which point convergence is reached. Further numerical details of the algorithm are given elsewhere.[4,5]

Illustration of Numerical Methodology

To illustrate how one approaches binding problems in a numerical context we will consider the case of a dimeric protein which binds two ligands in a cooperative manner. First, we will demonstrate the procedure for setting up the model for this system, which constitutes the input to the BIOEQS program. We will then simulate three-dimensional data sets (with

TABLE I
SPECIES AND CHEMICAL POTENTIALS

Species	Description	μ_0 (kcal/mol)
ML	Once liganded monomer	−5.00
M_2	Dimer	−6.00
M_2L	Once liganded dimer	−11.00
M_2L_2	Twice liganded dimer	−19.00
M	Free monomer	0.0
L	Free ligand	0.0

5% error) corresponding to titrations of the protein by ligand, observing both the degree of ligation and the degree of dimer dissociation. The simulated data will be analyzed, and the recovered free energy parameters will be compared to those used in the simulation. In the last section of this chapter we will consider the question of unique recovery of the free energy parameters from the examination of such complex systems.

The Model

The first step consists of setting up the macromolecular model. There are six possible species for a system composed of two elements (monomeric protein M and ligand L) in which each subunit can bind one ligand and the subunits can dimerize. The species are listed in Table I. A free energy diagram for this system may be found in Fig. 3. It can be seen from Fig. 3 that ligand binding in this model is cooperative and favors dimerization. It can also be seen that this system can be fully described by four independent free energy equations $[R (4) = N (6) − M (2)]$. The formula matrix for this particular system can be written as follows:

FORMULA MATRIX

	Species					
Element	M	L	ML	M_2	M_2L	M_2L_2
M	1	0	1	2	2	2
L	0	1	1	0	1	2

As labeled above, the rows represent the elements, whereas the columns correspond to the species. Thus, the once liganded dimer (M_2L) contains 2 mol of monomer and 1 mol of ligand, and its column vector is thus 2, 1. The reader will note the particular order in which the columns and rows have been arranged. The general trend in the species (column) ordering is to begin with the elemental species and to continue in increasing complexity. If the elements are ordered in the rows as they appear in the columns, then one has, at the left-hand side of this matrix, an $M \times M$ identity matrix.

The formula matrix is then used to construct the stoichiometric matrix mentioned in the previous section and which contains one of the possible sets of independent reactions. To construct the stoichiometric matrix one proceeds as follows. The identity matrix on the left-hand side of the formula matrix is removed, the remaining matrix is multiplied by -1, an $R \times R$ identity matrix is placed below it, and finally the rows of this new matrix are labeled as were the columns of the original formula matrix. The resulting matrix (shown below) is referred to as the stoichiometric matrix. This matrix would have been different had the formula matrix been set up with different ordering of the columns and rows.

STOICHIOMETRIC MATRIX

Species	Reaction			
	1	2	3	4
M	-1	-2	-2	-2
L	-1	0	-1	-2
ML	1	0	0	0
M_2	0	1	0	0
M_2L	0	0	1	0
M_2L_2	0	0	0	1

As it stands, the reactions represented in this stoichiometric matrix correspond to the particular solution to Eq. (6) in which the set of four independent free energies are those of formation of the nonelemental species from the elements, themselves. One need not of course write down these matrices for each system. Rather, one simply sets the chemical potentials of the elements to zero and defines the relative chemical potentials of the remaining species by adding up the free energies of the various reactions necessary for the formation of the species from the elements. When fitting data from such a system, these chemical potential values would serve as initial parameter guesses for the analysis program. Referring to Table I and Fig. 3, the chemical potentials of the elemental species (M and L) are set to 0. The free energy for dimerization to M_2 is -5.0 kcal, whereas that for ligation of the monomer to give ML is -6.0 kcal. The partial ligation of the dimer to give M_2L is -6.0 kcal, and thus the chemical potential of the once liganded dimer is $-(5.0 + 6.0) = -11.0$

TABLE II
OBSERVABLE QUANTITIES OF BIOEQS PROGRAM

Keyword	Description
alig1	Fraction of protein not bound by ligand1
alig2	Fraction of protein not bound by ligand2
blig1	Fraction of ligand1 not bound by protein
blig2	Fraction of ligand2 not bound by protein
aprot	Degree of oligomer dissociation
frac XX	Fractional population of any given species, XX

kcal/mol. Ligand binding is cooperative; thus, the binding of the second ligand to the dimer is more favorable than the binding of the first, -8.0 compared to -6.0 kcal/mol. The relative chemical potential of the twice liganded dimer, M_2L_2, is thus $-(11.0 + 8.0) = -19.0$ kcal/mol.

For any system one wishes to consider, it is convenient to draw such a free energy diagram. It can be seen that besides the zero level, there are only four chemical potential values specified. These four values are entered as the initial guess parameter values for the four independent free energies describing the system. In macromolecular systems, because the elements are also species, it is always true that the number of free energy equations (the reactions correspond to the formation of the species from the elements) is equal to the number of species minus the number of elements (in this case, $6 - 2 = 4$). For a more complete treatment, see Smith and Missen[4] and Royer and co-workers.[3]

For the fitting of the data, one must also create an observable mapping matrix. This matrix specifies how the calculated concentrations of the individual species, provided by the numerical solver, must be combined to yield the observable function, for example, degree of ligation or degree of protein oligomer dissociation. These observable quantities are normalized binding data. By normalized data we refer to dimensionless numbers between 0 and 1.0 which correspond to the observable quantities described in Table II. These values must be independently obtained from the actual experimentally measured quantity using the relationship predetermined by the investigator. For example, in the case of the *lac* repressor protein, the degree of inducer dissociation (alig1 in Table II) maybe calculated from measurements of the fluorescence emission energy, since inducer binding causes the tryptophan spectrum to shift to shorter wavelengths.[10] In the mapping matrix shown here there is one observable parameter per element, which corresponds to its degree of dissociation. The observable

---

[10] S. L. Laiken, C. A. Gross, and P. H. Von Hippel, *J. Mol. Biol.* **66,** 143 (1972).

mapping matrix corresponding to the system under consideration here is given below.

OBSERVABLE MATRIX

| | Observable | |
Species	α_1	α_2
ML	1.0	0.0
M_2	0.0	1.0
M_2L	0.0	0.5
M_2L_2	0.0	0.0
M	1.0	1.0
L	1.0	0.0

The rows in this matrix denote the species, while the columns denote the fractional contribution of a species to an observable. Thus, the first represents the degree of protein oligomer dissociation. Because the highest order oligomer in this system is the dimer, the monomeric species are 100% dissociated while the dimeric species are 0% dissociated. The second column represents the fraction of protein not bound by ligand. Thus, the species ML is 0% dissociated because the binding sites of the monomers are saturated. The once liganded dimer is 50% unsaturated while the unliganded monomer is 100% unsaturated. Other observables are possible, including the fractional population of a particular species. Table II gives a list of those currently available. In principle any combination of the species concentrations may be readily added to the mapping algorithm to provide additional observables as experimentation warrants. The observable keywords, the independent axis variable, the number and names of the data files, and the total concentrations of each element in each experiment must be supplied to the program. Further information and instructions concerning the specific ASCII input file format of the BIOEQS program are available from the authors.

In the simulations carried out in this example, 20 experiments are to be simulated and then back-analyzed. The observable quantity for the first 10 experiments is the degree of dissociation of ligand 1 (alig1), whereas that for the second 10 experiments is the degree of dimer dissociation (aprot). The independent axis in both data sets is the ligand concentration. The protein concentration ranges from 0.1 to 398 μM. In general a model must be associated with each experiment. However, in the example given here all of the experiments are carried out with the same two elements (M and L), and a single model describes all of the experiments. The framework, however, is general, such that combinations of experiments involving different (yet overlapping) models may be considered. For example, one can imagine an oligomeric protein which binds two types of ligands. Some experiments could be carried out in presence of one or the other ligand, some with both ligands present, and some with only the protein itself. Evidently, the model would differ for all of these types of experiments since the number and type of elements and species are not the same. In addition, although some of the chemical potential parameters would be shared among experiments, others would be unique to a particular type of experiment. Thus, there may be local and global fitting parameters. This permits the simultaneous analysis of all of the experiments on a particular system. For more information about local and global fitting parameters, see Ref. 9.

In Fig. 4 are shown the three-dimensional (3-D) data surfaces generated by simulating binding curves for the model in Fig. 3. In this case no error was added to the simulated data. The 3-D surface in Fig. 4A corresponds to the first 10 ligand titration experiments in which the observable quantity (the z axis in the Fig. 4A) corresponds to the degree of ligand dissociation. The x axis is the concentration of ligand, and the y axis points correspond to the 10 protein concentrations at which the ligand titration was performed. In Fig. 4B, the x and y axes are the same as in Fig. 4A, but now in these 10 experiments the observable quantity (or z axis) corresponds to the degree of dimer dissociation. It is important to note the protein concentration dependence of the curves in Fig. 4A as well as the ligand concentration dependence of those in Fig. 4B.

Simulations using the same free energy scheme were then carried out with a 5% random error associated with the data, and the resulting binding surfaces (again, a total of 20 titration curves containing 10 data points each) were analyzed using initial guesses for the four chemical potentials which differed by approximately 2 kcal from the values used to simulate the data. The minimization was obtained in nine iterations, with a final global χ^2 value of 0.918. It took approximately 12 min on an IBM PC compatible 80386 computer. The actual recovered chemical potential values are quite close to those used for the simulation of the data and are summarized in Table III. The first value differed by 0.028 kcal/mol, the second by 0.041 kcal/mol, the third by 0.3 kcal/mol, and the fourth by 0.1 kcal/mol. The results from 2 of the 20 individual binding curves are shown in Fig. 5 in which the experimental data, the calculated data from the fit, and the residuals are plotted. The BIOEQS program constitutes a complete data analysis environment. Because it is based on general ASCII model information input/output, it can be easily used for many different applications. The speed of the fitting algorithm is comparable to other global data

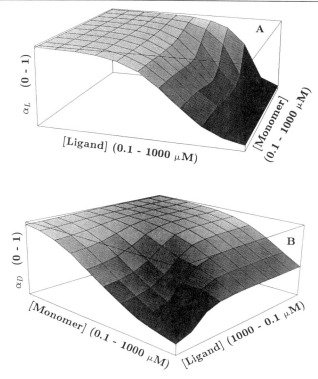

FIG. 4. Three-dimensional data surfaces of the data simulated for the twice liganded dimer system shown in Fig. 3. In both plots, the x axis corresponds to the concentration of protein expressed as monomer in μM units, and the y axis is the concentration of ligand, also in μM. In (A) the z axis corresponds to the degree of ligand dissociation (or fraction of protein not bound by ligand), whereas in (B) the z axis corresponds to the degree of dimer dissociation. The scale is in both cases 0–1.

analysis programs. Finally, the ASCII output allows for ease of interface with commercially available graphics packages.

Unique Parameter Recovery

It is of course a simple matter to evaluate the parameters recovered from a data analysis procedure when the actual values are known, as is the case with simulated data. However, this exercise serves as a valuable method of testing whether the analysis program functions correctly. More importantly, generating and subsequently analyzing data surfaces allows

TABLE III
RESULTS OF BIOEQS ANALYSIS

Parameter	Target value (kcal/mol)	Recovered value[a] (kcal/mol)	± Error (kcal/mol)	Recovered value[b] (kcal/mol)
μ_1	−5.00	−4.97	+0.11/−0.04	−5.58
μ_2	−6.00	−5.96	±0.10	−5.90
μ_3	−11.00	−11.30	—	−12.47
μ_4	−19.00	−18.90	+0.18/−0.30	−19.27

[a] Simultaneous analysis of 10 data sets mapping degree of ligand dissociation along with 10 data sets mapping degree of dimer dissociation.

[b] Analysis of the 10 data sets mapping the degree of ligand dissociation only.

one to ask the crucial question of how much information concerning binding models can one expect to obtain from a particular set of data. The fact that the BIOEQS program has the capability of performing rigorous confidence interval calculations allows the examination of the information content of multidimensional data. In brief, to perform an absolutely rigorous error estimate on the ith fitting parameter, one can systematically fix this parameter at a series of values, then perform an entire nonlinear minimization allowing the remaining $n − 1$ parameters to vary to minimize the χ^2 or variance. One can then record the series of minimum χ^2 values found over a particular range of the ith fitting parameter. Although this method requires a whole series of nonlinear analyses to be carried out, it is very rigorous because it takes into account all of the higher order correlations which may exist between fitting parameters (i.e., all other parameters are allowed to compensate for each particular change imposed on the parameter of interest). See the chapter on confidence interval estimation ([2], this volume) for a more detailed examination of this methodology. An asymmetric confidence interval on the ith parameter is then calculated by determining the appropriate statistically significant increase in the minimum χ^2 (or variance) given the number of degrees of freedom of the problem.

Rigorous confidence limit calculations for the four free energy parameters of this model were carried out using the data in Fig. 4. The results of these tests can be found in Figs. 6A–9A (top graphs). In these plots the y axis corresponds to the χ^2 value while the x axis corresponds to the value of the particular parameter. It can be seen, for example, in Fig. 6A, that a minimum in the χ^2 versus parameter value curve is found at 5 kcal/mol, which was in fact the target value. A horizontal line is drawn at the χ^2 value corresponding to the 67% confidence level of this fit. Drawing vertical lines

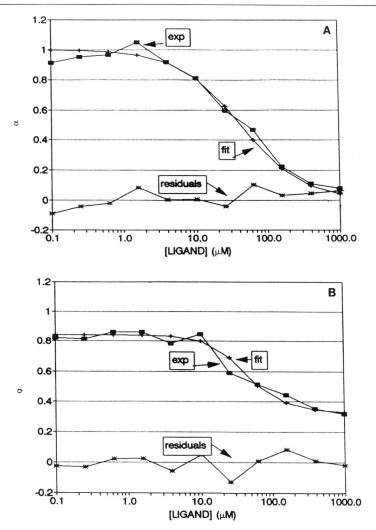

FIG. 5. Two curves of experimental data from each of the simulated three-dimensional binding surfaces are shown. (A) The experimental data correspond to the degree of ligand dissociation versus ligand concentration at a protein concentration of 0.1 μM in monomer; (B) the experimental data correspond to the degree of dimer dissociation versus ligand concentration at a protein concentration of 63 μM in monomer. Also plotted in both graphs are the calculated data and the residuals of the fit. A random error of 5% was imposed in these simulations.

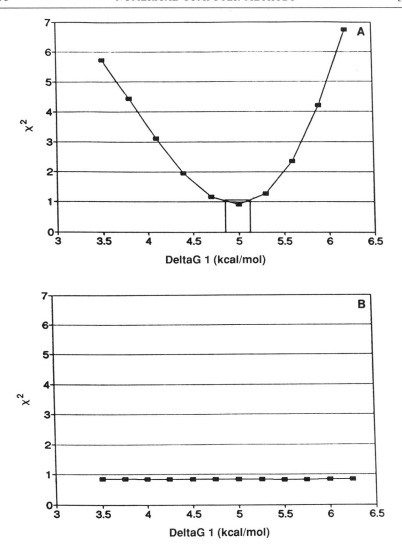

FIG. 6. χ^2 confidence plots for the DeltaG1 (ΔG_I) parameter resulting from rigorous confidence interval testing of (A) the combined ligation and dimerization data sets and (B) the ligation data sets only. The $\pm$ values for this parameter may be found in Table III.

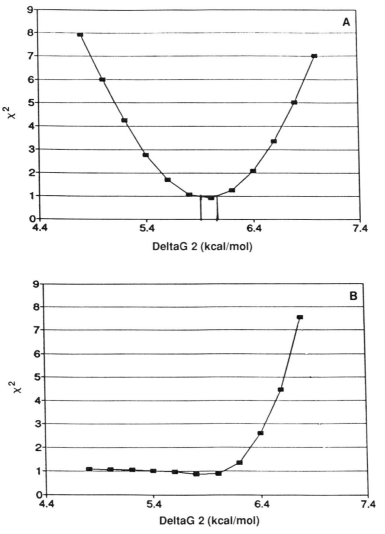

FIG. 7. χ^2 confidence plots for the DeltaG2 (ΔG_2) parameter resulting from rigorous confidence interval testing of (A) the combined ligation and dimerization data sets and (B) the ligation data sets only. The $\pm$ values for this parameter may be found in Table III.

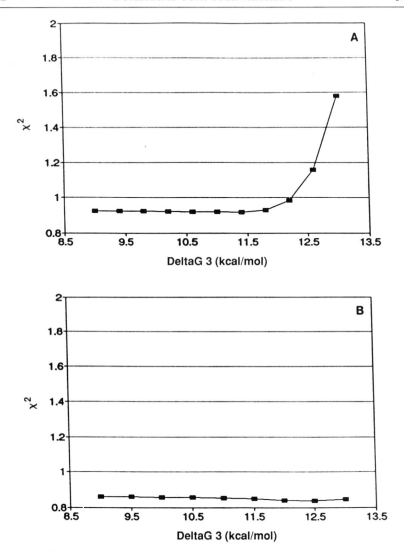

FIG. 8. χ^2 confidence plots for the DeltaG3 (ΔG_3) parameter resulting from rigorous confidence interval testing of (A) the combined ligation and dimerization data sets and (B) the ligation data sets only. This parameter could not be uniquely recovered.

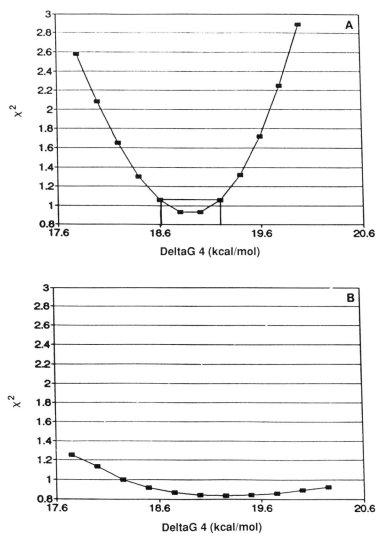

FIG. 9. χ^2 confidence plots for the DeltaG4 (ΔG_4) parameter resulting from rigorous confidence interval testing of (A) the combined ligation and dimerization data sets and (B) the ligation data sets only. The ± values for this parameter may be found in Table III.

down from the intersection of the horizontal 67% level and the χ^2 curve onto the x axis gives the ± error values (at the 67% confidence level) for that particular parameter. All of the parameters of this particular fit are relatively well resolved, with the exception of the third free energy (or chemical potential) value. The target values and the recovered parameter values with their confidence limits are given in Table III.

Next, the data surface from Fig. 4A alone, corresponding to information on ligand binding, was analyzed for the same set of four free energy parameters. In this case, no information relating to the protein or ligand concentration dependence of the dimerization properties of this protein were supplied to the analysis program. The recovered values from this fit are also given in Table III. When rigorous confidence limit tests were performed on the parameters recovered from this fit, the results were quite strikingly different and are shown in Figs. 6B–9B (bottom graphs). Without the additional data axis corresponding to protein–protein interactions, none of the four free energy parameters could be uniquely recovered.

Two important and related conclusions can be drawn from the results of this exercise. First, the binding energies for complex systems can be reasonably well determined. However, and second, this is only true if the number of data axes is sufficiently high as to actually contain the information necessary for this recovery. As a general rule, it would appear that the number of data axes must be at least equal to the number of elements in the system. Additionally, the chemical potentials of intermediate species will only be resolved with difficulty. Key to their resolution will be the design of experiments which will report on their existence. Certain experimental techniques (i.e., spectroscopy, DNase footprinting, gel shift mobility assays) allow for the direct observation of particular intermediates, greatly decreasing the errors in the recovered free energies. Again, the information which can be gained about a particular system should not be limited by our ability to analyze binding data. In many cases, investigators may discover that more than one model is consistent with the data sets provided to the program. Elimination of alternative models then depends on the design of new and imaginative experiments.

Conclusions

We have attempted in this chapter to convince the reader both of the wide applications of a numerical binding analysis methodology and of the ease of use of the particular BIOEQS program. By employing such a tool, the scientist can concentrate more time on model testing and experimental design, as opposed to confronting both the drudgery and constraints of developing and applying analytical expressions to describe the data. Although BIOEQS, as currently configured, can handle a large number of complex problems, we consider it to be an ongoing software development project. There are a number of features which are currently being incorporated into the next version. Some of these include completion of the incorporation of hydrostatic pressure as an additional independent axis, integrated graphical output, and a more versatile input interface. In addition, we envision expanding the capabilities of BIOEQS to fit not only for the macroscopic chemical potentials as it is currently configured, but also for intrinsic site affinities within a given complex stoichiometry. Recovery of such parameters of course depends on being able to identify the relative population of the particular site isomers with a particular stoichiometry. However, techniques do exist, such as DNase footprinting, which provide such capabilities.

We have taken pains to build the BIOEQS program on a highly generalized base such that the incorporation of new features can be accomplished with relative ease. We anticipate that the ongoing development of numerical analysis capabilities for the study of biological interactions will greatly assist in the understanding of the important role of free energy couplings in biological systems.

Acknowledgments

This work was carried out with funding from the National Institutes of Health (Grant GM39969 to C.A.R.) and from a Lucille P. Markey Scholar award to J.M.B. Copies of BIOEQS, as well as a user guide, can be obtained from either of the two authors.

[24] Deconvolution Analysis for Pulsed-Laser Photoacoustics

By JEANNE RUDZKI SMALL

Introduction

Pulsed-laser, time-resolved photoacoustics is a technique which provides information on nonradiative channels of deactivation of molecular excited states. It is thus complementary to radiative techniques such as fluorescence. This chapter describes time-resolved, pulsed-laser photoacoustics and how its data can be deconvolved to give kinetic and volumetric information by using two analysis methods common to time-resolved fluorescence decay, namely, nonlinear least-squares iterative reconvolution and the method of moments. The analysis techniques are illustrated with experimental and synthetic data.

Photoacoustic Experiments

The processes involved in the pulsed-laser photoacoustic experiment have been outlined previously.[1-3] A photon from a laser pulse is absorbed by a molecule in solution. A portion of that absorbed energy is released as heat to the solvent, which, at room temperature, induces a thermal volumetric expansion of the solution in the illuminated region. For some molecules (especially photoactive protein molecules), the absorption of light may induce a rapid conformational volume change in the molecule, which also induces a volumetric change in the solution. The volume changes from both of these sources combine to generate ultrasonic pressure waves, which reflect the volumetric changes in magnitude and time. A pressure-sensitive transducer, such as a piezoelectric transducer, is clamped to the side of the sample cuvette and detects pressure waves, generating a voltage signal. The voltage signal is amplified and digitized to give the photoacoustic waveform.

In a typical photoacoustic experiment, using aqueous solvent at 25°, the heat release to the solution per 10-μJ laser pulse is approximately 1 μJ, resulting in a temperature increase of about 10^{-5}° and a change in solution volume of about 10^{-10} ml. A rapid photoinduced volume change in a molecule of the order of 10 ml per mole would also give rise to an observable photoacoustic waveform.[2] The time response of the photoacoustic experiment is determined by the laser beam shape, as well as by the laser pulse width, intrinsic transducer frequency, and digitization channel width (0.3 to 0.5 nsec, 1 to 2 MHz, and 10 nsec, respectively, in our laboratory).[3] Typically, signals faster than 10 nsec are detected in amplitude but not resolved in rate, other than that they were fast; signals slower than 10 μsec are not detected at all; and in the intermediate time regime, both the rate and amplitude of the signals may be resolved. In the case where only very fast and very slow processes are occurring, the amplitude of the waveform is directly related to the magnitude of only the fast signal-inducing events.[4] However, when the lifetimes of the photoinduced processes are of intermediate value, more complex analysis methods (i.e., deconvolution) are often required.

[1] A. C. Tam, *Rev. Mod. Phys.* **58**, 381 (1986).
[2] J. R. Small, J. J. Hutchings, and E. W. Small, *Proc. SPIE Int. Soc. Opt. Eng.* **1054**, 26 (1989).
[3] J. R. Small and S. L. Larson, *Proc. SPIE Int. Soc. Opt. Eng.* **1204**, 126 (1990).
[4] J. E. Rudzki, J. L. Goodman, and K. S. Peters, *J. Am. Chem. Soc.* **107**, 7849 (1985).

Deconvolution of Photoacoustic Data: An Overview

The purpose of photoacoustic waveform deconvolution is to obtain information on the rate and magnitude of volume changes (e.g., owing to heat release) following the absorption of a photon. Just as fluorescence decay deconvolution involves a measured excitation and a measured fluorescence decay, photoacoustic waveform deconvolution involves two waveforms. These waveforms are the reference waveform, representing the instrument response to rapid (<1 nsec), complete conversion of absorbed photon energy to heat, and the sample waveform, which contains the information of interest. The deconvolution methods developed in our laboratory[4–7] give results in terms of ϕ_i and τ_i values for simultaneous or sequential exponential decays. In general, one to three decays are resolved. The first will always be fast, as there is always some vibrational relaxation occurring quickly after photon absorption. There may be other processes occurring quickly as well, such as fluorescent decay to the ground state. For the purposes of this chapter, the fast decay is defined as less than 10 nsec, the channel width of the digitizer. In practicality, it is not possible to distinguish rates which occur faster than the channel width; it is possible to detect their presence, however, and the ϕ_i value associated with the rate. For longer lifetimes, up to about 10 μsec, the rate and magnitude of heat release (or other processes) may be discerned by deconvolution.[5]

Some features of photoacoustic waveform analysis which should be noted are as follows[5]:

1. The assumed decay form for exponential decays is $\phi(1/\tau)\exp(-t/\tau)$. The $(1/\tau)$ preexponential factor is a weighting factor required for photoacoustic waveform analysis, so that ϕ is then a measure of enthalpy and volume change and is independent of τ. The ϕ preexponential factor is very important in photoacoustics, as it yields thermodynamic information distinct from the kinetic information (τ values). Even for a process too fast to be measured accurately by the photoacoustic technique (e.g., vibrational relaxation), it is still important to obtain an accurate ϕ value for that process.

[5] J. R. Small, L. J. Libertini, and E. W. Small, *Biophys. Chem.* **42**, 29 (1992).

[6] E. W. Small, L. J. Libertini, D. W. Brown, and J. R. Small, *Proc. SPIE Int. Soc. Opt. Eng.* **1054**, 36 (1989).

[7] J. R. Small, S. H. Watkins, B. J. Marks, and E. W. Small, *Proc. SPIE Int. Soc. Opt. Eng.* **1204**, 231 (1990).

2. The range of τ values commonly analyzed is very broad. With photoacoustic data, it is typical to analyze τ from 1 to 10^4 nsec, with a channel width of 10 nsec, using a 1-MHz transducer. Thus, the photoacoustic experiment requires analysis of lifetimes less than the digitization channel width, which is seldom done in other methods such as fluorescence. For these fast lifetimes, the objective is to determine that they were fast, not necessarily to determine their rate accurately.

3. The convolution of the instrumental response with a delta function is very important in photoacoustics, as it represents fast processes, for example, heat release due to vibrational relaxation. This point has governed the modifications required for adapting nonlinear least-squares and method of moments computer programs to photoacoustic data.

Experimental Requirements for Deconvolution

The data required for an analysis are photoacoustic waveforms for the reference and sample compounds, average laser pulse energy measured concomitantly with those waveforms, absorbances of all of the solutions at the laser light wavelength, and an average waveform baseline collected with no laser light incident on the sample cuvette.[2]

Data Preprocessing

Preprocessing of photoacoustic data is necessary before waveform deconvolution is performed.[5] The waveform baseline is subtracted from each photoacoustic waveform to yield a net waveform having positive and negative excursions. As suggested by Eq. (1) below and described previously,[4] each point of the net waveforms must be divided by the sample absorbance factor $[(1 - 10^{-A})$, where A is the absorbance at the excitation wavelength] and by the average laser pulse energy, E_0, as measured by an energy meter. The reference waveform thus processed is then normalized to 1.0 at its maximum amplitude, and the same normalization factor is applied to the sample waveform. At this stage, the waveforms are ready for deconvolution: the sample waveform has been processed to have a rigorous relationship to the reference waveform, and the reference waveform has been processed to characterize fast, complete ($\phi = 1.0$) heat release.

Deconvolution Programs

Deconvolution is achieved by an iterative nonlinear least-squares technique[5] or by the method of moments.[6-8] The reference and sample waveforms are input into the computer program, and a portion of the waveforms

---

[8] E. W. Small, this volume [11].

is selected for analysis. Generally, this includes the first few oscillations, where the signal-to-noise ratio is highest and where there are no contributions from later acoustic reflections in the cuvette. The data are deconvolved for one or two simultaneous exponential decays, with the results expressed as τ_i (lifetime) and ϕ_i (preexponential factor) values.

Interpretation of ϕ_i Values

The ϕ_i values may be interpreted in a variety of ways, depending on the experimental system and the choice of model for the system.[3,4,9] Perhaps the simplest interpretation is that ϕ_i is the fraction of the absorbed photon energy released as heat for the ith decay process. Thus, if a nitrogen laser (337 nm, $E = h\nu = 84.8$ kcal/mol) is used to excite a sample leading to a photoacoustic waveform with $\phi_1 = 0.19$, $\tau_1 < 10$ nsec, then the rapid decay can be said to occur with heat release of $\phi_1 h\nu = 16$ kcal/mol. More complex systems require other interpretations of ϕ_i.[3,9]

Photoacoustic Waveform Generation

In general, the measured photoacoustic voltage signal with time, $V(t, T)$, is given by[5]:

$$V(t, T) = k \sum_i \left\{ \phi_i \frac{\beta(T)}{C_p\rho} + \frac{m_i}{h\nu} \right\} E_0(1 - 10^{-A}) M(t) * q_i(t) \qquad (1)$$

where the variables are as follows (common units are in brackets): i, index for the transient decay of interest; ϕ_i, fraction of photon energy released as heat for the ith transient decay; $\beta(T)$, thermal volumetric expansion coefficient of the solvent [K^{-1}]: for water, $\beta(T)$ is strongly dependent on temperature; T, temperature [K]; C_p, heat capacity of solvent at constant pressure [cal g^{-1} K^{-1}]; ρ, density of solvent [g liter^{-1}]; m_i, volume change per mole of photoexcited molecules for the ith decay [liter mol^{-1}]; $h\nu$, photon energy [cal mol^{-1}]; E_0, laser pulse energy [cal]; A, sample absorbance at the photon frequency ν; t, time [sec]; $M(t)$, instrumental response, determined primarily by characteristics of the transducer; $q_i(t)$, time-dependent impulse response function, usually involving τ_i variables, with τ_i being the characteristic relaxation time of the ith decay; and k, proportionality constant relating $V(t, T)$ to the transient volume changes in solution [V liter^{-1}]; dependent on instrumental design. In addition, the symbol $*$ denotes a convolution:

[9] K. S. Peters, T. Watson, and K. Marr, *Annu. Rev. Biophys. Biophys. Chem.* **20**, 343 (1991).

$$M(t) * q_i(t) = \int_0^t M(u) \, q_i(t - u) \, du \tag{2}$$

The form of $q_i(t)$ cannot be determined *a priori,* and the researcher must choose a model based on what is expected theoretically. Two models which are commonly encountered for photoacoustic applications are simultaneous exponential decays, $q_i(t) = (1/\tau_i) \exp(-t/\tau_i)$, and sequential exponential decays (also characterized by a sum of exponentials), defined later in this chapter and elsewhere.[5,10]

Let R and S represent $V(t, T)$ for the reference and sample, respectively.[5] Both waveforms are governed by Eq. (1). Care must be taken to measure R and S under identical experimental geometry (i.e., same k), solvent conditions [i.e., same $\beta(T)$, C_p, and ρ], and temperature, T. The sample absorbance, A, and incident laser pulse energy, E_0, should be similar; however, small corrections can be made for differences in A and E_0 for the sample and reference waveforms. The reference compound is chosen to be one which relaxes back to the ground state (lifetime τ), with unit efficiency on a time scale very fast compared to the transducer response (frequency ν_{tr}). This feature means that, for a reference compound, $q_i(t)$ approximates a Dirac delta (δ) function, and the $M(t) * q_i(t)$ term in Eq. (1) is indistinguishable from $M(t)$:

$$M(t) * q_i(t) \cong M(t) \tag{3}$$

for $\tau_i \ll 1/\nu_{tr}$. Note that this feature eliminates the convolution from Eq. (1). The instrumental response, $M(t)$ can then be calculated from

$$M(t) = R(t)/k' \tag{4}$$

where k' is a constant under the conditions of identical k, $\beta(T)$, C_p, ρ, A, and E_0 for the R and S waveforms. For simplicity, m_i is taken to be zero for this discussion so that the source of the photoacoustic signal is solely due to heat release. It follows that

$$S(t) = k' \sum_i \phi_i M(t) * q_i(t) = \sum_i \phi_i R(t) * q_i(t) \tag{5}$$

From Eq. (5), then, the sample waveform S may be deconvolved using the reference waveform R to yield the desired parameters, ϕ_i and τ_i.

By defining the function $h(t)$ as

$$h(t) = \sum_i \phi_i q_i(t) = \sum_i \phi_i (1/\tau_i) \exp(-t/\tau_i) \tag{6}$$

[10] J. A. Westrick, K. S. Peters, J. D. Ropp, and S. G. Sligar, *Biochemistry* **29,** 6741 (1990).

where the expression given is for simultaneous exponential decays, then Eq. (5) becomes

$$S(t) = R(t) * h(t) \tag{7}$$

Equation (7) is equivalent to the expression $C_{exptl}(t) = E(t) * T(t)$ described earlier,[4] with $C_{exptl}(t)$, $T(t)$, and $E(t)$ equivalent to $S(t)$, $R(t)$, and $h(t)$, respectively.

Photoacoustic Waveform Simulation

Photoacoustic waveforms can be modeled according to the following equation[4]:

$$V(t) = \sum_{i=1}^{n} K' \, \phi_i \frac{\nu/\tau_i}{\nu^2 + (1/\tau_i')^2} \left\{ \exp(-t/\tau_i) \right.$$

$$\left. - \exp(-t/\tau_0) \left[\cos(\nu t) - \frac{1}{\nu\tau_i'} \sin(\nu t) \right] \right\} \tag{8}$$

where $V(t)$ is the detector response, K' is a constant (taken to be unity), ν is the characteristic oscillation frequency of the transducer (referred to earlier as ν_{tr}), τ_0 is the relaxation time of the transducer, ϕ_i is the amplitude factor for the ith decay, τ_i is the lifetime of the ith decay, t is time, and $1/\tau'_i = 1/\tau_i - 1/\tau_0$. [Equation (8) is the corrected version of Eq. (3) in Ref. 4, which was missing one set of brackets.] As described previously,[4] Eq. (8) represents the mathematical convolution of n simultaneous transient decays with a model transducer response function (which may be described by $n = 1$, $\tau_1 \ll 1/\nu$ and $\tau_1 \ll \tau_0$). Equation (8) may be used to simulate reference and sample waveforms of known ϕ_i and τ_i values. The waveforms may then be deconvolved using the analysis programs described in this chapter.

Simulated noise can be added to waveforms by a computer program which computes a normally distributed random number with a given mean (the original, noiseless data point) and standard deviation (specified by the user), and thus generates a new, noisy waveform based on the original noiseless data. It is assumed that the standard deviation is independent of time.

Nonlinear Least-Squares Iterative Reconvolution Program

A standard nonlinear least-squares iterative reconvolution program utilizing the Marquardt algorithm, used for the analysis of fluorescence decay, was modified for use with pulsed-laser photoacoustics. The modifications were required because of the wide range of transient decay times which must be addressed by the photoacoustic technique.[5] This new program (the quadratic-fit convolution program) is available from the author in both FORTRAN and ASYST languages. The equations utilized for modifying the original Marquardt program for photoacoustic waveform analysis are given here. A more detailed derivation is presented elsewhere.[5]

Equations for Quadratic-Fit Convolution Program

The quadratic-fit program was first modified to account for the fact that photoacoustic waveform analysis assumes the decay form $\phi(1/\tau) \exp(-t/\tau)$, rather than $\alpha \exp(-t/\tau)$ as used in fluorescence. Then, $R(t)$ was approximated as a continuous function (quadratic fits to sets of discrete points, done three at a time), permitting $R(t) * h(t)$ in Eq. (7) to be computed by direct integration.

Specifically, $R(u)$ is represented by a series of numbers (R_n, photoacoustic wave amplitude at channel n) evenly spaced on the time axis by the interval δ. The value of $R(u)$ between points R_n and R_{n+2} is given by a quadratic which passes through the three points (including R_{n+1}). Given the convolution to point n (C_n), the object is to calculate the value of the convolution at point $n + 1$ (C_{n+1}) and $n + 2$ (C_{n+2}). For clarity, the ϕ preexponential factor, a simple constant in the convolution expression, has not been included in the following equations:

$$\begin{aligned}
\mathbf{C_{n+1}} = {} & \mathbf{C_n}\, e^{-\delta/\tau} + \mathbf{R_n}[-e^{-\delta/\tau} + (1/2)(1 - 3\, e^{-\delta/\tau})(\tau/\delta) + (1 - e^{-\delta/\tau})(\tau/\delta)^2] \\
& + \mathbf{R_{n+1}}[1 + 2\, e^{-\delta/\tau}(\tau/\delta) - 2(1 - e^{-\delta/\tau})(\tau/\delta)^2] \\
& + \mathbf{R_{n+2}}[-(1/2)(1 + e^{-\delta/\tau})(\tau/\delta) + (1 - e^{-\delta/\tau})(\tau/\delta)^2] \quad (9)
\end{aligned}$$

$$\begin{aligned}
\mathbf{C_{n+2}} = {} & \mathbf{C_n}\, e^{-2\delta/\tau} + \mathbf{R_n}[e^{-2\delta/\tau} - (1/2)(1 + 3\, e^{-2\delta/\tau})(\tau/\delta) + (1 - e^{-2\delta/\tau})(\tau/\delta)^2] \\
& + \mathbf{R_{n+1}}[2(1 + e^{-2\delta/\tau})(\tau/\delta) - 2(1 - e^{-2\delta/\tau})(\tau/\delta)^2 \\
& + \mathbf{R_{n+2}}[1 - (1/2)(3 + e^{-2\delta/\tau})(\tau/\delta) + (1 - e^{-2\delta/\tau})(\tau/\delta)^2] \quad (10)
\end{aligned}$$

The following derivatives with respect to τ are necessary for the Marquardt algorithm:

$$\begin{aligned}
\partial\mathbf{C_{n+1}}/\partial\tau = {} & \mathbf{C_n}(\delta/\tau^2)\, e^{-\delta/\tau} + \mathbf{R_n}[1/2\delta + 2\tau/\delta^2 + (-5/2\delta - 2\tau/\delta^2 - 3/2\tau \\
& - \delta/\tau^2)\, e^{-\delta/\tau}] + \mathbf{R_{n+1}}[-4\tau/\delta^2 + (4/\delta + 4\tau/\delta^2 + 2/\tau)\, e^{-\delta/\tau}] \\
& + \mathbf{R_{n+2}}[-1/2\delta + 2\tau/\delta^2 + (-3/2\delta - 2\tau/\delta^2 - 1/\tau)\, e^{-\delta/\tau}] \quad (11)
\end{aligned}$$

$$\begin{aligned}
\partial\mathbf{C_{n+2}}/\partial\tau = {} & \mathbf{C_n}(2\delta/\tau^2)\, e^{-2\delta/\tau} + \mathbf{R_n}[-1/2\delta + 2\tau/\delta^2 + (-7/2\delta - 2\tau/\delta^2 \\
& - 3/\tau - 2\delta/\tau^2)\, e^{-2\delta/\tau}] + \mathbf{R_{n+1}}[2/\delta - 4\tau/\delta^2 + (6/\delta + 4\tau/\delta^2 \\
& + 4/\tau)e^{-2\delta/\tau}] + \mathbf{R_{n+2}}[-3/2\delta + 2\tau/\delta^2 + (-5/2\delta \\
& - 2\tau/\delta^2 - 1/\tau)\, e^{-2\delta/\tau}] \quad (12)
\end{aligned}$$

Incorporation of the preexponential factor, ϕ, is done by multiplying

Eqs. (9)–(12) through by ϕ. Thus, for the purposes of the Marquardt algorithm,[11] the required derivatives with respect to ϕ will be simply C_{n+1} and C_{n+2}. Equations (9)–(12) were substituted into the original Marquardt program utilized for fluorescence analysis. The program was modified to prevent any value of τ from becoming negative; τ is constrained to be greater than or equal to 1 fsec. No changes were made in the α and β matrices,[11] so that the methods for minimizing χ^2 were unchanged.

Goodness-of-Fit

The goodness-of-fit is judged by reduced χ^2 values, by visual inspection of the residuals, and by the autocorrelation of the residuals. Reduced χ^2 is defined as usual by[12]

$$\chi^2 = \frac{1}{(n_2 - n_1 + 1) - n_{\text{var}}} \sum_{i=n_1}^{n_2} [C(i) - S(i)]^2 \tag{13}$$

where n_1 and n_2 are, respectively, the starting channel and ending channel of the analysis, n_{var} is the total number of variables being sought (ϕ's and τ's), and $C(i)$ and $S(i)$ are the values of the convolved wave and the sample wave at channel i, respectively. The autocorrelation of the residuals is given by[12]

$$\text{cor}(j) = \frac{\dfrac{1}{(n_2 - n_1 + 1) - j} \sum\limits_{i=n_1}^{n_2-j} [C(i) - S(i)][C(i + j) - S(i + j)]}{\dfrac{1}{(n_2 - n_1 + 1)} \sum\limits_{i=n_1}^{n_2} [C(i) - S(i)]^2} \tag{14}$$

where j is an index which runs from 0 to $(n_2 - n_1 + 1)/2$. A "good" autocorrelation is represented by $\text{cor}(0) = 1.0$, $\text{cor}(j) \cong 0.0$ for $j \neq 0$.

Although the photoacoustic nonlinear least-squares analyses presented here are similar to fluorescence analyses, there are some differences in the details of the evaluation of goodness-of-fit. We have not yet weighted the residuals in χ^2 calculations[12-14] to yield a more statistically correct analysis and giving χ^2 values which tend toward unity for good fits. (The use of weighting factors would require the estimation of variance for

[11] P. R. Bevington, "Data Reduction and Error Analysis for the Physical Sciences." McGraw-Hill, New York, 1969.
[12] D. A. Holden, *in* "CRC Handbook of Organic Photochemistry" (J. C. Scaiano, ed.), Vol. 1, p. 261. CRC Press, Boca Raton, Florida, 1989.
[13] A. Grinvald and I. Z. Steinberg, *Anal. Biochem.* **59**, 583 (1974).
[14] A. Grinvald, *Anal. Biochem.* **75**, 260 (1976).

photoacoustic data, a function of the experimental noise. The experimental noise is constant with time, so there is probably no need for weighting.) Thus, the χ^2 values reported here have a wide range of magnitudes, all less than 1.0. In addition, there frequently are strong oscillations in the autocorrelation function for reasonably good fits of photoacoustic waveforms, owing to the oscillations intrinsic to the waveforms.[9] In practice, this sometimes limits the diagnostic value of the autocorrelation function for photoacoustics.

Extension of Method to Sequential Decays

In the quadratic-fit convolution program, simultaneous decays are assumed, with $q_i(t) = 1/\tau_i \exp(-t/\tau_i)$ for decay i, $i = 1, 2, 3, \ldots$ What if the decays are sequential rather than simultaneous?

$$A^* \xrightarrow{\tau_1} B \xrightarrow{\tau_2} C \xrightarrow{\tau_3} \cdots \tag{15}$$

It can be shown that analyses of data arising from sequential decays can still be done with the quadratic-fit convolution program.[5] In principle, the correct τ values will be returned regardless of which kinetic model, simultaneous or sequential, is valid. However, interpretation of the preexponential factors obtained will depend on the model assumed. For example, given a three-component fit by the analysis program and assuming a three-step sequential process, one finds

$$\phi_3 = \phi_3{}^{app} \frac{(\tau_3 - \tau_1)(\tau_3 - \tau_2)}{\tau_3{}^2} \tag{16}$$

$$\phi_2 = \phi_2{}^{app} \frac{\tau_2 - \tau_1}{\tau_2} + \phi_3 \frac{\tau_2}{\tau_3 - \tau_2} \tag{17}$$

$$\phi_1 = \phi_1{}^{app} + \phi_2 \frac{\tau_1}{\tau_2 - \tau_1} + \phi_3 \frac{\tau_1{}^2}{(\tau_3 - \tau_1)(\tau_1 - \tau_2)} \tag{18}$$

where $\phi_i{}^{app}$ are the amplitude values reported by the quadratic-fit convolution program for the exponential decays. (In the case of only two sequential exponential decays, ϕ_3 may be set to zero.) It is apparent from Eqs. (16)–(18) that if τ_1, τ_2, and τ_3 are well separated (e.g., $\tau_1 = 1$ nsec, $\tau_2 = 1$ μsec, and $\tau_3 = 1$ msec), then the apparent ϕ values are equivalent to the actual ϕ values.

This extension to sequential decays can be incorporated into the nonlinear least-squares program[4] or can reside in a separate program into which the user enters the ϕ and τ parameters returned by the analysis program. Thus, this extension to sequential decays is applicable also to the output of the method of moments program.

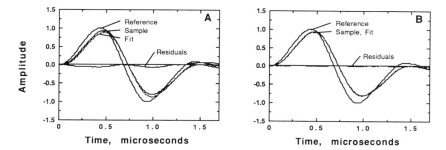

FIG. 1. Nonlinear least-squares analysis of photoacoustic waveforms from horse skeletal muscle myoglobin in 0.1 M phosphate buffer, pH 7.6, at 31°. Photolysis was performed at 337 nm. The pulsed-laser photoacoustic apparatus has been described elsewhere [J. R. Small and S. L. Larson, *in* "Time-Resolved Laser Spectroscopy in Biochemistry II" (J. R. Lakowicz, ed.), SPIE Proc., Vol. 1204, p. 126. Society of Photo-Optical Instrumentation Engineers, Bellingham, Washington, 1990]. The reference and sample waveforms were generated from metmyoglobin and carboxymyoglobin, respectively, and were analyzed by the nonlinear least-squares program to give (A) one-component and (B) two-component fits. The recovered parameters are as follows: (A) $\phi = 0.91$, $\tau = 49$ nsec, $\chi^2 = 2.0 \times 10^{-3}$; (B) $\phi_1 = 0.80$, $\tau_1 = 22$ nsec, $\phi_2 = 0.40$, $\tau_2 = 0.58$ μsec, $\chi^2 = 5.6 \times 10^{-5}$.

Example of Usage with Experimental Data

Figure 1 illustrates the usage of the nonlinear least-squares program with photoacoustic waveforms generated by 337-nm photolysis of met-myoglobin (reference waveform) and carboxymyoglobin (sample waveform). The region of the waveforms with highest signal-to-noise ratio was selected for analysis. It is usual to start the analysis with a single component fit (Fig. 1A), with starting guesses suggested by the appearance of the waveforms.[4] For example, starting guesses of $\phi = 0.8$, $\tau = 10^{-7}$ sec would be reasonable for the data in Fig. 1. The poor fit and high χ^2 of the single component analysis (Fig. 1A) suggest that at least two components are necessary to fit the data (Fig. 1B). Typically, for double component fits, all four parameters (ϕ_1, τ_1 ϕ_2, τ_2) are allowed to vary unconstrained, with starting guesses such as $\phi_1 = 0.5$, $\tau_1 = 10^{-9}$ sec, $\phi_2 = 0.5$, $\tau_2 = 10^{-7}$ sec. It is possible to constrain τ_1 to be fast, for example, $\tau_1 = 10^{-9}$ sec, but we have not found this to be necessary; the program quickly returns a τ_1 value less than the digitization channel width, if in fact a fast rate is discernible. The final χ^2 of 6×10^{-5} (Fig. 1B) is typical of the best fits obtained with the instrumentation used to generate Fig. 1. Deviations from a more ideal fit are most likely due to nonrandom errors in the data or to the intrinsic limitations of resolutions using single curve nonlinear least-squares analysis.[5]

Comments

The nonlinear least-squares program easily handles two decay components. Three decays, with six unconstrained variables, are more difficult to resolve, yet increasingly seen owing to improved instrumentation. More powerful nonlinear least-squares techniques, such as global analysis,[15] should be helpful in the resolution of multiple decays.

Method of Moments Program

The method of moments has been developed primarily for the deconvolution and analysis of fluorescence decay data.[6,8] The method of moments is a transform method. It is not a method for determining which parameters best fit the data of interest; rather, it asks a different question: Given the intrinsic errors in the experiment, what decay parameters are most likely to have generated the data of interest? The method of moments has two main features which distinguish it from other approaches: (1) it is robust with respect to certain nonrandom data errors usually encountered in fluorescence data, and (2) it has built into it a series of tests which verify the correctness of the model that is being applied to the data. The method of moments has been adapted to the oscillating waveforms of time-resolved, pulsed-laser photoacoustics.[7] A version of the program, in FORTRAN, is available from the author.

Equations for Method of Moments Program

The method of moments, with moment index displacement (*MD*) and λ-invariance, is described in detail elsewhere in this volume.[8] The adaptation of the method to pulsed-laser photoacoustics has been published.[6,7] Here, we point out the notable modifications which must be made when the method is used for photoacoustic waveform deconvolution.

First, we assume that the impulse response, $q(t)$, consists of an instantaneous heat release and a sum of exponential decays:

$$q(t) = \xi\, \delta(t) + \sum_{i=1}^{n} \frac{\phi_i}{\tau_i} e^{-t/\tau_i} \tag{19}$$

The "scatter" parameter ξ is the amplitude of the delta function component. Second, the reduced moments of the impulse response function, G_k, are given by

---

[15] J. M. Beechem, this volume [2].

$$G_1 = \sum_{i=1}^{n} \phi_i + \xi$$

$$G_2 = \sum_{i=1}^{n} \phi_i \tau_i$$

$$\vdots \tag{20}$$

$$G_k = \sum_{i=1}^{n} \phi_i \tau_i^{k-1}$$

The delta function component of $q(t)$ affects only the first G. Third, in order to find the lifetimes, τ_i, one inverts the following equations to find d_i:

$$\begin{bmatrix} G_{MD+1} & G_{MD+2} & \cdots & G_{MD+n} \\ G_{MD+2} & G_{MD+3} & \cdots & G_{MD+n+1} \\ \vdots & \vdots & & \vdots \\ G_{MD+n} & G_{MD+n+1} & \cdots & G_{MD+2n-1} \end{bmatrix} \begin{bmatrix} d_0 \\ d_1 \\ \vdots \\ d_{n-1} \end{bmatrix} = \begin{bmatrix} G_{MD+n+1} \\ G_{MD+n+2} \\ \vdots \\ G_{MD+2n} \end{bmatrix} \tag{21}$$

$MD = 0, 1, 2, 3, \ldots$. The d_i are the coefficients of the polynomial

$$P_n = x^n + d_{n-1}x^{n-1} + \cdots + d_0 = 0 \tag{22}$$

the roots of which are the lifetimes. The method of moments solves for the lifetimes in photoacoustic data in the same way that it does for fluorescence data.

After the lifetimes have been recovered, they are substituted back into Eq. (20), which are then inverted to recover the amplitudes. For photoacoustic data, the amplitudes recovered by the method of moments program must then be multiplied by their corresponding lifetimes to yield ϕ values. This is because photoacoustic waveform analysis assumes the decay form $\phi(1/\tau) \exp(-t/\tau)$, rather than $\alpha \exp(-t/\tau)$ as used in fluorescence.

Fourth, the "scatter" parameter ξ, the amplitude of the delta function component, is calculated by subtracting the sum of the ϕ values from the measured G_1 [see Eqs. (19)–(20)].

Rules for Accepting Result

The criteria for accepting a result from the method of moments are described fully elsewhere[7,8] but are reviewed here:

Rule 1—λ-Invariance: For a given MD and number of components, the parameter values (ϕ's and τ's) should be taken from a region of the λ-

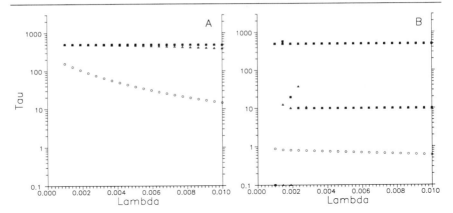

Fig. 2. λ-invariance plots for analyses of noiseless, synthetic photoacoustic waveforms. The axes are τ (in nsec) versus λ (in nsec^{-1}). The waveforms were generated using Eq. (8), with $\phi = 1.0$, $\tau = 1$ psec for the reference waveform and $\phi_1 = \phi_2 = 0.50$, $\tau_1 = 1$ nsec, $\tau_2 = 500$ nsec for the sample waveform. The digitization channel width was 10 nsec. For (A) and (B), the open circles, open triangles, filled squares, and filled triangles represent $MD0$, $MD1$, $MD2$, and $MD3$, respectively. The scatter coefficient option was set for all analyses. (A) and (B) represent one- and two-component analyses, respectively. The values obtained from these λ-invariance plots are recorded in Table I. [Adapted from J. R. Small, S. H. Watkins, B. J. Marks, and E. W. Small, *in* "Time-Resolved Laser Spectroscopy in Biochemistry II" (J. R. Lakowicz, ed.), SPIE Proc., Vol. 1204, p. 231. Society of Photo-Optical Instrumentation Engineers, Bellingham, Washington, 1990.]

invariance plot (τ, in nsec, versus λ, in nsec^{-1}) which is locally flat (i.e., $d\phi_i/d\lambda \cong 0$; $d\tau_i/d\lambda \cong 0$). Examples of λ-invariance plots are shown in Fig. 2, to be discussed below.

Rule 2—*MD* Agreement: Results obtained from scans at different *MD* values should agree.

Rule 3—Component Incrementation: An analysis for $n + 1$ components should indicate the same parameter values as the n-component analysis.

Protocol for Analyzing Data

The method of moments has been tested extensively with synthetic data,[6,7] which has led to the following protocol for analyzing photoacoustic data:

1. Select a λ range of 0.001 to 0.01. This is a reasonable range for typical photoacoustic data.
2. Select $MD0$ to $MD3$.
3. Select the option of calculating the scatter coefficient, ξ (operational with $MD1$ to $MD3$).

Interpret the results as follows:

1. For each MD in the range $MD1$ to $MD3$, sum the scatter coefficient with ϕ_1 (the ϕ value returned corresponding to a τ value of less than the digitization channel width), then use this sum as the total magnitude (ϕ_1') of the fast decay.

2. For $MD0$, there is no scatter coefficient. Therefore, use the ϕ_1 value only as the total magnitude of the fast decay.

3. Judge the returned values by the criteria of λ-invariance, MD agreement, and component incrementation outlined above.

Example of Usage with Synthetic Data

Photoacoustic waveforms were simulated using Eq. (8) as described previously.[4] The transducer was modeled as a 1-MHz transducer with a 1-msec intrinsic relaxation time. The digitization channel width was set at 10 nsec, and the reference waveform was generated with one decay, $\phi = 1.0$ and $\tau = 1$ psec (any decay faster than 1 nsec would have sufficed). The sample waveform was synthesized with two decays, $\phi_1 = \phi_2 = 0.50$, $\tau_1 = 1$ nsec, $\tau_2 = 500$ nsec.

The data were analyzed using the method of moments with moment index displacement and λ-invariance. The λ-invariance plots are shown in Fig. 2. Figure 2A shows good flatness for $MD1$–$MD3$ and good MD agreement for $MD1$–$MD3$. In Fig. 2A $MD0$ is not flat, and all $MD0$ values are greater than 10 nsec. In Fig. 2B, all MD values show flatness. There is again good MD agreement with the apparent exception of $MD0$. This is somewhat misleading, as explained by Table I. Table I shows that for τ_1, all MD values report lifetimes of 10 nsec or less. Thus, the $MD0$ τ_1 value of 0.7 at $\lambda = 0.0055$ is truly in agreement with the higher MD values, since all lifetimes of 10 nsec or less are equivalent.

Further examining the results in Table I, we see that all MD values report τ_2 to be 500 nsec, with $\phi_2 = 0.50$. $MD0$ accurately recovers ϕ_1 as 0.50. There is no scatter coefficient for $MD0$, but there is for $MD1$–$MD3$. Because the higher MD values report τ_1 as 10 nsec or less, it is appropriate to sum the scatter coefficient with ϕ_1 and designate this answer as ϕ_1', a modified ϕ_1. It is clear that ϕ_1' accurately returns the modeled value of 0.50. Thus, this two-component analysis seems to have been successful. It is only necessary to check for component incrementation. In this case, the most interesting comparison is with a one-component analysis (Fig. 2A). Because of the steep slope in the λ-invariance plot for $MD0$, the results for $MD0$ are not valid. For $MD1$–$MD3$, on the other hand, Table I reports one component of about 500 nsec, $\phi = 0.50$, and a scatter coefficient of 0.50. Because the scatter coefficient intrinsically reports on

TABLE I

RECOVERED PARAMETERS FROM λ-INVARIANCE PLOTS IN FIG. 2[a]

MD	ϕ_1	τ_1 (nsec)	ϕ_2	τ_2 (nsec)	ξ	$\phi_1 + \xi$[b]
From Fig. 2A						
0	0.76	38	—	—	—	—
1	—	—	0.50	469	0.49	—
2	—	—	0.50	498	0.50	—
3	—	—	0.50	500	0.50	—
From Fig. 2B						
0	0.50	0.7	0.50	498	—	—
1	0.04	10.0	0.50	500	0.46	0.50
2	0.04	9.9	0.50	500	0.46	0.50
3	0.05	9.8	0.50	500	0.45	0.50

[a] The values were taken from the midpoint, λ = 0.0055. Adapted from J. R. Small, S. H. Watkins, B. J. Marks, and E. W. Small, in "Time-Resolved Laser Spectroscopy in Biochemistry II" (J. R. Lakowicz, ed.), SPIE Proc., Vol. 1204, p. 231. Society of Photo-Optical Instrumentation Engineers, Bellingham, Washington, 1990.

[b] The recovered scatter coefficient (ξ) is added to ϕ_1 when τ_1 is less than or equal to the digitization channel width of 10 nsec.

a fast lifetime, this one-component analysis in Fig. 2A is actually reporting on two components. The results from Fig. 2A thus agree exactly with those in Fig. 2B and accurately return the modeled values.

Comments

An important feature of the method of moments is that it is robust with some types of nonrandom error.[8] The effects of two nonrandom errors in photoacoustic data have been examined elsewhere.[7] These errors include time origin error, in which the sample and reference waveforms are displaced in time with respect to each other, and the use of a reference waveform which results from a decay rather than instantaneous heat release. It was found that the method of moments is robust with respect to small amounts of time jitter and reference waveform lifetime errors.

When the photoacoustic data in Fig. 1 are analyzed by the method of moments, the following results are obtained: $\phi_1 = 0.83$, $\tau_1 < 10$ nsec, $\phi_2 = 0.47$, $\tau_2 = 0.37$ μsec.

Conclusions

Both nonlinear least-squares iterative reconvolution and the method of moments can be used to analyze time-resolved, pulsed-laser photoacoustic waveform data accurately. Both techniques adequately return the magnitude of fast components, as well as the magnitude and lifetime of slower components. The nonlinear least-squares method is somewhat easier to conceptualize and interpret, but the method of moments has strength in being robust with respect to the nonrandom errors undoubtedly present in experimental photoacoustic data. Each method has an important place in pused-laser photacoustics laboratories.

Acknowledgments

I am indebted to Dr. Ludwig Brand for enlightening me to the problems of data analysis and guiding me in my initial excursions with nonlinear least-squares techniques. I thank Dr. Enoch W. Small for 6 years of exploring the method of moments, and moments less methodical. Dr. Louis J. Libertini and Shane L. Larson were instrumental in developing the nonlinear least-squares computer code for our laboratory. Barbara J. Marks and Stephen H. Watkins adapted the method of moments code for pulsed-laser photoacoustics. Shane L. Larson provided expert assistance with figures. This work was supported by the National Institutes of Health (GM-41415).

[25] Parameter Estimation in Binary Mixtures of Phospholipids

By E. E. Brumbaugh and C. Huang

Background

The basic information regarding the mixing behavior of two component phospholipids in the two-dimensional plane of the lipid lamella at various temperatures is contained in the temperature–composition binary phase diagram, or simply the phase diagram, for the binary lipid system. It has been well documented that many types of phase diagrams can be exhibited by binary membrane lipids.[1,2] Some of the most commonly observed phase diagrams are shown in Fig. 1. These various phase diagrams reflect the miscibility and/or immiscibility of the component lipids in the gel, the liquid–crystalline, and the two coexisting phases over a certain range of the lipid composition.

The miscibility of the two phospholipids in the bilayer plane must depend on the lateral lipid–lipid interactions which, in turn, can be expected to depend on the structural similarity between the component lipids. Furthermore, the structural parameters of phospholipid molecules

[1] A. G. Lee, *Biochim. Biophys Acta* **472**, 285 (1977).
[2] B. G. Tenchov, *Prog. Surf. Sci.* **20**, 273 (1985).

METHODS IN ENZYMOLOGY, VOL. 210

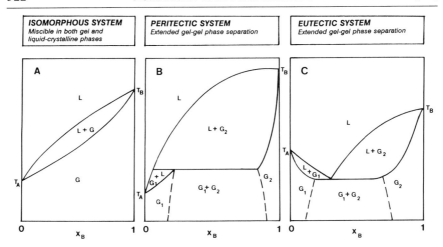

FIG. 1. Phase diagrams for (A) an isomorphous system with complete miscibility in both gel and liquid–crystalline phases, (B) a peritectic system showing extensive gel–gel separation, and (C) a eutectic system showing partial gel–gel phase separation combined with liquid–crystalline phase miscibility.

in the bilayer may change significantly as a function of temperature, particularly around the gel to liquid–crystalline phase transition temperature. The mixing behavior of binary lipid systems may, therefore, be strongly temperature dependent. For instance, the thickness of the lipid bilayer of $C(14):C(14)PC$ $[C(X):C(Y)PC$, saturated L-α-phosphatidylcholine having X carbons in the sn-1 acyl chain and Y carbons in the sn-2 acyl chain] decreases from 43 to 35 Å as the temperature increases from 10° to 30°.[3] In contrast, the bilayer thickness of $C(18):C(10)PC$ lamellas increases, albeit only slightly, with increasing temperature, changing from 33 Å at 10° to 35 Å at 30°.[4] Based on the structural information, one can expect that $C(14):C(14)PC$ and $C(18):C(10)PC$ are most likely immiscible in the bilayer plane at 10° owing to the large difference (10 Å) in the bilayer thickness between these two lipid systems. However, $C(14):C(14)PC$ and $C(18):C(10)PC$ are expected to be miscible at 30° owing to their identical thickness. These expectations are indeed borne out by the eutectic phase diagram (Fig. 1C) observed for $C(14):C(14)PC/C(18):C(10)PC$ mixtures.[5]

The simplest mixing behavior of a binary phospholipid system is the isomorphous system in which the component lipid A and component lipid

[3] M. J. Janiak, D. M. Small, and G. G. Shipley, *Biochemistry* **15**, 4575 (1976).

[4] S. W. Hui, J. T. Mason, and C. Huang, *Biochemistry* **23**, 5570 (1984).

[5] H.-N. Lin and C. Huang, *Biochim. Biophys. Acta* **496**, 178 (1988).

B are completely miscible in both gel and liquid–crystalline states over the entire composition range (Fig. 1A). For an ideal binary system which exhibits the simplest mixing behavior, the lateral lipid–lipid interactions of A–B pairs are equal to those of A–A and B–B pairs in the bilayer plane, and, in addition, the entropy of mixing is negligible. For such an ideal binary system, the following thermodynamic equations relating the chemical composition and calorimetric parameters of the system can be derived:

$$\ln(X_B^G/X_B^L) = \frac{\Delta H_B}{R}\left(\frac{1}{T} - \frac{1}{T_B}\right) \tag{1}$$

$$\ln[(1 - X_B^G)/(1 - X_B^L)] = \frac{\Delta H_A}{R}\left(\frac{1}{T} - \frac{1}{T_A}\right) \tag{2}$$

where $(1 - X_B)$ and X_B are the mole fractions of A and B, respectively, X_B^G and X_B^L are the mole fractions of B in the gel and liquid–crystalline states, respectively, ΔH_A and ΔH_B are the transition enthalpies of the pure component lipids A and B, respectively, T_A and T_B are the transition temperatures of A and B in kelvins, respectively, R is the universal gas constant, and ln is logarithm to the base e.

It was first demonstrated calorimetrically by Mabrey and Sturtevant[6] that binary lipid mixtures in which one component of the pair differs from the other by only two methylene units in each of their long acyl chains such as C(14):C(14)PC/C(16):C(16)PC do exhibit complete miscibility in both gel and liquid–crystalline phases. However, the phase diagram of such a mixture constructed based on the calorimetric data does not agree completely with the theoretical curves calculated based on Eqs. (1) and (2) for an ideal mixture. Specifically, the solidus and liquidus curves obtained calorimetrically lie below the respective curves calculated for an ideal binary mixture. This deviation from ideal mixing indicates that the transition behavior of the component lipid A (or B) is affected by the presence of the second component lipid B (or A) in the binary mixture. Mabrey and Sturtevant also demonstrated that as the chain length difference between A and B is further increased, the experimental phase diagram deviates progressively more from the calculated ideal behavior. In the case of C(12):C(12)PC/C(18):C(18)PC, the shape of the phase diagram has changed into the one characteristic of a typical peritectic system (Fig. 1B). Consequently, nonideal mixing behavior between the component lipids in the two-dimensional plane of the lipid bilayer in both gel and liquid–crystalline states must be taken into consideration when one analyzes the shape of phase diagrams for binary lipid mixtures. This nonideal mixing behavior

[6] S. Mabrey and J. M. Sturtevant, *Proc. Natl. Acad. Sci.U.S.A.* **73,** 3862 (1976).

can be attributed primarily to the difference in the lateral repulsive lipid–lipid interactions between the A–B pairs and those between the A–A and B–B pairs.

A term called the nonideality parameter, ρ, has been introduced by Lee to describe the nonideality of mixing for binary lipid mixtures.[1] This parameter is phase dependent and related to the lateral pair-interaction energy between unlike pairs and between like pairs in the two-dimensional plane of the bilayer as follows: $\rho = Z[E_{AB} - \frac{1}{2}(E_{AA} + E_{BB})]$, where Z is the number of nearest neighbors and E_{AB}, E_{AA}, and E_{BB} are the molar interaction energies of A–B, A–A, and B–B pairs of nearest neighbors, respectively. If the value of ρ is zero, it reflects a complete ideal mixing, implying zero enthalpy of mixing of the two component lipids. If $\rho > 0$, it reflects the immiscibility between unlike lipids, resulting in the lateral phase separation and domain formation of lipids of the same types; in the case of $\rho < 0$, it reflect a "chessboard" type arrangement of the component lipids.[2]

Using the regular solution theory and incorporating the nonideality parameter of ρ, Eqs. (1) and (2) can be modified to yield Eqs. (3) and (4), as shown below, which can be employed to simulate phase diagrams for various binary lipid mixtures[1,7,8]:

$$\ln(X_B^G/X_B^L) = \frac{\Delta H_B}{R}\left(\frac{1}{T} - \frac{1}{T_B}\right) + \frac{1}{RT}[\rho^L(1 - X_B^L)^2 - \rho^G(1 - X_B^G)^2] \quad (3)$$

$$\ln[(1 - X_B^G)/(1 - X_B^L)] = \frac{\Delta H_A}{R}\left(\frac{1}{T} - \frac{1}{T_A}\right) + \frac{1}{RT}[\rho^L(X_B^L)^2 - \rho^G(X_B^G)^2] \quad (4)$$

where ρ^G and ρ^L are nonideality parameters in the gel and liquid–crystalline phases, respectively.

Simulated phase diagrams for a given binary mixture can be generated using Eqs. (3) and (4), provided that the calorimetric data (ΔH_A, ΔH_B, T_A, and T_B) of the pure component lipids are known and that the values of ρ^G and ρ^L are assigned. If the phase diagram is determined experimentally, then the various simulated phase diagrams obtained systematically by varing ρ^G and ρ^L can be used to compare with the experimental one. The best-fit phase diagram will allow us to identify the basic system parameters of ρ^G and ρ^L which, in turn, can provide quantitative information regarding the degree of nonideality in the gel and liquid–crystalline states for the binary lipid mixture under study. However, it should be

[7] P. J. Davis and K. M. W. Keough, *Chem. Phys. Lipids* **25**, 299 (1984).
[8] R. Mendelsohn and C. C. Koch, *Biochim. Biophys. Acta* **598**, 260 (1980).

mentioned that Eqs. (3) and (4) are transcendental in nature; hence, they require numerical methods for their solutions.

Prior to our discussion of the numerical methods which have been successfully applied to simulate the phase diagrams for various binary lipid systems,[9] it is pertinent to mention that experimental phase diagrams of phospholipid systems are most commonly obtained using high-resolution differential scanning calorimetry (DSC).[10] Aqueous dispersions of binary lipid mixtures of various molar ratios are first incubated extensively at a given low temperature; the lipid samples are then subjected to DSC heating scans at slow scan rate of $10°-15°/hr$. The phase transition curves in these DSC scans are used to construct the solidus and liquidus curves of the phase diagram. Specifically, the transition peak in each DSC curve obtained at a given mole fraction of one of the lipid mixtures has characteristic onset and completion temperatures positioned at the beginning and ending of the transition peak, respectively. The phase diagram is constructed by plotting the onset and completion temperatures, after proper correction for the finite width of the transition peaks of the pure components,[6] as a function of the relative concentration of the higher melting component. The onset and completion temperature points then define the solidus and liquidus curves, respectively, of the temperature–composition phase diagram of the binary lipid mixture.

Numerical Methods

Given data in the form of onset and completion temperatures at various mole fractions ranging between 0.0 (pure lower-melting component) and 1.0 (pure higher-melting component), the goal is to find a mathematical model which accurately reproduces the phase behavior of the system. Based on previous work it appeared that the model of Lee [Eqs. (3) and (4)] using regular solution theory could provide a reasonable fit to moderately complex phase behavior. Davis and Keough,[7] for example, developed a computer program based on these equations which allowed them to adjust interactively the two parameters ρ^L and ρ^G in the equations until a visual fit to a given phase diagram was achieved. We could not find in the literature, however, any examples of parameter estimation carried out in such a way as to provide, along with values of the parameters, accompanying estimates of the associated errors. We therefore decided to explore the technique of nonlinear least-squares parameter estimation for analyses of phase diagrams.

[9] E. E. Brumbaugh, M. L. Johnson, and C. Huang, *Chem. Phys. Lipids* **52**, 69 (1990).

[10] S. Mabrey-Gaud, *in* "Liposomes: From Physical Structure to Therapeutic Applications" (C. G. Knight, ed.), p. 105. Elsevier/North-Holland, Amsterdam and New York, 1981.

Nonlinear least-squares parameter estimation methods and algorithms designed to carry out those methods are described in detail earlier in this volume. The remainder of this chapter describes a specific application of a program (NONLIN) developed by Michael L. Johnson.

The structure of NONLIN requires for each application a FORTRAN subroutine which evaluates the mathematical model being tested as a description of the experimental data. This model generally contains, in addition to independent and dependent variables, other parameters, at least one of which is not experimentally available but is to be estimated numerically. When the model consists of a single equation (as, for example, in the analysis of kinetic data using sums of exponential decays), the subroutine is simply a FORTRAN coding of the equation itself, along with a bit of bookkeeping to designate which parameter(s) are to be estimated. Starting with values supplied by the user (some of which must be guesses) the main program systematically varies the parameter(s), then uses values of the function returned by the subroutine for each case of the independent variable (giving predicted values of the corresponding dependent variables) to determine estimates of the parameter(s) which minimize the variance between the predicted and experimental values.

Transformation of Model

Equations (3) and (4), as written, do not readily lend themselves to coding as a function for the least-squares minimization program: the dependent variables (onset and completion temperatures) do not appear explicitly as functions of concentration; instead, a single temperature appears as a function of both concentrations in both equations. We therefore used the following strategy in creating a function evaluator for the NONLIN program.

Equations (3) and (4) can be recast as follows:

$$T_1 = \frac{\Delta H_B + \rho^L(1 - X_B^L)^2 - \rho^G(1 - X_B^G)^2}{\Delta H_B/T_B - R \ln(X_B^L/X_B^G)} \tag{5}$$

$$T_2 = \frac{\Delta H_A + \rho^L(X_B^L)^2 - \rho^G(X_B^G)^2}{\Delta H_A/T_A - R \ln(1 - X_B^L)/(1 - X_B^G)} \tag{6}$$

In this form, predicted temperature has been isolated on the left-hand side of both equations. This temperature (which is either an onset or a completion temperature) must be the same in both equations when evaluated at a given data point.

For any given data point, only one of the two concentrations is known. The other concentration, however, can be determined using the above temperature equality as a goal: the evaluation routine must find a value for the unknown concentration which gives a consistent temperature in both equations. That temperature can then be returned to the main program.

Our function evaluator, then, is really a subprogram which is considerably more complicated than a one-line equation. However, as long as a single dependent variable prediction is produced for each experimental point, no modifications need be made in the NONLIN main program. A slight amount of bookkeeping is needed to inform the subroutine whether X_G or X_L is the known concentration (i.e., whether the temperature falls on the solidus or liquidus). This is readily accomplished by including for each point in the data file an integer which tells the subroutine to use one of two code segments.

Search Algorithm

In our evaluation routine, a search for the unknown concentration proceeds in a binary fashion. If, for example, X^L is the known concentration (i.e., the data point falls on the liquidus curve), the difference (ΔT) between Eqs. (5) and (6) is determined for a value of X^G which very close to zero (10^{-8}); then X^G is given a value of 0.5, and the resulting ΔT is compared to the previous difference. If the sign has changed, then there must be a concentration in the range $0 < X^G < 0.5$ which causes the two equations to become equal. If the sign is the same, ΔT is determined at $X^G = 0.75$, etc. At each stage the test interval is divided in half, successively bracketing the common solution into smaller intervals. The search ends when the interval becomes less than a specified tolerance (usually 10^{-8}), and the resulting value of T is then returned to the main program.

This search algorithm will occasionally fail to find the correct temperature, but these cases can be detected visually since the main program also uses the subroutine to plot the resulting fitted curve through the data points (in addition to its use in finding the variance). The symptom of search failure is a discontinuity in one of the plotted curves. Examination of the functions represented by Eqs. (5) and (6) shows that, in these cases, there are multiple points at which the two curves cross (see Fig. 2). The discontinuities occur when an incremental change in the independent variable (X^L in the above example) causes one of the intersections to move to the opposite side of the 0.5 initial guess; the interval-dividing routine then moves to the other side of the boundary, and if there is a second crossing between 0.5 and 0.99999999 the algorithm will "jump" to that other (spurious) solution. A small amount of intelligence has been incorporated into the subroutine to detect this type of situation, and the search is

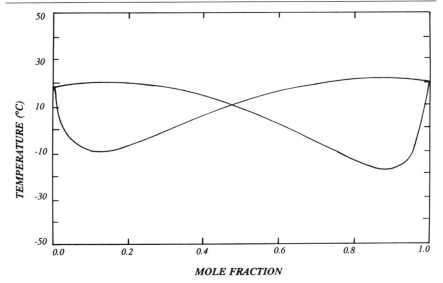

FIG. 2. Curves produced by Eqs. (3) and (4) using $\rho^L = 1.5$, $\rho^G = 2.5$, $\Delta H_A = 5$, $\Delta H_B = 8$, $T_A = 298$, $T_B = 292$. X^L was fixed at a value of 0.8, while $0.0 < X^G < 1.0$.

forced to start again in the opposite direction (where it then finds the correct solution). The criteria for detection were discovered by exhaustive examination of the separate curves for several "pathological" cases. (In Fig. 2 the two curves move vertically in relation to each other as the "fixed" concentration is changed, causing the intersections to move along the concentration axis.)

Estimation of Parameters

In the above equations there appear, in addition to the independent (X^L, X^G) and dependent (T) variable and the gas constant (R), a total of six parameters in three pairs: (a) T_A and T_B are the melting temperatures for the pure lipid species (with A as the higher-melting component), (b) ΔH_A and ΔH_B are the transition enthalpies for the pure components, and (c) ρ^L and ρ^G are "nonideal" energies which represent the interactions between the lipid components in the mixture. Pairs (a) and (b) are typically determined during the experiment, and it is pair (c) which is estimated from nonlinear minimization. Because the NONLIN program accommodates an arbitrary number of fitted parameters, we allow for the possibility that any of the above six parameters be permitted to "float" during the minimization process. This can provide independent verification of the experimental

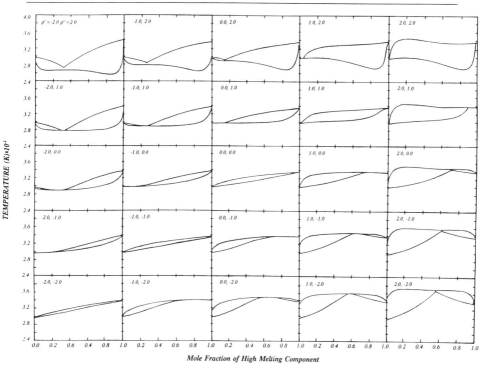

FIG. 3. Phase diagram grid showing the effect of systematic changes in the parameters ρ^L and ρ^G. Horizontally, ρ^L varies from -2 to $+2$ in steps of 1 kcal, whereas ρ^G varies from $+2$ (top) to -2 (bottom) with the same increment. The ideal case ($\rho^L = \rho^G = 0$) is in the center.

values (e.g., by beginning the fitting process with a "bad" guess for one of the temperatures) and allows one to use data where for some reason enthalpies and/or pure melting temperatures are not available.

The overall program is interactive in the sense that the effect of a given combination of parameters "floating" may be tested first visually and then more critically by examining the goodness-of-fit information supplied by NONLIN. The most useful criterion is usually the variance: we look for combinations of parameters which give a markedly lower variance after minimization.

With a new set of phase data, an analysis typically begins by allowing only the nonideal energies to be varied. Then, when the program has found the best values for those parameters, T_A and T_B are varied also. Those four "best" values are then used as new starting guesses, and the enthalpies also are permitted to change. An excellent test for local minima is

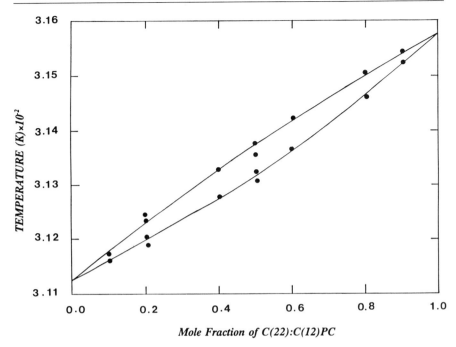

Mole Fraction of C(22):C(12)PC

FIG. 4. Onset and completion temperatures for a mixture of C(10):C(22)PC and C(22):C(12)PC are plotted along with the curves predicted from least-squares parameter estimation. The parameters are given in Table I.

provided by a comparison of these results with the estimates obtained when all of the values are allowed to float from the six starting values (the experimental values for T and ΔH and the original guesses for ρ). If various strategies all appear to converge to the same estimated values, one gains additional confidence in their validity.

TABLE I
FITTED PARAMETERS AND CONFIDENCE INTERVALS FOR DATA OF FIG. 4[a]

Parameter or interval	ρ^L	ρ^G	ΔH_A	ΔH_B	T_A	T_B
Upper limit	0.675	0.685	—	—	315.93	311.28
Parameter value	0.517	0.531	13.1	12.5	315.79	311.23
Lower limit	0.363	0.385	—	—	315.65	311.07

[a] If no limits are given, that parameter was held constant at the value shown. The variance of the fit was 0.009 with 14 degrees of freedom.

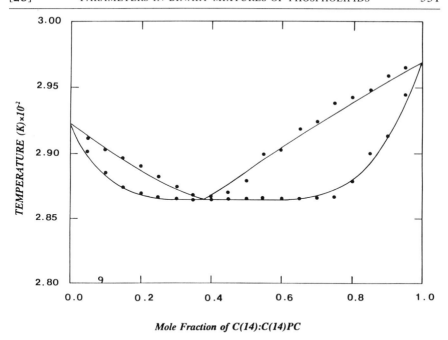

FIG. 5. Fitted phase data for a mixture of C(14):C(14)PC and C(18):C(10)PC. The estimated parameters and confidence intervals are given in Table II.

Because the program provides error estimates in the form of confidence intervals for each varied parameter, these ranges can be also used as criteria for fitting strategies. (If, for example, new estimates are consistently falling within these error limits, then one may as well quit trying similar combinations.) These confidence limits provide the most intuitive

TABLE II
FITTED PARAMETERS AND CONFIDENCE INTERVALS FOR DATA OF FIG. 5[a]

Parameter or interval	ρ^L	ρ^G	ΔH_A	ΔH_B	T_A	T_B
Upper limit	0.815	1.273	11.80	7.29	297.70	293.20
Parameter value	0.369	1.134	7.94	5.70	297.13	292.38
Lower limit	−0.064	0.998	3.96	4.07	296.54	291.49

[a] The variance of the fit was 0.244 with 32 degrees of freedom.

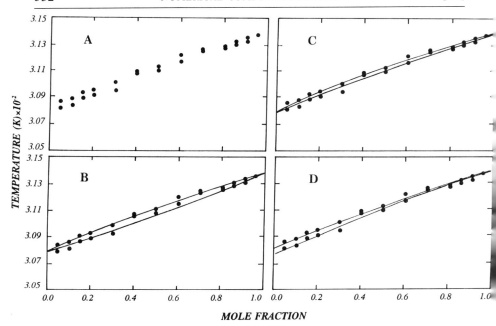

FIG. 6. (A) Onset and completion temperatures for a mixture of C(20) : C(12)PC and C(10) : C(22)PC. Prior to the parameter estimation process, these data might appear to exhibit ideal behavior. (B) The curves for $\rho^L = \rho^G = 0$ (ideality) are superimposed on the data of Fig. 6A, showing that the phase behavior is not in fact ideal. Reference to the lower left-hand corner of Fig. 2 gives a clue as to the correct values. (C) The result of allowing only ρ^L and ρ^G to vary during the fitting process is shown by the two curves. The values for these parameters are given in Table III. (D) If, in addition to ρ^L and ρ^G, the enthalpies ΔH_A and ΔH_B are also allowed to vary, the resulting curves seem to be an even better fit to the data. However, as seen in Table IV, the estimated values for these parameters are physically unrealistic.

form of error information, and in most cases they are the best means of deciding on the worth of the results. The program NONLIN also provides detailed information about relationships among the fitted parameters in the form of covariances which can also guide the investigator by indicating pairs of parameters that are strongly correlated and may therefore be a source of suspicion concerning the validity of the fitting process.

Results

Effect of Parameters

Because the main goal of parameter estimation for phase data is to obtain values for the nonideal energies (ρ^L and ρ^G) which are not directly accessible from experiment, it is of interest to examine the impact of these

TABLE III
FITTED PARAMETERS AND CONFIDENCE INTERVALS FOR DATA OF FIG. 6C[a]

Parameter or interval	ρ^L	ρ^G	ΔH_A	ΔH_B	T_A	T_B
Upper limit	0.854	0.821	—	—	—	—
Parameter value	−1.049	−1.084	10.22	9.40	313.82	307.92
Lower limit	−2.894	−2.929	—	—	—	—

[a] If no limits are given, that parameter was held constant at the value shown. The variance of the fit was 0.048 with 24 degrees of freedom.

variables on the shape of the phase diagram according to regular solution theory.

In the analysis of Lee, ρ^L and ρ^G appear as constants multiplying concentration terms which are added to the expressions for ideal behavior: thus, setting these two parameters equal to zero yields the classic "airplane wing" phase diagram. The effect of other combinations of ρ^L and ρ^G may be seen in Fig. 3. In general, the more complex shapes are produced when there is a larger difference in the values of these two parameters. Furthermore, the shapes which are most typically observed in binary lipid systems tend to be those produced when $\rho^G > \rho^L$ (upper left part of Fig. 3). It is apparent from Fig. 3 that a large variety of phase diagrams can be produced by varying only these two parameters. Refinements on these basic shapes for a particular lipid system are produced by differences in the two melting temperatures (at $X = 0, 1$) and by variations in ΔH_A and ΔH_B which tend to compress the basic shape of the phase diagram (or, in some cases, distort it horizontally).

Experimental Results

Figure 4 shows a phase diagram for a binary lipid mixture consisting of a mixture of C(10) : (22)PC and C(22) : C(12)PC. Both from the relatively simple shape itself and from the estimated values for ρ^L and ρ^G given in

TABLE IV
FITTED PARAMETERS AND CONFIDENCE INTERVALS FOR DATA OF FIG. 6D[a]

Parameter or interval	ρ^L	ρ^G	ΔH_A	ΔH_B	T_A	T_B
Upper limit	0.00	0.00	00.00	00.00	—	—
Parameter value	−281.40	−308.47	613.81	1740.09	313.82	307.92
Lower limit	0.00	0.00	00.00	00.00	—	—

[a] No confidence limits were available for the estimated parameters because of a zero eigenvalue.

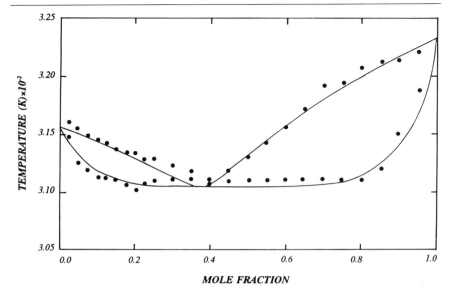

FIG. 7. Data and fitted curves for C(17) : C(17)PC/C(22) : C(12)PC mixtures. This was the best result allowing all six parameters to float during the estimation process. Note that the fit is reasonable except for the liquidus for $0.8 < X < 1.0$ where the data become almost horizontal. See Table V for the parameter values.

Table I, it is clear that this lipid pair behaves in a relatively ideal fashion. Contrast this with the data shown in Fig. 5 and Table II which exhibit a classic eutectic phase behavior. [In all figures, the solid curves are drawn from Eqs. (5) and (6) using the parameter estimates given in the corresponding tables.]

A less straightforward example is shown in Fig. 6. The experimental data (shown without a fitted curve in Fig. 6A) at first appear to exhibit relatively ideal behavior. However, superimposing the predicted curves

TABLE V
FITTED PARAMETERS AND CONFIDENCE INTERVALS FOR DATA OF FIG. 7[a]

Parameter or interval	ρ^L	ρ^G	ΔH_A	ΔH_B	T_A	T_B
Upper limit	0.584	1.438	17.41	14.41	324.32	316.36
Parameter value	−0.723	1.237	12.13	11.73	323.58	315.75
Lower limit	−0.708	1.043	6.68	9.09	322.81	315.11

[a] The variance of the fit was 0.378 with 42 degrees of freedom.

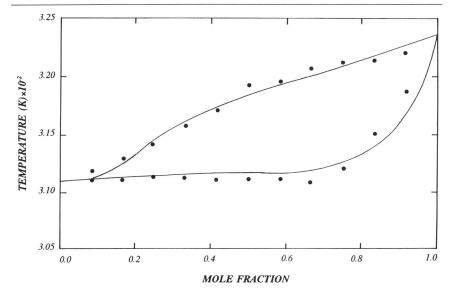

FIG. 8. Results of fitting the portion of the data shown in Fig. 7 for $0.4 < X < 1.0$ (data points to the right of the eutectic, but not including the eutectic point itself). Note by comparing Tables V and VII that the melting temperature of pure component A [C(17):C(17)PC] is closer to the experimental value (322.5 K) in this fit than when the full data set is used, and that the liquidus curve is more nearly horizontal. Concentrations were rescaled from $0.4 < X < 1.0$ to $0.0 < X < 1.0$.

for $\rho^L = \rho^G = 0$ (Fig. 6B) shows clearly that the onset and completion temperatures are much closer together than they should appear if the system were ideal. Figure 6C shows the curve obtained by allowing only ρ^G and ρ^L to float during the estimation process. The estimated values (Table III) are in this case equal within experimental error but are quite high (about 1 kcal). Note also that the errors in this case are large, with a

TABLE VI
FITTED PARAMETERS AND CONFIDENCE INTERVALS FOR DATA OF FIG. 8[a]

Parameter or interval	ρ^L	ρ^G	ΔH_A	ΔH_B	T_A	T_B
Upper limit	0.998	1.258	25.09	324.49	—	—
Parameter value	0.693	1.162	13.87	19.60	323.70	310.90
Lower limit	0.179	1.064	7.18	322.54	—	—

[a] If no limits are given, that parameter was held constant at the value shown. The variance of the fit was 0.678 with 18 degrees of freedom.

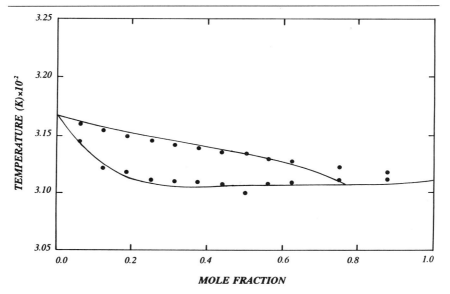

FIG. 9. Corresponding results for the remaining data from Fig. 7, this time using the data from the left-hand side of the eutectic concentration (again, not including the eutectic point). Except for the leftmost two points (where the solidus and liquidus merge) the fit is quite representative of the data. Concentrations were rescaled from $0.0 < X < 0.4$ to $0.0 < X < 1.0$.

confidence range 3 times the value itself, resulting from the high scatter in the data.

These data also provide an instructive example of the dangers lurking in indiscriminate curve fitting. If, in addition to the nonideal energies, we also allow the enthalpies to vary during the fitting process, we obtain a

TABLE VII
FITTED PARAMETERS AND CONFIDENCE INTERVALS FOR DATA OF FIG. 9[a]

Parameter or interval	ρ^L	ρ^G	ΔH_A	ΔH_B	T_A	T_B
Upper limit	0.888	1.377	—	—	—	—
Parameter value	0.818	1.315	19.60	16.40	310.90	316.76
Lower limit	0.692	1.202	—	—	—	—

[a] If no limits are given, that parameter was held constant at the value shown. The variance of the fit was 0.300 with 22 degrees of freedom.

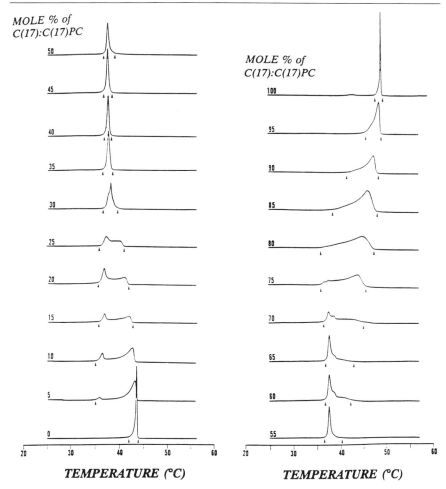

FIG. 10. Calorimeter scans for various concentrations of C(17):C(17)PC in C(22):C(12)PC. Note the similar shapes of the three peaks at 34–45 mol % which lie along the flat sections of the fitted curves in Figs. 8 and 9.

better visual fit (Fig. 6D) and a reduction in variance (0.048 to 0.030). However, the fitted values for these parameters (Table IV) are physically unrealistic (in the range of hundreds of kilocalories!). In addition, the program provided the information that a zero value for one of the eigenvalues had been obtained during the calculations, and that it was probable that too much information was being sought from a limited amount of data. In cases such as this, even in the absence of any indication that there is a mathematical problem, one can draw on a knowledge of the physical system as a guide in rejecting estimated results.

Possible Variations in Model

During our investigation of the many possible pairs of lipids of various hydrocarbon chain length, we have found several binary systems which exhibit phase diagrams having shapes that are not readily explained by regular solution theory. Figure 7 shows one of these examples. In Fig. 7, we show the predicted curves corresponding to the estimated values given in Table V. The general shape of the curve matches the data, but on the rightmost limb of the liquidus the fitted curve fails to reproduce the relative flatness of the experimental data for $0.7 < X < 1.0$. One consequence of this is that the melting temperature T_A for pure component A must be allowed to float, and the estimated value for this parameter falls outside of any realistic experimental error.

Shapes similar to that seen to the right of the eutectic temperature in these data can be produced by regular solution theory. We therefore decided to try analyzing the data as if the eutectic species were a pure component; that is, separate the data into two groups (one on each side of the eutectic point) and analyze each data set independently. Results of this procedure are shown in Figs. 8 and 9 (Tables VI and VII). The concentrations were rescaled to fall between 0.0 and 1.0 in both data sets to reflect the fact that the eutectic was being treated as a pure component. For this technique to be internally consistent, we used the same (nonfloating) values for both the eutectic temperature and transition enthalpy (both of which are available from the experimental data). Although the improvement in fit was not dramatic, this procedure did result in a lower melting temperature on the right-hand side and a somewhat better match over the horizontal segment of the liquidus.

One interesting aspect of the fitted curves was the horizontal segment which appeared on both sides of the eutectic concentration. Over this range of concentration the fitted horizontal curves appear to be an inadequate representation of the data points at 0.30, 0.35, and 0.45 (but obviously passes through the point at 0.40—the eutectic concentration). Examining the original calorimeter scans (Fig. 10) we noted that the scans for concentrations of 0.35–0.45 might well be interpretable as showing similar (and identical) onset and completion temperatures after correction. It is possible, then, that the fitted curves are not as much in error as they first appear to be in this region. Although this type of analysis is still in its preliminary stages, we have included it in order to show the flexibility of the overall procedure.

Future Directions

We also plan to explore the effect of modifications to regular solution theory. For example, one might substitute a different function of concentration for the added (nonideal) term. Different functional forms for existing terms are preferable to additional terms since added parameters (which probably must be estimated) would require a substantial amount of additional data to keep the same level of significance in the parameter estimation statistics. It might also be feasible to include within the evaluation routine additional logic which imposes physically reasonable constraints on the parameters in order to avoid situations such as that shown in Fig. 6D.

Acknowledgments

This research was supported by National Institutes of Health Grant GM-17452.

[26] Deconvolution Analysis of Hormone Data

By JOHANNES D. VELDHUIS and MICHAEL L. JOHNSON

Introduction and Definition of Deconvolution

Introduction

Deconvolution is a mathematical technique that has been utilized in the physical, applied, and natural sciences as a method for estimating the particular behavior of one or more component processes contributing to an observed outcome. For example, in seismology, the intensity of the geologic shock wave recorded at some distant site is a function of the initial disturbance and various dissipation kinetics acting en route to the monitoring point. Deconvolution would attempt to reconstruct the initial shock impulse from time records of the remote signal.[1] In the field of spectroscopy, the intensity of an emitted wavelength of light measured at some remote point from the light source is influenced by the energy properties of the initial fluorescent signal and attenuation of the emitted signal as it travels to the point of observation. Deconvolution attempts to recover estimates of the intensity of the original fluorescent discharge.[2]

[1] J. J. Kormylo and J. M. Mendel, *IEEE Trans. Geosci. Remote Sensing* **GE-21,** 72 (1983).
[2] P. A. Jansson, "Deconvolution with Applications in Spectroscopy," p. 11. Academic Press, New York, 1984.

In the life sciences, many biological phenomena observed in the time domain are the result of several constituent processes that are regulated and operate independently, or nearly independently, but contribute jointly to the outcome of interest. A typical example is the behavior of metabolite, substrate, or hormone concentrations sampled in the blood compartment over time. Spontaneous variations in measured concentrations of a metabolite, substrate, or hormone over time are controlled by at least two distinct processes (in addition to effects of "noise" in the system): (1) the rate of entry of the constituent into the blood compartment and (2) the magnitude and type of elimination kinetics serving to remove the compound from the circulation. Both the rate of entry and the rate of removal of the substance from the sampling compartment (as well as experimental uncertainty) specify the estimated concentration at any given instant.[3] The "convolution" or entwining of these two contributing processes designates the overall outcome.

Although the concept of deconvolution has been useful in a large range of physical and natural sciences, here we evaluate methods of deconvolution analysis that have been applied to a specific topic in biology, namely, the temporal behavior of hormones, metabolites, and substrates in a sampled fluid compartment, such as blood. Such analyses take on significance in this area of biology, because the time structure of the hormone, metabolite, or substrate concentration signal conveys important information to the target tissue, yields insights into systems regulation, and subserves the homeostasis of an organism.

Convolution Integral

To relate two or more relevant processes to an overall outcome quantitatively, the specific mathematical functions that are algebraic descriptors of those processes can be "convolved" as a mathematical dot product, the integral of which yields the net output of the system at any given instant in time. Such an integral of two explicit functions, which are said to be convolved (intertwined, or jointly contributory), is designated a convolution integral. Solution of the convolution integral provides an important basis for estimating the individual constituent processes, given some assumptions about the overall behavior of the data, the structure of the phenomenon, and the response characteristics of the measurement system. For example, measurements of plasma hormone concentrations serially over time can be considered to result from finite entry rates (secretion impulses) of the hormone molecules into the bloodstream, measurable kinetics of hormone removal, and confounding experimental uncertainty

[3] J. D. Veldhuis and M. L. Johnson, *Front. Neuroendocrinol.* **11,** 363 (1990).

contributed by an array of factors (e.g., biological variations as well as those introduced by sample collection, sample processing, assay, and interpolation from standards). The convolution integral relates the combined contributions of these processes to the observed output of interest, namely, the measured blood concentration of the hormone or substrate at some instant, as given below:

$$C(t) = \int_0^t S(z)E(t - z)\,dz + \varepsilon \tag{1}$$

In the above formulation, $C(t)$ gives the concentration of the hormone, metabolite, or substrate at some time, t. $S(z)$ denotes the secretion function or the input function, which represents the rate of entry of the molecules of interest into the sampling compartment over time. The definite integral of $S(z)$ specifies the mass of substance delivered into the system per unit distribution volume over the time interval of interest. The form of the secretion function is of particular interest in hormone data, since various physiological states and/or pathological conditions can modify the secretory behavior of an endocrine gland (*vide infra*). In addition, the concentration of analyte at time zero can be defined as an appropriate constant within $S(z)$. $E(t - z)$ describes the elimination function, or the kinetics of removal of the hormone, metabolite, or substrate from the sampled space. Epsilon denotes random experimental variance in the system, which contributes to and confounds the measurement.

We shall discuss two particular methods for solving such a convolution integral, and thereby estimating the underlying secretory behavior of the endocrine gland and/or the dissipation kinetics that result in effective clearance of the substance from the system. Because the rate of entry and/or the rate of exit of a hormone, metabolite, or substrate from the sampling space can be altered in physiology and pathophysiology, deconvolution techniques have been designed to evaluate from the convolution integral the value(s) of either rate alone or both rates simultaneously. Here we present two particular approaches to deconvolution analysis, which can now be defined practically as solution of the convolution integral [Eq. (1)] to obtain quantitative estimates of constituent processes, given an observed outcome. This is illustrated schematically in Fig. 1. Some advantages of deconvolution analysis are given in Table I.

Special Features of Endocrine Data

Biological Constraints

In the life sciences, the collection of experimental data can be significantly constrained by the biological system. Specific difficulties encountered in the accumulation of endocrine data include the adverse effects of

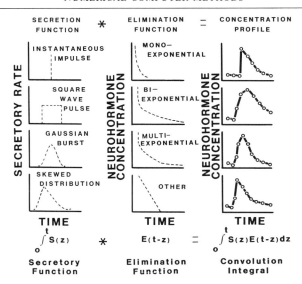

FIG. 1. Schematic illustration of the concept of a convolution integral. A convolution integral is used to specify the simultaneous effects of two functions operating to control the overall observed outcome. For example, the convolution of a secretion function (left-hand graphs) and a clearance or elimination function (middle graphs) when integrated with respect to time will describe a profile of hormone, metabolite, or substrate concentration over time (right-hand graphs). A variety of potential secretion functions could be employed to represent the temporal behavior of hormone release rates in a particular circumstance. For example, a secretion episode may be represented as a nearly instantaneous impulse, a "square-wave" event, or a symmetric Gaussian burst. In addition, the kinetic function to describe the amount of hormone eliminated per unit time could consist of one or more exponential terms, a concentration-dependent function, a partially saturable process, etc. [Adapted with permission from J. D. Veldhuis and M. L. Johnson, *J. Neuroendocrinol.* **2,** 755 (1990).]

TABLE I

INFORMATION CONTENT PROVIDED BY KNOWLEDGE OF ENDOCRINE SECRETORY SIGNAL

1. Secretory events may be quite dissimilar from plasma hormone concentration pulses, e.g., in duration, amplitude, shape, and timing
2. Increased analytical power can be achieved by detecting secretory bursts, when the confounding effects of subject-, condition-, and hormone-specific metabolic clearance are removed from the plasma hormone concentration data
3. Cosecretion of two or more hormones can be evaluated despite differences in endogenous hormone half-life, baseline, plasma peak duration, etc.
4. If secretion rates are known or can be estimated from the data, then endogenous hormone clearance rates can also be calculated

TABLE II
LIMITATIONS AND PITFALLS IN EVALUATING *in Vivo* SECRETION INDIRECTLY

1. Hormone concentration series are often sparse, noisy, and contain an unknown true signal configuration
2. True metabolic clearance rates of endogenous hormone isoforms are often unknown or difficult to estimate independently in the specific study context of interest
3. Multiple sources of experimental variance contribute to statistical uncertainty in the data, e.g., sample processing and assay noise and uncertainty in half-life

experimental observation on the processes being studied. For example, placement of intravenous catheters in an animal to collect blood samples at frequent intervals to measure plasma concentrations of a substrate, hormone, or metabolite can elicit significant stress responses from the animal, result in substantial blood withdrawal (which itself could alter homeostatic responses), and require infusion of anticoagulants or fluids that might dilute the substance of interest. Moreover, the number of sample observations that can be obtained per unit time is sometimes severely limited in biological models. The total duration over which observations can be carried out without disturbing the system can also be undesirably short. In addition, biological data typically show substantial diversity among different organisms within ostensibly homogeneous groups, which requires that the observation be replicated over a relatively large number of individual animals or subjects. Moreover, in the case of endocrine data, the regulation of hormone output by a secretory gland is subject to multifactorial control by a host of variables, such as time of day, nutrient status, associated stress, ambient hormone milieu, and prior endocrine changes.

The "noise" or experimental uncertainty in many biological systems is relatively large and/or its nature is not precisely defined. For example, one does not always know whether experimental variance is distributed as a random normal deviate with zero mean or conforms to some other statistical distribution. As importantly, in many circumstances the exact form of the biological signal [e.g., $S(z)$ in Eq. (1)] is not known *a priori* and/or is difficult to document directly.

Consequently, biological data series pose special problems in that they are typically short, noisy, and sparse, contain an unknown signal, and are confounded by uncertain sources and patterns of experimental variation (see Table II). Because deconvolution techniques are typically dependent on the total number and time density of observations as well as the nature and amount of experimental uncertainty, deconvolution approaches in biology have been challenged by the above restrictions.[4]

[4] J. D. Veldhuis and M. L. Johnson, *BioTechniques* **8**(6), 634 (1990).

Assay Characteristics

The measurement of a hormone, metabolite, or substrate in a body fluid is always attended by some degree of assay imprecision, is restrained by some level of assay sensitivity (minimal detectable or limiting hormone concentration), and is subject to some particular specificity (the extent to which the assay correctly reports amounts of the substance of interest, and conversely the extent to which it does not falsely report the presence of other substances). For example, analytical precision in the measurement of endocrine phenomena will correspond typically to a coefficient of variation (ratio of standard deviation of the measurement to the mean measurement value) between 3 and 25%. Often, higher precision (a lower coefficient of variation) is achieved at intermediate hormone, substrate, or metabolite concentrations, and lower precision (higher coefficients of variation) is observed at very low or very high concentrations.[5] Thus, there is dose-dependent experimental uncertainty in the measurement.

To determine the distribution of measurement error, a given relevant sample containing a known concentration of a hormone should be submitted to assay in multiple replicates (e.g., 20–100), and the distribution of measurement values observed. The replication procedure should be repeated as a function of hormone dose to encompass the full range of relevant substance concentrations. This is illustrated for luteinizing hormone (LH) in Fig. 2. Note that at intermediate and high plasma LH concentrations, the distribution of replicate estimates is random and normal about the sample mean, whereas at the lowest concentrations of hormone there is significant departure from normality. This observation is of more than trivial importance, because many deconvolution techniques assume that the distribution of experimental uncertainty in the dependent variable is random and normal (Gaussian).

In addition to evaluating the precision of endocrine assays, one must ascertain the sensitivity and specificity characteristics. Sensitivity can be determined by replicate assays of biological samples containing no hormone. The preparation of such samples requires special precautions, since the chemical composition of the "blank" sample should be identical to that encountered in other experimental samples except for absence of the hormone, substrate, or metabolite of interest. In some cases, serum obtained under pathological conditions that are associated with complete absence of the hormone will be suitable. In other circumstances, antibody-mediated precipitation or chemical or physical methods for removing the analyte are appropriate. Given an appropriate "blank" sample, the sensi-

---

[5] R. J. Urban, W. S. Evans, A. D. Rogol, D. L. Kaiser, M. L. Johnson, and J. D. Veldhuis, *Endocr. Rev.* **9,** 3 (1988).

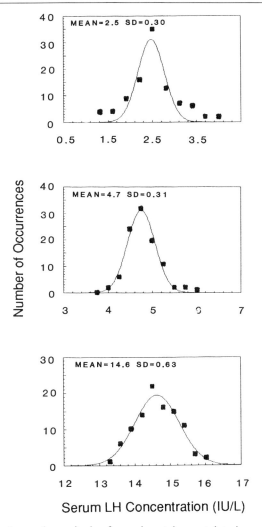

FIG. 2. Distribution and magnitude of experimental uncertainty in a particular hormone assay applied to the measurement of serum luteinizing hormone (LH) concentrations. The frequency distribution of serum LH estimates is given for each of three concentrations of this glycoprotein hormone. Each histogram distribution was created experimentally by assaying 100 replicates of each of three pools of blood containing a specific dose of LH. The distribution of the 50 mean LH sample concentration values calculated from the assay duplicates was random and normal (Gaussian) for the two higher doses, but it departed from normality at the lowest concentration, which approximates assay sensitivity. The squares give the observed data (number of LH measurements within a given interval), and the solid curve the best-fit Gaussian.

tivity of the assay can then be defined as the amount of added pure compound that exceeds by 2 or 3 standard deviations the mean value estimated in the zero-dose tubes.

Once the sensitivity of the endocrine assay is defined, one must then deal with the problem of values that fall at or below assay sensitivity. By definition, the concentration of hormone in such samples is indeterminate. Moreover, the measurement imprecision of such indeterminate samples is unknown. Accordingly, if a substantial fraction of the experimental samples exhibit hormone concentrations that are at or below assay sensitivity, then significant bias could be introduced into the analysis of hormone secretion and clearance. Specifically, one cannot estimate the number and amplitude of secretory events occurring within the undetectable range, and the propagation of error associated with these samples in the deconvolution algorithm becomes problematic. How this problem is best dealt with analytically is not absolutely clear. However, if an arbitrary value is assigned to such undetectable sample concentrations (such as the sensitivity limit of the assay per se), as well as a minimum standard deviation (such as that interpolated to the sensitivity limit by some dose-dependent variance function or one-half the sensitivity value), then the remaining endocrine data that are within the measurable range can be submitted to deconvolution analysis. Irrespective of the approach used, the extent of bias in the fit that is produced by the low-end samples should be evaluated.

Preferably the problem of poor sensitivity should be dealt with experimentally; that is, improved assays with enhanced sensitivity should be used in circumstances in which a significant fraction of the samples would otherwise yield undetectable values. Improved assays can offer significant new insights into the patterns of hormone secretion. For example, in the case of growth hormone (GH) measurements in young, healthy, nutritionally replete individuals, approximately 60–80% of the serum GH concentrations in conventional assays can fall below assay sensitivity. The use of an enhanced measurement technique recently demonstrated that GH secretory pulses continue to occur at a significant frequency even at extraordinarily low plasma hormone concentrations.[6] In short, no mathematical tool will overcome inherent deficiencies in the quality of the data. Rather, such deficiencies must be addressed by enhancing experimental techniques.

The specificity of an assay is commonly assumed, but it must be rigorously demonstrated. For example, in measurements of LH, radioimmunoassays are usually sensitive to one or more discrete epitopes in the polypeptide component of the glycoprotein molecule. The epitopes

[6] L. M. Winer, M. A. Shaw, and G. Baumann, *J. Clin. Endocrinol. Metab.* **70**, 1678 (1990).

recognized by the particular antiserum utilized may or may not relate to relevant regions of the hormone molecule that actually specify biological responses to target tissues.[7] Alternatively, antisera may detect one or more isoforms of a hormone and/or its partially degraded fragments that retain only sparse biological activity. Moreover, uncombined subunits of a glycoprotein hormone such as the common free α subunit of thyroid-stimulating hormone, LH, follicle-stimulating hormone, and human chorionic gonadotropin can be measured in some radioimmunoassays, leading to poor specificity and potentially erroneous inferences under various conditions. The investigator must recognize that refinements of analytical tools (such as deconvolution) cannot overcome any primary deficiencies in specificity inherent in the particular assay system chosen.

Specific Deconvolution Techniques

Given that serial measurements of a hormone, metabolite, or substrate over time have been carried out accurately and specifically and with good sensitivity, the resultant temporal profile of circulating hormone concentrations can be examined mathematically so as to describe the relative contributions of input and output (secretion and metabolic clearance) that generated the observations. Many methods for accomplishing such dissection of plasma hormone concentration profiles into constituent secretion and/or clearance processes have been developed, and they were reviewed recently elsewhere.[3,4,8]

Here we evaluate two particular methods that either assume or do not assume a specific waveform for the secretion function [$S(z)$ in Eq. (1)]. In the first case, when a particular algebraic form of the secretion function is assumed or known, quantitative values for both secretion and elimination can be deduced by deconvolution analysis of serial plasma hormone concentration measurements in appropriate circumstances. Several possible waveforms for a secretion function, $S(z)$, could be assumed, some of which are illustrated in Fig. 1, such as the following: a nearly instantaneous impulse; a square-wave increase in secretion (on/off switch); a symmetric Gaussian; or one of a variety of asymmetric trajectories of secretion rate over time. In most circumstances in endocrine research, the exact *in vivo* waveform of the secretion event (plot of secretion rate against time as enacted by the endocrine gland) has not been established. However, direct monitoring of glandular effluent blood has been accomplished occasionally

[7] J. D. Veldhuis, R. J. Urban, I. Beitins, R. M. Blizzard, M. L. Johnson, and M. L. Dufau, *J. Steroid Biochem.* **33,** 739 (1989).

[8] J. D. Veldhuis and M. L. Johnson, *J. Neuroendocrinol.* **2,** 755 (1990).

and has revealed either skewed or symmetric secretion bursts that occur as apparently discrete punctuated episodes of release, for example, for GH, LH, follicle-stimulating hormone, adrenocorticotropic hormone (ACTH), and others.[9,10] Wherever possible, the investigator should obtain such direct information about the time course of *in vivo* secretion by direct but minimally disturbing measurements of glandular secretion in the specific experimental context of interest. Alternatively, a reasonable waveform for the secretion event can be assumed provisionally in the deconvolution analysis, and relevant predictions of such assumptions tested experimentally in the animal.[11]

Waveform-Defined Method

The multiple-parameter deconvolution technique provides one means to estimate secretion and clearance rates simultaneously from observed serial plasma hormone concentrations measured over time. This methodology requires that some specific waveform for the secretion function be assumed or known (e.g., as illustrated in Fig. 1). In addition, estimates of experimental uncertainty in the data are important, and a sufficient number of observations over a reasonably extended interval should be available. Estimates of hormone secretion and metabolic clearance rates can then be accomplished in individual animals and specific experimental settings.[11] As a necessary but not sufficient condition for validity of such a model, independently validated estimates of hormone secretion and clearance rates should be obtained; this has indeed been accomplished for a large number of hormones.[12-16]

The multiple-parameter deconvolution methodology is based on Eq. (1) above. Although various algebraic forms for the secretion function, $S(z)$, can be considered, we have found that the following secretion func-

[9] L. A. Frohman, T. R. Downs, I. J. Clarke, and G. B. Thomas, *J. Clin. Invest.* **86,** 17 (1990).
[10] S. L. Alexander and C. H. Irvine, *J. Endocrinol.* **114,** 351 (1987).
[11] J. D. Veldhuis, M. L. Carlson, and M. L. Johnson, *Proc. Natl. Acad. Sci. U.S.A.* **84,** 7686 (1987).
[12] J. D. Veldhuis, A. Iranmanesh, G. Lizarralde, and M. L. Johnson, *Am. J. Physiol.* **257,** E6 (1989).
[13] J. D. Veldhuis, M. L. Johnson, and M. L. Dufau, *Am. J. Physiol.* **256,** E199 (1989).
[14] M. L. Hartman, A. C. S. Faria, M. L. Vance, M. L. Johnson, M. O. Thorner, and J. D. Veldhuis, *Am. J. Physiol.* **260,** E101 (1991).
[15] J. D. Veldhuis, A. Iranmanesh, M. L. Johnson, and G. Lizarralde, *J. Clin. Endocrinol. Metab.* **71,** 452 (1990).
[16] J. D. Veldhuis, A. Faria, M. L. Vance, W. S. Evans, M. O. Thorner, and M. L. Johnson, *Acta Paediatr. Scand.* **347,** 63 (1988).

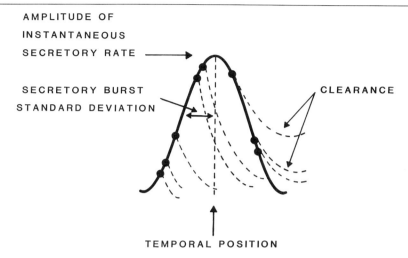

AMPLITUDE OF
INSTANTANEOUS
SECRETORY RATE ──────→

SECRETORY BURST   CLEARANCE
STANDARD DEVIATION

TEMPORAL POSITION

FIG. 3. Hypothetical hormone secretory burst and its relevant characteristics. The schema illustrates a hypothetical secretion burst of symmetric form that can be modeled as a Gaussian event with a distinct amplitude, some centered position in time, and an individual *SD* or half-duration (time in minutes elapsing at half-maximal amplitude; the half-duration equals 2.354*SD*). Descriptive features of an algebraically definable secretion burst can then be estimated by nonlinear methods of multiple-parameter deconvolution analysis (see text). All secreted molecules are assumed to be subject to hormone elimination or clearance, that is, they exhibit finite disappearance kinetics. [Adapted with permission from J. D. Veldhuis, M. L. Carlson, and M. L. Johnson, *Proc. Natl. Acad. Sci. U.S.A.* **84,** 7686 (1987).]

tion (a series of Gaussian events) provides a good fit of many experimental endocrine data series and yields good estimates of irreversible metabolic clearance rates and endogenous secretion rates for a large number of different hormones:

$$S(z) = \sum_{i=1}^{n} A_i \, e^{-(1/2)[(pp_i - z)/SD]^2} \qquad (2)$$

where A_i is the amplitude of the ith secretory event occurring at time pp_i, and *SD* represents the standard deviation, which is proportional to the half-duration (*HD*, duration at half-maximal amplitude) of the secretory bursts. Thus, in this formulation a secretory burst is modeled as a Gaussian waveform with finite standard deviation (where $HD = SD \times 2.354$), amplitude (maximal value), and location in time. These characteristics of a hypothetical secretion burst are illustrated in Fig. 3. We chose this general approximation because even a square-wave or rectangular impulse delivered into a turbulent stream would be subject to smoothing at the leading and trailing edges and result in a roughly symmetric waveform.[17] The overall secretion function, $S(z)$, is composed of a finite number of secretory bursts, n, each with an individual finite, real amplitude, A_i, and a corresponding location or position in time centered at pp_i. In this waveform-defined method, one can develop deconvolution-based estimates of individual secretion burst amplitudes, half-duration (duration at half-maximal amplitude), and locations in time.

The elimination fraction, $E(t - z)$, can be approximated in many biological circumstances as a mono-, bi-, or pluriexponential decay function:

$$E(t - z) = A\,e^{-k_1(t-z)} + B\,e^{-k_2(t-z)} + C\,e^{-k_3(t-z)} \qquad (3)$$

The rate constants k_1, k_2, and k_3 are elimination constants in inverse time units, and they are related to the half-life as $t_{1/2} = \ln 2/k$. Note also that the total amplitude of the decay process is partitioned in relation to the different rate constants, as relative amplitudes A, B, and C.

The overall form of Eq. (1) for monoexponential decay as used in the multiple-parameter deconvolution technique can then be given as:

$$C(t) = \int_0^t \left[\sum_{i=1}^n A_i\,e^{-(1/2)[(pp_i - z)/SD]^2} \right] A\,e^{-k_1(t-z)}\,dz + \varepsilon \qquad (4)$$

We have recently shown that the convolution product of an error function (the expression for the secretion event represented algebraically by a Gaussian) and an elimination function (taken, for example, as an exponential disappearance process) is susceptible to an analytical solution.[18] Use of the analytical solution to Eq. (4) in parameter estimation obviates the computational burden that numerical approximation of the above integral otherwise imposes.

When using the foregoing waveform-defined deconvolution technique, an investigator can obtain specific estimates of hormone secretion burst number (n), individual amplitudes (A_i), half-duration (HD), and the associated half-life of removal of the hormone, metabolite, or substrate from the sampling compartment. The values of four classes of parameters can be estimated by appropriate nonlinear least-squares methods of curve fitting and error propagation, in which solutions of the convolution integral are

---

[17] One can also introduce a skewness factor to create a range of asymmetries, for example, so as to induce leftward or rightward skew of the secretion impulse.[18] Note also that one can carry out deconvolution analysis without any assumptions about the secretion waveform (see section "Waveform-Independent Method").

[18] J. D. Veldhuis, A. B. Lassiter, and M. L. Johnson, *Am. J. Physiol.* **259**, E351 (1990).

TABLE III
FEATURES OF MULTIPLE-PARAMETER DECONVOLUTION ALGORITHM

1. All experimental data and their variances are considered when estimating optimal parameter values
2. Mean estimate and corresponding statistical confidence limits are calculated for each secretion and clearance parameter. Only positive nonzero real estimates are obtained
3. Endogenous hormone disappearance rates are computed from a model of combined secretion and clearance. Disappearance rate constants are thus estimated in each individual and/or experimental context
4. High data densities (e.g., high sampling frequencies) enhance rather than detract from parameter estimation
5. Temporal locations as well as the amplitudes and half-durations of secretory impulses can be estimated. This allows assessment of secretory concordance for two or more different hormone series

compared to the observed data. These features of this multiple-parameter deconvolution algorithm are summarized in Table III.

To obtain numerical estimates of secretion and clearance, the measured concentration of a hormone, substrate, or metabolite in blood at each instant can be related mathematically by way of Eq. (4) to four distinct, finite, determinable, real, and nonnegative parameters embracing (1) the number and locations in time, (2) the individual amplitudes and (3) the half-duration(s) of all prior secretion episodes, as well as (4) subject- and condition-specific elimination rate constants and their corresponding amplitudes. The observed serial blood concentrations of the hormone, substrate, or metabolite are then arrayed in a family of equations, which can be given as follows:

$$
\begin{aligned}
C(t_1) &= \int_0^{t_1} \left[\sum_{i=1}^{n} A_i\, e^{-(1/2)[pp_i - z)/SD]^2} \right] A\, e^{-k_1(t_1 - z)}\, dz\, +\, \varepsilon \\
C(t_2) &= \int_0^{t_2} \left[\sum_{i=1}^{n} A_i\, e^{-(1/2)[pp_i - z)/SD]^2} \right] A\, e^{-k_1(t_2 - z)}\, dz\, +\, \varepsilon \\
&\;\vdots \\
C(t_j) &= \int_0^{t_j} \left[\sum_{i=1}^{n} A_i\, e^{-(1/2)[pp_i - z)/SD]^2} \right] A\, e^{-k_1(t_j - z)}\, dz\, +\, \varepsilon
\end{aligned}
\tag{5}
$$

Specifically, we write a family of j simultaneous integral equations, each of which contains $C(t_j)$, the concentration of the hormone at some particular time, t_j. Each convolution integral relates the hormone concentration at that time to all prior and concurrent secretion events and the simultaneous effects of metabolic clearance acting on all secreted molecules. For example, if blood is sampled at 10-min intervals for 24 hr, and the subsequent sera subjected to immunoassay to measure growth hormone concentrations, then the available 145 serum GH concentration measurements permit one to create a family of 145 equations. Each equation contains a relevant convolution integral, in which the number, amplitude, location in time, and duration of all prior secretion events and subject-specific kinetic rate constants are incorporated as unknown parameters. Typically, one will then need to estimate approximately 20–40 unknown parameters (individual secretory burst locations, amplitudes, and/ or durations as well as several kinetic terms) from the corresponding 145 equations. The solution of such sets of simultaneous integral equations is facilitated by analytical solution to the integral[18] and high-speed memory-adequate computational devices. To constrain parameters to positive real numbers, we fit to the logarithm of the unknowns (*vide infra*). Stable, efficient, and robust computer convergence subroutines are required to obtain rapid and reliable estimates of the multiple secretory (and clearance) parameters that are often highly correlated.

If one assumes that the form of secretion bursts is homogeneous within any given subject and experimental condition, then a single term for the secretory burst half-duration (duration of the secretion event at half-maximal amplitude) can be estimated for a data set. In addition to this parameter value, one must compute the individual best-fit location and amplitude of each secretory burst that significantly reduces variance of the fit, and simultaneously calculate relevant parameters of hormone metabolic clearance (one or more rate constants and their fractional amplitudes). This feat is made more challenging because some of the parameters in the convolution integral are interdependent or correlated. For example, any given estimate of secretory burst half-duration would require a corresponding inverse adjustment in hormone half-life, since prolonging the duration of the secretion event would allow an almost equivalent fit of the data with a shorter half-life (more hormone needs to be removed per unit time if the secretion event is extended) (Fig. 4). Consequently, very carefully selected computer-based convergence methods are required to achieve best-fit parameter estimates in such potentially highly correlated yet multiple parameter spaces.[19] A modified Gauss–Newton quadratic convergence routine is one such method,[11,19] as discussed elsewhere in this volume. Joint asymmetric variance spaces are also generated to offer statistical confidence limits for multiple (correlated) parameters.[19]

Figure 5 illustrates the application of the above waveform-specific method of deconvolution. We depict two types of data with widely differing

[19] M. L. Johnson and S. G. Frasier, this series, Vol. 117, p. 301.

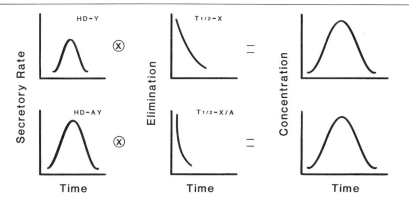

FIG. 4. Interrelationship between secretory burst half-duration and hormone half-life. As shown by the right-hand graphs, a given profile of plasma hormone concentrations can result from either a long secretion burst half-duration (*HD*) and a short half-life ($t_{1/2}$), or from a brief secretion burst half-duration and a long half-life. Thus, terms of secretion and clearance are intercorrelated when estimated in relation to any single data set.

quality (in one case blood was sampled every 1 min for 400 min, and in the other every 20 min for 300 min). Note that the above multiple-parameter deconvolution algorithm is relatively well behaved, since it provides only positive real-number estimates of secretion rates, yields a good fit of the observed data, and performs well over as much as a 20-fold range in data density. Moreover, as required, the above method yields reasonable estimates of steady-state hormone metabolic clearance rates (or half-lives) for a range of anterior pituitary hormones, as illustrated in Fig. 6. Correspondingly, this deconvolution technique correctly predicts the endogenous hormone secretory rates under various pathophysiological conditions.[11–16] Accurate predictions of endogenous hormone half-life and production rates are necessary but not sufficient criteria for methodological validity (*vide infra*).

Assumptions of the waveform-defined methodology of deconvolution include all the assumptions inherent in nonlinear methods of parameter estimation, as discussed in detail elsewhere in this volume. If these assumptions are satisfied, the method is a maximum likelihood procedure. In brief, pertinent assumptions require the following: (1) error is restricted to the dependent variable (*y* axis); (2) the distribution of experimental uncertainty is random and normal; (3) the experimental observations are independent; (4) no systematic bias exists in the data; and (5) the algebraic structure of the convolution model is relevant to the biology of the system. In relation to these assumptions, we have demonstrated that the independent variable (timing of blood sample collection) is relatively error free;

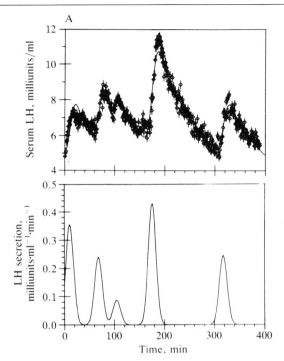

FIG. 5. Waveform-specific deconvolution analysis applied to two illustrative hormone data series. In (A), serum luteinizing hormone concentrations were measured in blood collected at 1-min intervals for approximately 400 min. In (B), serum concentrations of LH were measured in blood collected at 20-min intervals for 300 min. Note that in each circumstance the multiple-parameter convolution model predicts a fitted function (top graphs) that corresponds closely to the observed data, and yields an estimate of underlying hormone secretion burst number, amplitude, half-duration, and location (bottom graphs). [Reprinted with permission from J. D. Veldhuis, M. L. Carlson, and M. L. Johnson, *Proc. Natl. Acad. Sci. U.S.A.* **84,** 7686 (1987).]

that is, the majority of blood samples are collected within 5–20 sec of the intended time (*x* axis). By replication of assay measurements, we have determined that the distribution of experimental variance in the dependent variable (hormone concentration, *y* axis) is approximately Gaussian over the range of detectable and physiological hormone concentrations. Of course, in any experiment, one must anticipate and define as well as possible all sources of systematic bias in the data.

Finally, the relevance of the algebraic structure of the model to the biology of the process should be examined. In this regard, if the model predicts erroneous hormone half-lives and production rates, then a signifi-

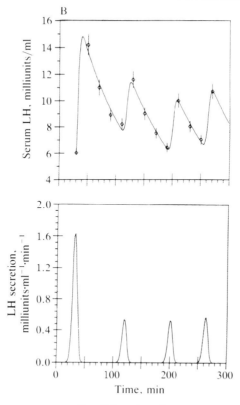

FIG. 5 (*continued*)

cant error is likely to exist in either the model or the independent tests of that model. Positive evidence for the relevance of the Gaussian waveform deconvolution technique has been obtained via direct measurements of pituitary effluent blood in sheep to observe GH secretion and in horses to define the waveform of LH and follicle-stimulating hormone secretion.[8-10] In these circumstances, a generally symmetric pituitary hormone release episode is evident, and this profile can be approximated by a series of Gaussian secretion bursts.

In summary, waveform-specific deconvolution utilizes an assumed algebraic approximation to the secretion function to depict the behavior of secretion over time and one or more kinetic rate constants to describe hormone, metabolite, or substrate removal from the system. A family of equations is written to relate the measured hormone concentrations to the convolution parameters of interest, namely, secretion burst location,

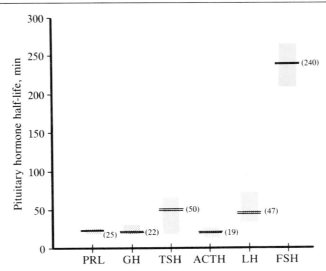

FIG. 6. Deconvolution-predicted *in vivo* half-lives for various anterior pituitary hormones studied in the human. The vertical axis gives the computer-predicted half-time of disappearance from plasma of several distinct individual anterior pituitary hormones: luteinizing hormone (LH), follicle-stimulating hormone (FSH), growth hormone (GH), adrenocorticotropic hormone (ACTH), prolactin (PRL), and thyroid-stimulating hormone (TSH). The double horizontal lines (and numerical values) denote mean literature estimates of the half-lives, and the stippled boxes give the statistical confidence limits for deconvolution-based estimates. [Reprinted with permission from J. D. Veldhuis, M. L. Carlson, and M. L. Johnson, *Proc. Natl. Acad. Sci. U.S.A.* **84**, 7686 (1987).]

amplitude, and duration and condition-specific kinetic terms. The set of simultaneous integral equations is used to solve for values of the parameters of interest by nonlinear least-squares methods of parameter estimation, with appropriate statistical propagation of joint variance (error) spaces (discussed below). Because the amplitude of secretion bursts and the hormone half-life should be zero or positive real numbers, available methods for constraining parameters to such a universe are applied; for example, one can fit the data to the logarithm of the amplitude, so that whether the argument is negative or positive the amplitude will be a positive real number. Other methods for constraining the domain of parameter estimates have been discussed.[2,20]

---

[20] J. D. Veldhuis and M. L. Johnson, *in* "Advances in Neuroendocrine Regulation of Reproduction" (S. S. C. Yen and W. Vale, eds.), p. 123. Plenum, New York, 1990.

Waveform-Independent Method

In the absence of any assumptions or knowledge concerning the secretion-event waveform, one or more methods of deconvolution that are waveform-independent can be employed. In brief, one such method utilizes the following equation:

$$C(t) = \sum_{i=1}^{n} S_i E(t - t_i) H(t - t_i)$$

$$H(t - t_i) = \begin{cases} 1, t - t_i \geq 0 \\ 0, t - t_i < 0 \end{cases}$$

(6)

where Δt is the sampling interval (or time between consecutive samples), which is assumed to be uniform, S_i is the sample secretion rate (mass units of hormone secreted per unit time per unit distribution volume) associated with the ith observation, n is the number of samples, and $C(t)$ is the concentration of the hormone, substrate, or metabolite in the sampling compartment at time t. The elimination function, $E(t - z)$, typically can be represented by an exponential disappearance function [Eq. (3)].

Because secretion and elimination rates are highly correlated parameters for any given data set, one cannot easily estimate individual secretion rates for all samples and the subject-specific hormone half-life simultaneously without further information. For example, given any particular half-life (or elimination rate constant), a particular set of sample secretion rates will provide a good fit of the observed serial hormone concentrations over time. However, if the half-life value changes, the sample secretion rate values will change in the opposite direction to fit the same concentration data. Consequently, such correlated parameters cannot be estimated uniquely in a ready fashion. However, if the half-life is known *a priori* or assumed, then estimates of individual sample secretion rates, S_i, can be achieved by writing a family of n equations each containing an individual sample hormone concentration that is related to the sum of all prior secretion acted on by relevant metabolic clearance[21] [Eq. (6)]. The results of this technique (with appropriate error propagation[19] as discussed below) are illustrated in Fig. 7.

Error propagation in waveform-independent (as well as waveform-specific) deconvolution techniques is essential in order to estimate whether individual sample secretion rates are significantly nonzero and/or significantly different from flanking values (i.e., a "burst" of secretion has occurred). We believe that error propagation should include the transfer

[21] J. D. Veldhuis, A. Iranmanesh, M. L. Johnson, and G. Lizarralde, *J. Clin. Endocrinol. Metab.* **71**, 1616 (1990).

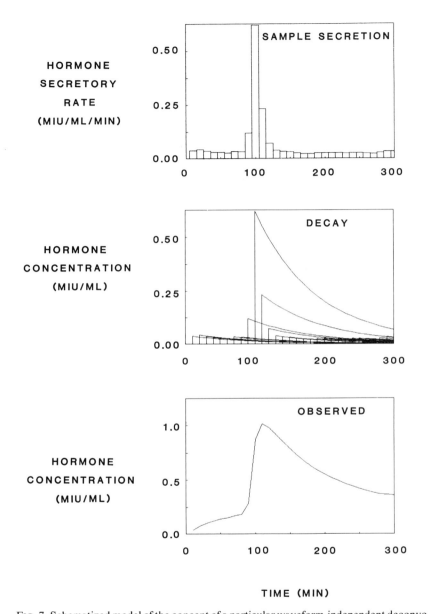

FIG. 7. Schematized model of the concept of a particular waveform-independent deconvolution technique. Observed sample hormone concentrations (bottom graph) are considered to arise from the combined effects of sample secretion (top) and clearance kinetics or decay (middle). Error intrinsic in the hormone half-life estimate and dose-dependent experimental variance inherent in the hormone assay are both used to propagate conjoint error spaces and thereby estimate the statistical confidence limits for the calculated secretion values. In this model, some *a priori* knowledge or assumption regarding the half-life of hormone disappearance (decay) must be available. (Arbitrary hormone mass units of mIU are shown.)

of experimental variance to the secretion estimate from all plausible sources of experimental uncertainty. Such sources for error include the investigator-specified and/or independently determined estimates of hormone half-life and the imprecision in the measurement introduced by sample collection, processing, and assay.[5,18-21]

Some sources of error are difficult to estimate, including, for example, error introduced by sample collection and processing. To estimate such error, we recommend that similar sampling procedures be carried out in both validating and experimental sessions. The concentrations of a highly stable blood constituent should be measured over time using samples collected in the validating session. If a highly precise physical method of measurement is used to estimate concentrations of this stable analyte (e.g., total protein or calcium), then the imprecision in such serial measurements can be assumed to reflect largely the variability in sample collection and processing.[22] Such variation can be introduced by the position of the intravenous catheter used in sample collection, the amount of dilution that has occurred during sample collection, the amount of sublimation of the sample when frozen, the relative dehydration of the sample in the thawing process, etc. On the other hand, error in the measurement instrument is easier to quantitate and as discussed above can be estimated at variable concentrations of the substance of interest using appropriate multiple replicates of properly prepared standards and quality controls. Lastly, errors in the half-life estimate can be approximated by independent calculations of hormone half-life in suitably prepared animals or subjects administered the pure hormone under the same conditions as those anticipated in the experiment. Of course, each of these validation steps has the potential for introducing additional error in either its design, conduct, or interpretation, and accordingly must be carried out and reviewed critically.

Application of the waveform-independent method to a series of serum cortisol concentration measurements is illustrated in Fig. 8. Note that the sample observations (serum cortisol concentrations assayed in duplicate at each time point) are represented with error bars indicating the dose-dependent experimental uncertainty in the measurement system. These data were analyzed assuming a monocomponent cortisol half-life of 65 ± 5 min, so as to allow for intrinsic error in the estimate of hormone removal rate. The resultant calculated sample secretion rates with their individual statistical confidence intervals are shown. We emphasize that the statistical confidence intervals for the individual sample secretion rates are derived from the joint contributions of error in the measurement system (radioimmunoassay) and the half-life estimate. Because each sample secre-

[22] J. D. Veldhuis, W. S. Evans, M. L. Johnson, and A. D. Rogol, *J. Clin. Endocrinol. Metab.* **62,** 881 (1986).

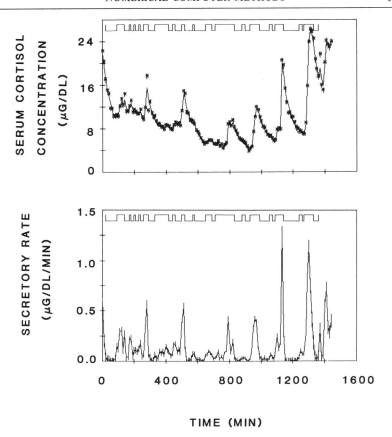

FIG. 8. Application of waveform-deconvolution analysis to serial plasma cortisol concentration measurements. As shown at top, the fitted curve predicted by the waveform-independent deconvolution model corresponds well to the observed serial serum cortisol concentration measurements. The bottom graph depicts the computer estimates of individual sample cortisol secretion rates. The vertical bars associated with each sample secretion estimate give the 67% statistical confidence limits for the secretion calculation. These error estimates derive from the combined influence of dose-dependent experimental error intrinsic in the cortisol assay and errors in the estimate of cortisol half-life per se. Schematized deflections above the data denote deconvolution-identified pulses.

tion estimate has an associated standard deviation, one can compute a z score (ratio of secretion estimate to the standard deviation) for each calculated secretory rate, as well as its first derivative. Such calculations are subject to all the assumptions noted above for other nonlinear parameter estimation techniques (see also other chapters in this volume).

Therefore, using this waveform-independent method, one can estimate whether secretion is significantly nonzero in individual samples without requiring any specific assumption about the waveform of the secretion process, for example, whether it is tonic (time invariant), burstlike, or impulsive. Moreover, a significant increase in secretion rate (i.e., a significantly positive first derivative of secretion rate) can be used to mark the onset of an individual secretory peak in the data, and a significant decrease in secretion rate, the offset of a pulse.

Others

As reviewed elsewhere, various other numerical and analytical approaches to deconvolution can also be considered.[1-5,8,20]

Current Problems, Limitations, and Suggested Solutions in Deconvolution Approaches to Endocrine Data

Error Definition and Propagation

An essential component of statistically valid deconvolution approaches is the propagation of experimental uncertainty. Error propagation is accomplished by identifying the various sources of experimental variation in data and assessing their conjoint influence on the predicted values of the parameters.[19] Various techniques are available to aid in systematic error propagation, assuming symmetric or asymmetric highly correlated or independent variance spaces. One useful class of techniques includes Monte Carlo procedures, in which random perturbations are made in the apparent parameter values and the effects on the resultant "fitted variance" assessed. (The fitted variance is the square root of the sum of the squares of the differences between the observed data and the fitted curve.) By varying two or more parameters simultaneously, one can create a *joint* variance space that describes the impact of small changes in multiple parameters on the fit of the predicted curve. As reviewed elsewhere in this volume, error propagation is a sine qua non of appropriate deconvolution techniques, since proper error propagation permits the investigator to evaluate statistical confidence intervals for the secretion and clearance terms of interest.

The absolute values of the statistical confidence limits for individual secretion and clearance estimates will depend on many factors including the quality of the data (the degree of experimental variation in the sample measurements, etc.), the applicability of the model to the data, the number of observations (which will control the number of degrees of freedom of

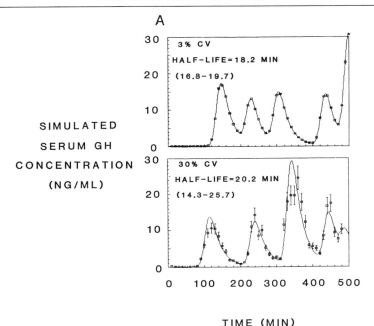

TIME (MIN)

FIG. 9. Computer simulations using the multiple-parameter convolution model to illustrate the effects of varying amounts of experimental noise (random variance) (A) and different sampling intensities (B) on the statistical confidence limits of the half-life estimates. In (A), simulated data are shown for episodic GH secretion assuming 3% intrasample variance (coefficient of variation, CV) or 30% intrasample variance. The synthetic data were subjected to deconvolution analysis and the apparent half-life calculated. Note that the joint statistical confidence limits for the deconvolution-based estimate of the half-life depend on the amount of random experimental variance inherent in the data. In (B), deconvolution was performed for the simulated series representing either 1- or 30-min sampling with corresponding predictions of hormone half-life. In both (A) and (B), the actual hormone half-life used by the convolution procedure to create the simulated data was 18.0 min.

the fitted model), etc. This concept is illustrated in Fig. 9, where simulated endocrine data with known experimental variance are subjected to deconvolution fitting. The corresponding confidence intervals for the estimated hormone half-life are given, and as anticipated they vary inversely with sample number and directly with sample measurement uncertainty. We believe that deconvolution methods which fail to provide error estimates are largely uninformative, since one cannot determine whether a putative secretory event or sample secretion rate is statistically significantly different from zero, or whether two estimates of half-lives are statistically different from each other, etc.

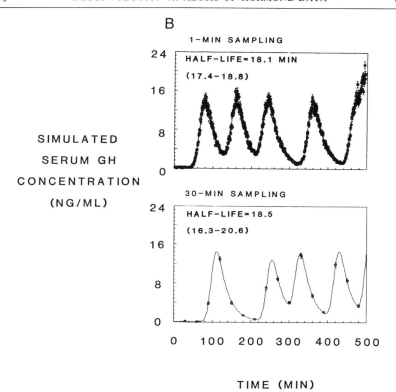

FIG. 9 (*continued*)

Adequacy of Fit

An adequate fit typically requires as a minimum that each added parameter significantly reduce the fitted variance (e.g., as assessed by an F-test[19] or the Akaike information content index[23]) and that the distribution of residuals—differences between fitted curve and observed data—be random (e.g., as evaluated by autocorrelation, the runs test, or the Kolmogorov–Smirnov statistic[24]). Moreover, any consistent discrepancies between the reconvolution fit and the observed data should be investigated for systematic bias. Finally, predictions of the deconvolution analysis and the validity of peak detection should be tested for accuracy by independent experimental means (see section on determining peak number, below).

[23] J. C. Spall ed., "Bayesian Analysis of Time Series and Dynamic Models," p. 25. Dekker, New York, 1988.

[24] J. S. Bendat and A. G. Piersol, "Random Data," 2nd Ed., p. 120. Wiley, New York, 1986.

Weighting Functions Used in Reconvolution Curve Fitting

Because endocrine and many other forms of experimental data in the natural sciences are typically susceptible to significant measurement variability (e.g., coefficients of variation of 3–20%), and because such variability is commonly dependent on dose, assay, and condition, accurate estimates of measurement precision are necessary for valid deconvolution-based calculations of secretion and clearance parameters (see section on error propagation). For example, in addition to their utility in error propagation, the intrasample variances can be used in an inverse weighting function in the iterative fits when computing the parameter values. Measurements associated with large experimental uncertainty receive reduced weight in the fit compared to data with greater precision. Currently, we require as a minimum that dose-dependent experimental variance be defined for the assay system utilized, and that an inverse weighting of variance be employed in the nonlinear fitting of the data.

Determining Peak Number

With endocrine data, investigators are particularly interested in determining the frequency and amplitude characteristics of the secretory phenomena. As noted above, if appropriate methods of error propagation are used, then the amplitudes of sample secretion rates and/or of secretory events can be estimated with statistical confidence intervals. Amplitudes not significantly different from zero would be rejected as nonpeaks.

Determining the number of secretory peaks is still difficult with biological data, which are potentially susceptible to either Type I or Type II statistical errors. A Type I statistical error involves a false-positive estimate, in which one falsely rejects the null hypothesis of no peak being present (i.e., a peak is inserted where none occurs in fact). A Type II statistical error is a false-negative estimate, or false acceptance of the null hypothesis that no peak is present (i.e., a true peak is missed) (see Fig. 10). Ideally, an investigator would determine both the false-positive and false-negative error rates associated with the methodology and conditions he is employing. This is important since statistical confidence limits do not necessarily equate with biological truth; for example, data generated in poor-quality assays with large experimental uncertainties and/or data consisting of few observations offer less power in the statistical identification of biological events.

To date, relatively limited data are available characterizing Type I and Type II statistical errors in the performance of deconvolution techniques. This issue is particularly complex because the means of collecting biological data, the number of samples obtained, the precision of the measure-

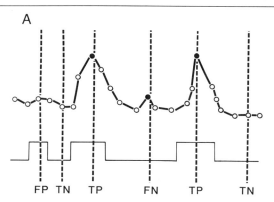

FP TN   TP   FN   TP   TN

FIG. 10. Illustration of Type I and Type II statistical errors consisting, respectively, of false-positive (FP) and false-negative (FN) errors in peak identification (A). In addition, true-positive (TP) and true-negative (TN) peak identification can occur. The convolution integral [Eq. (1)] can be used to create simulated hormone data series, which in turn can be utilized to test the discriminative value of specific algorithms for peak detection (B). These algorithms are susceptible to false-positive and false-negative errors, which influence the overall sensitivity and specificity of the procedure (see text). In (B), the top "observed LH profile" is a synthetic series constructed by combining the convolved secretion and clearance functions (middle) with Gaussian noise (bottom). Deflections above the graphs depict algorithm-identified peaks.

ments, the relevance of the functional form and structure of the equations, etc., all contribute to the probability of either falsely rejecting or falsely accepting a possible secretory peak in the data. Because of these complexities, we recommend combined *in vivo* biological validation and computer simulations to evaluate Type I and Type II statistical errors in the deconvolution approach employed.[3–5,8,18,20]

Estimating Basal Secretion

A question of particular interest to endocrinologists is the extent to which a secretory gland maintains a low level of invariant hormone secretion over time, namely, basal or tonic secretion. As illustrated in Fig. 11, GH concentrations measured in certain pathological conditions (e.g., in patients with GH-secreting pituitary tumors) may exhibit an apparently constant high level of activity with relatively minimal superimposed variations (pulses). This pattern contrasts with that typically encountered in normal physiological states, in which only small or near-zero amounts of basal secretion seem to occur with superimposed burstlike episodes of hormone discharge.[11–16]

Consequently, whether basal secretion is nonzero is of potential impor-

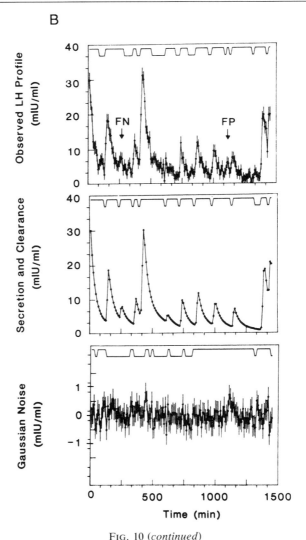

FIG. 10 (*continued*)

tance in separating pathological and physiological states. To make such a distinction, the deconvolution methodology must estimate sample secretion rates with relevant statistical confidence intervals. A current challenge in this area is the propagation of statistical confidence intervals that include all relevant sources of experimental uncertainty. As noted above, estimates of imprecision introduced by sample collection, processing, and

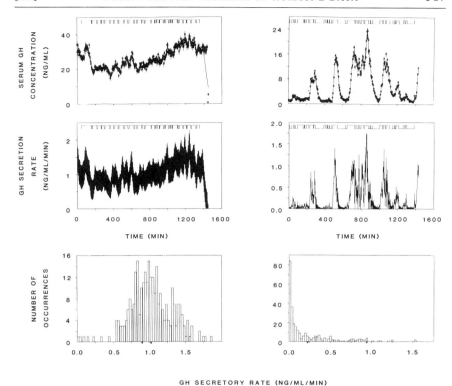

FIG. 11. Apparent increase in basal growth hormone (GH) secretion in a patient with a GH-secreting pituitary tumor (left) compared to normal (right). The top graphs give the observed plasma GH concentration data and the calculated reconvolution fit. The profiles in the middle graphs are calculated sample secretory rates (and their 63% confidence limits) based on a waveform-independent deconvolution algorithm. The bottom graphs give the distributions (histograms) of sample GH secretory rates. Note that the tumor patient exhibited a markedly different distribution of individual sample secretory rates (middle). The pathological GH series also exhibits a high baseline or tonic secretory mode with small pulsatile episodes superimposed. For deconvolution purposes, a biexponential kinetic function was assumed to consist of GH half-lives of 3.5 ± 1.0 min (first component) and 21 ± 1.1 min (second component), with the second component constituting 67% of the total amplitude. Data were provided by Drs. Mark Hartman and Michael Thorner and derive from blood sampled every 5 min for 24 hr for the subsequent immunoradiometric assay of serum GH concentrations in duplicate.

assay are possible, and estimates of error in the half-life calculation are practicable. However, any additional error present in the independent variable (e.g., time of sample collection) is not considered by most nonlinear methods of parameter estimation, and the exact extent to which the biological processes depart from the algebraic structure utilized in the convolution integral is difficult to assess under some conditions of study. Consequently, some experimental uncertainty remains in appraising secretion rates absolutely even when statistical confidence limits are generated in an appropriate manner.

Integrating a Systems View

Another major area for further development is the integration of systems behavior in convolution expressions. For example, certain hormones when secreted into the bloodstream associate strongly with various transport proteins in the plasma compartment (sex steroid hormones and their binding globulins, GH and its binding proteins, etc.). Other hormones, substrates, and metabolites are subject to metabolic interconversion, distribution into extravascular compartments, postsecretion modification, etc. Appropriate convolution formulations will need to include kinetic statements and additional association and dissociation rate constants to accommodate the full range of *in vivo* biological behavior and corresponding fate of circulating effector and product molecules.

Other Applications and Implications of Deconvolution Techniques

Model Synthesis

The convolution integral [Eq. (1)] provides a description of two or more processes that interact over time to specify an overall outcome. With endocrine data, a wide range of outcome phenomena are susceptible to experimental study if specific hypotheses of a quantitative nature can be stated. For example, the mechanism subserving the massive increase in plasma LH concentrations that occurs before ovulation in various species in principle could be attributed to (1) an increase in the number of spontaneous hormone secretion bursts, and/or their duration and amplitude; (2) an increase in the half-time of hormone disappearance; and/or (3) an increase in basal secretion rate, on which are superimposed secretion events. As shown in Fig. 12, any of these mechanisms would promote a remarkable increase in circulating hormone concentrations. Most effective in generating a rapid increase in hormone levels in the blood would be a combined augmentation of hormone secretory burst number, duration, and amplitude with or without a corresponding prolongation in hormone half-life.

Single Pulse Generator

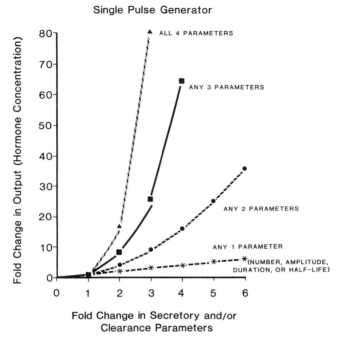

FIG. 12. Multiplicative interactions among multiple secretion and clearance parameters in controlling output (hormone concentrations). Note that in principle a marked increase in plasma hormone concentrations could occur by means of increased secretory burst frequency, amplitude, or duration and/or a prolonged half-life of hormone disposal. [Adapted with permission from J. D. Veldhuis, A. B. Lassiter, and M. L. Johnson, *Am. J. Physiol.* **259,** E351 (1990).]

Validity Testing

The convolution integral [Eq. (1)] is useful for the mathematical construction of synthetic endocrine data series, in which the number, amplitude, and duration of simulated hormone secretion bursts can be specified precisely and the apparent half-life of hormone removal stated explicitly. As shown in Fig. 13, parameters of hypothetical hormone secretion and clearance can be used to generate a synthetic series of hormone data, on which is superimposed a particular distribution and quantity of "noise" (experimental variance). Such synthetic series are then useful for testing the performance not only of various deconvolution techniques per se, but also other procedures for time series analysis and/or the detection of discrete episodes of hormone release. Various (simplified) forms of Eq.

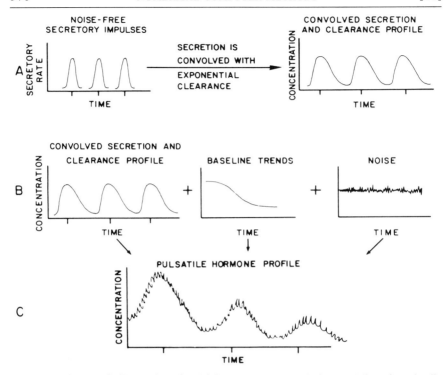

FIG. 13. Schematic illustration of multiple-parameter convolution modeling of a pulsatile hormone time series. The schema indicates in summary form the sequential process for building a simulated pulsatile hormone time series using a convolution of random secretory bursts with clearance (A). The combined secretion and clearance profile is superimposed on systematic baseline trends and perturbed by random experimental noise (B). The final synthetic hormone time series (C) results from algebraic summation of these constituents. [Reprinted with permission from J. D. Veldhuis and M. L. Johnson, *Am. J. Physiol.* **255,** E749 (1988).]

(1) have been used to this end.[3–5,7,18,20,25–29] For example, depending on the particular algebraic form assumed for $S(z)$, a number of different synthetic hormone concentration series can be generated and submitted to waveform-independent deconvolution analysis. Ideally, the waveform-independent deconvolution technique should recover the originally synthesized

[25] R. J. Urban, M. L. Johnson, and J. D. Veldhuis, *Am. J. Physiol.* **257,** E88 (1989).

[26] R. J. Urban, M. L. Johnson, and J. D. Veldhuis, *Endocrinology (Baltimore)* **124,** 2541 (1989).

[27] J. D. Veldhuis, A. Iranmanesh, I. Clarke, D. L. Kaiser, and M. L. Johnson, *J. Neuroendocrinol.* **1,** 185 (1989).

[28] V. Guardabasso, G. De Nicolao, M. Rochetti, and D. Rodbard, *Am. J. Physiol.* **255,** E775 (1988).

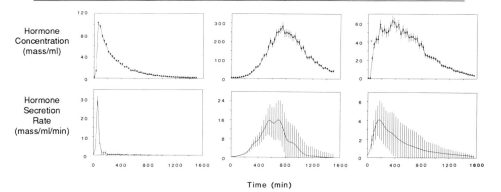

FIG. 14. Waveform-independent deconvolution analysis as a means to recover the underlying original secretion waveform. The convolution model described in Fig. 13 was used to create either a nearly instantaneous impulse (left), a symmetric secretion burst (middle), or an asymmetric secretion burst (right). In each case the top graph gives the synthetic hormone concentration profile, which is a convolution of secretion and clearance. The bottom graph gives the recovered secretion pulse as a function of time.

"secretory" waveform and provide statistical confidence limits to indicate the precision of the sample secretion estimates. This is illustrated in Fig. 14 for three simulated hormone secretion bursts: an instantaneous impulse, a symmetric Gaussian secretion event, and a skewed secretion burst. Because the exact structure of the synthetic data is known *a priori,* comparisons can be made between the known secretion and clearance values and those estimated by the analytical tool of interest.

Multiple-Pulse Generators

In many physiological circumstances, the output of a hormone, metabolite and/or substrate from a secretory gland or target tissue is specified by two or more physiological factors acting either independently or interdependently (see Fig. 15). For example, the episodic release of a GH is controlled not only by release of an inhibitory tetradecapeptide, somatostatin, but also by the availability of a stimulatory polypeptide, growth hormone-releasing hormone.[30] In such circumstances, the overall output of GH is dependent on simultaneous withdrawal of an inhibitory agent and the emergence of a stimulatory compound. Moreover, multiple stimulatory agents may act in parallel or in series to specify secretion. Independent pulse generators acting concurrently over time have a finite probability of

[29] J. D. Veldhuis and M. L. Johnson, *Am. J. Physiol.* **255,** E749 (1988).

[30] P. M. Plotsky and W. Vale, *Science* **230,** 461 (1985).

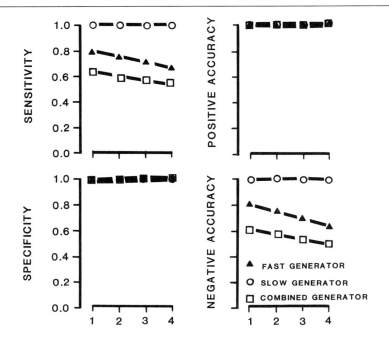

THRESHOLD T-STATISTIC

FIG. 15. Influence of two-pulse generators on the discriminative value of a discrete peak-detection method [Cluster analysis, J. D. Veldhuis and M. L. Johnson, *Am. J. Physiol.* **250,** 334 (1986)]. Synthetic hormone "peaks" were generated as described in Fig. 13, but two-pulse generators were simulated. Data were evaluated for each peak generator alone and for the combined output of both generators. Discriminative indices of test performances were evaluated as a function of instrument (Cluster) threshold. [Reprinted with permission from J. D. Veldhuis, A. B. Lassiter, and M. L. Johnson, *Am. J. Physiol.* **259,** E351 (1990).]

firing at the same moment or within the same (or a lagged) sample observation interval. Computer simulations indicate that the probability of such coincident occurrences can be given by the hypergeometric probability density function for two series[27] and by more complex formulations for three or more pulse generators (see Appendix). Consequently, an examination of pulse generators that are multiple and/or are multipotential (the same generator controlled by two or more noninteracting pathways) will become important in further deconvolution analysis of hormone data. In addition, how independent pulse generators might jointly influence false-positive and false-negative error rates requires further study.

TABLE IV
IDEAL FEATURES OF DECONVOLUTION TECHNIQUES

1. Estimate endogenous secretory and/or clearance rates by model-specific as well as waveform-independent methodologies
2. Provide formal statistical error propagation to encompass experimental uncertainty in both data measurements and half-life/secretion estimates
3. Avoid negative secretion and/or half-life calculations, and avoid "ringing" (oscillations of secretory rate zero)
4. Compare fitted (predicted) function with observed data
5. Operate robustly over a range of sampling intensities; signal shape, frequency, and amplitude; hormone half-lives; and signal-to-noise ratios
6. Be validated by both computer simulations and *in vivo* biological data
7. Allow for more complex systems behavior (e.g., multiple pulse generators and multicompartmental kinetics)

Summary and Conclusion

Deconvolution analysis of hormone data poses special problems in view of the sparse, noisy, and short data series typically available for analysis; the unknown true nature of the underlying secretory event; and potentially large variations in dissipation or clearance kinetics in different settings. Consequently, deconvolution techniques, which concern themselves with the estimation of hormone secretion and/or clearance based on serial circulating hormone concentration measurements, face a particular challenge. Ideal features of deconvolution algorithms are summarized in Table IV.

Specific deconvolution techniques available to analyze hormone data include both waveform-defined procedures and waveform-independent algorithms. These approaches should be viewed as complementary rather than antagonistic. All deconvolution techniques are subject to individual limitations and specific strengths. Independently of the method employed, error propagation is necessary so as to define the statistical uncertainty intrinsic to the estimate of secretion and clearance. Such calculations of experimental uncertainty should include error inherent in the sample collection, processing, and assay as well as error in the kinetic constants and/or anticipated departures of the biological process from the algebraic structure of the convolution formulation. Moreover, more complex convolution statements will be required to describe the full range of behavior of hormone data in a systems view. The applications of such newer convolution methods as well as currently available techniques include model synthesis, model testing, and analysis of the interactions among multiple pulse generators.

Appendix: Coincident Firing of Two or More Independent Hormone
  Pulse Generators

If pulse generators A and B are firing independently at respective rates such that m and n individual bursts are identified within z sample observations, and if m and n are small compared to z, then the probability of observing exactly x coincident events is given from the hypergeometric distribution by $P(x)$:

$$P(x) = \frac{\binom{m}{x}\binom{z-m}{n-x}}{\binom{z}{n}}; \quad x = 0, 1, \ldots, \min(m,n) \quad \text{(A1)}$$

The expected number of purely random coincidences can be estimated as x_e:

$$x_e = \frac{mn}{z} \quad \text{(A2)}$$

which has a variance of x_{var}:

$$x_{var} = \frac{mn}{z}\left(1 - \frac{m}{z}\right)\left(1 - \frac{n-1}{z-1}\right) \quad \text{(A3)}$$

The probability of observing at least x (i.e., x or more), coincidences on the basis of chance associations alone is $P(\geq x)$:

$$P(\geq x) = 1 - \sum_{i=0}^{x-1} P(i) \quad \text{(A4)}$$

As shown elsewhere [J. D. Veldhuis, A. Iranmanesh, I. Clarke, D. L. Kaiser, and M. L. Johnson, *J. Neuroendocrinol.* **1,** 185 (1989)], this formulation provides an accurate representation of computer simulations of the behavior of two independent pulse generators. A more complex combinatorial algebra than Eq. (A1) is required to stipulate the expected behavior of 3 or more pulse generators acting independently, as summarized elsewhere [J. D. Veldhuis, M. L. Johnson, and E. Seneta, *J. Clin. Endocrinol. Metab.* **73,** 569 (1991)]. However, for j (*multiple*) generators, which have independent pulse frequencies of m, n, o, $\ldots$, the expected mean number of random coincidences can be estimated reasonably from a general form of Eq. (A2), namely,

$$x_{e(j-1)} = \frac{mno \cdots}{z^{(j-1)}} \quad \text{(A5)}$$

The variance of this mean is approximately equal to the mean, according to our computer simulations.

Acknowledgments

We thank Patsy Craig for preparation of the manuscript and Paula P. Azimi for the artwork. This work was supported in part by National Institutes of Health Grant RR 00847 to the Clinical Research Center of the University of Virginia, Grant RCDA 1 KO4 HD00634 (J.D.V.), Grant GM-28928 (M.L.J.), the Diabetes and Endocrinology Research Center Grant NIH DK-38942, the NIH-supported Clinfo Data Reduction Systems, the Pratt Foundation, the University of Virginia Academic Enhancement Fund, and the National Science Foundation Science Center for Biological Timing (DIR-8920162).

[27] Dynamic Programming Algorithms for Biological Sequence Comparison

By William R. Pearson and Webb Miller

Introduction

The concurrent development of molecular cloning techniques, DNA sequencing methods, rapid sequence comparison algorithms, and computer workstations has revolutionized the role of biological sequence comparison in molecular biology. Twenty years ago, protein sequence determination was the last step in the characterization of a protein. Today, we clone first, sequence the clone, search the databases, and then, armed with hypotheses about structure and function that are based on sequence similarity, proceed with experiments to characterize the biochemical properties of a protein. This strategy—clone first and experiment later—is likely to become even more popular as sequence databases grow larger and thus more comprehensive.

In this chapter, we discuss several dynamic programming algorithms that have been applied to biological sequence comparison problems.[1] Such algorithms are the basis of most methods currently used to identify distantly related proteins by sequence similarity. We do not discuss the application of these algorithms, or of their heuristic approximations, to sequence database searches. For reviews of the problems encountered in

---

[1] "Dynamic programming" refers to a broad class of algorithms that have been applied to problems from a number of areas. For other examples, see T. H. Cormen, C. E. Leiserson, and R. L. Rivest, "Introduction to Algorithms," p. 301. McGraw-Hill, New York, 1990.

identifying distantly related sequences by sequence similarity, see the papers by Pearson[2] and Doolittle.[3]

A dynamic programming algorithm was first used to align protein sequences in 1970[4]; these algorithms are now used widely not only for sequence alignment[5-9] (for reviews, see Sankoff and Kruskal[10] and Waterman[11]), but also for sequence library searches,[12] evolutionary tree construction,[5,13,14] multiple sequence alignment,[15,16] and even for alignment of restriction maps[17-19] and protein structures.[20] Despite the widespread use of these algorithms, the computational requirements (the number of steps and the amount of memory required) for biological sequence comparison are often misunderstood. As a result of this confusion, several widely available packages of sequence analysis programs limit the lengths of the sequences that can be aligned, because they implement an algorithm that requires an amount of memory proportional to the product of the sequence lengths $[O(N^2)]$.[21] For example, as many as 4 million words of memory

[2] W. R. Pearson, this series, Vol. 183, p. 63.

[3] R. F. Doolittle, this series, Vol. 183, p. 99.

[4] S. Needleman and C. Wunsch, *J. Mol. Biol.* **48,** 444 (1970).

[5] P. H. Sellers, *SIAM J. Appl. Math.* **26,** 787 (1974).

[6] T. F. Smith and M. S. Waterman, *J. Mol. Biol.* **147,** 195 (1981).

[7] O. Gotoh, *J. Mol. Biol.* **162,** 705 (1982).

[8] S. F. Altschul and B. W. Erickson, *Bull. Math. Biol.* **48,** 603 (1986).

[9] E. W. Myers and W. Miller, *Comput. Appl. Biosci.* **4,** 11 (1988).

[10] D. Sankoff and J. B. Kruskal, eds. "Time Warps, String Edits, and Macromolecules: The Theory and Practice of Sequence Comparison." Addison-Wesley, Reading, Massachusetts, 1983.

[11] M. S. Waterman, *in* "Mathematical Methods for DNA Sequences" (M. S. Waterman, ed.), p. 53. CRC Press, Boca Raton, Florida, 1989.

[12] M. Gribskov, R. Lüthy, and D. Eisenberg, this series, Vol. 183, p. 146.

[13] D. Sankoff and R. J. Cedergren, *in* "Time Warps, String Edits, and Macromolecules: The Theory and Practice of Sequence Comparison" (D. Sankoff and J. B. Kruskal, eds.). Addison-Wesley, Reading, Massachusetts, 1983.

[14] D. Feng and R. F. Doolittle, *J. Mol. Evol.* **25,** 351 (1987).

[15] M. Murata, J.S. Richardson, and J. L. Sussman, *Proc. Natl. Acad. Sci. U.S.A.* **82,** 3073 (1985).

[16] D. J. Lipman, S. F. Altschul, and J. D. Kececioglu, *Proc. Natl. Acad. Sci. U.S.A.* **86,** 4412 (1989).

[17] M. S. Waterman, T. F. Smith, and H. L. Katcher, *Nucleic Acids Res.* **12,** 237 (1984).

[18] E. W. Myers and X. Huang, *Bull. Math. Biol.* in press (1991).

[19] W. Miller, J. Barr, and K. Rudd, *Comput. Appl. Biosci.* **7,** 447 (1991).

[20] M. Zuker and R. L. Somorjai, *Bull. Math. Biol.* **51,** 55 (1989).

[21] We use the notation $O(X)$ to denote a quantity that is proportional to X. Somewhat imprecisely, we can reason that a method which requires $O(N^3)$ operations will take N times longer to run than one that requires $O(N^2)$ time. (We shall generally assume that the two sequences are about the same length.)

might be required to align two sequences that are each 2000 residues long; alignment of two 50,000-nucleotide genes would be impossible with such an algorithm. However, dynamic programming algorithms that require space proportional to the sum of the lengths of the two sequences $[O(N)]$ have been described,[9] and complex local sequence comparisons of two globin gene clusters (44,000 and 73,000 nucleotides) have been performed.[22,23] In this chapter, we describe $O(N^2)$-time and $O(N)$-space implementations of the dynamic programming algorithm and discuss generalizations of the algorithm to alignments based on multiple sequence profiles.[12,24] Our goal is to provide examples of efficient dynamic programming algorithms that can be applied to most biological sequence comparison problems.

Dynamic programming algorithms for two protein or DNA sequences calculate alignment scores that allow for substitutions, insertions, and deletions. The alignment score can be calculated as a similarity score (closely related sequences have very high scores while distantly related sequences have lower scores, and unrelated sequences have scores that are lower still) or a distance score (identical sequences have a distance score of zero, closely related sequences have a distance score that is low, and distantly related sequences have scores that are much higher). Distance scores are frequently used to construct evolutionary trees,[25,26] whereas inferences of homology are more commonly based on similarity scores. It is straightforward to transform a distance into a sequence similarity score, but similarity scores can be calculated in ways that cannot be converted to distances.[11,27] Dynamic programming methods guarantee the calculation of the best possible, or optimal, similarity score or distance for the set of match, mismatch, insertion, and deletion scoring parameters that are used.

The demonstration of sequence homology (common evolutionary ancestry) based on sequence similarity has been extremely fruitful, perhaps because the dynamic programming algorithm calculates similarity in a way that accurately reflects the process of molecular evolution. Similarity searches of protein sequence libraries routinely identify protein sequences that diverged from a common ancestor 500–2500 million years in the

[22] X. Huang, R. C. Hardison, and W. Miller, *CABIOS* **6,** 373 (1990).
[23] X. Huang and W. Miller, *Adv. Appl. Math.* **12,** 337 (1991).
[24] G. J. Barton and M. J. E. Sternberg, *J. Mol. Biol.* **212,** 389 (1990).
[25] W. M. Fitch and E. Margoliash, *Science* **155,** 279 (1967).
[26] J. Felsenstein, *Annu. Rev. Genet.* **22,** 521 (1988).
[27] T. F. Smith, M. S. Waterman, and W. M. Fitch, *J. Mol. Evol.* **18,** 38 (1981).

past.[28,29] Mutations cause amino acid substitutions, insertions, and deletions; the dynamic programming algorithm is guaranteed to produce scores that reflect a minimum number of these events. In contrast, we do not know exactly how the transcription machinery of a cell interacts with a promoter, nor how the splicing machinery identifies intron/exon boundaries, so we do not have computer methods that accurately reflect these processes.

Some of the confusion about dynamic programming algorithms for DNA and protein sequence comparison probably stems from the number of permutations that have been described over the past 20 years. Any family of algorithms that can calculate several types of scores (global similarity, distance, local similarity) with so many variations (unpenalized end gaps, a penalty for a gap of any length, a penalty for each residue in a gap, or one penalty for opening a gap and a second for each residue in the gap) is initially somewhat confusing. However, the algorithms are in fact very simple; only minor changes are required to convert an algorithm that calculates a global distance score to one that calculates a local similarity score. In the next sections, we present efficient dynamic programming algorithms for calculating global and local similarity scores, and for comparing a sequence "profile" or pattern to a sequence. In the final section, we discuss variations on these algorithms that can greatly decrease, or increase, the amount of computation required.

Global Sequence Similarity and the Needleman–Wunsch Algorithm

Needleman and Wunsch were the first to use dynamic programming to compare biological sequences.[4] These authors described a method for calculating the "largest number of amino acids of one protein that can be matched with those of another protein while allowing for all possible deletions" in either sequence.[4] Their original formulation of the algorithm has since been improved; we can now compute exactly the same scores and alignments far more efficiently. For simplicity, we first present an algorithm that uses a different gap penalty than that considered by Needleman and Wunsch. Later in this section we describe an algorithm that calculates exactly the same similarity score.

For an introduction to the dynamic programming method for aligning sequences, let us consider the problem of aligning $seq1 = FFKL$ with $seq2 = FLL$. We award $+1$ for matching residues, -1 for mismatching

[28] R. F. Doolittle, *Science* **214**, 149 (1981).
[29] R. F. Doolittle, D. F. Feng, M. S. Johnson, and M. A. McClure, *Cold Spring Harbor Symp. Quant. Biol.* **51**, 447 (1986).

residues, and -1 for each insertion or deletion. For now, we are concerned only with computing the score of an optimal alignment; later we describe ways to produce an explicit optimal alignment.

To find the best alignment score for *FFKL* and *FLL,* we form a 4×3 matrix, label each row with the corresponding entry of *seq1,* and label each column with the corresponding entry of *seq2.* At the intersection of

	F	L	L
F	1	0	-1
F	0	0	-1
K	-1	-1	-1
L	-2	0	0

row i and column j, we place the score of an optimal alignment of the first i residues of *seq1* with the first j residues of *seq2;* we denote this entry $S(i, j)$. For example, $S(3,1)$ is the optimal score of aligning *FFK* with *F,* namely, -1, which is the score of

$$\text{either:} \quad \begin{array}{ccc} F & F & K \\ F & — & — \end{array} \qquad \text{or:} \quad \begin{array}{ccc} F & F & K \\ — & F & — \end{array}$$

This illustrates that there can be several optimal alignments, though of course there is always a unique optimal score. Thus we are careful to say *an* optimal alignment instead of *the* optimal alignment.

The fundamental observation of Needleman and Wunsch, which lies at the heart of dynamic programming algorithms, was that to calculate the alignment score $S(i, j)$, one need only enumerate and score all the ways that one aligned pair can be added to a shorter alignment to produce an alignment of the first i residues of *seq1* with the first j residues of *seq2.* Specifically, an alignment can be extended in three ways: (1) by aligning the next two residues in both sequences (thus causing a match if the two are identical and a mismatch if they are not); (2) by aligning a residue in the first sequence with a gap in the second (making a deletion of the second sequence with respect to the first); or (3) by aligning a residue in the second sequence with a gap in the first (making an insertion in the second sequence with respect to the first).

For the sequences, *FFKL* and *FLL,* consider the computation of $S(4,3)$ from $S(3,2)$, $S(3,3)$, and $S(4,2)$. If the last pair (which might be an insertion or deletion) is removed from an optimal alignment of *FFKL* and *FLL,* then there are three possibilities: (1) the removed pair is $\begin{bmatrix} L \\ L \end{bmatrix}$ and what remains is an optimal alignment of *FFK* and *FL* [thus $S(4,3) = S(3,2) + 1$]; (2) the

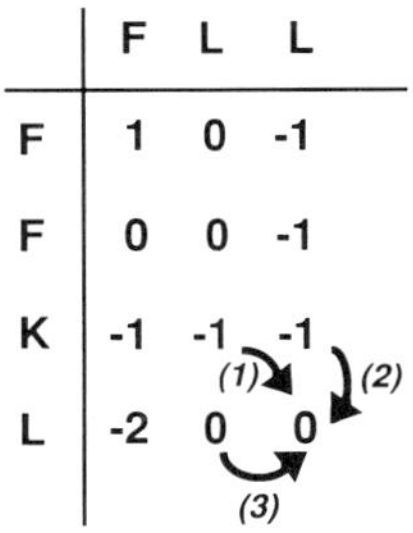

FIG. 1. The three alternatives for $S(4, 3)$.

removed pair is $\begin{bmatrix} L \\ - \end{bmatrix}$ and what remains is an optimal alignment of FFK

and FLL [thus $S(4,3) = S(3,3) - 1$]; or (3) the removed pair is $\begin{bmatrix} - \\ L \end{bmatrix}$ and

what remains is an optimal alignment of $FFKL$ and FL [thus $S(4,3) = S(4,2) - 1$]. Thus $S(4,3)$ is the maximum of these three possible scores, which is obtained by path (1).

These alternatives are depicted in Fig. 1. There the arrows labeled (1)–(3) correspond to the options given above. For example, arrow (1)

corresponds to appending $\begin{bmatrix} L \\ L \end{bmatrix}$ to an optimal alignment of FFK and FL,

such as $\begin{bmatrix} FFK \\ FL- \end{bmatrix}$.

The same strategy works for aligning arbitrary sequences $seq1$ and $seq2$ using arbitrary scores for aligned pairs. Let the length of $seq1$ be M, denote its residues by $a_1, a_2, \ldots, a_M$, and let A_i denote the prefix of $seq1$ consisting of its first i residues. Similarly, let $seq2$ be $b_1 b_2 \cdots b_N$ and let $B_j = b_1 b_2 \cdots b_j$. Let $S(i, j)$ be the score of an optimal alignment of A_i with

B_j. We refer to the score for aligned pair $\begin{bmatrix} a \\ b \end{bmatrix}$ as $\sigma\left(\begin{bmatrix} a \\ b \end{bmatrix}\right)$, where a and b

are either residues or the symbol —. (An insertion or deletion can be considered equivalent to aligning a residue with —.) In the example above,

$$\sigma\left(\begin{bmatrix} a \\ a \end{bmatrix}\right) = 1 \text{ and } \sigma\left(\begin{bmatrix} a \\ b \end{bmatrix}\right) = \sigma\left(\begin{bmatrix} a \\ - \end{bmatrix}\right) = \sigma\left(\begin{bmatrix} - \\ b \end{bmatrix}\right) = -1.$$ To get the

computation rolling, we add a row to the top of the score matrix S; $S(0, j)$ gives the score of aligning the first 0 symbols of $seq1$ with the first

j symbols of $seq2$. This entry is simply the sum of the $\sigma\left(\begin{bmatrix} - \\ b \end{bmatrix}\right)$ as b ranges

$S(0,0) \leftarrow 0$
for $j \leftarrow 1$ **to** N **do**
    $S(0,j) \leftarrow S(0,j-1) + \sigma(\begin{bmatrix} - \\ b_j \end{bmatrix})$
for $i \leftarrow 1$ **to** M **do**
    $\{$ $S(i,0) \leftarrow S(i-1,0) + \sigma(\begin{bmatrix} a_i \\ - \end{bmatrix})$
      for $j \leftarrow 1$ **to** N **do**
        $S(i,j) \leftarrow \max\{S(i-1,j-1) + \sigma(\begin{bmatrix} a_i \\ b_j \end{bmatrix}), S(i-1,j) + \sigma(\begin{bmatrix} a_i \\ - \end{bmatrix}), S(i,j-1) + \sigma(\begin{bmatrix} - \\ b_j \end{bmatrix})\}$
    $\}$

write "Similarity score is" $S(M,N)$

FIG. 2. The basic dynamic programming method for sequence alignment.

over the first j symbols of *seq2*. Similarly, we add a "column 0" to S. Outside of row 0 and column 0, there are only three ways to align A_i with B_j optimally, namely,

1. Add the pair $\begin{bmatrix} a_i \\ b_j \end{bmatrix}$ to an optimal alignment of A_{i-1} with B_{j-1}, yielding the score $S(i-1, j-1) + \sigma\left(\begin{bmatrix} a_i \\ b_j \end{bmatrix}\right)$.

2. Add the pair $\begin{bmatrix} a_i \\ - \end{bmatrix}$ to an optimal alignment of A_{i-1} with B_j, yielding the score $S(i-1, j) + \sigma\left(\begin{bmatrix} a_i \\ - \end{bmatrix}\right)$.

3. Add the pair $\begin{bmatrix} - \\ b_j \end{bmatrix}$ to an optimal alignment of A_i with B_{j-1}, yielding the score $S(i, j-1) + \sigma\left(\begin{bmatrix} - \\ b_j \end{bmatrix}\right)$.

These observations give the algorithm of Fig. 2.

Our program in Fig. 2 can be characterized as calculating a global sequence similarity that penalizes end-gaps and uses an additional penalty for each residue in a gap. A global sequence similarity is one that extends from one end of both sequences to the other; it does not matter whether it is calculated from beginning to end or vice versa. The algorithm calculates sequence similarity, rather than evolutionary distance, because it increases the similarity count for every amino acid match, rather than increasing a distance count for every mismatch. Also, the method penalizes end-gaps; that is, the alignment

```
S(0,0) ← 0
for j ← 1 to N do
    S(0,j) ← q
for i ← 1 to M do
{   S(i,0) ← q
    for j ← 1 to N do
        S(i,j) ← max{ S(i−1,j−1)+σ([ᵃⁱ_bⱼ]), max{S(0,j)..S(i−1,j)}−q, max{S(i,0)..S(i,j−1)}−q }
}
```

write "Similarity score is" $S(M,N)$

FIG. 3. The Needleman–Wunsch algorithm.

$$M P\ M A C D F\ K L$$
$$— — — A\ M D F\ R M$$

has a lower score than

$$A C D F\ K L$$
$$A M D F\ R M$$

(Assume that deletions have negative σ.) This algorithm increases the penalty for each residue in a gap, so that

$$A C D G S\ M F\ K L \quad \text{and} \quad A C D S\ G M F\ K L$$
$$A M D — S — F\ R M \qquad\quad A M D S — — F\ R M$$

have the same score. In contrast, the Needleman–Wunsch algorithm utilizes a gap (insertion or deletion) penalty q that is independent of the length of the gap. Thus,

$$A C D G S\ M F\ K L \quad \text{and} \quad A C D S\ F\ K L$$
$$A M D — — — F\ R M \qquad\quad A M D — F\ R M$$

would have the same score with the Needleman–Wunsch algorithm, because they both have one gap. The algorithm in Fig. 2 would score the latter alignment higher than the former.

When the dynamic programming principle is applied to optimize alignments with a constant penalty for a gap of any length, gaps can be extended in more than just the three ways shown in Fig. 1. Consider an optimal alignment of A_i with B_j that ends in the deletion of k successive residues of *seq1*, whence $S(i, j) = S(i − k, j) − q$. Here, we must account for gaps of every length k from 1 to i.[30] Similarly, when determining $S(i, j)$ the algorithm must account for all alignments ending with a multiresidue insertion. The resulting algorithm is given in Fig. 3.

---

[30] Although Needleman and Wunsch described their calculation by proceeding from the end of both sequences to the beginning, for consistency with this paper and the literature we describe an implementation that starts at the beginning of the two sequences and proceeds forward.

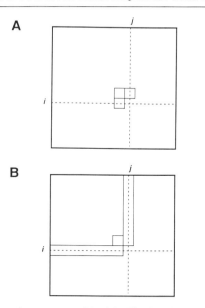

FIG. 4. (A) Locations used to compute $S(i, j)$ in Fig. 2. (B) Locations used to compute $S(i, j)$ in Fig. 3.

It is useful to discuss the computational complexity of the algorithm in Fig. 2 and to compare it with the Needleman–Wunsch algorithm (Fig. 3) and later implementations. The algorithms in Figs. 2 and 3 fill a scoring matrix S of $M + 1$ rows and $N + 1$ columns; thus the memory requirement is $O(MN)$, or $O(N^2)$ if the two sequences are about the same length. In Fig. 2, the value of $S(i, j)$ is calculated by comparing the three alternatives listed above; each of these cases depends only on an adjacent value of $S(i, j)$. Because $3N$ alternatives are considered on each of the M rows, the number of operations required is $O(MN)$ or $O(N^2)$. In contrast, calculation of the value of $S(i, j)$ in Fig. 3 involves consideration of all values of S that precede position j in row i and all values that precede position i in column j, or $O(N)$ operations on average. This difference between Fig. 2 and Fig. 3 is depicted in Fig. 4. Thus, the Needleman–Wunsch alignment procedure requires the calculation of $O(N^2)$ matrix entries, each of which requires $O(N)$ operations, for a total of $O(N^3)$ steps, and takes roughly N times longer than the algorithm in Fig. 2.

It is possible to calculate a similarity score that uses the Needleman–Wunsch penalty-per-gap in $O(N^2)$ operations by using a small amount of additional memory to keep track of where the best match-scores have been found in each row and column.[7] Current implementations of

```
S(0) ← 0
for j ← 1 to N do
    SS(j) ← SS(j−1) + σ([ ⁻/b_j ])
for i ← 1 to M do
  {  s ← SS(0)
     SS(0) ← c ← SS(0) + σ([ a_i/⁻ ])
     for j ← 1 to N do
        {  c ← max{s + σ([ a_i/b_j ]), SS(j) + σ([ a_i/⁻ ]), c + σ([ ⁻/b_j ])}
           s ← SS(j)
           SS(j) ← c
        }
  }

write "Similarity score is" SS(N)
```

FIG. 5. Linear-space version of Fig. 2.

the dynamic programming algorithm for similarity score calculations often offer a pair of parameters, q and r, so that the penalty for a gap is $g = q + rk$, where k is the number of residues in the gap. Figure 2 handles the case $q = 0$ and $r = 1$ if we set $\sigma\left(\begin{bmatrix} a \\ \_ \end{bmatrix}\right) = -1 = \sigma\left(\begin{bmatrix} \_ \\ b \end{bmatrix}\right)$ for all a and b. For the Needleman–Wunsch algorithm, q is the gap penalty and $r = 0$. Similarity scores that allow gap penalties of the form $q + rk$ can also be calculated in $O(N^2)$ time.[7] Although more complex gap penalties that require $O(N^3)$ operations have been proposed,[31] comparisons using penalties of the form $q + rk$ perform well in practice. However, because of the additional bookkeeping involved, algorithms that calculate a gap-penalty function with $q \neq 0$ are more difficult to describe; hence our examples require that $q = 0$.

At first glance, these algorithms for calculating sequence similarity scores might appear to require $O(N^2)$ space (i.e., the S matrix). However in Fig. 2, values in the ith row of S depend only on values in rows i and $i - 1$.[32] This means that two row-sized vectors are adequate to compute the score of an optimal alignment. In fact, the algorithm of Fig. 5 uses only one vector, SS, and two temporary values, c and s. Figure 5 follows readily from Fig. 2 if we keep in mind that at the top of the inner **for** loop we have

[31] M. S. Waterman, T. F. Smith, and W. A. Beyer, *Adv. Math.* **20**, 367 (1976).
[32] This is not the case for the Needleman–Wunsch algorithm, but the bookkeeping tricks that reduce its time complexity from $O(N^3)$ to $O(N^2)$ also reduce the memory requirement to $O(N)$.

$$SS(k) = \begin{cases} S(i, k) & \text{if } k < j \\ S(i - 1, k) & \text{if } k \geq j \end{cases}$$

$$c = S(i, j - 1)$$
$$s = S(i - 1, j - 1)$$

In addition to the algorithm being far more space-efficient than Fig. 2, an implementation of Fig. 5 will run somewhat faster because the array-indexing operations are simpler. Thus, it is possible to calculate the optimal global similarity score for a very general set of match/mismatch penalties and gap penalties in $O(N^2)$ time and $O(N)$ space.

Constructing Sequence Alignments

The algorithms described above calculate an optimal similarity or distance score without specifying the alignment or alignments that produce that score. By alignment, we mean the explicit mapping of the residues in one sequence to the residues in the other. For many sequence comparison problems, the score is more important than an alignment. For example, when one searches a protein or DNA sequence database, a similarity score is calculated for every sequence in the database, but alignments are shown typically for the top-scoring sequences only. Likewise, it is possible to construct an evolutionary tree for K sequences based on the $K(K - 1)/2$ pairwise distances[25] without specifying a multiple alignment of all the sequences. Because rigorous multiple alignment of K sequences requires $O(N^K)$ time and $O(N^{K-1})$ space,[33] evolutionary trees that are based on $O(K^2)$ distances calculations, each of which requires $O(N^2)$ operations, are far more feasible.

Space-Efficient Alignments

Not only is it possible to calculate the optimal similarity score between two sequences in $O(N)$ (linear) space, it is also possible to calculate an optimal alignment in linear space.[9] This has been surprising to many biologists who practice sequence alignment; almost all of the methods described for constructing an alignment use the full S matrix,[4-6] or some approximation of it that requires $O(N^2)$ space.[7,8] These investigators had noted that if the S matrix is available, one could start at the end of the alignment [position (M, N)] and retrace the steps required to calculate the optimal score. For the algorithm shown in Fig. 2, one would examine three

[33] H. Carrillo and D. J. Lipman, *SIAM J. Appl. Math.* **48,** 1073 (1988).

matrix entries, (1) on the upper-left diagonal, (2) above, and (3) to the left, and identify the entry yielding the best score. The entry yielding the best score indicates the direction from which the alignment was extended: (1) a match/mismatch, (2) a deletion, or (3) an insertion. Often there would be a tie, so that the same score could be obtained by proceeding in more than one direction; ties could be broken arbitrarily. By repeating this process, the alignment can be reconstructed in $O(N)$ time.

Computer scientists discovered that applying the divide-and-conquer programming paradigm produces a much more space-efficient method for reconstructing an optimal alignment.[34] To discuss explicit determination of an optimal alignment, we find it helpful to reformulate the S matrix in terms of an "edit graph." For sequences $seq1 = a_1a_1 \cdots a_M$ and $seq2 = b_1b_2 \cdots b_N$ consider the set of "vertices" (i, j) where $0 \le i \le M$ and $0 \le j \le N$. The following edges and only these edges are in the edit graph:

1. If $1 \le i \le M$ and $1 \le j \le N$, there is a substitution edge $(i - 1, j - 1) \rightarrow (i, j)$ labeled $\begin{bmatrix} a_i \\ b_j \end{bmatrix}$.

2. If $1 \le i \le M$ and $0 \le j \le N$, there is a deletion edge $(i - 1, j) \rightarrow (i, j)$ labeled $\begin{bmatrix} a_i \\ - \end{bmatrix}$.

3. If $0 \le i \le M$ and $1 \le j \le N$, there is an insertion edge $(i, j - 1) \rightarrow (i, j)$ labeled $\begin{bmatrix} - \\ b_j \end{bmatrix}$.

Figure 6 provides an example of the construction. A path from $(0, 0)$ to (i, j) in this graph corresponds to the alignment created by concatenating the labels that lie along the path.

For each edge of the edit graph we assign the weight $\sigma(\lambda)$, where λ is the label of the edge. The score of a path in this graph is defined as the sum of its edge weights. Then the value $S(i, j)$ computed by Fig. 2 is simply the maximum score of a path from $(0, 0)$ to (i, j), and the problem of constructing an optimal alignment is equivalent to displaying a highest-scoring path. Edit graphs also are particularly useful for reasoning about alignments with gap penalties of the form $q + rk$. Reference 23 shows the edit graph and algorithm for this more complex gap penalty. This more complex edit graph[23] includes nodes for insertion initiation and deletion initiation.

The first step in understanding how to compute optimal alignments in linear space is to note that the algorithms of Figs. 2 and 5 can be "inverted"

[34] D. S. Hirschberg, *Commun. ACM* **18**, 341 (1975).

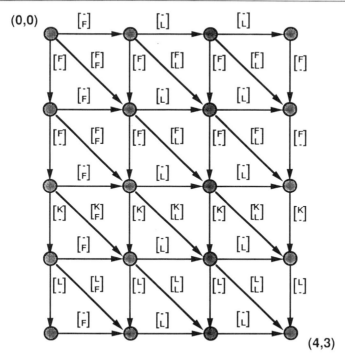

FIG. 6. The edit graph for *seq1* = FFKL and *seq2* = FLL.

to treat rows from last to first, sweeping right to left within a row. Here, the value computed at vertex (i, j) is the score of an optimal path from (i, j) to (M, N), which is the same as the score of a path from (M, N) to (i, j) if we invert the direction of edges; we call this value $T(i, j)$. To compute an alignment, we begin by applying Fig. 5 to evaluate the first $\lceil M/2 \rceil$ rows of S (the brackets tell to round downward if M is odd); thus it determines $SS(j) = S(\lceil M/2 \rceil, j)$ for $0 \leq j \leq N$. Then, with an inverted version of Fig. 5, we compute $TT(j) = T(\lceil M/2 \rceil, j)$ for $0 \leq j \leq N$.

An optimal path (i.e., alignment) can be bisected into an optimal path from $(0, 0)$ to node $(\lfloor M/2 \rfloor, j)$ for some j between 0 and N [score = $SS(j)$] followed by an optimal path from $(\lfloor M/2 \rfloor, j)$ to (M, N) [score = $TT(j)$]. Thus, we can find the midpoint on an optimal path by determining a j_{opt} that maximizes $SS(j) + TT(j)$; the cost of doing so is essentially identical to that of running Fig. 5. There may be several columns j where the maximum is attained; ties can be broken arbitrarily. This midpoint divides the problem into two smaller problems, namely, aligning $a_1 a_2 \cdots a_{\lfloor M/2 \rfloor}$ with $b_1 b_2 \cdots b_j$ and aligning $a_{\lfloor M/2 \rfloor + 1} \cdots a_M$ with $b_{j+1} \cdots b_N$. For each

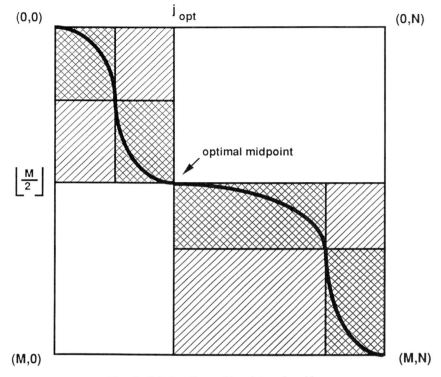

(0,0) j_{opt} (0,N)

optimal midpoint

$\left\lfloor \dfrac{M}{2} \right\rfloor$

(M,0)   (M,N)

FIG. 7. Splitting the problem into subproblems.

subproblem, we compute an optimal midpoint as before. Notice that for any j, the (singly or doubly) hatched regions in Fig. 7 are exactly half the entire area. This implies that determining the next two points on an optimal path costs half as much as the first point. At the next step, we are left with determining optimal subpaths through four regions of total area $\frac{1}{4} \times MN$, as in Fig. 7. Continuing in this manner, we fill in the vertices on an optimal path. If t denotes the time taken by Fig. 5, then the time for this divide-and-conquer procedure is t for the first midpoint, plus $t/2$ for the next two points, plus $t/4$ for the next four, and so on, for a total of $(1 + \frac{1}{2} + \frac{1}{4} + \cdots)t \cong 2t$. In other words, using only space for two rows of the S matrix and twice the execution time of Fig. 5, we can explicitly produce an optimal alignment.

Local Sequence Similarity and Smith–Waterman Algorithm

The Needleman–Wunsch algorithm calculates a global similarity score, that is, one that reflects an alignment which starts at one end of both sequences and stops at the other end. This similarity score is very useful for ranking the similarities—or distances—of members of a family of proteins, such as the hemoglobin family. Table I shows a set of global and local similarity scores calculated for several members and nonmembers of the globin superfamily.

A minor variant of the global similarity calculation solves the problem of aligning $seq1 = a_1 a_2 \cdots a_M$ with only a region of $seq2 = b_1 b_2 \cdots b_N$ optimally.[35] For example, suppose that M is much smaller than N and we want to find the region $b_p b_{p+1} \cdots b_q$ of $seq2$ that is most like $seq1$. Pictorially, the alignment matrix is short and wide, and we seek a highest scoring alignment that begins somewhere in row 0 and ends somewhere in row M. Equivalently, the problem can be formalized as aligning $seq1$ with all of $seq2$, where an insertion pair $\begin{bmatrix} - \\ b_j \end{bmatrix}$ is not penalized if it occurs either before or after all substitution or deletion pairs; in other words, we give the weight zero to horizontal edges in the 0th and Mth rows, then seek an optimal path from $(0, 0)$ to (M, N). This variation is equivalent to a global alignment without penalizing end gaps in $seq2$. Figure 2 can be modified to solve this problem by setting $S(0, j) = 0$ for every j and replacing $S(i, j - 1) + \sigma\left(\begin{bmatrix} - \\ b_j \end{bmatrix}\right)$ with $S(i, j - 1)$ when the maximum is found in row $i = M$.

Global similarity scores are most useful when the sequence similarity extends from one end of the protein to the other, as is the case for the globins. However, many proteins have a more complex evolutionary pedigree. Some proteins have evolved by duplication of a subset of the sequence, whereas others contain portions of two or more distinct families. The family of proteins that contain the EF-hand calcium-binding domain is one such example. This family includes the calmodulins, troponins, tropomyosins, myosin catalytic and regulatory chains, and a variety of other calcium-binding proteins.[36] Members of this family have evolved by duplications of the EF-hand domain; in some proteins, the EF-hand domain is associated with other functional domains. The members of this family have from two to eight calcium-binding domains. Proteins in one branch of the family, the calpains, contain four calcium-binding domains and a neutral protease domain. All of the sequences in the middle of Table I contain EF-hand calcium-binding domains. However, the global sequence similarity calculation is not appropriate for identifying the relationship between calmodulin (148 amino acids) and calpain (714 amino

[35] P. H. Sellers, *J. Algorithms* **1**, 359 (1980).
[36] N. D. Moncrief, M. Goodman, and R. H. Kretsinger, *J. Mol. Evol.* **30**, 522 (1990).

TABLE I
GLOBAL AND LOCAL SEQUENCE SIMILARITY SCORES

PIR entry		Compound	Source	Similarity score[a]			Distance[b]
				Global			
				End penalty	No end penalty	Local	
HBHU versus	HBHU	Hemoglobin, β chain	Human	725	725	725	0
	HAHU	Hemoglobin, α chain	Human	314	320	322	152
	MYHU	Myoglobin	Human	121	164	166	212
	GPYL	Leghemoglobin I	Yellow lupine	8	28	43	239
	LZCH	Lysozyme c precursor	Chicken	−107	16	32	220
	NRBO	Pancreatic ribonuclease	Bovine	−124	16	31	280
	CCHU	Cytochrome c	Human	−160	10	26	321
MCHU versus	MCHU	Calmodulin	Human	671	671	671	0
	TPHUCS	Troponin C, skeletal muscle	Human	395	430	438	161
	PVPK2	β-Parvalbumin	Northern pike	−57	103	115	313
	CIHUH	Calpain I heavy chain	Human	−2085	89	100	2463
	CIHUL	Calpain light chain	Human	−358	79	87	682
	AQJFNV	Aequorin precursor	Jellyfish	−65	48	76	391
	KLSWM	Calcium-binding protein	Scallop	−89	45	52	323
QRHULD[c] versus	EGMSMG	Epidermal growth factor precursor	Mouse	−591	475	655	2549

[a] Similarity scores were calculated using the PAM250 matrix with a penalty of −8 for opening a gap and a penalty of −4 for each residue in a gap.

[b] Distance scores were calculated using a minimal mutation distance scoring matrix with a penalty of −8 for opening a gap and a penalty of −4 for each residue in a gap.

[c] Human LDL receptor.

acids); the similarity in the EF-hand domain is masked by the gap and mismatch penalties from the 529 unrelated amino acids. Table I also includes a more complex example: the local similarity between the LDL receptor (PIR code QRHULD) and the mouse epidermal growth factor precursor (EGMSMG). These two proteins share several regions of similarity as the result of exon shuffling.[37]

Similar domains that span a portion of two sequences are best identified by calculating the optimal local sequence similarity.[6] Local similarity means that the alignment need not extend from one end of the sequence to the other; it must simply be an alignment with the best similarity score, starting and ending at any point in either sequence. Thus, a local similarity calculation might be used to identify the exons in a gene by comparing the mRNA to the genomic DNA sequence, or to identify the multiple EF-hand domains in calpain. In contrast to the global similarity score, which can have positive or negative values, the local similarity score must be greater than zero if there are any identical residues in the two sequences (it may be equal to zero if there is no sequence identity). Although it might seem that an algorithm which calculates the best alignment score, regardless of where the alignment starts or ends, should be more complicated than one which starts and stops at the ends of the two sequences, the two calculations are almost identical.

The Smith–Waterman algorithm[6] computes an optimal alignment between regions of *seq1* and *seq2*. Here we seek a highest scoring path in the edit graph where the end points are not fixed in advance, but are chosen to maximize the score. Letting $S'(i, j)$ denote the highest score of a path ending at (i, j) (with an arbitrary starting point), there are now four possibilities. The path can have zero edges or, as before, it can come through one of (i, j)'s three immediate predecessors. Assuming that deletion and insertion pairs have nonpositive weights, Fig. 8 computes S'.

Including the alternative that the similarity score be zero in the expression for $S'(i, j)$ allows the local alignment to restart at any pair of aligned residues. In addition, it makes the calculation much more sensitive to the precise match and mismatch scores and gap penalties. This is not the case with the global similarity calculation, where adding a constant to the scores for matches, mismatches, and gaps will not change the alignment. In contrast, if all of the match and mismatch values are greater than zero, then every local similarity score will start at the beginning of at least one, and stop at the end of at least one, of the two sequences. Likewise, if all of the match and mismatch scores are zero or less, the best local similarity score will be zero.

[37] T. C. Sudhof, J. L. Goldstein, M. S. Brown, and D. W. Russell, *Science* **228,** 815 (1985).

```
best ← 0
for j ← 0 to N do
    S'(0,j) ← 0
for i ← 1 to M do
{   S'(i,0) ← 0
    for j ← 1 to N do
    {   S'(i,j) ← max{0, S'(i-1,j-1) + σ([ᵃⁱ_bⱼ]), S'(i-1,j) + σ([ᵃⁱ_-]), S'(i,j-1) + σ([⁻_bⱼ])}
        best ← max{S'(i,j), best}
    }
}
```

write "The score of the best local alignment is" *best*

FIG. 8. An algorithm for computing the score of an optimal local alignment.

Thus for local alignments, it is essential that σ be chosen so that the score of a "random" path is negative. Attempts to define the local alignment problem in terms of minimizing a difference score have resulted in excessively complex algorithms.[38,39] Protein sequence comparisons frequently use the PAM250 matrix,[40] which reflects the frequency of amino acid replacement over a long period of evolutionary time. In the PAM250 matrix, the aligning of identical residues yields scores ranging from 2 (Ala–Ala) to 17 (Trp–Trp); the scores for alignment of nonidentical residues range from -8 (Cys–Trp) to 4 (Met–Leu).[41]

Explicit determination of an optimal local alignment, not just its score, can be performed in $O(N)$ space using the linear-space global alignment procedure. Figure 8 can be modified so that at each (i, j) we determine both the score, $S'(i, j)$, and the starting point (g, h) of an optimal path ending at (i, j). [Alternatively, the starting point can be found by a backward sweep from the point (i, j) where $S'(i, j)$ is maximal.] Then finding an optimal local alignment ending at (i, j) is equivalent to determining an optimal global alignment of the sequence $a_{g+1}a_{g+2} \cdots a_i$ with $b_{h+1}b_{h+2} \cdots b_j$, so the earlier linear-space procedure can be invoked.

Often one wants to determine several local alignments between two sequences. For example, there are 3 calcium-binding domains in parval-

[38] W. B. Goad and M. I. Kanehisa, *Nucleic Acids Res.* **10,** 247 (1982).
[39] P. H. Sellers, *Bull. Math. Biol.* **46,** 501 (1984).
[40] M. Dayhoff, R. M. Schwartz, and B. C. Orcutt, *in* "Atlas of Protein Sequence and Structure" (M. Dayhoff, ed.), Vol. 5, Suppl. 3, p. 345. National Biomedical Research Foundation, Silver Spring, Maryland, 1978.
[41] Thus, the PAM250 matrix suggests that the alignment of two alanine residues is less significant than a leucine–methionine alignment. This is based on the observation that alanines are very common in proteins, and thus their alignment may be expected frequently by chance. Methionine and leucine are less common, but interchange rapidly in evolutionary time.

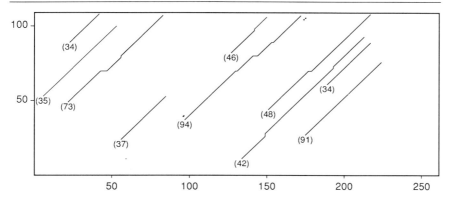

FIG. 9. Local alignments of calbindin and parvalbumin. The ten best local alignments between human calbindin (PIR code S00234, 109 amino acids) and *Amphiuma* parvalbumin (PVNESA, 260 amino acids) are plotted. The PAM250 matrix, with -12 for opening a gap and -4 for each residue in a gap, was used for scoring. The numbers in parentheses indicate the local similarity score. The "..." indicate potential subalignments of the best alignment that have similarity scores greater than the second best local alignment.

bumin and 6 in calbindin,[36] so there are 18 potential alignments between calcium-binding domains in these two proteins. Graphical presentations of local sequence alignments resemble "dot-matrix" similarity plots[42] but are more informative because they provide an optimal local alignment that has a similarity score.

Although it is easy to determine a highest scoring local alignment, given a proper choice of σ, it is not so clear how to proceed to subsequent alignments. It will not work merely to report the second largest value of S' and a corresponding path. The score at a point that is one edge beyond the end of an optimal path equals the optimal score plus the cost of the edge, so it is very nearly optimal. For example, one can trim one residue from the start or one or two residues from the end of the best local alignment in Fig. 9 and still have an alignment with a score that is 91 (the score of the second best local alignment shown) or better. More generally, the second-highest scoring alignment will typically differ only slightly from the highest scoring alignment. Thus, for the "second best" alignment, we take the highest scoring path that shares no diagonal edge (i.e., match or mismatch) with the first path, and in general deliver the optimal path that is similarly disjoint from the first $k-1$ reported paths.[43]

Thus, to calculate or display the kth local alignment, we must recalcu-

[42] J. Maizel and R. Lenk, *Proc. Natl. Acad. Sci. U.S.A.* **78,** 7665 (1981).
[43] M. S. Waterman and M. Eggert, *J. Mol. Biol.* **197,** 723 (1987).

late those parts of the S' graph in the "shadow" of the $k - 1$ earlier alignments. The program in Fig. 8 can locate the end of an optimal local alignment, and a traceback procedure can be applied to S' to produce the entire alignment. (For now, we are assuming that S' is retained.) Diagonal edges on that path can be removed from the graph, and the process can be repeated to find a highest scoring path in the reduced graph. However, this complete recomputation of S' is extremely wasteful, and a more efficient procedure has been described.[43] Let $S_1'(i, j)$ denote the value at node (i, j) on the initial execution of Fig. 8, and let $S_2'(i, j)$ denote the value computed without using diagonal edges from the first path; note that $S_2'(i, j) \leq S_1'(i, j)$. Typically, $S_2'(i, j)$ will equal $S_1'(i, j)$ for all (i, j) except those in some neighborhood of the first path. Thus, computing time can be drastically reduced by computing S_2' only for vertices where it differs from S_1'.[43]

In practice, there are two further complications. First, one wants to use gap penalties of the form $q + rk$ rather than just rk. As discussed earlier, these penalties can be handled at modest additional cost (perhaps a factor of 2 in both time and space requirements), but the necessary bookkeeping complicates the implementation. More difficult is the problem of using only linear space, which prohibits saving all but a tiny fraction of the S values. These hurdles can be overcome,[23] but the solution is too complicated to describe here.

Dynamic Programming Algorithms for Profiles and Patterns

One characteristic of dynamic programming algorithms for biological sequence comparison is that they are very pessimistic, that is, their average execution time is exactly the same as their worst-case execution time. In some sense, it is this pessimism—$O(N^2)$ operations are always performed—that allows the potentially more complex local similarity calculation to be performed in the same time as a global similarity calculation. This pessimism can be avoided in some calculations of distance scores; these shortcuts are discussed in the next section. It also reflects the fact that a dynamic programming algorithm can do much more than compare two sequences; it can also compare a sequence to a variety of sequence patterns.[12,24,44,45]

Very distantly related members of a protein sequence family can sometimes be identified more clearly with a pattern that summarizes the family

[44] M. Gribskov, A. D. McLachlan, and D. Eisenberg, *Proc. Natl. Acad. Sci. U.S.A.* **84**, 4355 (1987).
[45] E. W. Myers and W. Miller, *Bull. Math. Biol.* **51**, 5 (1989).

rather than with individual members of the family. Thus, rather than compare a plant leghemoglobin to a mammalian β-globin (which has a very low similarity score, see Table I), one compares the leghemoglobin sequence to a pattern that reflects the number of times a particular amino acid has occurred in each position along the sequence in aligned members of the protein family. For example a pattern for the serine protease family would reflect the highly conserved GDSGGP sequence in the active site of these enzymes.

The strategy of comparing sequences to a sequence "profile," rather than a single sequence, was first proposed by Gribskov *et al.*[12,44] Here, a profile for a sequence family is constructed by examining the number of times each amino acid or a gap occurs at each position in a multiple alignment of several members of the protein family. The result of this calculation is a set of $\sigma_i(b)$ that describe the score for aligning residue b with position i of the profile. In addition, the multiple alignment can be used to set δ_i, the penalty for each gap in the sequence with respect to the profile, and ι_i, the penalty for inserting residue b between positions i and $i + 1$ of the profile. An algorithm that calculates the optimal similarity score using σ_i, δ_i, and ι_i can be derived from Fig. 2 by substituting $\sigma_i(b)$

for $\sigma\left(\begin{bmatrix} a \\ b \end{bmatrix}\right)$, δ_i for $\sigma\left(\begin{bmatrix} a \\ - \end{bmatrix}\right)$, and ι_i for $\sigma\left(\begin{bmatrix} - \\ b \end{bmatrix}\right)$.

A different type of profile, termed a "flexible pattern,"[24,46] specifies a set of amino acid residues and a range of gap sizes that can be accommodated at different positions in the sequence, rather than a variable set of gap penalties. Remarkably, a pattern that specifies only 28% of a globin sequence can rank 343 of 346 globin sequences above the highest-scoring unrelated sequence in the PIR protein sequence database.[24,46] "Flexible patterns" performed considerably better than sequence profiles for Barton and Sternberg; in their hands the latter identifies only 318 sequences before the highest-ranking unrelated sequence.

A flexible pattern has the form

$$r_0\langle l_1, u_1\rangle r_1\langle l_2, u_2\rangle r_2 \cdots r_{P-1}\langle l_P, u_P\rangle r_P$$

Each r_i is a "residue matcher"; the score awarded for aligning r_i with residue b is given by $\sigma_i(b)$, and the score for aligning the pattern with a sequence is the sum of these σ's. A term $\langle l_i, u_i\rangle$ specifies that when the pattern is aligned with a sequence, there must at least l_i, and at most u_i, residues of the sequence lying between the residue matched by r_{i-1} and the residue matched by r_i. (Although we use the term "pattern," we are calculating a similarity score; some pattern matching algorithms look only

---

[46] G. J. Barton, this series, Vol. 183, p. 403.

```
for j ← 1 to N do
    R (0, j) ← σ₀(bⱼ)
start ← 1
for i ← 1 to P do
  { start_prev ← start
    start ← start + 1 + lᵢ
    for j ← start to N do
        R (i, j) ← max{R (i−1, j−1−t) where lᵢ ≤ t ≤ min{uᵢ, j−1−start_prev }} + σᵢ(bⱼ)
  }
```

write "Similarity score is" max{R (P, j) where start ≤ j ≤ N }

FIG. 10. A dynamic programming algorithm for pattern matching.

for an exact match.) A simple sequence $a_1 a_2 \cdots a_M$ is equivalent to a flexible pattern where $l_i = u_i = 0$ for all i.

To calculate an optimal score for a pattern, we consider a sequence $b_1 b_2 \ldots b_N$ (with a prefix denoted by B_j) and let A_i denote the pattern prefix

$$r_0 \langle l_1, u_1 \rangle r_1 \langle l_2, u_2 \rangle r_2 \cdots r_{i-1} \langle l_i, u_i \rangle r_i$$

We consider an optimal alignment of A_i and B_j that ends by aligning r_i with b_j and denote that optimal score by $R(i, j)$. Because r_{i-1} must be aligned with b_{j-1-t} where $l_i \leq t \leq u_i$, and $R(i, j) = R(i - 1, j - 1 - t) + \sigma_i(b_j)$, we have

$$R(i, j) = \max\{R(i - 1, j - 1 - t) \text{ where } l_i \leq t \leq u_i\} + \sigma_i(b_j)$$

(We note that since $u_i \geq l_i \geq 0$, we need not consider deletions in the pattern A.) In practice, $R(i, j)$ must be evaluated carefully. If r_i can be aligned with b_j, then $j \geq start_i = i + 1 + \Sigma_{h=1}^{i} l_k$, since i previous b's must be aligned to residue matchers and at least $\Sigma_{h=1}^{i} l_h$ previous b's must appear in gaps. Moreover, when defining $R(i, j)$ in terms of values $R(i - 1, j - 1 - t)$, we must consider only values t where $j - 1 - t \geq start_{i-1}$, that is, $t \leq j - 1 - start_{i-1}$. The algorithm of Fig. 10 uses these observations to optimally align a pattern with a sequence.

Comparison of Fig. 11 and Fig. 4 suggests that the computational complexity of matching flexible patterns (Fig. 10) lies between that of simple sequence alignment (Fig. 2) and the Needleman-Wunsch algorithm (Fig. 3). Let F denote the maximum "gap flexibility" $u_i - l_i + 1$. Using a basic data structure called a "priority queue," the successive values $R(i, 1), R(i, 2), \cdots$ for fixed i can be computed in $O(\log_2 F)$ time each. If

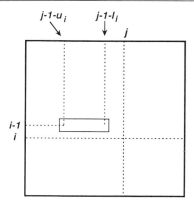

FIG. 11. Locations used to compute $R(i, j)$ in Fig. 10.

Fig. 10 is implemented this way, then its total running time is $O(PN \log F)$. Since we may assume $F \leq N$, the time is $O(PN \log N)$.

Table II summarizes several of the major dynamic programming algorithms for biological sequence alignment that have been described above.

Variations on Dynamic Programming: Shortcuts and Detours

Program developers have worked hard to produce methods that improve on dynamic programming. However, such efforts must recognize the following fact: any procedure that can align arbitrary sequences optimally under an arbitrary scoring scheme must take $O(N^2)$ time in the worst case.[47] A variety of strategies have been employed to cope with this obstacle. The most successful approaches have sacrificed the guaranteed optimality of the computed alignments, while still producing alignments that work well in practice.[2,48,49] However, here we review only approaches that retain optimality. For a more comprehensive review, see the excellent survey paper by Myers.[50]

One approach has been to live with $O(N^2)$-time algorithms, but to implement them to run on computers that can perform many operations in parallel.[51,52] That way, the algorithm takes less than $O(N^2)$ actual time,

[47] X. Huang, *Inf. Process. Lett.* **27**, 319 (1988).

[48] W. R. Pearson and D. J. Lipman, *Proc. Natl. Acad. Sci. U.S.A.* **85**, 2444 (1988).

[49] S. F. Altschul, W. Gish, W. Miller, E. W. Myers, and D. J. Lipman, *J. Mol. Biol.* **215**, 403 (1990).

[50] E. W. Myers, *in* "Mathematic and Molecular Biology" (E. Lander and W. Gilbert, eds.), 1991.

[51] E. W. Edmiston, N. G. Gore, J. H. Saltz, and R. M. Smith, *Int. J. Parallel Program.* **17**, 259 (1988).

TABLE II
DYNAMIC PROGRAMMING IMPLEMENTATIONS

Alignment	Gap penalty	Time required	Space required	Ref.
Global similarity	Penalty/gap	$O(N^3)$	$O(N^2)$	a
(Global) distance	Penalty/residue	$O(N^2)$	$O(N^2)$	b
(Global) distance	Penalty/residue	$O(N^2)$	$O(N)$	c
(Global) distance	Penalty/residue	$O(ND)^d$	$O(ND)$	e, f
(Global) distance	Arbitrary	$O(N^3)$	$O(N^2)$	g
Global similarity				
Score only	$q + rk$	$O(N^2)$	$O(N)$	h–j
Alignment		$O(N^2)$	$O(N^2)$	
Global similarity (score and alignment)	$q + rk$	$O(N^2)$	$O(N)$	k
Unweighted end gaps in sequence 2	$q + rk$	$O(N^2)$	$O(N^2)$	l
One local similarity	$q + rk$	$O(N^2)$	$O(N^2)$	m
K local similarities	$q + rk$	$O(N^2K)^o$	$O(N^2)$	n
K local similarities	$q + rk$	$O(N^2K)^o$	$O(N)$	p, q
Global similarity to sequence profile	Profile	$O(PN)^u$	$O(N)^r$	s, t
Global similarity to sequence pattern	$\langle l_i - u_i \rangle^v$	$O(PN \log N)^u$	$O(N)$	w, x

[a] S. Needleman and C. Wunsch, *J. Mol. Biol.* **48**, 444 (1970).

[b] P. H. Sellers, *SIAM J. Appl. Math.* **26**, 787 (1974).

[c] D. S. Hirschberg, *Commun. Assoc. Comput. Mach.* **18**, 341 (1975).

[d] D is the maximum distance between the two sequences.

[e] E. Ukkonen, *Inf. Control* **64**, 100 (1985).

[f] J. W. Fickett, *Nucleic Acids Res.* **12**, 175 (1984).

[g] M. S. Waterman, T. F. Smith, and W. A. Beyer, *Adv. Math.* **20**, 367 (1976).

[h] O. Gotoh, *J. Mol. Biol.* **162**, 705 (1982).

[i] S. F. Altschul and B. W. Erickson, *Bull. Math. Biol.* **48**, 603 (1986).

[j] W. Miller and E. W. Myers, *Bull. Math. Biol.* **50**, 97 (1988).

[k] E. W. Myers and W. Miller, *CABIOS* **4**, 11 (1988).

[l] P. H. Sellers, *J. Algorithms* **1**, 359 (1980).

[m] T. F. Smith and M. S. Waterman, *J. Mol. Biol.* **147**, 195 (1981).

[n] M. S. Waterman and M. Eggert, *J. Mol. Biol.* **197**, 723 (1987).

[o] In practice these methods require $O(N^2 \sum_{k=1}^{k} L_k^2)$ time.

[p] X. Huang, R. C. Hardison, and W. Miller, *CABIOS* **6**, 373 (1990).

[q] X. Huang and W. Miller, *Adv. Appl. Math* **12**, 337 (1991).

[r] The authors do not explicitly state the complexity.

[s] M. Gribskov, A. D. McLachlan, and D. Eisenberg, *Proc. Natl. Acad. Sci. U.S.A.* **84**, 4355 (1987).

[t] M. Gribskov, R. Lüthy, and D. Eisenberg, this series, Vol. 183, p. 146.

[u] P is the number of residue matchers in the profile or pattern.

[v] See text.

[w] G. J. Barton, this series, Vol. 183, p. 403.

[x] G. J. Barton and M. J. E. Sternberg, *J. Mol. Biol.* **212**, 389 (1990).

though it still requires $O(N^2)$ operations. The dynamic programming algorithms discussed in this chapter lend themselves very well to parallel computation; all the variations discussed above, including computation of k best local similarities in linear space, can be "parallelized."[52]

Alternatively, we can beat $O(N^2)$ time for pairs of input sequences that are likely to be interesting because they have low distance scores. For sequences that are similar to each other, one need evaluate only the portion of the dynamic programming matrix near the main diagonal.[53,54] For alignment problems expressed as minimizing a difference score D, this approach leads to $O(ND)$-time methods; that is, the computation is fast if D is small.

Another approach has been to work with a limited class of alignment scores. For example, if the gap parameters q and r and all substitution scores are integers, then there exits an alignment program that runs in average time $O(N + D^2)$.[55] This limited class of weights includes those usually employed for DNA or protein comparisons. It is important to remember how large D can be in practice, however. One is frequently interested in similarities between DNA sequences that are as much as 40% different ($D \cong N/2$); moreover, alignments between protein sequences that are 80% different ($D \cong N$) can be used to demonstrate common ancestry. In practice, complex modifications of the algorithm of Wu *et al.*[56] to include general scoring penalties have not been significantly faster than dynamic programming (W. Miller, unpublished, 1986).

A different line of research has been to work with more general scoring schemes, in particular, more general gap penalties. For example, the penalty might rise quickly with gap length for short gaps, then level off to be almost constant for long gaps. The hope here is that by allowing more flexible scores, we can produce more useful alignments. Figure 3 is readily modified to handle the case that gaps of length k are penalized w_k, where the w's are arbitrary,[31] but the time complexity is $O(N^3)$. Notice that gap penalties of the form $w_k = q + rk$ put the w_k's on a straight line. A piecewise linear w with P pieces can be handled in $O(N^2 \log P)$ time.[57] Thus, there is very little increased time required for penalties of the form $w_k = q + rk$ for $k < l$ and $w_k = c$ for $k \geq l$, that is, a constant penalty c for gaps greater than l. The case where $w_{i+1} - w_i \geq w_{j+1} - w_j$ if $i < j$, that is, the additional penalty for extending a gap by one residue is less for

[52] X. Huang, W. Miller, S. Schwartz, and R. C. Hardison, Parallelization of a local similarity algorithm. Submitted (1990).

[53] E. Ukkonen, *Inf. Control* **64**, 100 (1985).

[54] J. W. Fickett, *Nucleic Acids Res.* **12**, 175 (1984).

[55] E. W. Myers and W. Miller, *ACM Trans. Prog. Languages Syst.* **11**, 33 (1989).

[56] S. Wu, U. Manber, E. W. Myers, and W. Miller, *Inf. Process. Lett.* **35**, 317 (1990).

[57] W. Miller and E. W. Myers, *Bull. Math. Biol.* **50**, 97 (1988).

longer gaps, has also been solved. Such "concave" gap penalties can be handled in $O(N^2 \log N)$ time.[57-59] The divide-and-conquer strategy for producing an alignment in linear space works for this class of weights,[57] though now the linear space bound holds only in the average case but not in the worst case.[60]

All of the algorithms we discuss above seek to maximize a similarity score (or minimize a distance). However, the alignment with the highest score may be neither the most statistically significant nor the most biologically relevant. In particular, this is a problem with local similarity scores that are calculated during searches of sequence databases. Protein sequence databases often contain fragments of a protein that are derived from direct amino acid sequencing; DNA sequence databases may contain limited regions of high identity because of exon structure. In library searches, it is common to find a short (15–25 amino acid) domain that shares 85% sequence identity to a query sequence and has the same similarity score as a 50–80 amino acid domain with 30% identity. The former alignment is much more likely to be biologically relevant than the latter, but the relationship between statistical significance, similarity score, and alignment length is complex. For example, in Fig. 9, the highest scoring local alignment is 75 residues long and has a score of 94, the second highest scoring local alignment is 49 residues long with a score of 91, while the longest local alignment (82 residues) is ranked sixth with a score of 42. Which alignment is the most significant? Analytical solutions to this problem are available only for alignments that do not allow gaps.[61,62] Altschul has described an $O(N^3)$ algorithm that finds the most statistically significant alignment where gaps are not allowed.[63]

Summary

Efficient dynamic programming algorithms are available for a broad class of protein and DNA sequence comparison problems. These algorithms require computer time proportional to the product of the lengths of the two sequences being compared [$O(N^2)$] but require memory space proportional only to the sum of these lengths [$O(N)$]. Although the requirement for $O(N^2)$ time limits use of the algorithms to the largest computers

[58] M. S. Waterman, *J. Theor. Biol.* **108,** 333 (1984).
[59] Z. Galil and R. Giancarlo, *Theor. Comput. Sci.* **64,** 107 (1989).
[60] Y. Rabani and Z. Galil, On the space complexity of some algorithms for sequence comparison. *Theor. Comput. Sci.* in press (1992).
[61] R. Arratia, L. Gordon, and M. S. Waterman, *Ann. Math. Stat.* **14,** 971 (1986).
[62] S. Karlin and S. F. Altschul, *Proc. Natl. Acad. Sci. U.S.A.* **87,** 2264 (1990).
[63] S. F. Altschul and B. W. Erickson, *Bull. Math. Biol.* **48,** 617 (1986).

when searching protein and DNA sequence databases, many other applications of these algorithms, such as calculation of distances for evolutionary trees and comparison of a new sequence to a library of sequence profiles, are well within the capabilities of desktop computers. In particular, the results of library searches with rapid searching programs, such as FASTA[2,48] or BLAST,[49] should be confirmed by performing a rigorous optimal alignment. Whereas rapid methods do not overlook significant sequence similarities,[64] FASTA limits the number of gaps that can be inserted into an alignment, so that a rigorous alignment may extend the alignment substantially in some cases. BLAST does not allow gaps in the local regions that it reports; a calculation that allows gaps is very likely to extend the alignment substantially. Although a Monte Carlo evaluation of the statistical significance of a similarity score with a rigorous algorithm is much slower than the heuristic approach used by the RDF2 program,[2] the dynamic programming approach should take less than 1 hr on a 386-based PC or desktop Unix workstation.

For descriptive purposes, we have limited our discussion to methods for calculating similarity scores and distances that use gap penalties of the form $g = rk$. Nevertheless, programs for the more general case ($g = q + rk$) are readily available.[9,23] Versions of these programs that run either on Unix workstations, IBM-PC class computers, or the Macintosh can be obtained from either of the authors.

Acknowledgments

This work was supported by grants from the National Library of Medicine (LM04969 to W.R.P. and LM05110 to W.M.).

[64] W. R. Pearson, *Genomics* **11**, 635 (1991).

[28] Programs for Symbolic Mathematics in Biochemistry

By HERBERT R. HALVORSON

Introduction

The intent of this chapter is to present an introduction to symbolic computing (computer mathematics) in contrast to numerical computing (computer arithmetic). It might seem that such a chapter would be out of place in a volume otherwise devoted to numerical methods of analysis. However, any problem in numerical analysis should be presented to the computer in a form that is both accurate and efficient. To assure accuracy in a numerical answer it is necessary to avoid formulations that involve taking the difference between quantities of similar magnitude (round-off error). Efficiency is enhanced by reducing the operation count, either by factorizing polynomials or by writing them in nested (Horner) form. Carelessness in a derivation can introduce extraneous roots, with disastrous effects on a Newton–Raphson root-finding procedure. Many problems have a closed-form solution that can eliminate the need for an iterative routine or the numerical evaluation of an integral. Others have a Taylor series expansion that can be truncated and solved analytically to provide an adequate approximation that is easy to calculate.

The speed of the computer limits the extent to which it is profitable to explore some of these alternatives. That is, with reasonable assurance that round-off error has been avoided, the computer can generate a great many inefficient solutions in the time it would take to optimize the problem for faster operation. "Fine-tuning" is customarily reserved for problems that run extraordinarily slowly or that are run repetitively in a production environment.

It is not widely appreciated that a class of readily available computer programs exists, specifically suited for these problems. Tedious derivations can now be accomplished much more quickly. The market is competitive, and the vendors have devoted effort to make their offerings "user-friendly." Versions are available for mainframes, workstations, and personal computers. Particular attention here is given to the smallest of the available programs simply to indicate the mathematical power that is now accessible even to a laptop computer.

What Is Computer Mathematics?

Computer mathematics is an application of the techniques of artificial intelligence to the programmed manipulation of symbols (tokens, lists, or strings), according to the rules of mathematics.[1] Although the computer is ultimately doing binary arithmetic, the concept of number need not enter the problem. For most people, this introduction is the major conceptual hurdle to be overcome. As a simple example, consider $\int_0^1 \sin(x)\, dx$. The symbolic solution to this is $1 - \cos(1)$, and the numerical solution is $0.459697694...$, which are equivalent, of course. But the integral

---

[1] An article that illustrates how rapidly the field has developed within the past decade is R. Pavelle, M. Rothstein, and J. Fitch, *Sci. Am.* **December,** 136 (1981). One section of an ACS Symposium Report is devoted to the topics of this chapter; see T. H. Pierce and B. A. Hohne, eds., "Artificial Intelligence Applications in Chemistry," ACS Symp. Ser. 306, American Chemical Society, Washington, D.C., 1986.

$\int_0^y \sin(x)\ dx$ does not have a numerical solution unless y is bound to a numerical value, even though the symbolic solution of $1 - \cos(y)$ is obvious. On the other hand, the integral $\int_0^1 \sin[\sin(x)]\ dx$ is approximately 0.430606103... although it does not have a symbolic solution.

A computer program that performs these symbolic operations can be regarded as a symbolic counterpart to a scientific calculator. [Indeed, some scientific calculators already have limited symbolic capabilities.] Alternatively, such a program can be regarded as a mathematical counterpart to a text editor.

How Can One Start?

A dedicated hacker might be tempted to write such a program. It is not all that difficult to prepare a program that will encompass a small set of differentiation rules, for example. The programming could be done in any language that handles ASCII variables (BASIC), but the task would be made easier by taking the time to learn one of the newer list-oriented (LISP) or logic programming (PROLOG) languages. A major difficulty is coping with the mathematical subtleties of expressions that seem simple $(1/x)$, but that behave differently depending on whether x is real or complex, and on whether the domain includes the origin. This is not a reasonable approach.

There are now many commercial "equation solving" software packages for the personal computer. The prospective customer should be aware that these come in either symbolic or numerical flavors. Although this is not a product review, it is appropriate to list the more readily available programs whose emphasis is symbolic.[2] These would be Macsyma (Symbolics Inc., Burlington, MA), Mathematica (Wolfram Research Inc., Champaign, IL), Maple (Waterloo Maple Software, Waterloo, Ontario, Canada), and Derive (Soft Warehouse, Honolulu, HI). The system requirements range from an 8088 with 512 KByte RAM (1 floppy) to an 80386 with 4 MBytes of memory and 30 MBytes of disk space (code and work area), with prices that span a factor of 10.

The actual performance capabilities of these programs do not differ so markedly in terms of the mathematics that they can do. The differences come in the size of the problem that can be solved, in the quality and speed of the graphics routines, and in their ability to recognize the presence

---

[2] New versions of these programs with augmented features appear in a steady stream, but two relatively recent comparative product reviews can be found in P. Coffee, K. Moser, and J. Frentzen, *PC Week* **October 30** (1989); K. R. Foster and H. B. Bau, *Science* **243,** 679 (1989).

of a numeric coprocessor (80x87) for faster (approximate) numerical calculations.

Intermediate stages in evaluating expressions can make astonishing demands on computer memory. Programs that can access the high memory of 32 bit personal computers will naturally perform better, but a workstation is probably a more reasonable platform for the middle-level problem. The prospective user with a truly large problem should explore the availability of mainframe versions at the nearest computing center.

Most of these programs are not intrinsically numerical, although they all provide some capability for numerical evaluation in approximate or "real" mode. Generating two- or three-dimensional graphics images is another numerically intensive activity. Programs that can exploit the presence of a numerical coprocessor (80x87) perform these tasks much more quickly. The symbolic operations use integer arithmetic and are limited by the CPU clock speed.

Other kinds of programs, although outside the topic of this chapter, might be more appropriate to the needs of some readers. The number of numerical equation solvers is growing rapidly. These programs provide quick and accurate numerical solutions without requiring any programming effort. Well-known examples would be MathCAD (MathSoft Inc., Cambridge, MA) and TK Solver Plus, but the list could be very long. Other readers might be interested in programs such as Automated Programmer (KGK Automated Systems, Hartsdale, NY), which specialize in converting a mathematical description of a problem to computer code (FORTRAN). (The symbolic programs referred to previously all provide some capabilities for both of these actions.)

What Can One Do?

The most profitable application of symbolic computing is to eliminate the "grunt work" from a problem. The level at which mathematics is tedium depends on the mathematical sophistication of the user, of course. But who dares to expand a 4 × 4 determinant without rechecking all the signs? Roughly speaking, the available software includes a decent undergraduate mathematics curriculum: two years of calculus, including vector calculus and the special functions, linear algebra (vectors, matrices, determinants, eigenvalues), and some ordinary differential equations (the latter being variable in content). Set theory, group theory, topology, graph theory, and the like are not included. It should be mentioned that this software is not a particularly good way to learn new mathematics, and even familiar mathematics can produce surprises. For example, $\ln(x^2 - x) - \ln(x)$ is not the same as $\ln(x - 1)$ if it is possible for x to take a negative real value or if x is complex, even though the logarithm function is well defined in these circumstances.

These considerations make it difficult to present concrete examples oriented to the needs of the biochemist. The simple problems have well-known simple solutions, and being able to develop them with a computer program is not particularly impressive. The complicated problems (which can be as simple as the competitive binding of two ligands to an enzyme) have complicated solutions (the roots of a cubic), and publishing such a solution is neither compelling nor informative. (However, one can determine quickly which of the solutions should be used.) Mathematically impressive demonstrations tend to be remote from biochemistry and can appear to be contrived.

A reasonable compromise is provided by the supposition of an experiment that monitors the formation of a complex PX between a protein P and a ligand X. Suppose that it is desired to study the sensitivity of complex formation to some thermodynamic intensive variable (T or p) in order to determine the conjugate extensive variable (ΔH or ΔV). That is, how does $\partial \ln([PX])/\partial T$ relate to ΔH? Because $\partial \ln(K)/\partial T$ is $\Delta H/RT^2$, the problem is to determine $\partial \ln([PX])/\partial \ln(K)$, but the answer needs to be expressed in terms of the stoichiometric concentrations $[P]_t$ and $[X]_t$.

The task here is the paradoxical one of trying to explain in detail that a procedure is very simple. Figure 1 shows a transcript of a session with the Derive program, although any commercial package could certainly handle a problem this simple. The meaning of the individual expressions should be clear, as should the sequence of operations necessary to get from one expression to the next. There might be some question as to whether expression #5 is the correct root of expression #4, and some readers might want to convince themselves that expressions #7 and #8 are truly equivalent. There now follows a complete step-by-step description of the actions required to produce this transcript. It ends up being a partial explanation of the user interface for the program, but the purpose is to show that it is sufficient to know what needs to be done; one need not personally do the intervening steps to reach the answer.

Procedure

The conservation of mass relations (#1 and #2) are first entered. This is done by selecting Author (as a verb) from a menu in order to activate an edit line, typing the text pt = p + k p x, followed by Enter ($\hookleftarrow$), and repeating the sequence for the other expression. Expressions #1 and #2 appear in the work window. The program understands that adjacent variables without an intervening explicit operator ($+$, $-$, $*$, $/$, or $^\wedge$) are to

1: $\quad$ pt = p + k p x

2: $\quad$ xt = x + k p x

3: $\quad$ $x = \dfrac{xt}{k\ p\ +\ 1}$

4: $\quad$ $pt = p + k\ p\ \dfrac{xt}{k\ p\ +\ 1}$

5: $\quad$ $p = \dfrac{\sqrt{(k^2\ (pt\ -\ xt)^2\ +\ 2\ k\ (pt\ +\ xt)\ +\ 1)}\ +\ k\ (pt\ -\ xt)\ -\ 1}{2\ k}$

6: $\quad$ $pt - p = -\ \dfrac{\sqrt{(k^2\ (pt\ -\ xt)^2\ +\ 2\ k\ (pt\ +\ xt)\ +\ 1)}\ -\ k\ (pt\ +\ xt)\ -\ 1}{2\ k}$

7: $\quad$ $k\ \dfrac{d}{dk}\ LN\ \left[\ -\ \dfrac{\sqrt{(k^2\ (pt\ -\ xt)^2\ +\ 2\ k\ (pt\ +\ xt)\ +\ 1)}\ -\ k\ (pt\ +\ xt)\ -\ 1}{2\ k}\ \right]$

8: $\quad$ $\dfrac{1}{\sqrt{(k^2\ (pt\ -\ xt)^2\ +\ 2\ k\ (pt\ +\ xt)\ +\ 1)}}$

FIG. 1. Transcript of a session with the Derive program, showing the solution to a simple binding problem. Expressions 1 and 2 are the conservation of mass relations for protein and ligand. The final expression is $\partial \ln([PX])/\partial \ln(K)$ in terms of the stoichiometric concentrations P_t and X_t. The entire session took less than 5 min.

be multiplied. (This comprises most of the typing of expressions. The remaining steps are done by selecting highlighted expressions or subexpressions, in a fashion entirely analogous to word processing, and by executing commands from a menu. The commands appear here with their identifying capital letter.)

The expression for total ligand (#2) is solved for free ligand (#3) by choosing soLve from the menu, accepting #2 for the expression number and accepting the default variable x as the one to be solved for. The result is substituted into the expression for total protein (#1) to give an expression involving only free protein (#4) by choosing Manage-Substitute from the menu, responding with a #1 to the prompt for expression number. The program steps through the variables, and at x move the highlight to the right side of expression #3 with the cursor arrows and press the function key F3, copying that subexpression into the edit line as a replacement for x. After the final Enter, expression #4 appears in the work window. This relation is then soLved for free protein (p) and expression #5 appears, along with the conjugate root. (The extraneous root is quickly identified and rejected with a Manage-Substitute of 0 for Xt and Simplifying to a nonphysical result. The correct root Simplifies to p = pt. The scratch work is Removed.) Subtracting free protein from total (choose Simplify and enter pt − #5 at the prompt for expression) gives the amount of complex (#6). The Derive program does not permit $dY(x)/d \ln(x)$ at the command level, although there is a utility function for implicit differentiation.

Expression #7 was created by highlighting the right side of expression #6, choosing Author, and typing k dif(ln(F3), k) (first-order is the default). The desired derivative is formed (#7) and then Simplified (#8). This last reduction is the most time-consuming step, taking about 65 sec on an 80286, about 145 sec on an 8088. [Just as in doing this problem by hand, the work is accelerated if pt and xt are replaced by variables representing (xt + pt) and (xt − pt).] Although this problem is only at the level of introductory calculus, simplifying the derivative by hand can be tedious. For cosmetic purposes I went back and executed Factor-Rational commands on the subexpressions under the radicals in expressions 5, 6, 7, and 8. (One can scroll through the entire work window to review previous work, delete expressions, rearrange the sequence of expressions, or select a subexpression for a current task.) A Transfer-Print-File command saved the material for printing later. The entire exercise took less than 5 min, much less time than it took to describe the steps.

Perhaps the first documented use of a symbol-manipulating program in a biochemical context was when Stafford and Yphantis[3] expressed the concentration dependence of the weight average molecular weight for an ideal self-associating system as a virial expansion in total concentration. The problem and its answer are deceptively simple; the intermediate stages are agonizingly complicated. The desired expansion is

$$1/M_w(c) = 1/M_1 + 2B_2c + 3B_3c^2 + 4B_4c^3 + \cdots \tag{1}$$

where M_1 is the molecular weight of the monomer and the B_j are the desired virial coefficients. For a monomer–N-mer association the weight-average molecular weight is

$$M_w = M_1(1 - \alpha) + NM_1\alpha \tag{2}$$

where α is the weight fraction of polymer. This can be rearranged to

[3] W. F. Stafford III and D. A. Yphantis, *Biophys. J.* **12,** 1359 (1972).

$$f = M_1/M_w = 1/[1 + (N - 1)\alpha] \tag{3}$$

Formally, Eq. (3) can be represented as

$$f = \Sigma_j \, d^j f/dc^j \, c^j/j! \tag{4}$$

and the coefficient of c^j in the power series will be $(j + 1)B_{j+1}M_1$. The difficulty arises because f is not known as an explicit function of c. The mass-action expression

$$K = \alpha/[(1 - \alpha)^N c^{N-1}] \tag{5}$$

relates α and c, but this cannot be solved without knowledge of N. Instead, Eq. (3) is solved for α, and the resulting expression is substituted into Eq. (5). After introducing a new independent variable $q = Kc^{N-1}$, the result simplifies to

$$q = (1/f - 1)(N - 1)^{N-1}/(N - 1/f)^N \tag{6}$$

The necessary derivatives are then developed by implicit differentiation: Eq. (6) is differentiated with respect to q, and the result is solved for df/dq in terms of f and q. Higher-order derivatives are obtained by differentiating the current result, but it is necessary to keep substituting the expression for df/dq in order to keep the result in terms of f and q. The desired expressions are then found by taking the limit as $q \rightarrow 0$ ($f \rightarrow 1$).

The implicit expressions for the derivatives grow explosively with increasing order. The solution was found by using the mainframe program FORMAC (REDUCE, ALTRAN, SCRATCHPAD, SMP, and MACSYMA are similar). It is simply

$$1/M_w = 1/M_1\{1 + \Sigma_j(-q)^j(jN - 1)!/j! \, [j(N- 1) - 1]!\} \tag{7}$$

Although this result is significant to the ultracentrifugal analysis of self-associating systems, as discussed in the original work, it is probably fair to say that the work would not have been done without the aid of the program for computer mathematics. To indicate how much things have changed in the 20 years since the publication of this work, the result was checked with the Derive program, using a supplied utility function for implicit differentiation and a built-in limit command. Determining the first three coefficients of the series took about 20 min. Because evaluating the third derivative took most of the time (15 min), higher-order terms were not attempted, but it is likely that they would exceed the memory capacity of the personal computer. However, the first three derivatives are ample to infer the pattern. If this were a new problem, one would want to predict the fourth derivative and then test that prediction. This could be done. The memory demands are particularly severe because the program uses a recursive algorithm, generating all of the previous solutions and keeping them in memory simultaneously until it can perform the final simplification. The approach would be to single-step through the algorithm manually (but letting the program do all the real work), saving intermediate expressions to disk so that the current task can be performed in available memory. The intermediate expressions would be individually recalled and reduced until the final expression could be constructed. The limit operation could be introduced at several intermediate points, greatly reducing the size of the expressions. This procedure might take a couple of hours, a reasonable investment for a real and significant problem.

Another kind of activity where these programs are particularly valuable is in introducing nonidealities or perturbations into a derivation in a "what-if" fashion. The effects can be propagated through the problem to see if the perturbations are damped or amplified. The physical sources of the changes remain identifiable. Alternatively, it is sometimes possible to identify parts of a complete description of a system that contribute significantly to the complexity of the problem but that are always numerically small for the interesting range of parameter values. Eliminating the inconsequential details can lead to a better understanding of the problem.

Conclusion

One motivation for writing this chapter was the perception that few biochemists use this technology or are even aware that it exists. The lack of familiarity produces an underutilization of a powerful set of tools. Another factor may be resistance to the program's knowledge: "I can perform that operation without the aid of a computer." But how many of us are willing to do long division, much less extract a square root, without reaching for the calculator?

This kind of software places an extraordinary amount of mathematical knowledge on the desktop (or laptop). Whatever the limitations of the programs, none of them are toys. The tasks that they are capable of doing are performed accurately and amazingly rapidly on the human time scale, Use of this software is not only a great timesaver to anyone doing mathematics (for love or necessity); it also permits a concentration on the mathematics itself rather than on the mechanical manipulations. The capabilities of the least expensive of these programs are sufficiently impressive that one could seriously consider purchasing it solely to determine what more advanced features might be desirable. Because a major use of programs for symbol manipulation is within the field of computer science for the purpose of algorithmically developing algorithms, it is impossible to predict the pace at which more mathematical knowledge will be embedded in these packages.

Acknowledgments

This work has been supported by a grant from the Office of Naval Research.

[29] Artificial Neural Networks

By W. T. KATZ, J. W. SNELL, and M. B. MERICKEL

Introduction

Historically, artificial neural networks have been studied for years in the hope of solving complex real-world problems with humanlike performance. Some appreciation of the magnitude of this task can be obtained by considering the "simple" household fly.[1] The fly can simultaneously process information from multiple sensors and make complex decisions involving the coordination of a myriad of motor tasks, such as avoiding your fly swatter as it converges on your picnic lunch. This is a particularly impressive task since the neurons of a fly's nervous system have a frequency response of approximately 100 Hz which is 100,000 times slower than the microprocessor components in a home computer. Even today's supercomputers are unable to effectively solve relatively "simple" problems such as the fly scenario presented above. The reasons that biological neural networks have such impressive performance are just beginning to be understood.

Modern digital computers built with traditional designs have a fundamental limitation, the so-called von Neumann bottleneck. Traditional computation requires a problem to be broken down to a set of operations which are performed in serial fashion, that is, one instruction at a time. Typically, each instruction must be completed before the next instruction is executed.

Artificial neural networks represent a fundamentally different approach to computation. They are explicitly designed to mimic the basic organizational features of biological nervous systems: parallel, distributed processing. It is not surprising then that artificial neural networks (ANNs) have also been called parallel distributed processing, connectionist, and neuromorphic systems. ANNs consist of a large number of simple inter-

---

[1] J. F. Shepanski, "Quest Technology Report," p. 19. TRW Space and Defense Sector, Winter 1987–1988.

connected processing elements, where the processing elements are simplified models of neurons and the interconnections between the processing elements are simplified models of the synapses between neurons. Each processing unit or "neuron" can process some piece of information at the same time as other units. The processing of information in such networks therefore occurs in parallel and is distributed throughout each unit composing the network. This approach allows networks of relatively slow, simple processing elements to solve complex, difficult problems with inexact solutions. The rationale behind such a computational model stems in part from the desire to have computers deal gracefully with various real-world problems, namely, situations which require perception and "common sense," two stumbling blocks of the traditional, symbolic approaches to artificial intelligence.

Interest in ANNs dates back to at least 50 years ago with some of the early work of investigators in neuroanatomy, neurophysiology, and psychology who were interested in developing models of human learning. An important early model of the biological neuron was proposed in 1943 by McCulloch and Pitts.[2] This McCulloch–Pitts "neuron" is a relatively simple model which assumes the output of the neuron to be binary (i.e., all-or-none) and due to the combined action of inhibitory and excitatory inputs. In this model, the action of inhibitory inputs is absolute such that any inhibitory input completely inhibits the firing of a neuron. In the absence of inhibitory input, the neuron adds all of its excitatory inputs and compares the sum to a threshold to determine whether it should fire.

The development of a learning rule which could be used for neural models was pioneered by D. O. Hebb, who proposed the now famous Hebbian model for synaptic modification.[3] This model basically states that the connection (i.e., the synapse) between two neurons is strengthened if one neuron repeatedly participates in firing the other. This Hebbian synaptic modification rule does not express a quantitative relationship between pre- and postsynaptic neurons, and therefore many alternative quantitative interpretations have been developed. However, the Hebbian model for synaptic modification remains important to this day and serves as the reference point for all other learning rules.

Rather than tersely cover the breadth of ANN models developed after Hebb's seminal work, this chapter concentrates on two classes of widely used artificial neural networks: the perceptron–back-propagation and the Hopfield–Boltzmann machine models. First, the characteristics of a simple feedforward ANN model is explored in more detail. Then, the per-

[2] W. S. McCulloch and W. Pitts, *Bull. Math. Biophys.* **5,** 115 (1943).
[3] D. O. Hebb, "The Organization of Behavior." Wiley, New York, 1949.

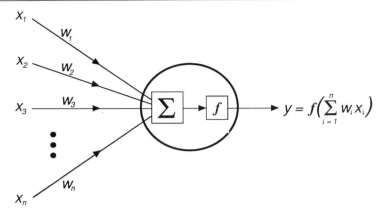

FIG. 1. Typical "neuron" or processing unit in an artificial neural network.

ceptron–back-propagation model is presented in an intuitive applications-oriented style. The chapter concludes with a description of the Hopfield–Boltzmann machine ANN models.

Basic Artificial Neural Network Model

In Fig. 1, we show the basic structure of the simple processing element or "neuron" in the artificial neural network. The processing unit receives some number of input signals, $x_1, \ldots, x_n$, through weighted links, sums the weighted inputs, and then passes the resulting sum or activation level through an output function f. The weights on the input lines, $w_1, \ldots, w_n$, represent the strength of the connections to a unit, and learning rules (such as the Hebbian rule and the back-propagation algorithm) alter these weights in order to create a desired input/output response from an artificial neural network. In other words, the "knowledge" or functionality of an ANN is encoded in the values of its weights.

In many ANNs, the processing units are arranged in layers (Fig. 2). The first layer receives a number of input signals and produces some output which is then fed to the next layer of processing elements, and so on. The input signals constitute some input vector **x,** whereas the resulting signals from the final or output layer form an output vector **y.** The cascade of layers can be thought of as a black box which maps input vectors to output vectors.

In supervised ANN models, a desired mapping can be obtained by presenting the ANN with training samples, that is, providing the desired output vector $\mathbf{y}_d$ for a given input vector **x.** The ANN then computes

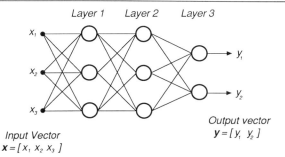

FIG. 2. Artificial neural network with its processing units divided into layers.

some measure of the error between the actual and desired output, using a learning algorithm to adjust the weights on the interconnections to reduce the error.

Self-organizing ANN models are unsupervised in the sense that no training samples need be provided; the mapping is created after presentation of input vectors only. Therefore, a self-organizing ANN produces similar output vectors when given similar input vectors, the interpretation of "similarity" varying with the particular ANN model.

ANN models differ in the manner in which they adjust their weighted interconnections (the learning algorithm), the processing performed by the individual "neurons," and the overall architecture of processing unit interconnection. At the level of the individual processing element, three possible output functions are shown in Fig. 3. The first is linear while the last two are the nonlinear step and sigmoid functions. As mentioned before, each processing unit takes a weighted sum of the input signals and passes this value through the output function. It can be shown that if a linear output function is used, a single layer can be constructed to have

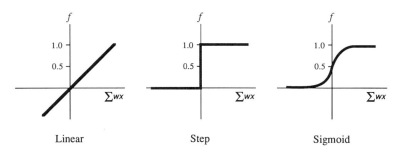

FIG. 3. Linear, step, and sigmoid output functions. The output is plotted against the net input $\Sigma\ wx$ to a processing unit.

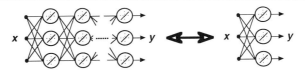

FIG. 4. If units with linear output functions are used, multilayer ANNs can be replaced with single layer ANNs.

the same mapping effect as any number of cascaded layers (Fig. 4). Consequently, in order to benefit from additional layers, ANN models usually have nonlinear output functions. Two of the most popular are the step and the sigmoid functions.

The Perceptron

Classification Ability of the Perceptron

In the late 1950s, Frank Rosenblatt introduced a neural network model called the perceptron,[4] a name which is used for the individual units as well as the overall layered network of units (Fig. 5). The perceptron follows the basic model described above; it accepts inputs through weighted links, sums the inputs, and passes the sum through a step function. The perceptron also has a bias term θ which serves as a threshold; if the sum from the weighted inputs is greater than $-\theta$, the unit outputs a "1," otherwise it outputs a "0." One way of implementing this bias term is to use an additional constant input 1 with its corresponding weight, w_0, set to θ.

[4] F. Rosenblatt, *Psychol. Rev.* **65,** 386 (1958).

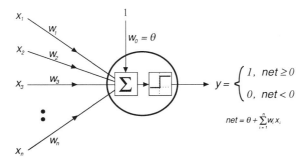

FIG. 5. Perceptron processing unit. Because it uses a hard-limiting step output function, the perceptron gives binary output, "0" or "1," depending on the values of the weighted inputs and the threshold term θ.

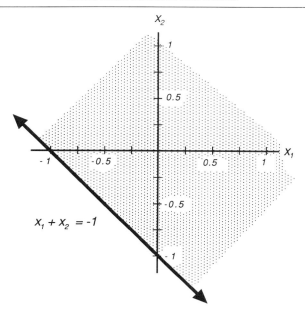

FIG. 6. Decision region of a two-input perceptron. The shaded area, the region above the line $x_1 + x_2 = -1$, describes those input values for which the perceptron is "on" (i.e., gives a "1" output).

Despite the simplicity of the model, a perceptron network was shown to be capable of recognizing simple characters as well as other interesting patterns. To get a more intuitive feel for what the perceptron is computing, we will be looking at its processing using geometry. For example, consider a simple case of two inputs and the bias term. The input forms a two-dimensional input vector, and the space of possible input vectors (the input or feature space) can be shown on a two-dimensional graph. If we map the area for which the perceptron outputs a "1," we find that the border of this "on" area (the decision region) is formed by a line (the decision boundary) described by the equation $x_0 w_0 + x_1 w_1 = -\theta$. By varying the values of the weights w_0 and w_1, we can move the decision boundary and partition the two-dimensional input space into any two parts as long as the parts are linearly separable. And by modifying the sign of the weights, we can choose which side of the decision boundary forms the "on" area or decision region.

A simple example is shown in Fig. 6. We have chosen $w_0 = 1$ and $w_1 = 1$ with $\theta = 1$. The decision boundary of this perceptron is a line which runs through $(0, -1)$ and $(-1, 0)$; the "on" area is the region above the

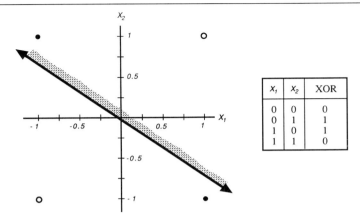

FIG. 7. The XOR problem. Given two inputs, the perceptron must be "on" if the two inputs are not identical (filled circles) and "off" if the two inputs are identical (empty circles). However, as can be seen, there is no orientation of the decision line which will separate the filled and empty circles.

line. To reverse the labeling, that is, to make the "on" area the region below the line, we only have to switch the signs of the weights and bias so that $w_0 = -1$ and $w_1 = -1$ with $\theta = -1$.

The classic XOR problem shows the limitation of a single perceptron (Fig. 7). The XOR (exclusive-or) function is a simple logical function which returns a "1" if the two inputs are not identical and a "0" if they are identical. As can be seen from simple inspection of the input space, the required mapping cannot be produced by any orientation of a single decision line, that is, it is linearly inseparable. Therefore, the XOR function cannot be implemented by a single perceptron.

If more input signals are allowed, the input vector and the corresponding feature space grow in dimensionality. For example, if we have a perceptron with three inputs, the resulting decision boundary is a plane in the three-dimensional feature space. For a perceptron with four or more inputs, the decision regions are bounded by a n-dimensional hyperplane which splits the feature hyperspace. But we return to the two-dimensional case to visualize the mapping ability of multilayer perceptrons.

Figure 8 shows some of the decision regions that can be formed by using just three perceptrons in two layers. The first layer consists of two perceptrons which partition the feature space using two decision lines. Each perceptron in this first layer divides the feature space into an "on" area and an "off" area. The weights of the final perceptron can be set so the unit emulates any of a number of logical functions, thereby allowing

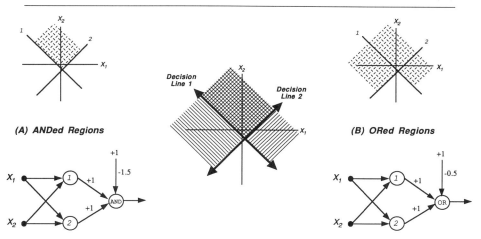

FIG. 8. Use of a two-layer perceptron with two-dimensional input vectors $[x_1 \, x_2]$. Each of the two units in the first layer (numbered *1* and *2*) partition the input space into two parts (see Fig. 6). The single unit in the second layer combines these resultant decision regions depending on the connection weights between the first and second layer. (A) The final unit implements an "AND" function, and the final output of the two-layer perceptron is the intersection between the two "on" areas of the first-layer units. (B) The final unit implements an "OR" function, and the final decision region is the union of the two "on" areas of the first-layer units.

the combination of decision regions resulting from the first layer. Thus, the intersection of the two "on" areas can be obtained by setting the weights so that the final perceptron acts as an "AND" unit. Alternatively, the union of the "on" areas can be found by using the perceptron as an "OR" unit.

By increasing the number of units in the first layer, we can add edges to our decision boundary and create more complex decision regions. In fact, we can come arbitrarily close to making a decision region for any convex and many concave connected areas, bounded or unbounded. Figure 9 provides an example of how a complex nonconvex decision region can be formed using a two-layer perceptron. The six units in the first layer divide the two-dimensional feature space into 19 different regions with their six decision lines. The perceptron in the final layer selects those areas which lie in at least four "on" regions. The resulting decision region is quite complex despite the use of only 7 perceptrons.

The next step in our geometric analysis is to add a final layer to form a three-layer perceptron. The perceptron in the final layer can combine the results from several units in the second layer, and in so doing, extend a decision region to incorporate disjoint areas of arbitrary shape in the

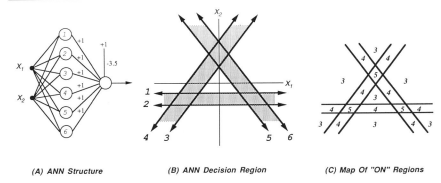

(A) ANN Structure   (B) ANN Decision Region   (C) Map Of "ON" Regions

Fig. 9. Formation of a complex nonconvex decision region using a two-layer perceptron with six units in the first layer and a single second layer unit. (A) The perceptron architecture. (B) The output decision region (shaded). Each numbered line corresponds to the decision boundary for the identically numbered unit in the first layer. Even-numbered units have their "on" regions covering the origin, whereas odd-numbered units have their "on" regions facing away from the origin. (C) The input space is divided into 19 regions by the six decision boundaries. The total number of first-layer units which are "on" in each region is shown. Note that the output decision region in (B) corresponds to regions which have four or more "on" units.

input space. In the example shown in Fig. 10, a triangular donut is formed by subtracting a smaller triangular area from a larger region.

There are two conclusions which can be drawn from our geometric analysis of the simple two-dimensional case. First, three layers are sufficient to represent any decision regions, provided they consist of a finite number of disjoint areas.[5] Second, *a priori* information regarding the complexity of the desired decision region can directly influence the required number of units in each network layer. For example, if we are using two-dimensional input vectors, and it is known that the desired decision region consists of two disjoint areas, we will probably need at least six units in the first layer and two units in the second layer. Closed boundaries in two-dimensional space require a minimum of three sides (a triangle); therefore, two sets (for the two disjoint areas) of three units are required in the first layer. An additional two units are required in the second layer to combine (via "AND," "OR," or other boolean functions) the first layer results and create the disjoint decision regions. This example can be extrapolated to a general heuristic: a minimum of $m(n + 1)$ first layer perceptrons and m second layer perceptrons are needed to construct a decision region encompassing m disjoint areas in a n-dimensional feature space.

[5] G. J. Gibson and C. F. N. Cowan, *Proc. IEEE* **78,** 1590 (1990).

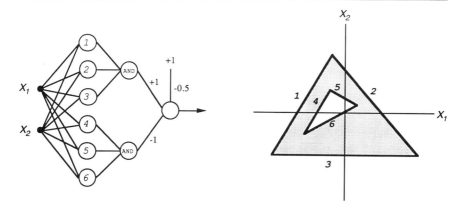

(A) ANN Structure   **(B) ANN Decision Region**

Fig. 10. Decision region formed by a three-layer perceptron. (A) The six first-layer units form six decision lines in the input space. The first three (1–3) form the sides of the outer triangle, and the second three (4–6) form the sides of the smaller inner triangle. The "on" areas for all six units face the interior of the triangles, so the "AND" units in the second layer form triangular decision regions. The single third-layer unit subtracts the inner triangle from the outer triangle to form the final triangular donut decision region. (B) The final decision region (shaded) is shown with the numbered edges corresponding to the decision lines of identically numbered first-layer units shown in (A).

Perceptron Learning Algorithm

In the previous section, we analyzed the types of decision regions which could be represented by multilayer perceptrons. Although a three-layer perceptron can generate arbitrarily complex decision regions, in the 1950s there were no known ways in which to adaptively set the weights in these multilayer networks given a training set of input vectors and the corresponding desired outputs.

There were, however, methods for learning the weights for a one-layer perceptron. Rosenblatt published the perceptron learning theorem which gave both a learning algorithm and a proof that any decision region that could be represented by a single layer perceptron could be learned using his learning algorithm. As we have seen in the section above, single-layer perceptrons can represent (and therefore learn by the perceptron learning theorem) any linearly separable decision regions. Unfortunately, if presented with nonlinearly separable problems, the perceptron learning algorithm may continue to oscillate between the decision boundaries indefinitely.

Shortly after Rosenblatt's perceptron learning rule was presented, Widrow and Hoff independently introduced the least mean squares (LMS) or Widrow–Hoff delta (δ) rule.[6,7] This algorithm attempts to minimize the mean squared error between the desired and actual output. Whereas the LMS method is not guaranteed to separate linearly separable classes, it will converge on reasonable decision boundaries in linearly inseparable problems.

LMS or Widrow–Hoff Delta Rule

0. Initialize the weights and bias terms to small random values.
1. Compute the net input, $net = \Sigma\, wx$, to each perceptron unit using some input vector $\mathbf{x}$ from the training set.
2. Calculate the error, $\delta = d - net$, where d is the desired output.
3. Adjust each of the weights of the unit in proportion to the error and the input signal coming in over that line. The change in weight w_i is given by the equation $\Delta w_i = \alpha \delta x_i$ where α is a learning rate or gain term (usually set to the range $0.1 < \alpha < 1.0$) controlling the stability and speed of convergence to the correct weight values.
4. Go to Step 1.

The key difference between the perceptron and LMS training algorithms is in Step 1: the perceptron learning algorithm passes the weighted sum of inputs, net, through the output step function so that $\delta = d - y$ where $y = f(net)$ as in Fig. 5. In other words, the perceptron learning algorithm uses the actual unit output y, whereas the LMS algorithm uses the net input in the derivation of the error term. This means that the adaptation of weights can continue ($\Delta w_i \neq 0$) in the LMS case even if the actual output agrees with the desired output. In human learning, such continual adaptation is typical as in the case of medical problem solving. For example, first-year medical students may correctly make a diagnosis ($y = d$) although they lack confidence in their judgment ($\delta \neq 0$). On the other hand, trained clinicians with repeated exposure to a number of similar cases can make the same diagnosis more rapidly and with more certainty (smaller δ).

The perceptron and LMS learning procedures work for a single layer perceptron. But what about the multilayer perceptron? In this case, the weights for the final layer can be modified because the error δ between the final layer and the desired outputs can still be computed. But the adjustment for weights in "hidden" layers (i.e., all layers but the final one) are

[6] B. Widrow and M. E. Hoff, "1960 IRE WESCON Convention Record," p. 96. IRE, New York, 1960.
[7] B. Widrow and M. A. Lehr, *Proc. IEEE* **78**, 1415 (1990).

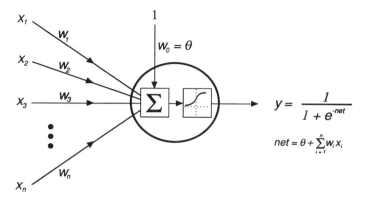

FIG. 11. "Neuron" or processing unit in the back-propagation ANN.

not so easily computed since there is no direct error measurement δ between a hidden layer units and the given desired output.

Back-propagation

Network and Geometry of Decision Regions

The training of multilayer neural networks proved to be a thorny problem. Without the prerequisite learning algorithm, training was limited to single layer perceptrons with all of the inherent limitations in representing only linearly separable decision regions. It was no surprise, then, that the introduction of the back-propagation algorithm in the 1980s proved a great boon to the fledgling artificial neural network movement. Back-propagation or "backprop" is a simple, easily implemented training algorithm for multilayer feedforward networks.[8] The units in the backprop network are identical to perceptrons except for the replacement of the step output function with a sigmoid function (Fig. 11). The sigmoid is used because the error measurements require the output function to be differentiable.

The conclusions obtained from the geometric analysis of the perceptron can be extended to the back-propagation ANN model. The use of the sigmoid as the output function complicates the geometric analysis, but the overall results are similar. Instead of lines, planes, and hyperplanes forming the decision boundaries, the sigmoid output function generates curves,

[8] D. E. Rumelhart, G. E. Hinton, and R. J. Williams, "Parallel Distributed Processing," p. 318. MIT Press, Cambridge, Massachusetts, 1986.

curved surfaces, and hypersurfaces which form the decision boundaries. One caveat should be given regarding the theoretical analysis of minimum ANN requirements. As mentioned in the discussion of perceptrons, a three-layer ANN can represent any arbitrarily complex decision region. However, it may be advisable to use more than three layers, because in practice, the size (number of hidden units) and training time requirements of a minimal three-layer network may be much greater than that of a network consisting of more layers.

Back-propagation Learning Algorithm

Back-propagation Learning Algorithm

0. Initialize the weights and bias terms to small random values (e.g., −0.5 to 0.5).
1. Compute the actual output vector **y** by propagating some input vector **x** from the training set forward through each layer.
2. Start at the output layer and calculate the error $\delta_i = y_i(1 - y_i)$ $(d_u - y_i)$, where δ_i is the error term for output unit i, y_i is the output for unit i (the ith component of the output vector **y**), and d_i is the desired output for unit i. The $(d_i - y_i)$ term relates the magnitude of the error while the remaining terms are the derivative of the sigmoid output function.
3. Adjust each of the weights of the output units in proportion to the error and the input signal coming in over that line. The change in weight w_i is given by the equation $\Delta w_i = \alpha \delta_i x_i$ where α is a learning rate or gain term (usually set to the range $0.1 < \alpha < 1.0$) controlling the stability and speed of convergence to the correct weight values.
4. After completing the weight changes for the output layer, work backward layer by layer to the first hidden layer. At each layer:
4a. Calculate the error

$$\delta_i = x_i(1 - x_i) \sum_j \delta_j w_{ij}$$

where δ_i is the error term for unit i of the current layer, x_i is the output for unit i, δ_j is the error term for unit j in the layer after the current layer, and w_{ij} is the weight between unit i and unit j. The sum in the error term basically back-propagates the errors from the units j to which the current unit i is connected.
4b. Adjust the weights (as in Step 3) leading to unit i using the δ_i calculated in Step 4a.
5. Go to Step 1.

Note that the only difference between the back-propagation and LMS

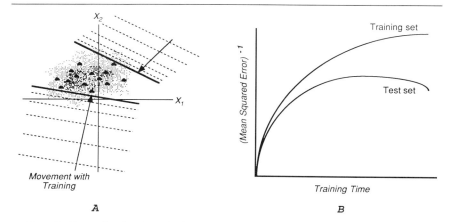

FIG. 12. Generalization degradation due to overtraining. (A) Decision lines move closer to training points (large triangles) during training. Possible test input points in the same class (small points) are distributed over a larger area. As the decision lines move closer to the training points, they may pass over the test input points. (B) The result is an eventual degradation of test point classification even though performance on the training set continues to rise.

learning algorithms is the definition of the error term δ. In LMS, the error term was simply the difference between the actual output and the sum of weighted inputs. In backprop, we have two different δ terms (Steps 2 and 4 above) depending on whether the unit is in a hidden layer or output layer.

The back-propagation procedure can require thousands of presentations of each input and desired output vector pair in the training set. To speed the process, a momentum term can be added to the weight change (Step 3). Then, the change in weight is $\Delta w_i = \alpha \delta_i x_i + \eta \Delta w_i(n - 1)$ where η is the momentum (usually less than 0.9) and $\Delta w_i(n - 1)$ is the weight change from the previous training input.

Overtraining

It is possible to overtrain an artificial neural network so that it generalizes poorly to novel input vectors. This problem can be seen geometrically (Fig. 12). The training set is a sample from the set of all possible input/output vector pairs. If training on the samples continues for too long, the ANN will move its decision boundaries arbitrarily close to the sample points. Also, if too many units are used in the network relative to the number of samples in the training set, the ANN may "memorize" the samples by forming tight, disjoint decision regions around each sample point. In either case, the resultant decision regions will exclude input vectors similar to those in the training set: generalization will be poor. One method of preventing overtraining is simply to halt training as soon as all training samples are correctly identified rather than trying to obtain zero error or training for a predetermined number of passes.

Error Surfaces

Learning algorithms attempt to find some set of weight values which produce the desired mapping (or, in geometric terms, represent the required decision regions). If we had n connection weights in a neural network, we could think of an $(n + 1)$-dimensional space with n axes corresponding to the different weights and one axis representing the overall error between desired and actual outputs.[9] In this space, we could describe an error surface giving the error for each possible combination of weights. The back-propagation learning algorithm implements a form of gradient descent on this error surface. It changes the weight values in such a manner as to follow the error surface slope downward to a minimum. Usually, the error surface is highly convoluted with many areas of shallow slope (owing to the small effect of weight changes when unit outputs are very large), pockets of varying depth (local minima), and possibly many equally deep holes (global minima). By following the steepest descent, it is possible that back-propagation could get stuck in one of the local minima or progress with extreme slowness through relatively flat terrain. Although the latter scenario (slow convergence) may be prevalent in many real-world applications, the local minima problem is rarely encountered, and a good minimum, if not a global minima, is usually discovered. Extensions of the backprop learning algorithm have been developed specifically to combat the slowness of training.[10]

Back-propagation Applications

The back-propagation algorithm has been applied extensively, covering a gamut of problems from backgammon playing to medical diagnosis. In each case, the researcher has (1) chosen a suitable encoding scheme for the problem at hand, (2) specified the number of units and layers as well as the interconnection pattern, and (3) acquired a good training set.

The first task is to choose a suitable encoding scheme so that one can represent the requisite information in a form suitable for ANN processing. Therefore, we need to transform our inputs and desired responses into

[9] R. Hecht-Nielsen, "Neurocomputing," p. 128. Addison-Wesley, Reading, Massachusetts, 1990.
[10] R. A. Jacobs, *Neural Networks* **1**, 295 (1988).

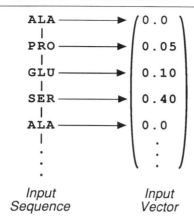

FIG. 13. One method for representing amino acid sequences as numerical vectors. Each residue position corresponds to one component of the input vector, and the types of amino acids are mapped to unique numbers (e.g., ALA = 0.0, PRO = 0.05).

numerical input and output vectors. For example, suppose we want to start with a sequence of amino acids and map this sequence into secondary structure information (α helix, β sheet, or coil). How would we represent the input (the sequence of amino acids) and the output (the secondary structure) in numerical terms? One possible input representation is to assign a number to each amino acid: alanine would be represented by 0.0, proline by 0.05, glutamate by 0.10, etc. Then, a single input value (i.e., one component of the input vector) could represent one amino acid position in the sequence (Fig. 13). Likewise, the secondary structure types could be assigned values so that α helix is represented by 0.0, β sheet by 0.5, and coil by 1.0. Using this type of encoding, the ANN would receive a n-dimensional input vector where n is the number of amino acids in the sequence to be input. The ANN output would be a single value corresponding to a type of secondary structure.

Although this encoding scheme transforms the training information to numerical vectors efficiently, it is likely that the ANN will be unable to perform a useful mapping because our input and output representations do not mirror the real world. For example, one would expect the biological properties of proline to be a similar to alanine since their encodings, 0.0 and 0.05, are very close numerically. In reality, proline is an atypical amino acid that tends to be an α-helix breaker. This property of proline will be difficult to encode using any such "analog" encoding scheme.

An alternative encoding scheme uses "local" representation, a form of binary representation. In this scheme, each component of the input vector represents a particular amino acid; therefore, if we only had to deal with 5 amino acids (instead of the 20 in nature), one position in the sequence could be represented by a five-dimensional input vector. For example, (1 0 0 0 0) could represent proline while (0 1 0 0 0) could represent alanine. This form of representation allows the ANN greater differentiation among the various amino acids. However, this type of encoding requires a $5n$-dimensional input vector where n is the number of amino acids in the sequence to be input. The same type of encoding can be used for the output vector. With local representation, we use a three-dimensional output vector (produced by 3 units in the output layer) to signify the secondary structure type. An output of (1 0 0) indicates α helix while the vector (0 1 0) indicates β sheet. The encoding scheme can greatly influence the success of an ANN application.[11]

NETtalk. In 1987, Sejnowski and Rosenberg described a back-propagation application which converted English text into intelligible speech.[12] This application demonstrated the power of ANNs and accelerated the use of the back-propagation algorithm in particular.

Sejnowski and Rosenberg organized their ANN into two layers, a hidden and output layer, that accepted a group of sequential letters and output the phonetic symbol corresponding to the central letter. The input and output data were encoded using local representation. By using additional equipment, the researchers were able to convert the ANN output phonetic symbols into sounds. Initially, the ANN "speech" is like baby babble, but as training progresses, words become properly enunciated.

Protein Secondary Structure Prediction. Shortly after the NETtalk publication, two independent groups, Qian and Sejnowski[13] and Holley and Karplus,[14] applied a similar ANN architecture to the problem of protein secondary structure prediction, replacing the input string of letters with amino acids and the output phonetic symbols with secondary structural types. The researchers obtained a training set of primary and secondary structure information derived from known protein structures listed in the Brookhaven National Laboratory Protein Data Bank. A test set of protein segments was created after removal of candidate proteins with homologies in the training set.

As shown in Fig. 14, the ANN receives an input vector representing a segment of primary amino acid sequence. Because the encoding scheme uses local representation, the input vector is a $21n$-dimensional vector

[11] P. J. B. Hancock, "Proceedings of the 1988 Connectionist Models Summer School," p. 11. Morgan Kaufmann, San Mateo, California, 1989.
[12] T. J. Sejnowski and R. R. Rosenberg, *Complex Syst.* **1**, 145 (1987).
[13] N. Qian and T. J. Sejnowski, *Mol. Biol.* **202**, 865 (1988).
[14] L. H. Holley and M. Karplus, *Proc. Natl. Acad. Sci. U.S.A.* **86**, 152 (1989).

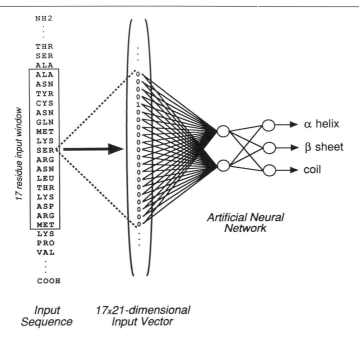

FIG. 14. Protein secondary structure prediction using an ANN. A small number of contiguous amino acids are transformed to an input vector. To reduce the complexity of the diagram, only the input vector components corresponding to the central residue position (SER in this case) are shown. The input vector is fed through the ANN, and the output units give the likelihood of the three secondary structure types.

where n is the number of positions (usually 13 to 17) in the amino acid segment. (There are 21 different inputs at each segment position: 20 amino acids and 1 spacer.) After propagating the input vector through the layers, the ANN outputs a vector giving the secondary structure type (α helix, β sheet, or coil) of the amino acid at the center of the input segment. This result is compared to the actual secondary structure determined through atomic coordinate analysis using the method of Kabsch and Sander.[15] Errors are used to correct the weights in the ANN via the back-propagation algorithm.

Both research groups had similar findings: predictive accuracy was approximately 63% on test proteins nonhomologous with proteins used in the training set. This result indicated that the ANN approach was more accurate than many other methods. An analysis of the trained ANN

[15] W. Kabsch and C. Sander, *Biopolymers* **22**, 2577 (1983).

showed that the network weights had indeed captured a number of real-world properties. For example, weights connecting "proline" components of the input vector with the "α helix" output unit were strongly negative, correctly indicating that proline tends to prevent α-helix formation. (Note that use of the encoding scheme in Fig. 13 would hinder the weight representation of the antihelix property of proline.)

Hopfield Networks and Boltzmann Machine

Introduction to Optimization Problem

There are many tasks in science and engineering which are referred to as optimization problems, that is, problems that have many valid solutions, but one or more of these solutions is considered to be best. The search for an optimum solution is often a difficult and time-consuming task. The game of chess is an example. If you are the white, there are a number of play sequences leading to the optimum solution: the eventual capture of the black king. But brute force evaluation of all valid moves is computationally intractable owing to the combinatorial explosion of possible lines of play.

Conventional approaches to solving optimization problems (e.g., random search or gradient descent techniques) are often slow and yield poor solutions. It is obvious that biological systems rapidly solve these types of problems with a high degree of success. Besides higher cognitive functions like game playing, many routine tasks can be considered optimization problems. Such tasks include the construction of depth field from two monocular scenes by the visual system, finding the proper orientation of the arm for grasping an object given a set of impeding obstacles, and the rapid recognition of a familiar face or object. The Hopfield and Boltzmann models are neurally inspired computational techniques which attempt to deal effectively with optimization problems.

Two-State Neuron Model

In 1982, John Hopfield showed that a network of densely interconnected two-state units or "neurons" can exhibit emergent collective computational properties.[16] These networks differ from the perceptron/backprop ANN models in that each neuron is potentially connected to every other neuron; therefore, cycles in the network may form. The resulting feedback endows the ANN with dynamic properties and may prevent the network from reaching a stable state.

---

[16] J. Hopfield, *Proc. Natl. Acad. Sci. U.S.A.* **79**, 2554 (1982).

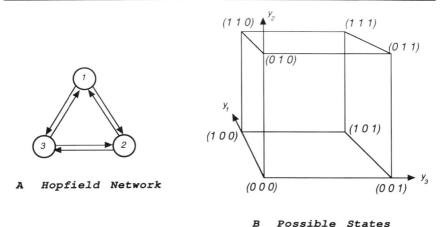

A   Hopfield   Network

B   Possible   States

FIG. 15. Three-unit Hopfield network and its possible output states.

Each unit in the Hopfield network is essentially a perceptron with zero threshold, $\theta = 0$ (see Fig. 5). It sums the weighted inputs and then outputs a "0" if the result is negative or "1" if the result is greater than or equal to zero. The network is updated asynchronously. Each neuron randomly and independently evaluates its inputs and updates its output signal. This is very different from the systematic layer-by-layer evaluation of units using the back-propagation algorithm.

At any time, the state of the network is defined by the collection of unit outputs, the output vector. Because each unit in the network can be in only two states, "0" or "1," the output vector can be thought to represent a vertex of a n-dimensional hypercube, where n is the number of units comprising the network. Figure 15 shows a Hopfield network with three units and the possible values of its output vector.

Initially, a Hopfield network is in a state corresponding to some vertex in the n-dimensional hypercube. As each unit evaluates its input and updates its output,the output vector can change by only one component value. Graphically, this means that the point on the hypercube corresponding to the network state can travel only one edge during a single unit update. As time elapses and the output vector changes, the network state can be described by a point moving along the edges of the hypercube.

One of the most interesting aspects of Hopfield's paper was its description of the energy (E) of the network state:

$$E = -\frac{1}{2}\sum_{j \neq i}\sum w_{ij}y_iy_j$$

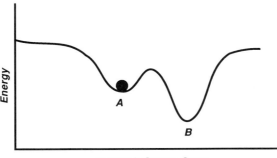

FIG. 16. "Energy" surface of a Hopfield network. The ball represents the output state of the network. As time progresses, the ANN moves from an initial state to either a local minimum (A) or a global minimum (B).

where w_{ij} is the connection weight between units i and j, and y_i is the output of unit i. The result of each unit update is to lower the overall energy of the network. This can be seen by considering the change in energy due to the change in the output of a single unit:

$$\Delta E = \Delta_{y_i} \sum_{j \neq i} w_{ij} y_j$$

The summation (net input) and Δy_i in the equation are either both positive or negative since the output y_i of a unit depends on the sign of the net input; therefore, the energy of the network state is always decreasing. Depending on the values of the connection weights, some vertices of the hypercube will have lower energy than other vertices. So as the state of the system evolves, the output vector will settle onto one of the vertices. This stable state is reached when all of the edges, along which the network state can move, lead to vertices with higher energy.

It has been shown that a network comprised of two-state units will eventually reach a stable, fixed state if the connection weights between any two units are identical (i.e., symmetric connections) and no unit excites or inhibits itself. The state which is eventually reached may be a local minimum instead of a global minimum (Fig. 16). Even though some vertex and its corresponding network state has the smallest energy value, the current network state may not be able to reach that global minimum because it first runs into a local minimum. The initial state of the network decides which of the minima the network will eventually settle into.

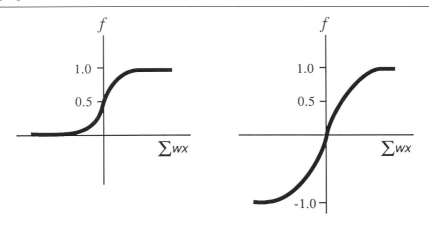

(A) Back-propagation   (B) Graded response
                                  Hopfield network

FIG. 17. The sigmoid used in back-propagation (A) allows output from 0 to 1 while the sigmoid used in graded response Hopfield networks (B) allows output from -1 to 1.

Graded Response Neurons

In 1984, Hopfield replaced the two-state units of the network with continuous output units like those in the back-propagation model.[17] The only difference between the back-propagation unit and Hopfield's graded response unit is the nature of the output sigmoid function (Fig. 17). In the Hopfield model, the output of the sigmoid changes sign when the net input equals zero. This adjustment is necessary to ensure that the energy of the network state always decreases.

The graded response of the units allows the output vector components to be continuous values. Therefore, the point corresponding to the network state can lie in the interior of the hypercube, and the constraint of traveling along the hypercube edges is removed. This freedom of movement makes it less likely (although not impossible) that the network state will fall into a local minimum.

Boltzmann Machine: Simulated Annealing

In an effort to avoid local minima, a number of researchers began to experiment with the technique of simulated annealing.[18,19] Simulated annealing describes a process similar to the annealing of metal. First, the

---

[17] J. Hopfield, *Proc. Natl. Acad. Sci. U.S.A.* **81,** 3088 (1984).
[18] S. Kirkpatrick, C. D. Gelatt, Jr., and M. P. Vecchi, *Science* **220,** 671 (1983).
[19] S. Geman and D. Geman, *IEEE Trans. Pattern Anal. Machine Intelligence* **6,** 721 (1984).

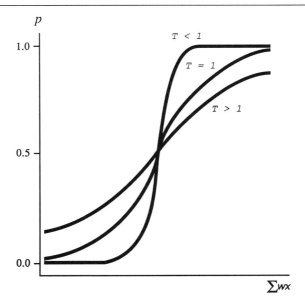

FIG. 18. Effect of temperature T on the probability that a unit in the Boltzmann machine will become active. As T approaches absolute zero, the sigmoid approaches a step function. At higher temperatures, the sigmoid becomes flatter, and net input becomes less of a factor in determining whether a unit becomes active.

metal is heated to high temperatures, agitating the atoms. Then, as the metal is slowly cooled, the atoms become less agitated and the metal eventually falls into a low energy state.

In the case of Hopfield networks, a temperature term T was added to the sigmoid output function and the output function was made probabilistic instead of deterministic (Fig. 18). The probability p_i that a given unit is activated (i.e., it gets turned on with output "1") is given by

$$p_i = \frac{1}{1 + e^{net_i/T}}$$

where net_i is the sum of weighted inputs. At high temperatures, each unit has a similar chance of being activated regardless of its input. Then, the network is allowed to iterate while the temperature is slowly reduced. At some low temperature, the network output vector will "freeze" to a final, unchanging state.

This procedure prevents the network from being trapped in a local minimum since there is always a finite probability of assuming a higher energy state than the current one. In fact, convergence of the system to the global minimum is guaranteed as long as the temperature is reduced slowly enough. However, the cooling time is not necessarily finite.

Simulated annealing can be described graphically with the help of Fig. 16. Initially, the high temperatures can be thought of as violently shaking the state of the network so the state has an equal probability of falling in either well. As the agitation decreases, the probability of the state being in well B (the global minimum) begins to exceed that for well A. The state may jump from A to B, but it is less likely that a jump from B to A is possible since this represents a greater energy difference. Eventually, the state becomes trapped in the desired global minimum and is unable to jump back to the local minimum.

Setting Weights of Hopfield–Boltzmann Machine Artificial Neural Network

To solve a particular problem, a researcher must determine appropriate values of the connection weights so that the stable states of the network correspond to desired memories or valid solutions. Whereas the perceptron and back-propagation ANN models were rooted in clearly defined weight learning algorithms, the Hopfield–Boltzmann machine ANNs models rely more on the ability of the researcher to translate a problem into a suitable energy function so that low energy states correspond to optimal solutions of the problem. A brief overview of techniques for weight setting is provided below.

Content-Addressable (Associative) Memories. Associative memories are capable of recalling a complete memory given a noisy or incomplete input pattern. This is characteristic of human memory and is extremely useful in pattern recognition applications. Typical computer memory is dependent on knowledge of particular storage locations rather than what is actually stored in those locations. Associative memories, on the other hand, retrieve information by content rather than by location.

Recurrent networks, such as the Hopfield networks, can implement these associative memories. If the correct set of weights is used, the network will output a complete desired memory given an incomplete or distorted input. In terms of energy states, the stored memories correspond to ANN states which are local or global minima. With incomplete or distorted input, the ANN starts in a high energy state and, with time, changes its output to the nearest low energy state which hopefully corresponds to the completed input pattern.

Given a set of binary encoded memories to be stored, the appropriate values for the weights are given by the following relation:

$$w_{ij} = \sum_{p=1}^{m} y_{i,p} y_{j,p}$$

where w_{ij} is the weight between units i and j, m is the number of memories to be stored, and $y_{i,p}$ is the desired output of unit i for memory p. To retrieve a memory, the network output is clamped to an input pattern. The output of the network is then freed and allowed to iterate into a low energy state. The resulting output will be the corresponding memory.

Associative memories simply find the nearest local minima to a given initial state. There are many situations in which the global minima is desired. This is the case with optimization problems. Here the determination of the connection weights is not nearly as straightforward as that for the associative memory problem. The constraints involved in the optimization must be mapped onto the energy function for the network.

An Optimization Problem. The traveling salesman problem (TSP) involves finding an optimum tour between a group of cities. The optimum tour is the shortest tour which visits each city only once and returns to the starting city. If there are n cities, then the number of possible valid tours is given by $n!/(2n)$. Obviously, for problems of more than a small number of cities, exhaustive search for the optimum tour is computationally intractable.

This problem can be mapped to a network by constructing an energy function which has minima for all valid tours.[20] These minima will have depths inversely proportional to the length of the tour they represent. As a result, a network with this energy function will tend to settle into one of the deeper minima, namely, a tour solution with minimal distance. An optimum solution is not guaranteed, but acceptable solutions are rapidly achieved.

Each city in an n-city tour is represented by a row of n units. Each unit in the row represents the order in the tour in which that city is visited. Therefore, a matrix of $n \times n$ units is formed to represent the entire tour (Fig. 19). Note that we reference each unit's output, y_{Xi}, in the ANN by two indices; the first designates the Xth city, whereas the second designates the position of the Xth city on the tour. For example, y_{32}, is the output from the unit associated with the hypothesis "City 3 is the second city in the tour."

We can devise a suitable energy function by considering (1) the constraints on valid solutions and (2) the criteria for "best" solutions. With regard to solution validity, there are two constraints. A city can only be visited once per tour; therefore, only one unit per row must be active. Because only one city can be visited at a time, only one unit per column must be active. Besides validity, we must also incorporate some measure of the goodness of the solution. In the case of the TSP, the total distance

[20] J. J. Hopfield and D. W. Tank, *Biol. Cybernetics* **52**, 141 (1985).

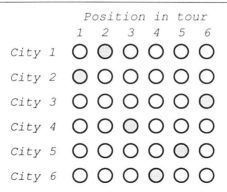

FIG. 19. Hopfield network solution to a six-city traveling salesman problem. Active units (shaded circles) describe a tour solution. The solution pictured in the diagram is a tour with visits to the cities 2, 1, 4, 6, 5, and 3, in that order. Each row represents a city while each column represents a position in the tour. All units are fully connected to each other.

required by a valid solution is the measure we must minimize. The following energy function satisfies all of these conditions:

$$E = A \sum_X \sum_i \sum_{j \neq i} \mathbf{y}_{Xi}\mathbf{y}_{Xj} + B \sum_i \sum_X \sum_{Y \neq X} \mathbf{y}_{Xi}\mathbf{y}_{Yi} + C\left(-n + \sum_X \sum_i \mathbf{y}_{Xi}\right)^2$$
$$+ D \sum_i \sum_X \sum_{Y \neq X} d_{XY}\mathbf{y}_{Xi}(\mathbf{y}_{Y(i-1)} + \mathbf{y}_{Y(i+1)})$$

where A, B, C, and D are some large constant values, n is the number of cities in a tour, $\mathbf{y}_{Xi}$ is the output of the unit associated with the hypothesis "City X is the ith visit on the tour," and d_{XY} is the distance between city X and city Y. The first two summations (A and B term) are zero only when there is a single active neuron in each row and column, respectively. The third summation (C term) is zero only when there are n active neurons. The fourth term is proportional to the tour length.

By relating the above TSP energy function to the general energy function (i.e., $E = a \sum\sum w_{ij}y_iy_j$), we can work out the required values for the connection weights between the ANN units. In this case, the weight between any two units is negative and proportional to the distance between the two cities they represent. The weights also include negative contributions from the validity constraints (A, B, and C terms in the energy function).

Boltzmann Machine Learning Algorithm. In addition to the two methods given above for the determination of the weights, Sejnowski and coworkers have developed a supervised training method for the Boltzmann machine ANN.[21,22] Although this method automatically configures the ANN weights for a particular problem, the learning algorithm tends to be extremely slow.

Discussion

Artificial neural networks are novel, robust computational tools for pattern recognition, data mapping, and other applications. The demonstrated success of ANN techniques has lead to an explosion of interest among scientific and engineering circles. In contrast to the paucity of introductory material a decade ago, ANN texts and journals are now readily available. Both commercial and public domain ANN software exists for a variety of computer systems ranging from small personal computers to large, massively parallel supercomputers. With these instructional and developmental resources, researchers are now able to apply ANNs to a wide range of scientific and engineering problems.

[21] D. H. Ackley, G. E. Hinton, and T. J. Sejnowski, *Cognit. Sci.* **9,** 147 (1985).
[22] G. E. Hinton and T. J. Sejnowski, "Parallel Distributed Processing," p. 282. MIT Press, Cambridge, Massachusetts, 1986.

[30] Fractal Applications in Biology: Scaling Time in Biochemical Networks

By F. Eugene Yates

Introduction

Aims of This Chapter

The purpose of this chapter is to present to a readership of biochemists, molecular biologists, and cell physiologists some of the terms and concepts of modern dynamical systems theory, including chaotic dynamics and fractals, with suggested applications. Although chaos and fractals are different concepts that should not be confounded, they intersect in the field of modern nonlinear dynamics. For example, models of chaotic dynamics have demonstrated that complex systems can be globally stable even though locally unstable and that the global stability reveals itself through the confinement of the motion of the system to a "strange attractor" with a microscopic fractal geometry. Some of the technical aspects of chaos, fractals, and complex dynamical systems are sketched in the Glossary at the end of this chapter, where appropriate references to the papers of experts can be found. The details in the Glossary permit me to use a freer style in the body of this chapter, where I shall explore the possible relevance of fractals to the understanding of both structure and function in biology.

The general motivation for biologists to examine chaotic dynamics and fractal geometries arises from recognition that most of physics addresses only simple systems, no matter how elaborate the mathematical apparatus seems to the uninitiated. For the purposes of theorizing, the simple systems examined, both objects and processes, are preferably rather uniform (all electrons are alike; there are only a few quarks), analytically smooth, conservative (nondissipative), time-symmetric (reversible), and linear. Biological systems, in contrast, are notably complex, nonlinear, dissipative, irreversible, and diverse. In the last 20 years, for the first time, we have witnessed the development of several branches of mathematics, both "pure" and applied, that attempt to confront nonlinearity and complexity in a systematic way. That development led to modern dynamical systems theory and vigorous extensions of nonlinear mechanics. These advances have transformed discussions of the shapes of common objects such as trees, clouds, coastlines, and mountains; of the stability of complex systems; of noise and apparent randomness; of the genesis and nature of turbulence and the Red Spot of Jupiter; of the intervals between drops in a dripping faucet; of the music of Bach. . . . Now we begin to see biological forms and functions examined from the same mathematical perspectives.

To provide a more specific motivation for biochemists to attend to the terms and concepts of modern dynamical systems theory, I ask the reader to accept on trial, as it were, the hypothesis that biological systems depend on a chemical network whose synthetic capabilities lead to many forms with fractal geometries (including bronchial and vascular trees and dendritic branchings of neurons), and many processes that organize time in a fractal manner. We further hypothesize that biological systems are not inherently noisy, but follow state-determined dynamics (motion and change) of the deterministic chaotic class. Their chaotic dynamics produce their marginal stability, many of their structures, and their fractal organization of time. Furthermore, biological systems with their very large numbers of degrees of freedom (i.e., high dimensionality) have biochemical traffic patterns as fluxes, transports, and transformations whose stability can be comprehended formally only through the confinement of their motions to low-dimensional, chaotic, strange attractors. (Note: These are all assumptions, for the sake of discussion.)

Because we lack commonplace terms or familiar metaphors to convey the essence of modern dynamical systems theory, I rely on the Glossary in either its alphabetical or indicated logical order to help the reader through the jargonistic jungle. I doubt that deterministic chaos and fractal time will ultimately prove to be the most useful or insightful models of the complexity of biological systems. Fractals and chaos have little in them of theoretical profundity, in spite of their attendant and fashionable mathematical pyrotechnics. But as models of data they may be the best we now have for complex systems.

Discovery of Fractals: Background Reading and Computing

Fractals can pertain to the organization of both space and time. It is astonishing, given that so much of our terrestrial surround has fractal shapes, that the concept can be attributed chiefly to one person: Benoit B. Mandelbrot. His first book on the subject, *Fractals: Form, Chance, and Dimension*, was published by W. H. Freeman and Company in 1977. A revised edition entitled *The Fractal Geometry of Nature* appeared in 1982–1983. This work is idiosyncratic, imaginative, and difficult for the layman to read. Mandelbrot showed that recursive phenomena can generate fractals, and one kind of basic fractal expression is a statement describing a recursion. For example, for a univariate system, the recursion is

$$x_{n+1} = f(x_n, c) \cdot \tag{1}$$

The $(n + 1)$th value of the recursive function is a function of the nth value plus a constant. The function for the recursion may be deterministic, stochastic, or even a combination. The example commonly used for a purely deterministic recursion is $Z_{n+1} = Z_n^2 + C$, where Z and C are complex numbers. Mandelbrot's insight was to examine recursions on the complex plane, instead of on the real number line. His recursion serves as a basis for the definition of both the Mandelbrot and Julia sets.[1] These remarkable sets are portrayed in the complex plane, where they show very irregular and stunning shapes with extremely elaborate boundaries. (Julia sets may be exactly self-similar at all scales of magnification, but a Mandelbrot set, consisting of all values of C that have connected Julia sets, has a fantastic fine structure at all scales, including repetition, and is not exactly self-similar across scales. The details change.)

In the case of generating fractal trees as computer images of this recursion, if the parameters remain constant through various generations of branching, a somewhat regular tree looking like a bracken fern is obtained, but if the parameters for the branchings are random, then very

[1] I. Peterson, "The Mathematical Tourist: Snapshots of Modern Mathematics," p. 157. Freeman, New York, 1988.

irregular shapes are obtained. In the mathematical generation of fractals by recursion rules, orders of magnitude from zero to infinity are permitted. However, in the real, physical world, there are cutoffs both at the lower and at the higher ends. The measurement of the length of a coastline would stop with a grain of sand; most physical and biological systems damp out or filter out very high frequencies.

There is a very comfortable fit between fractal geometries and the iterative characteristics of digital computers. As a consequence, fractal geometries have turned computer graphics into a new form of art. Software packages are available for the layman, and I mention several of them here because they provide a sense of immediacy about fractals impossible to convey by words or by formal definitions. Anyone with an IBM PC (or clone), enhanced graphics adapter, 640K of RAM, and DOS 2.0 or higher (a hard disk drive helps) can run "The Desktop Fractal Design System" by Michael F. Barnsley or James Gleick's "Chaos: The Software." Both systems are VGA and EGA compatible and are available as diskettes. A math coprocessor is recommended but not required. A book that gives programs, but does not provide the software on diskette, is *Exploring the Geometry of Nature: Computer Modeling of Chaos, Fractals, Cellular Automata and Neural Networks* by Edward Rietman (Windcrest Books, Blue Ridge Summit, Pennsylvania, 1989).

In book form fractals and the mathematical background (not easy) are available in the following: H.-O. Peitgen and P. H. Richter, *The Beauty of Fractals: Images of Complex Dynamical Systems,* Springer-Verlag, New York, 1986; M. Barnsley, *Fractals Everywhere,* Academic Press, New York, 1988; R. L. Devaney and L. Keen, eds., *Chaos and Fractals: The Mathematics Behind the Computer Graphics* (Proceedings of Symposia in Applied Mathematics, Vol. 39), American Mathematical Society, Providence, Rhode Island, 1989; R. L. Devaney, *Chaos, Fractals, and Dynamics: Computer Experiments in Mathematics,* Addison-Wesley, Menlo Park, California, 1990 (this book is for nonexperts); R. L. Devaney, *An Introduction to Chaotic Dynamical Systems,* Second ed., Addison-Wesley, Menlo Park, California, 1989 (written for the mathematical community).

Many videotapes vividly demonstrating fractal geometries in pseudocolors are available. A good example is C. Fitch's "Fractal Fantasy," produced by Media Magic (Nicasio, CA). Another is "Fractal Zooms" animation of the Mandelbrot set, obtainable from Art Matrix (Ithaca, NY). In addition to his beautiful book cited above, H.-O. Peitgen and colleagues have produced a video entitled "Frontiers of Chaos," also available from Media Magic. Robert L. Devaney has produced two very useful introductory videotapes that overlap somewhat, but are best studied as a pair.

They are entitled "Transition to Chaos: Orbital Diagrams and Mandelbrot Set" and "Chaos, Fractals and Dynamics: Computer Experiments in Mathematics," both available from Science Television, Dept. of Mathematics, Boston University (1990), ISBN 1-878310-08-9 and 1-878310-00-3, respectively.

From these sources the careful reader will find a deeper understanding of fractals than arises from merely supposing that a fractal is a geometric object having a fractional dimension. Some trivial fractal objects actually have integer dimension, but so do some that are not so trivial. In the discussion of biological structure and function in this chapter, I shall be emphasizing both structures that do in fact have fractional dimensions and also functions, revealed as time histories, that produce $1/f^m$ spectra typical of fractal time, where $m \cong 1$ or, more generally, some noninteger value. The key to this fractal view of both structure and function lies in two of the very strong features of fractals: lack of a characteristic scale and some self-similarity across all levels of magnification or minification. In the strongest cases self-similarity lies in the detailed shapes, or patterns, but in weaker cases it can be found only in statistical characteristics. Because most fractals are not homogeneous (i.e., not identical at every scale), the more closely you examine them, the more details you find, although there will not necessarily be infinite layers of detail.

The creation of fractal structures in the physical/biological world, as opposed to their generation in recursive computer models, may require the operation of chaotic dynamics. Chaotic processes acting on an environment such as the seashore, atmosphere, or lithosphere can leave behind fractal objects such as coastlines, clouds, and rock formations. I wish to emphasize that the mathematics of fractals was developed independently of the mathematics of chaotic dynamics. Although there were antecedents, I think it is fair to say that the general scientific communities became aware of chaotic dynamics through a 1971 paper on turbulence by Ruelle and Takens, and aware of fractals through Mandelbrot's 1977 book already cited. In biology some of these modern concepts were adumbrated in works by N. Rashevsky.

The common occurrence of fractal geometries in biological morphologies, also found in physical structures, suggests but does not prove that chaotic dynamics are very widespread in nature and should be found in biological morphogenetic processes. That search is a current and advanced topic for biological investigations and should provide, in my opinion, an enrichment of Edelman's "topobiology,"[2] which at present suffers from a lack of detailed, mathematical modeling. J. Lefevre in 1983 attempted

---

[2] G. M. Edelman, "Topolobiology: An Introduction to Molecular Embryology." Basic Books, New York, 1988.

to extend Mandelbrot's illustration of a fractal network, space-filling in two dimensions, into three dimensions for the bronchial tree. A fine account of the possible importance of fractal processes in morphogenesis has been provided by West.[3] It is now thought that many aspects of mammalian morphology involve fractal tree branchings, including bronchi, arteriolar networks, cardiac conduction systems, and neuronal dendrites.

Fractal Morphology in Mammals: Some Branchings

Bronchial Tree and Pulmonary Blood Flow

The bronchial tree meets all of the criteria for a fractal form (see Glossary). The conduits through which gases flow to and from the lungs branch repeatedly from the single trachea to the terminal structures, the alveoli. B. J. West et al. have recently reanalyzed lung casts of humans and several other mammalian species originally prepared in 1962 by Weibel et al. They found the type of scaling characteristic of a fractal geometry. The fractal dimension, D, $2 < D < 3$, may have provided a mechanism for converting a volume of dimension three (blood and air in tubes) into a surface area of dimension two, facilitating gas exchange. Goldberger et al.[4] have discussed these findings. Fractal branching systems greatly amplify surface area available for distribution or collection, or for absorption, or even for information processing.

Glenny and Robertson[5] have examined the fractal properties of pulmonary blood flow and characterized the spatial heterogeneity. Their data fit a fractal model very well with a fractal dimension (D_s) of 1.09 ± 0.02, where a D_s value of 1.0 reflects homogeneous flow, and 1.5 would indicate a random flow distribution in their model.

Vascular Tree

Elsewhere I have discussed the branching rules of the mammalian vascular tree.[6] As in the case of the bronchii, casts have been used (by Suwa) to discover the branching rules of the vascular tree, and that tree also turns out to be a fractal structure.

[3] B. J.West, in "Chaos in Biological Systems" (H. Degn, A. V. Holden, and L. F. Olsen, eds.), p. 305. Plenum, New York, 1987.

[4] A. L. Goldberger, D. R. Rigney, and B. J. West, Sci. Am. **262**, 44 (1990).

[5] R. W. Glenny and H. T. Robertson, J. Appl. Physiol. **69**, 532 (1990).

[6] F. E. Yates, in "The Resistance Vasculature" (J. A. Bevan, W. Halpern, and M. J. Mulvaney, eds.), p. 451. Humana Press, Clifton, New Jersey, 1991.

Morphology of the Heart

The branching of the coronary arterial network is self-similar, as is that of the fibers (chordae tendineae) anchoring the mitral and triscupid valves to the ventricular muscle. Furthermore, the irregular spatial branching of the conduction system of the heart (His–Purkinje system) sets up a fractallike conduction network, which has been interpreted as having important functional consequences because it forces a fractal temporal distribution on electrical impulses flowing through.[7]

Neurons (Dendrites)

By the criteria for fractals given in the Glossary and the preceding discussion, the branching of the dendritic tree of many neurons is fractal. Goldberger *et al.*[4] have suggested that this dendritic fractal geometry may be related to chaos in the nervous system, a feature of the nervous system suggested by the work of Walter Freeman.[8] Such conjectures are interesting but not yet proved to be correct. (See the comment of L. Partridge in the Glossary, section on chaos.)

The above examples represent a small sample of the data which seem to establish that branching treelike structures with fractal geometries abound in the biological realm, both for animal and plant life. However, it is less clear that biological processes are organized in fractal time. I consider that question next, cautioning the reader that I am not comfortable with the current fad for casually imposing chaotic dynamics and fractal concepts on models of physiological processes. These models should be examined skeptically, as provocative possibilities yet to be proved.

Fractal Time: Prelude to Fractal Function and Temporal Organization

The main purpose of this chapter is to show how fractals might help to describe and explain biochemical and physiological processes. A first step toward that goal is recognition of the general form of fractals; a second step is the examination of fractal structures in space. Now the third step is an examination of temporal organization from the fractal viewpoint. The final step will consist of applications to research in biochemistry and cellular physiology.

---

[7] A. L. Goldberger and B. J. West, *Biophys. J.* **48**, 525 (1985).
[8] W. F. Freeman, *Sci. Am.* **264**, 78 (1991).

Time History Analysis, Scaling Noises

Consider a record of the amplitude of a single biochemical or physiological variable over the duration of some experiment. If the variable is continuous, as in the case of blood pressure, it may be recorded continuously or intermittently (discrete sampling). For time history analysis in digital computers continuous records are ordinarily discretized at constant sampling interval and converted to a string of numbers. If the variable has the nature of an irregular event, such as the beat of the heart, it may be merely counted or treated as a point process. (As a first approximation these are usually expected to generate a Poisson exponential distribution of interevent intervals, as in radioactive decay of a specific type from a specific source, but recently attention has been directed to more complicated point processes, for example, in the analyses of neuronal firing in parts of the auditory system.)

Time histories of biological variables can be hypothesized to have been generated either by a deterministic process or by a random process. In the field of nonlinear topological dynamics under discussion in this chapter, varying deterministic processes can be periodic, quasi-periodic, or chaotic. The random processes include: (1) point processes with Poisson distributions, (2) processes generating intermittent or episodic pulses ("pulsatility"), and (3) processes that generate broad, rather featureless spectra such as white noise ($1/f^0$ spectrum) or Brownian motion noise ($1/f^2$ spectrum) (where f is the frequency against which an associated component of variance—roughly amplitude squared—is being plotted). But there is another case: Mandelbrot has called attention to stochastic fractals, an example of which is fractional Gaussian noise. The power spectrum in the low frequency range for such a univariate, real-valued function with fractal dimension D ($1 \leq D \leq 2$) can be shown to be proportional to f^{1-2H}, where $0 \leq H < 1$, and H is a constant related to D. In other words, fractional Gaussian noise corresponds to what is generically termed $1/f^m$ ($m \cong 1$) noise, where the power spectrum reveals concentration of low frequency energy, with a long higher frequency tail. Familiarly the spectrum of such processes is broadly referred to as "$1/f$." (The more general case where m is not an integer is discussed below.)

Some point processes can be modeled as stochastic fractals, such as the activity of primary discharge patterns of some neurons. However, the enthusiasm for searching for processes with $1/f$ spectra in biology lies not in the direction of stochastic processes, but in the domain of deterministic chaos, whose dynamics can produce both fractal time and fractal spatial structures. (The finding of a $1/f$ spectrum in a biological time history does not prove that the generator was deterministic chaos. It does, however, permit that hypothesis.) The "noise" classification according to the spectral sequence $1/f^0$, $1/f^1$, $1/f^2$ is implemented by a log–log plot in which log (amplitude squared) is plotted against log (frequency or harmonic number). The three kinds of noise in the sequence will produce log–log plots with a slope of 0 (white or Johnson noise), approximately -1 (Mandelbrot noise), or -2 (Brownian noise). Mandelbrot has called these various noises "scaling noises." If an investigator plots the power spectrum of the time history of a finite length record on a single variable, using the log–log transformation, and finds that the background, band-limited but broad, "noise" does not fall on a line with slope 0, -1, or -2, he could either reject the scaling noise model and assume some other basis for the background variations, or he might consider a harmonically modulated fractal noise model (if the data wander back and forth across a straight line with a slope of -1). Such a model has been provided by West[3] and is of the form

$$y(z) = [A_0 + A_1 \cos(2\pi \ln z/\ln \beta)]/z^\alpha \qquad (2)$$

where A_0, A_1, α, and β are parameters.

It should be noted that the log–log transformation of the power spectrum does not obliterate any spectral lines identifying periodic processes, which can occur on top of a background of scaling noise. In summary, an investigator believing that the time history of the variable he is looking at might have resulted from chaotic dynamics might well start with the search for the $1/f$ spectrum characteristic of fractal time, remembering that the presence of a $1/f$ spectrum is compatible with, but does not establish, the presence of a deterministic, chaotic generator of the time history. (In practice it is not always easy to decide on the value of a negative slope in a log–log plot of the power spectrum of real biological data.)

Because fractal time, like fractal space, has the features of heterogeneity, self-similarity, and absence of a characteristic scale, it follows that if recorded on a tape fractal noise always sounds the same when the tape speed is varied (making allowances for changes in loudness). In contrast, the pitch of any periodic process will be a function of tape speed (rising as the tape speed is increased, a familiar phenomenon when fast forward is used).

Schlesinger[9] offered a technical but clear account of fractal time and $1/f$ noise in complex systems. In what follows I paraphrase some points of his article, possibly relevant to biochemical and physiological data in which time appears as an independent variable. It should be an embarassment to biochemists that time rarely appears explicitly in plots of their

[9] M. F. Schlesinger, *Ann. N.Y. Acad. Sci.* **504,** 229 (1987).

data, their idea of kinetics too often being merely a relationship between a reaction velocity (time implicit) and a (steady) substrate concentration.

Suppose that a biologist is observing a process that seems to have some kind of "pulsatility" in which there is a variable time between events or peak concentrations. A starting point for the analysis of pulsatility is to suppose that there is some probability for time between events

$$\psi(t) \, dt = \text{Prob[time between events} \in (t, t + dt)] \tag{3}$$

There will be a mean time $\langle t \rangle$ and a median time, t_m, between events

$$\langle t \rangle \equiv \int_0^\infty t\psi(t) \, dt, \qquad \int_0^{t_m} \psi(t) \, dt = \frac{1}{2} \tag{4}$$

Shlesinger gives the references that provide the basis for these and other of his statements that follow. If $\langle t \rangle$ is finite, then we can say that some natural scale exists in which to measure time. For a very long series of data it will appear that events occur at the constant rate $\langle t \rangle^{-1}$. If $\langle t \rangle$ is very large, then events will occur at a slow rate, but we would not call such events rare. When $\langle t \rangle = \infty$, there is no natural time scale in which to gauge measurements, and events are indeed rare. Note that even under these conditions, t_m is finite, so events still occur.

Shlesinger then considers the case that we have three events in a row at times $t = 0$, $t = \tau$, and $t = T$, where the value of T is known. We want to know the probability of the middle event that occurs at $t = \tau$

$$f(t) = \frac{\psi(\tau)\psi(T - t)}{\displaystyle\int_0^T \psi(s)\psi(T - s) \, ds} \tag{5}$$

where "the denominator insures the proper normalization." For a purely random process $\psi(t) = \lambda \exp(-\lambda t)$, and $f(\tau)$ is a uniform distribution in the interval $(0, T)$. In that case the most likely time for the middle event is $\tau = T/2$; that is, on the average, events occur at rather regular intervals. However, if $\langle t \rangle$ is infinite, the process cannot resemble a constant rate renewal; so then $\tau = T/2$ is the least likely value of τ, and values of τ closer to $t = 0$ and $t = T$ are more probable. For such rare events (mean renewal rate of zero), the time sequence of events appears in self-similar clusters like points in a Cantor set (one of the best known demonstrations of self-similarity and fractal dimension). For proof of this by no means obvious result, see the references provided by Shlesinger.

One example of a physical mechanism that can generate such a fractal time distribution of events is hopping over a distribution of energy barriers. A small median jump time can exist and be consistent with an infinite mean jump time. In physics these ideas have become important in describing charge motion in some disordered systems. In that case $\psi(t)$ can represent the probability density for an electron's not moving, but it can also represent the probability density for remaining in a correlated motion.

It still has to be answered why $1/f$ noise is so prevalent in physical and biological systems. Shlesinger gives a plausible argument which says that this kind of noise is generic in the same sense that the Gaussian distribution arises from the central limit theorem which governs sums of independent random variables with finite second moments. Consider a process that is described by a product of random variables. Then, for an event to occur, several conditions have to be satisfied simultaneously or in sequence. If P is the probability for the event to occur, and if

$$P = p_1 p_2 \cdot \cdot \cdot p_N \tag{6}$$

then

$$\log P = \sum_{i=1}^{N} \log p_i \tag{7}$$

has a Gaussian distribution and P has a log-normal distribution:

$$P(\tau) = \frac{1}{\pi\sigma\tau} \exp(-[\log(\tau/\langle\tau\rangle)]^2/2\sigma^2) \tag{8}$$

where $\langle\tau\rangle$ and σ^2 are the mean and variance of the distribution, respectively. As more factors N participate, σ increases (see next section) and $P(\tau) \cong 1/\tau$ over a range of τ values. The greater the value for σ, the more extensive the range over which the τ^{-1} behavior persists, and thus the larger the range over which $1/f$ noise is found. Shlesinger concludes, "The underlying product of the random variables idea that leads to a log-normal distribution of relaxation times and naturally to $1/f$ noise provides a generic generation of this phenomenon. . . . The real message of $1/f$ noise is that a scale-invariant distribution of relaxation times has been generated." (See the following section for a similar account by Bassingthwaighte.)

An earlier paper by Careri et al.[10] provides a contrasting background for the analysis of Shlesinger; it examined statistical time events in enzymes. The great progress in understanding spatial aspects of enzyme action has not been matched by an equally important analysis of the temporal aspects. Carerri et al. explored elementary processes and assessed the microscopic mechanisms by comparative studies on representative model systems using the theory of random processes. They modeled relaxation and concentration fluctuation spectroscopy by use of the fluc-

[10] G. Careri, P. Fasella, and E. Gratton, Crit. Rev. Biochem. **3**, 141 (1975).

tuation-dissipation theorem. This theorem relates the average time for the decay of spontaneous microscopic fluctuations to the time course observed after a small perturbation around equilibrium. They focused on the correlation time (especially the autocorrelation time of a statistically stationary variable) as the basic quantity of interest because it gives a direct measure of the time interval over which a variable is behaving more or less regularly and predictably. For longer time intervals the behavior becomes progressively more random in this model. However, the authors caution that in their kind of analysis of random processes they must assume that the variables involved are statistically stationary and linearly superimposed, and that such an assumption may not hold for an enzyme, where the different classes of fluctuations may interact in a nonlinear way, merging into a new cooperative with nonstationary effects of great chemical interest—precisely the point of modern, nonlinear dynamics!

The presence of $1/f$ noise in a dynamic system does not imply any particular mechanism for its generation. For that one needs to draw on phase-space plots, Liapunov exponents, embedding plots, etc., for a more detailed investigation of dynamic behavior in complex biological systems. Recently H. H. Sun[11] considered the general case that a fractal system follows a generalized inverse power law equation of the form $1/f^m$, where m is any fractional number. He reexpresses such a system in fractional power pole form (not shown here) which, as he points out, has a much wider representation of natural and physiological phenomena than does a $1/f^1$ view because its low frequency magnitude is finite instead of infinite (as $f \to 0$). He then offers a time domain expression of such a fractal system consisting of a set of linear differential equations with time-varying coefficients. If the fractal dimensions m_i approach unity or any other integer numbers, his equations lead to regular time-invariant systems. Most importantly, he shows that certain familiar dynamic systems whose performance criteria are well known become more stable in the fractal domain. Thus he confirms the view expressed earlier by Bruce West that fractal systems are error tolerant and, in that sense, more stable than their nonfractal counterparts.

A clear and profound explanation of $1/f$ noise has been provided by Bak, Tang, and Wiesenfeld.[12] As they remark, "One of the classical problems in physics is the existence of the ubiquitous '$1/f$' noise which has been detected for transport in systems as diverse as resistors, the hour glass, the flow of the river Nile, and the luminosity of stars. The low-frequency power spectra of such systems display a power-law behavior

[11] H. H. Sun, *Ann. Biomed. Eng.* **18**, 597 (1990).
[12] P. Bak, C. Tang, and K. Wiesenfeld, *Phys. Rev. Lett.* **59**, 381 (1987).

$f^{-\beta}$ over vastly different time scales. Despite much effort, there is no general theory that explains the widespread occurrence of $1/f$ noise." They then argue and demonstrate numerically that dynamical systems with extended spatial degrees of freedom naturally evolve into self-organized critical structures of states which are barely stable. They propose that this self-organized criticality is the common underlying mechanism behind the common occurrence of $1/f$ noise and self-similar fractal structures. The combination of dynamic minimal stability and spatial scaling leads to a power law for temporal fluctuations. It should be emphasized that the criticality in their theory is fundamentally different from the critical point at phase transitions in equilibrium statistical mechanics, which can be reached only by tuning of a parameter. These authors refer to critical points that are attractors reached by starting far from equilibrium, and the scaling properties of the attractor are insensitive to the parameters of the model. In fact, this robustness is essential in supporting their conclusion that no fine tuning is necessary to generate $1/f$ noise and fractal structures in nature. They end their article with the remark, "We believe that the new concept of self-organized criticality can be taken much further and might be *the* underlying concept for temporal and spatial scaling in a wide class of dissipative systems with extended degrees of freedom."

The ubiquity of physical and biological processes in which some measure of "intensity" (such as frequency of usage or occurrence, physical power, or probability) varies inversely with f^m (where f is a frequency or a rank order) forces us to speculate about interpretations of different values of m, whether integer (0, 1, 2) or fractional. A rich and varied literature bears on such systems, real and model. The mathematical treatments are advanced, and we have no compact, reduced figure of thought to encompass all $1/f^m$ phenomena. There are, however, some informal inferences that may be drawn, and they are considered next.

Fractals and Scatter in Biological Data: Heterogeneity

Bassingthwaighte[13] has noted that there are spatial variations in concentrations or flows within an organ, as well as temporal variation in reaction rates or flows, which appear to broaden as the scale of observation is made smaller (e.g., smaller lengths, areas, volumes, or times), for the same constant total size or interval, composed of N_i units, where i indexes an observational scale (number of pieces or intervals in the population of samples). He then asks, "How can we characterize heterogeneity independently of scale?" This scale-dependent scatter in physiological and bio-

---

[13] J. B. Bassingthwaighte, *NIPS* **3**, 5 (1988).

chemical observations is a property inherent in the biological system and cannot be accounted for entirely by measurement error. Bassingthwaighte considers the example of channel fluctuations: "When the duration of openings of ion channels is measured, the variation is broader when the observations are made over short intervals with high-resolution instrumentation and narrower when made over long intervals with lower fidelity. While this problem seems obvious when considered directly, no standard method for handling it has evolved." Given any arbitrary choice for the size of the domain one wishes to consider, how could one describe the heterogeneity of the system in a fashion that is independent of the magnitude of the domain or the period of observation? The fractal concept provides the answer.

As an example, consider the measurement of variation in regional flows throughout some organ whose total blood flow is known. The mean flow per gram of tissue is the total flow divided by the mass of the organ. Next, assume that the flows everywhere are steady and that the organ is chopped up into weighed pieces and the flow to each piece is known (from the deposition of indicator or microspheres) so that we have an estimate of the flow per gram for each piece. The relative dispersion of the regional flows is given by the standard deviation divided by the mean (the usual coefficient of variation, CV). Empirically we find that the finer the pieces we chopped the organ into, the greater the relative dispersion or coefficient of variation. So, what is the true variation? From experimental data Bassingthwaighte obtained the result for relative dispersion (which he designated RD and I call CV):

$$CV = CV(N = 1)N^{D-1} \qquad (9)$$

where N is the number of pieces and D is the fractal dimension, a measure of "irregularity." In a plot of CV versus N, the real value of $CV(N = 1)$ can only be 0, and so it cannot actually be on the fractal curve. Therefore the value of $CV(N = 1)$ in these data had to be obtained by extrapolation of the log–log plot to its intercept. It was found that extrapolated $CV(N = 1) = 12.9\%$.

To generalize the expression for relative dispersion,

$$CV(w) = CV(w = 1 \text{ g})w^{1-D} \qquad (10)$$

where w is the mass of the observed pieces of tissues and D is the fractal dimension (which in the case of blood flow in the myocardium had an observed value of 1.18). (The exponent can be $D - 1$ or $1 - D$, depending on whether the measure is directly or inversely proportional to the measuring stick strength.) Thus it is the fractal dimension, not the coefficient of variation for any particular set of data, that expresses the "true" heterogeneity.

The same approach can be applied to temporal fluctuations, and the equation has the same form as that for spatial heterogeneity

$$CV(\tau) = CV(\tau = 1 \text{ unit})\tau^{1-D_\tau} \tag{11}$$

In this case we examine the standard deviation of flows over a given interval τ, divided by the mean flow determined over a much longer time. The standard deviation is broader the shorter the interval τ over which the flows are measured. Some arbitrary reference interval ($\tau = 1$) must be chosen. Preliminary analyses of capillary flow fluctuations gives a D_τ value of approximately 1.3. Similarly, Liebovitch et al.[14] obtained patch-clamp data on ion channel openings or closings in lens epithelial cells. They compared a fractal model with mono- or multiexponential rate constants and obtained better fits with the fractal model.

The important conclusion is that suggested by Bassingthwaighte: Fractals link determinism and randomness in structures and functions.

Chaos in Enzyme Reactions

To further illustrate possible applications of concepts of chaos and fractals for the interpretation of biochemical phenomena, I consider the historically important, brief report by Olsen and Degn[15] of perhaps the first direct experimental demonstration of chaos in a chemical reaction system. They studied the behavior of oxygen concentration as a function of time in a peroxidase-catalyzed oxidation of NADH in a system open to O_2 in a stirred solution, where the O_2 could enter by diffusion from the gas phase. They observed sustained "oscillations" by continuously supplying NADH to the reaction mixture. They discovered that the waveform of the variations depended on the enzyme concentration. At one concentration sustained, regular, true oscillations were observed; but at other concentrations the large variations showed no apparent periodicity. This irregular fluctuation was analyzed by plotting each amplitude against the preceding amplitude, and each period against the preceding period, according to the mapping technique introduced by Lorenz. (The study of such iterated maps originally arose from a desire to understand the behavior of solutions of ordinary differential equations, and they can be used to identify random, periodic, or chaotic behaviors.) Such maps diagram transition functions,

[14] L. S. Liebovitch, J. Fischbarg, J. P. Koniarek, I. Todorova, and M. Wang, *Biochim. Biophys. Acta* **896,** 173 (1987).
[15] L. F. Olsen and H. Degn, *Nature (London)* **267,** 177 (1977).

and, according to a theorem of Li and Yorke,[16] if the transition function allows the period 3 (which can be tested graphically on the map), then it allows any period and chaos exists. (Devaney credits this idea to an earlier source; see the videotapes already mentioned.) By this means Olsen and Degn demonstrated graphically that the fluctuations in the concentration of the reactant O_2 at a certain enzyme concentration produced a transition function admitting period 3. By the Li and Yorke criterion (not necessary but sufficient to identify chaos) the reaction kinetics were indeed chaotic. In the absence of chaos theory the data most likely would have been uninterpretable or regarded merely as showing a contaminating noise of unknown origin. With chaos theory it could be concluded that the data accurately conveyed the real (deterministic, nonlinear) dynamics of the reaction.

Practical Guide to Identification of Chaos and Fractals in
  Biochemical Reactions

Suppose that a biochemist is monitoring the concentration or thermo-dynamic activity of a reactant under conditions that the reaction is not at equilibrium. (It may be in a constrained steady state, or it may be tending toward an equilibrium, depending on the conditions.) Traditionally such kinetic systems are usually thought to have either monotonic or periodic behavior. Examples of nonmonotonic, nonperiodic behavior have long been known, but before the advent of deterministic chaos theory it was usually assumed that the variations had to be attributed to some kind of "noise." It was not easy to imagine where the noise came from during controlled *in vitro* studies. Today contamination of the reaction kinetics by some random process should not be the first choice for modeling; deterministic chaos deserves to be tried.

The first step in such an analysis consists of converting the concentration data into a string of numbers (usually at equally spaced time intervals). If the data were continuously recorded, as might be the case for hydrogen ion concentration using a pH electrode, then the data must be discretized. Here the choice of sampling interval depends on the frequency range the experimenter wishes to explore, recognizing restrictions imposed by the frequency response of the recording equipment. Once the string of pairs of numbers (concentration, time of observation) has been obtained, several practical considerations arise that I cannot treat formally here. These concern whether the data should be in any way smoothed or filtered according to what is known about measurement uncertainty. In addition,

[16] T.-Y. Li and J. A. Yorke, *Am. Math. Monthly* **82,** 985 (1975).

there must be an *a priori* policy for interpolation across missing points. Choices about the length of the record or the number of successive points to be included in the analysis can bear critically on the results of subsequent computations. Some of the most powerful techniques for exploring chaotic dynamics require more points (more than a thousand) than biochemists may have. If the conditions of the reaction are nonstationary, so that the statistical properties of segments of record are different in different pieces of the record, then it may be desirable to use less than the whole record length for analysis. But there is a caveat: As I have pointed out elsewhere in this chapter, one chaotic dynamic can produce different spectral results depending on the piece of the record used, even when the underlying dynamics are actually the same throughout. This is tricky business, and the investigator must proceed cautiously.

Having chosen a sampling interval and a length of record, the investigator may then apply spectral analysis and examine the power spectrum (see [15] in this volume) for evidence of periodicities (peaks) and broad-band, background "noise." If that background has a $1/f^m$ property, as revealed by a log–log plot of power (amplitude^2) versus frequency, yielding a straight line of negative slope $-m$, then the value of m can give some clue as to the nature of the process. m equals 2 is consistent with an assumption that a Brownian motion, diffusion process is dominant; an m of approximately 1 hints that chaotic dynamics may be operating (this is not a critical test).

The next step might be the graphical iteration technique in which each value x_{t+1} is plotted against the preceding value x_t, along the whole time series. The resulting iterated map gives the transition function for the reactant whose concentration was x. If the plot yields a random scattergram, the process is random. However, if the plot reveals regularities, a chaotic "attractor" may be present.

Next the "correlation dimension" of the data string may be computed using the Grassberger–Procaccia algorithm,[17] not further described here. If the computed dimension has a noninteger value greater than 2, then chaotic dynamics are suggested. (In my experience, if the correlation dimension ranges larger than 10, it is best just to assume that white noise is present.) There are many reservations about the interpretation of the correlation dimension as being the (fractal) dimension of a chaotic attractor (not reviewed here). It is safest just to regard it as some kind of "complexity" estimate on the dynamics of the reactant. As a practical matter this complexity will range from 2^+ to 10^+, the larger the number the greater

---

[17] P. Grassberger and I. Procaccia, *Physica* **9D,** 189 (1983).

the complexity of the dynamics. (In theory the correlation dimension can range from zero to infinity.)

There is a new measure of the complexity of dynamics that is mathematically robust and makes fewer assumptions than are required for the computation of the correlation dimension. Steven Pincus[18] has introduced a regularity statistic for biological data analysis that he calls "approximate entropy," abbreviated ApEn. It is a relative, not absolute, measure. Its rigorous use may require as many as 1000 data points, but its robustness permits valuable results from strings of time series data no longer than perhaps 200 points. This measure has been used to detect abnormal hormone pulsatility in physiological systems. Pincus developed ApEn as a corrective to blind applications of certain algorithms to arbitrary time series data. He commented (personal communication, 1991), "These algorithms include estimates of Kolmogorov–Sinai (K-S) entropy and estimates of system dimension. These algorithms were developed for application to deterministic systems, yet are frequently applied to arbitrary data with questionable interpretations and no established statistical results. Sometimes attempts are made to apply these algorithms to very noisy processes, but again, statistical understanding is lacking. Furthermore, correlated stochastic processes are almost never evaluated in this way. There are severe difficulties with such blind applications of these popular algorithms. One cannot establish underlying determinism via the correlation dimension algorithm." At the present time ApEn may be the best we can do to specify "complexity" of a time series, particularly given that biological systems may produce very messy, weakly correlated stochastic processes. However, like all other aggregated measures, ApEn loses information.

Summary, or What Does It All Mean?

I have tried to convey some of the new trends of thought regarding the performance of complex systems that have been provided by mathematical advances, particularly in topology. It would be impossible to try to fill in every step required for a clear understanding of the separate concepts of chaos and fractals that come together in the modern dynamic systems theory addressing the behavior of complex, nonlinear systems. I have provided only a few primary or secondary references out of the hundreds now available in this rapidly moving branch of applied mathematics. The bibliographies of the references I offer do give a foundation for understanding the new developments. Most of the material I have cited, and the references cited in them, are mathematically somewhat advanced and not

---

[18] S. M. Pincus, *Proc. Natl. Acad. Sci. U.S.A.* **88**, 2297 (1991).

easily accessible to a nonspecialist reader. Popular books such as James Gleick's *Chaos: Making a New Science* (Viking Press, New York, 1987) provide very attractive accounts of some of the excitement in the field, of some of the leading personalities, and of some of the claims. However, in my opinion a lay article inevitably must fail to convey the substance of chaos and fractals. Here, I have tried to indicate some of that substance, without figures and with only a few equations. This effort, too, must fail, but it should at least avoid misleading the reader about some of the achievements and the perplexities attendant on these new views of dynamics. A scholarly examination of the history of these trends in mathematics would show, as Ralph Abraham has done, that not all the fashionable and even faddish ideas are really new; there are strong antecedants going back to at least the early 1900s.

The study of chaos and fractals in a biological setting provides many valuable lessons, some of which I list below. Complex results do not imply complex causes. Simple systems need not behave in simple ways. Complex systems can have simple behavior. Randomness out does not imply randomness in (or inside). Small transient or distant causes can have large long-term effects. There can be instability at every point in phase space, while trajectories within it may be confined: aperiodic, never repeating, yet staying close together. Erratic behavior can be stable. Global behavior can differ from local behavior and may be predictable even when local behavior may be unpredictable. When a quantity changes, it can change arbitrarily fast. Deterministic, continuum dynamics can be intermittent. Trends can persist through seeming randomness. There can be scaling symmetry (i.e., self-similarity) across all levels of observation in systems having no characteristic scale of space or time. The region of phase space in which trajectories settle down can have noninteger dimension. Transitions from one dynamic mode of behavior to another can occur through bifurcations that can be subtle, catastrophic, or explosive. A nonlinear system, such as may be seen in a hydrodynamic field, may have regions with steady behavioral modes, separated by chaotic boundaries that are stable (e.g., the Red Spot of Jupiter, which may be a vortex stabilized by a chaotic surround, or which may be a soliton). Fluctuations may be enhanced and captured as a new form or function.

The list could go on, but the items above should convey the explanatory richness of the concepts of chaos and fractals and their potential for changing our viewpoints about determinism and randomness, about stability, evolution, speciation, noise. . . . Surely these are nontrivial yields to be had from a study of these subjects.

Perhaps for the first time we have a mathematics of complexity mature enough to describe motions and transformations within biochemical and physiological systems. But not all of those competent to understand the mathematical developments in detail are comfortable with the tendency, possibly a rampant distortion, to view more and more of biological processes through the spectacles of chaos and fractals. Are we ready to abandon all the old-fashioned models of deterministic systems contaminated by additive or multiplicative noise of various types? Is there really so much of chaotic dynamics in living systems? How can we tell?

Admittedly, deterministic chaotic dynamics can imitate many different kinds of what we previously thought of as random "noise." Above all there is a kind of efficiency or compactness about the concepts of chaos and fractals as applied to biological dynamics. They can dispel some of our confusion about properties of data, as I have tried to illustrate, and arguably they justify the claim that underneath many appearances of complexity lies some kind of deeper simplicity—an assumption that physicists have long made.

I have described some crude tests for the presence of fractal time, fractal space, and chaotic dynamics in living systems. There are no stand-alone, solid tests; however, there are indicators of the operation of chaotic dynamics and the presence of fractal space or time. The most important yield for biologists arising from the study of chaotic dynamics and fractals seems to me to pertain to the fundamental biological concept (originated separately by Sechenov, Bernard, and Cannon) of "homeostasis." Homeostasis says something about the stability, vitality, health, and persistence of a living system and its ability to accommodate perturbations without dying. The very nonlinear world of chaotic dynamics and fractal geometries seems to me to justify substitution of the term "homeodynamics" for homeostasis. Homeodynamics carries with it the potential for a deeper understanding of what it means for a complicated system to be stable and yet show very rich behaviors, including the possibilities of development of individuals and evolution of species. Our thinking itself is evolving as shown in the following diagram.

Biological Stability in Space and Time

Homeostasis $\xrightarrow{\text{Chaos}}$ Homeodynamics

(linear, limited)   Fractals   (nonlinear, general)

Glossary

In the setting of this book this glossary must necessarily be only semi-technical. The subject of fractals properly belongs to topology, whose contributions to dynamics cannot be appreciated without knowledge of sets, maps (linear and nonlinear), manifolds, metric spaces, and vectorfields. A brief reading list for some of the relevant mathematics might include the following:

1. *Encyclopedic Dictionary of Mathematics* (2 volumes), prepared by the Mathematical Society of Japan, MIT Press, Cambridge, Massachusetts, 1977.

2. M. Barnsley, "Metric Spaces, Equivalent Spaces, Classification of Subsets, and the Space of Fractals," in *Fractals Everywhere,* pp. 6–42, Academic Press, New York, 1988.

3. Robert L. Devaney, "Dynamics of Simple Maps," in *Chaos and Fractals: The Mathematics Behind the Computer Graphics* (Proceedings of Symposia in Applied Mathematics, Vol. 39) (R. L. Devaney and L. Keen, eds.), pp. 1–24, American Mathematical Society, Providence, Rhode Island, 1989.

4. Robert L. Devaney, *An Introduction to Chaotic Dynamical Systems,* Second ed., Addison-Wesley, New York, 1989.

5. H.-O. Peitgen and P. H. Richter, *The Beauty of Fractals: Images of Complex Dynamical Systems,* Springer-Verlag, New York, 1986.

6. Ian Stewart, *Does God Play Dice? The Mathematics of Chaos,* Basel Blackwell, Cambridge, Massachusetts, 1989.

7. R. Abraham and C. Shaw, *Dynamics: The Geometry of Behavior, Part One: Periodic Behavior; Part Two: Chaotic Behavior; Part Three: Global Behavior; Part Four: Bifurcation Behavior* (4-volume series), The Visual Mathematics Library, Aerial Press, Santa Cruz, California, 1982–1988.

Instruction to Reader

The following terms are partially defined and presented in alphabetical order: attractor (basin of attraction), bifurcation, chaos, complexity, cycle, dimension, dynamics, dynamical system (and state vector), fractals, limit cycle, linear, manifold, maps (mapping), noise, nonlinear, quasi-periodicity, spectral analysis, stability, state, vectorfield. Terms in italics in the definitions that follow are themselves defined elsewhere in the Glossary. The alphabetical order is not the logical order. Any reader wishing to use this Glossary as an introduction to modern nonlinear dynamics should read the entries in the following, more logical order: complexity, state, dynamics, dynamical system, manifold, maps, vectorfield, attractor, linear, nonlinear, bifurcation, chaos, fractals, dimension, limit cycle, quasi-periodicity, noise, spectral analysis, and stability.

Attractor (Basin of Attraction)

Modern *nonlinear dynamics* provides qualitative predictions of the asymptotic behavior of the system of interest in the long run. These predictions may hold even when quantitative predictions are impossible. In each *vectorfield* of a *dynamical scheme* there are certain asymptotic limit sets reachable sooner or later from a significant set of initial conditions. Each asymptotic set in each vectorfield of a dynamical scheme is an attractor. The set of initial states that tend to a given attractor asymptotically as time goes to +infinity comprise the basins of the attractor. Initial conditions from which the system will not reach the attractor belong to a separator. Attractors occur in three types: static (an attractor limit point); periodic (an attractive *limit cycle* or oscillation); and *chaotic* (meaning any other attractive limit set). If the attractor has a topological *dimension* that is not an integer it is considered "strange," and many *nonlinear* systems have such strange attractors (that are *fractals*). A strange attractor has weird geometry such as fractional dimension or nondifferentiability.

Bifurcation

Bifurcation theory asks how the equilibrium solution, or value, of interest for a dynamical system changes as some control parameter or variable is gradually changed. The equilibrium solution (itself a dynamical behavior or dynamical mode) can change in a subtle manner or in a very abrupt, catastrophic, or explosive manner at critical parameter values. Changes in states such as the freezing of liquid water into ice represent a familiar bifurcation. Changes in flows from laminar to turbulent as the velocity is increased (for a given geometry) also represent a familiar bifurcation.

Chaos

A readable introduction to the subject of chaos has been provided by Morrison.[19] The simplest view of chaos is that it is an irregular (aperiodic) fluctuation of a variable—unpredictable in the long term—generated by a fully deterministic process without noise. That view is all very well if one knows the deterministic system in advance, but if all one has is irregularly varying data, proof that the dynamics are chaotic is difficult, and may be impossible. It is important to note that even if long-term prediction of chaotic systems is impossible, accurate short-term prediction is possible. The chaotic orbit of the planet Pluto is an example.

Chaotic dynamics are deterministic, not random, although they may imitate various kinds of *noise*. Broad-band power spectra are often associated with chaotic dynamics. They may show the $1/f$ spectrum of Mandel-

---

[19] F. Morrison, "The Art of Modeling Dynamic Systems: Forecasting for Chaos, Randomness, and Determinism." Wiley (Interscience), New York, 1991.

brot noise. All of the known chaotic attractors have a fractal microstructure that is responsible for their long-term unpredictable behavior. The power *spectrum* of a periodic attractor is a discrete, or line, spectrum, but the power spectra of *chaotic motions on attractors* are usually continuous or "noisy-looking."

Farmer *et al.* computed the spectra for the Rössler chaotic attractor under six different parameter values for the simple equations generating the chaotic dynamics (see discussion by Schaffer[20]). For each of six parameter values the motion was indeed chaotic, but the spectra varied from sharp peaks (periodic behavior) to featureless spectra (broad band noise). Similar results from other chaotic systems give a warning: dividing a single chaotic trajectory into subsamples, so that the data sets are relatively short, causes the same chaotic process to reveal itself through very different spectra.

The onset of chaos is often preceded by an apparently infinite cascade of period doublings, so that by the time one reaches the chaotic region all cycles of period 2^n have gone unstable, even though they are still there. (Odd periods also appear; see Devaney videotapes for illustration.) As a result, the chaotic orbit wanders among former basins of attraction. We can conclude that *spectral analysis*, even of rather dense biological data, is not a reliable means to identify chaotic dynamics. However, the spectra are useful in their own right because they do identify such periodicities as may be present, regardless of origin.

Formal definitions of chaos require that systems show, as typical orbits on their attractors, trajectories with a positive Liapunov exponent. Chaos is characterized by the fact that adjacent initial conditions lead to exponentially diverging trajectories (i.e., there is extremely close dependence on initial conditions in the long run), and the exponent describing this divergence is the Liapunov exponent. In addition, the Kolmogorov–Sinai criterion requires that a chaotic system have nonzero entropy.

There is currently active research under way on the development of practical algorithms that can be used to compute numerically the *dimensions* and Liapunov numbers (which I do not define here), given the values of some variables as a function of time. The algorithms being used have many potential problems. There is debate on two aspects: (1) the requirements for the size of the data set being analyzed and (2) the effects of noise, large derivatives, and geometry of the attractor. Because of these difficulties, unambiguous interpretation of published reports is difficult. Any claim for "chaos" based solely on calculation of dimension or the

---

[20] W. M. Schaffer, *in* "Chaos and Biological Systems" (H. Degn, A. V. Holden, and L. F. Olsen, eds.), p. 233. Plenum, New York, 1987.

Liapunov numbers without additional supporting evidence such as well-characterized bifurcations must be viewed with extreme skepticism at present.[21]

Fractals and chaos are separate concepts and should not be confounded. "Chaos" refers to the dynamics of a system; "strange attractor" characterizes the (often fractal) geometry of an attractor. Chaotic dynamics can have attractors that are not strange; nonchaotic dynamics can display strange attractors, and not all strange attractors are chaotic!

Chaotic dynamics may describe the transient motions of a point in a phase space of high *dimension*, even infinity. The lower dimension region of phase space to which the motions tend asymptotically as time goes to infinity (the chaotic or strange *attractor*) will usually have a *fractal* dimension. It is thought that most strange or chaotic attractors have fractal dimensions, usually greater than 2^+. Even though in principle dimension can reach infinity, as a practical matter biological data rarely support a claim for the dimensionality of an attractor higher than about 6^+-10^+. (White noise, which has no attractor, can generate an apparent dimension of about 10 using the correlation dimension algorithms.)

Some physicists view the world algorithmically, saying that the existence of predictable regularities means that the world is algorithmically compressible. For example, the positions of the planets in the solar system over some interval constitute a compressible data set, because Newton's laws may be used to link these positions at all times to the positions and velocities at some initial time. In this case Newton's laws supply the necessary algorithm to achieve the compression. However, there is a wide class of theoretical and perhaps actual systems, the chaotic ones, that are not algorithmically compressible.

Whenever one has a mathematical model of a two-variable limit cycle oscillator (a nonlinear system) and adds just one more variable to it (such as a depletable source that regenerates at a finite rate), one is extremely likely to discover a chaotic regime after some fiddling with the parameters.[22] To produce chaos through nonlinear models based on ordinary differential equations, one needs at least three continuous independent variables. However, in the case of nonlinear discrete models (e.g., with finite difference equations) the dynamics of a single variable can have chaotic regimes.

Because random processes are often characterized by their interevent histograms, and because it is well known that in the Poisson process case

[21] L. Glass and M. C. Mackey, "From Clocks to Chaos: The Rhythms of Life," p. 53. Princeton Univ. Press, Princeton, New Jersey, 1988.
[22] O. E. Rössler, *Ann. N.Y. Acad. Sci.* **504**, 229 (1987).

the interevent histogram is an exponential function, a problem arises: namely, deterministic chaotic systems can also give rise to exponential interevent histograms. Thus, it is not a simple matter to distinguish between random *noise* and deterministic *chaos*, and it is always possible that irregular dynamics in many systems that have been ascribed to chaos may be noise, or vice versa. Glass and Mackey[21] remark, "The strongest evidence for chaotic behavior comes from situations in which there is a theory for the dynamics that shows both periodic and chaotic dynamics as parameters are varied. Corresponding experimental observation of theoretically predicted dynamics, including irregular dynamics for parameter values that give chaos in the deterministic equations, is strong evidence that the experimentally observed dynamics are chaotic." Other approaches to the identification of chaos include the power *spectrum*, Poincaré map, Liapunov number, and dimension calculations (but see the caveat, above, by Glass and Mackey).

If a chaotic attractor exists inside a basin of attraction in the phase space of a dynamical system, then globally the dynamics will be stable as time goes on, although microscopically, within the attractor, they may be locally unstable. However, if the parameters of the system change in such a way that a chaotic attractor collides with the boundaries of its attraction basin, all the chaotic trajectories will become only transients (metastable chaos), and eventually the system escapes the basin of chaos and evolves toward some other attractor in the phase space. When topological, unstable chaos occurs between two stable attractors, the basins acquire a *fractal* nature with self-similarity at their boundaries, which then possess fine structure at each scale of detail, creating a fuzzy border.

It is an open question whether low-dimension chaos and nonlinear mappings are relevant to the nervous system. Freeman[8] has argued strongly that they are. Lloyd Partridge (personal communication, 1991) points out that there seems to be no universally accepted definition of what constitutes deterministic chaos, but at least two things seem to appear in all definitions. First, chaos describes the continuously changing pattern of some variable in a system in which future values are determined by the operation of rules of change on the present value. Second, those rules are very sensitive to initial conditions. The generation of a chaotic response is always determined by the internal rules of the system while that system has only constant (including zero) input or uniform periodic input. Outputs that are considered chaotic continue to change in a never exactly repeating manner, yet they stay within specific bounds. Thus, while individual response values cannot be predicted, the bounds of possible responses can only be learned by observations. Partridge closes his presentation as follows: "I conclude that the contributions of nonlinearity, discontinuity, feedback, and dynamics to neural function demands serious examination. At the same time, the division between truly chaotic and nonchaotic effects that may result from these properties falls within the range that these studies should span and does not represent an important division. Thus, while understanding of the results of formal study of chaotic behavior can be a valuable background for neural science, distinguishing sharply between chaos and the variety of related phenomena, in particular neural function, may be relatively useless.''

Among some of those who understand chaotic dynamics very deeply, such as Stuart Kauffman, there is a lingering feeling that the emergence of order in self-organizing systems is in some sense "anti-chaotic."[23] Using Boolean networks as models, Kauffman tries to show that state cycles, as dynamic attractors of such networks, may arise because Boolean networks have a finite number of states. A system must therefore eventually reenter a state that it has previously encountered. In less formal terms I have tried to advance the same idea under the term "homeodynamics," discussed later. Kauffman concludes that parallel-processing Boolean networks poised between order and chaos can adapt most readily, and therefore they may be the inevitable target of natural selection. According to this view, evolution proceeds to the edge of chaos. Kauffman concludes, "Taken as models of genomic systems, systems poised between order and chaos come close to fitting many features of cellular differentiation during ontogeny—features common to organisms that had been diverging evolutionarily for more than 600 million years. . . . If the hypotheses continue to hold up, biologists may have the beginnings of a comprehensive theory of genomic organization, behavior and capacity to evolve." Although I am somewhat dubious that Boolean networks are the best form for modeling the dynamics of biological systems (Boolean networks can have arbitrary dynamics violating physical law), I think Kauffman has captured an important idea.

Complexity

There is no general agreement as to what constitutes a complex system. Stein[24] remarks that complexity implies some kind of nonreducibility: the behavior we are interested in evaporates when we try to reduce the system to a simpler, better understood one. In my opinion the best definition of

[23] S. A. Kauffman, *Sci. Am.* **265,** 78 (1991).
[24] D. L. Stein, *in* "Lectures in the Sciences of Complexity" (D. L. Stein, ed.), p. xiii. Addison-Wesley, Redwood City, California, 1989.

a complex system is that according to Rosen.[25] He notes that complex systems are counterintuitive, that is, they may do things that are unexpected and apparently unpredictable. Their behavior may show emergent properties, because complex systems do not appear to possess a single, universal, overarching description such as those postulated for simpler physical, mechanical, or thermodynamic systems. For example, an organism admits a multiplicity of partial descriptions, and each partial description, considered by itself, describes a simpler system, that is, one with a prescribed set of *states* and a definite *dynamical law* or state transition rule. Thus, an organism can present itself to different observers in various ways, each of which can be described simply according to a standard Newtonian-like paradigm, but the description will be different for different observers. A complex system is one that cannot be comprehensively described. A complex system is not effectively explainable by a superposition of the simple subsystem descriptions; it does not fit the Newtonian dynamical scheme.

Cycle

I use "cycle" and "rhythm" synonymously to describe a time history of a variable (usually referred to as a time series) in which there is a recurrent amplitude variation with statistical regularity (stationarity). The duration of one variation back to its original starting point is the length of one cycle, or the period. The reciprocal of period is frequency. A noise-free, sine-wave generator produces perfect cycles of constant period, amplitude, phase, and mean value. In the presence of certain kinds of *noise* there may be wobble, that is, there may be variation on the length of the period, as well as on amplitude, but if the dispersion is not so great as to obscure the underlying periodicity around some average period, we may still wish to claim the presence of a cyclic process. The shape of the recurring process can range from a train of spikes (as in nerve impulses, which have height but little width) separated by intervals, on one extreme, to a pure sine wave which is very rounded with no intervals between events on the other extreme. (Fourier-based spectral analysis deals well with the latter, but poorly with the former shapes.)

Time histories that are cyclic can be (1) periodic, (2) nearly periodic (this is an informal term, used when a little unexplained wobble is observed), or (3) *quasi-periodic*. In contrast is the acyclic (usually called aperiodic) time series in which there is no regularity in the occurrence of any amplitude value. Aperiodic time histories can be generated by

[25] R. Rosen, *in* "Quantum Implications" (B. J. Hiley and F. T. Peat, eds.), p. 314. Routledge and Kegan Paul, London, 1987.

deterministic chaotic dynamics, by some random processes, or by a mixture of both.

Dimension

The familiar three dimensions of Euclidean space give us an intuitive feeling for the meaning of the term. However, the concept of a topological dimension is more elaborate and not easy to explain in lay terms. As Peterson[1] remarks, "Experiments with soap films show the tremendous variability in the shapes of surfaces. Computer-generated pictures of 4-dimensional forms reveal unusual geometric features. The crinkly edges of coast lines, the roughness of natural terrain, and the branching patterns of trees point to structures too convoluted to be described as 1-, 2-, or 3-dimensional." Instead mathematicians express the dimensions of these irregular objects as decimal fractions rather than whole numbers. In the case of whole numbers, for example, any set of four numbers, variables, or parameters can be considered as a four-dimensional entity. In the theories of special and general relativity, three-dimensional space and time together make up a four-dimensional continuum. Going beyond the fourth dimension requires only adding more variables.

The *Encyclopedic Dictionary of Mathematics* states (Volume I, Section 121, "Dimension Theory): "Toward the end of the 19th century, G. Cantor discovered that there exists a 1-to-1 correspondence between the set of points on a line segment and the set of points on a square; and also, G. Peano discovered the existence of a continuous *mapping* from the segment onto the square. Soon, the progress of the theory of point-set topology led to the consideration of sets which are more complicated than familiar sets, such as polygons and polyhedra. Thus it became necessary to give a precise definition to dimension, a concept which had previously been used only vaguely." (The rest of the section gives a highly technical definition of the dimension of metric spaces.)

Glass and Mackey[21] offer this definition of one of the simplest meanings of (capacity) dimension. "Consider a set of points in N-dimensional space. Let $n(\varepsilon)$ be the minimum number of the N-dimensional cubes of side ε needed to cover the set. Then the dimension, d, of the set is

$$d = \lim_{\varepsilon \to 0} \frac{\log \eta(\varepsilon)}{\log(1/\varepsilon)} \tag{G1}$$

For example, to cover the length of a line L, $n(\varepsilon) = L/\varepsilon$, and d is readily computed to be 1. Similarly, for a square of side L, we have $n(\varepsilon) = L^2/\varepsilon^2$ and $d = 2$. Unfortunately, the many different views of dimension touch on mathematical issues much too deep for this chapter. A sense of those intricacies can be found in Mayer-Kress.[26] It will have to suffice for my purposes to invoke the technical capacity dimension or the intuitive notion that a dynamical system has a hyperspace defined by one dimension for each dynamical degree of freedom. These two views do not always coincide. For example, the capacity dimension of a limit cycle based on a two-dimensional nonlinear differential equation (e.g., the van der Pol equation) is 1 for the limit set of the orbit, but 2 for the basin of attraction feeding asymptotically onto the orbit. The number of degrees of freedom in the van der Pol oscillator is 2 in either the separated x and y form or the acceleration, velocity, position form for a single variable. (Both forms are presented elsewhere in the Glossary.) In spite of the difficulties, "dimension" can be used as a qualitative term corresponding to the number of independent variables needed to specify the activity of a system at a given instant. It also corresponds to a measure for the number of active modes modulating a physical process. Layne *et al.*[27] point out that it is therefore a measure of complexity. However, many also make use of the correlation dimension based on the algorithm proposed by Grassberger and Procaccia.[17] An application can be found in Mayer-Kress *et al.*[28]

Dynamics

The basic concepts of dynamics have been expressed by Abraham and Shaw[29] as follows: "The key to the geometric theory of dynamical systems created by Poincaré is the phase portrait of a *dynamical system*. The first step in drawing this portrait is the creation of a geometric model for the set of all possible states of the system. This is called the state space. On this geometric model, the dynamics determined a cellular structure of *basins* enclosed by separatrices. Within each cell, or basin, is a nucleus called the *attractor*. The *states* which will actually be observed in this system are the attractors. Thus, the portrait of the dynamical system, showing the basins and attractors, is of primary importance in applications. . . . The history of a real system [can be] represented graphically as a trajectory in a geometric state space. Newton added the concept of the instantaneous velocity, or derivative, of vector calculus. The velocity

[26] G. Mayer-Kress, ed., "Dimensions and Entropies in Chaotic Systems." Springer-Verlag, New York, 1986.

[27] S. P. Layne, G. Mayer-Kress, and J. Holzufs, *in* "Dimensions and Entropies in Chaotic Systems: Quantification of Complex Behavior" (G. Mayer-Kress, ed.), p. 248. Springer-Verlag, Berlin, 1986.

[28] G. Mayer-Kress, F. E. Yates, L. Benton, M. Keidel, W. Tirsch, S. J. Poppl, and K. Geist, *Math. Biosci.* **90,** 155 (1988).

[29] R. H. Abraham and C. D. Shaw, "Dynamics: The Geometry of Behavior, Part 1: Periodic Behavior," p. 11. Ariel Press, Santa Cruz, California, 1982.

vectorfield emerged as one of the basic concepts. Velocities are given by the first time derivative (tangent) of the trajectories. The prescription of a velocity vector at each point in the state space is called a velocity *vectorfield*."

Dynamical System and State Vector

A system with n degrees of freedom, that is, with n different, independent variables, can be thought of as living in an n-space. The n coordinates of a single point in the n-space define all the n variables simultaneously. If the motion of the point in n-space follows some rule acting on the positions (magnitudes) of the variables and their velocities, then that rule defines a dynamical system; the n variables are then more than a mere aggregate. The point in n-space that is moving is called a configuration or state vector. The rule defining the dynamical system expresses the law relating the variables and the parameters (Parameters are more or less constant or very slowly changing magnitudes that give the system its particular identity). Influences (controls) may act on the dynamical system either through the variables or the parameters. Controls change the motions of the system. A simple dynamical scheme is a function assigning a smooth *vectorfield* to the *manifold* of instantaneous states for every point in the manifold of control influences. The smooth function is a *mapping*.

Fractals

Fractal geometry is based on the idea that the natural world is not made up of the familiar objects of geometry: circles, triangles, and the like. The natural world of clouds, coastlines, and mountains cannot be fully described by the geometry of circles and squares. Fractals are structures that always look the same, either exactly or in a statistical sense, as you endlessly enlarge portions of them. According to a usage becoming standard, fractal forms ordinarily have three features: heterogeneity, self-similarity, and the absence of a well-defined (characteristic) scale of length. The first feature is not an absolute requirement, but the other two are. There must be nontrivial structure on all scales so that small details are reminiscent of the entire object. Fractal structures have both irregularity and redundancy and, as a result, they are able to withstand injury.

Fractals were first conceived by Benoit Mandelbrot. They are geometric fragments of varying size and orientation but similar in shape. It is remarkable that the details of a fractal at a certain scale are similar (although not necessarily identical) to those of the structure seen at larger or smaller scales. There is no characteristic scale. All fractals either have this look-alike property called self-similarity, or, alternatively, they may be self-affine. As Goldberger *et al.*[4] comment, "Because a fractal is composed of similar structures of ever finer detail, its length is not well defined. If one attempts to measure the length of a fractal with a given ruler, some details will always be finer than the ruler can possibly measure. As the resolution of the measuring instrument increases, therefore, the length of a fractal grows. Because length is not a meaningful concept for fractals, mathematicians calculate the *'dimensions'* of a fractal to quantify how it fills space. The familiar concept of dimension applies to the objects of classical, or Euclidean, geometry. Lines have a dimension of one, circles have two dimensions, and spheres have three. Most (but not all!) fractals have noninteger (fractional) dimensions. Whereas a smooth Euclidean line precisely fills a one-dimensional space, a fractal line spills over into a two-dimensional space." A fractal line, a coastline, for example, therefore has a dimension between one and two. Likewise a fractal surface, of a mountain, for instance, has a dimension between two and three. The greater the dimension of a fractal, the greater the chance that a given region of space contains a piece of that fractal.

Peterson[1] points out that for any fractal object of size P, constructed of smaller units of size p, the number, N, of units that fits into the object is the size ratio raised to a power, and that exponent, d, is called the Hausdorff dimension. In mathematical terms this can be written as

$$N = (P/p)^d \qquad \text{(G2)}$$

or

$$d = \log N/\log(P/p) \qquad \text{(G3)}$$

This way of defining *dimension* shows that familiar objects, such as the line, square, and cube, are also fractals, although mathematically they count as trivial cases. The line contains within itself little line segments, the square contains little squares, and the cube little cubes. (This is self-similarity, but without heterogeneity.) But applying the concept of Hausdorff dimension to other objects, such as coastlines, gives a fractional dimension. Fractals in nature are often self-similar only in a statistical sense. The fractal dimension of these shapes can be determined only by taking the average of the fractal dimensions at many different length scales.

The correlation dimension (not defined here, but see Grassberger and Procaccia[17]) serves as a lower bound for the fractal dimension. The fractal dimension itself is a number bounded below by the topological dimension (0 for a point, 1 for a line, 2 for a surface, etc.) and above by the Euclidean dimension in which the fractal is located (a point is on a line, Euclidean dimension 1; a line is on a surface, Euclidean dimension 2; etc.) The topological dimension approximately corresponds to the number of independent variables required for the definition of the function. The Euclidean dimension is the dimension of the range of the function.

Mandelbrot[30] discusses *dimension* in very advanced terms using the concept to define a fractal: "A fractal is by definition a set for which the Hausdorff–Besicovitch dimension strictly exceeds the topological dimension" (p. 15). But he thought this definition incomplete.

Chaotic attractors are all fractals, and usually they are of a dimension greater than 2, but not all systems with self-similar phase space (fractal properties) are necessarily ascribable to a chaotic system. I want to emphasize again that chaos and fractals should not be confused with each other. "Chaos" is about dynamics; "fractals" is about geometry. It happens that chaotic attractors often have a fractal microstructure geometry, but that does not make fractals and chaos synonymous.

In the case of processes, fractals can be stochastic or they can be deterministic. Stochastic fractals are an example of fractional Gaussian *noise* where the power *spectrum* contains large amounts of low frequency energy. Fractal noise ($1/f^m$ spectrum, $m \cong 1$, or, more generally, a noninteger) is very structured and is a long-time scale phenomenon. Measures that deal with short-time scales such as correlation functions and interspike interval histograms cannot assess fractal activity.

To identify fractals in objects, for example, in the case of branching structures common in biological objects, two types of scaling can be compared, one exponential and one fractal. If a tree structure follows the simple exponential rule: $d(z, a) = d_0 e^{-az}$, where $d(z, a)$ is the average diameter of tubes in the zth generation, d_0 is the diameter of the single parent trunk or vessel, and a is the characteristic scale factor, then a *semilog plot* of ln $d(z, a)$ versus z will give a straight line with negative slope, $-a$. In contrast, a fractal tree has a multiplicity of scales, and each can contribute with a different weighting or probability of occurrence that is revealed by an inverse power law. In that case, $d(z) \propto 1/z^\mu$, where μ is the power law index, $\mu = 1 - D$, and D is the fractal dimension.[12] Now a *log–log plot* of ln $d(z)$ versus ln z gives a straight line of negative slope $-\mu$. (There may be harmonic modulation of the data around the pure power law regression line without overcoming the fractal scaling.[3])

Limit Cycle

The limit cycle is a nonlinear cyclic time series creating, when abstracted, a closed orbit (on certain plots described below) of a wide variety of shapes (but not including that of a circle; the circle represents a pure,

---

[30] B. Mandelbrot, "The Fractal Geometry of Nature," Revised Ed. Freeman, New York, 1983.

harmonic, linear oscillation in these plots). In the single-variable picture of a limit cycle we plot the velocity of a variable against its magnitude. However, in some systems under nonholonomic constraints, velocities and magnitudes or positions are independently specifiable; thus the plot of velocity versus magnitude of a single variable is actually a two-variable plot. Poincaré studied nonlinear differential equations with two variables in which it is possible to have an oscillation that is reestablished following a small perturbation delivered at any phase of the oscillation. He called such oscillations stable limit cycles. One of the most thoroughly examined versions is the simple two-dimensional differential equation proposed by van der Pol to model nonlinear limit cycle oscillations:

$$\frac{d^2\mu}{dt^2} - \varepsilon(1 - \mu^2)\frac{d\mu}{dt} + \mu = \beta\cos(\alpha t) \tag{G4}$$

When $\beta = 0$ there is a unique, stable limit cycle oscillation. Alternatively, the van der Pol oscillator is given by the following pair of equations:

$$\frac{dx}{dt} = \frac{1}{\varepsilon}\left(y - \frac{x^3}{3} + x\right) \tag{G5a}$$

$$\frac{dy}{dt} = \varepsilon x \qquad \varepsilon > 0 \tag{G5b}$$

A one-dimensional, nonlinear finite difference equation (e.g., the logistic equation discussed under the section on maps) at certain parameter values can also generate stable limit cycles.

It has been a challenge to generalize the limit cycle concept beyond the two-dimensional (two degrees of freedom, two variable) case. However, generalization is essential for the understanding of homeodynamic stability of biological systems.[31] Complex modes of behavior almost always appear if two nonlinear oscillatory mechanisms are coupled, either in series or in parallel. Any interaction between two limit cycles can produce complex periodic oscillations, or chaos. (In biological systems the term "near-periodic" best describes the motions, and that this result is to be expected of homeodynamic systems of all kinds.)

Linear

A linear term is one which is first degree in the dependent variables and their derivatives. A linear equation is an equation consisting of a sum of linear terms. If any term of a differential equation contains higher

[31] F. E. Yates, *Can. J. Physiol. Pharmacol.* **60**, 217 (1982).

powers, products, or transcendental functions of the dependent variables, it is nonlinear. Such terms include $(dy/dt)^3$, $u(dy/dt)$, and $\sin u$, respectively. $(5/\cos t)(d^2y/dt^2)$ is a term of first degree in the dependent variable y, whereas $2uy^3(dy/dt)$ is a term of fifth degree in the dependent variables u and y.

Any differential equation of the form below is linear, where y is the output and u the input:

$$\sum_{i=0}^{n} a_i(t)\frac{d^i y}{dt^i} = u \qquad (G6)$$

If all initial conditions in a system are zero, that is, if the system is completely at rest, then the system is a linear system if it has the following property: (1) an input $u_1(t)$ produces an output $y_1(t)$, and (2) an input $u_2(t)$ produces an output $y_2(t)$, (3) then input $c_1u_1(t) + c_2u_2(t)$ produces an output $c_1y_1(t) + c_2y_2(t)$ for all pairs of inputs $u_1(t)$ and $u_2(t)$ and all pairs of constants c_1 and c_2.

The principle of superposition follows from the definition above. The response $y(t)$ of a linear system due to several inputs $u_1(t), u_2(t), \ldots, u_n(t)$ acting simultaneously is equal to the sum of the responses of each input acting alone, when all initial conditions in the system are zero. That is, if $y_i(t)$ is the response due to the input $u_i(t)$, then

$$y(t) = \sum_{i=1}^{n} y_i(t) \qquad (G7)$$

Any system that satisfies the principle of superposition is linear. All others are nonlinear.

Manifold

A manifold is a geometrical model for the observed states of a dynamical or experimental situation and is identical to the n-dimensional state space of a model of the situation. Each instantaneous state has a location in n-space, and all those locations achievable by the system following a rule for its motion constitute a manifold.

Maps (Mapping)

Functions that determine *dynamical systems* are called mappings or maps. This terminology emphasizes the geometric process of taking one point to another. The basic goal of the theory of dynamical systems is to understand the eventual (i.e., asymptotic) behavior of an iterative or ongoing process. In dynamical systems analysis we ask, Where do points go, and what do they do when they get there? The answer is a mapping.

A map is a rule. The rule can be deterministic or statistical, linear or nonlinear, continuous or discrete. In *dynamical systems* a mapping is the rule or law governing motion and change as a function of state configuration, initial velocities and positions, and constraints. An example of a discrete, nonlinear dynamical law of one dimension is

$$x_{n+1} = f(x_n), \qquad f \neq \text{a constant} \tag{G8}$$

where f is a function carrying out the mapping. A notation for this map is $f: x \rightarrow x$, which assigns exactly one point $f(x) \in x$ to each point $x \in x$, when (x, d) is a metric space. A common example is the nonlinear population growth model or logistic map:

$$x_{n+1} = f(x_n) = rx_n(1 - x_n) \tag{G9}$$

In this dynamic rule time appears as a generational step (discrete map). The level of the population, x, as a function of time (or iteration number) goes from some initial condition toward extinction (0), constancy, oscillation, period doubling oscillations, or chaos as r is extended in the range of positive values starting close to zero and increasing past 3. The switching of dynamical outcomes (extinction, constancy, oscillation, period doubling, chaos) occurs at successive critical values of r, at which the iterated dynamics undergo a bifurcation (change in behavioral mode).

For the logistic equation above as often displayed for illustration, plotting x_{n+1} against x_n, at a given value of parameter r, results in a parabola that opens downward. The parabola sits on the x_n axis between 0 and 1. The plotting of one value of x to the next creates a graph called a return map, which is also a transition function.

Maps can address real or complex numbers. An important example involving complex numbers is

$$z_{n+1} = z_n^2 + c \tag{G10}$$

where c is a complex constant and the mapping rule is

$$f_c: z \rightarrow z^2 + c \tag{G11}$$

This mapping results in quadratic Julia sets and the famous Mandelbrot set, which have led to astonishing computer graphical demonstrations.[32]

In summary, if a rule exists that assigns to each element of a set A an element of set B, this rule is said to define a mapping (or simply, map)

---

[32] A. Douady, *in* "The Beauty of Fractals" (H.-O. Peitgen and P. H. Richter, authors), p. 161. Springer-Verlag, Berlin, 1986.

function or transformation from A into B. (The term transformation is sometimes restricted to the case where $A = B$.) The expression $f: A \rightarrow B$ or $(A \rightarrow B)$ means that f is a function that maps A into B. If $f: A \rightarrow B$ and $A \in A$, then $f(a)$ denotes the element of B, which is assigned to A by f.

Noise

Noise is any unwanted (from the point of view of the investigator) variation in data. Noise may be stochastic (random) or deterministic; it is traditionally thought of as being random and pernicious. Curiously, the presence of low levels of noise can actually improve detection of weak signals in systems with stochastic resonance. In that case a bistable system operates as a detector when a sufficiently strong external force—a signal—provokes it into a change of state. If the force is too weak, the system stays in its original state and detects no signal. The addition of noise injects energy into the variations or fluctuations in each state and changes the probability, if the barrier between states is low, that the system might change state. Then, when a weak signal arrives, combined with the noise, a state change may be accomplished, whereas the signal, without the energizing noise, would not be sufficient to overcome the energy barrier.

One way to describe noise is through its *spectral analysis* or its distributional characteristics. The discrete events making up a Poisson distribution generate an exponential interval histogram. Brownian motion generates a $1/f^2$ *spectrum*; fractal, brown (Mandelbrot, Zipfian) noise generates a $1/f^1$ spectrum; band-limited, "white" (or uniform) noise generates a $1/f^0$ spectrum (where f is frequency). More generally, $1/f^m$ spectra, where m is not an integer, characterizes a fractal system.[11]

Nonlinear

Because nonlinearity is a more general concept than is linearity, it can be understood from the very restricted definition of linearity already given. There are many different kinds and causes of nonlinearity, some of them "hard," such as sharp thresholds or saturations, and some of them "soft," such as found in some memory elements of a system. For a good discussion of linearity and nonlinearity, see DiStefano *et al.*[33]

Quasi-periodicity

An *attractor* that is a torus (i.e., the surface of a doughnut) and that allows a trajectory winding around it an infinite number of times, filling its surface but never intersecting itself, describes quasi-periodicity. The

[33] J. J. DiStefano III, A. R. Stubberud, and I. J. Williams, "Feedback and Control Systems" (Schaum's Outline Series), 2nd Ed. McGraw-Hill, New York, 1990.

attractor has integral dimensions and is not "strange." It can be created in the case of two rhythms that are completely independent, whose phase relationships we examine. The phase relations between the two rhythms will continually shift but will never repeat if the ratio between the two frequencies is not rational. Glass and Mackey[21] note that the dynamics are then not periodic, but they are also not *chaotic* because two initial conditions that are close together remain close together in subsequent iterations. If two periodic motions have periods with a common measure, both being integer multiples of the same thing, and are added together, then their result is itself periodic. For example, if one motion has a period of 3 sec and the other 5 sec, their (linear) combination will repeat every 15 sec. But if there is no common measure, then the motion never repeats exactly, even though it does almost repeat—thus the term "quasi-periodic." Quasi-periodicity is often found in theoretical classic, conservative dynamics, although it is not considered to be a motion typical of a general dynamical system in the mathematical world. In the physical/biological world something that might be called "near-periodicity" is very typical of observational data. Whether this ubiquitous near-periodicity is formal quasi-periodicity is an open question. Some think it is *chaos,* but have not proved it for biological data sets. Elsewhere I have argued that near-periodicity is neither quasi-periodicity nor chaos, but the temporal organization to be expected of a homeodynamic, complex, system.[34]

Spectral Analysis

Classic Fourier-based spectral analysis consists of fitting time history data with a linear series of sines and cosines. Any recurrent process can be modeled by a Fourier series, but many of the terms in the series will simply adjust the shape of the fit. The Fourier series is harmonic, and all the terms have commensurate frequencies (the quotient of any two frequencies is a rational number). Spectral analysis consists of partitioning the variance of a variable undergoing a time history into frequency bands or windows or bins. In the absence of noise a linear, additive mixture of pure sines and cosines will lead to a line spectrum with extremely sharp "peaks" at the relevant frequencies. Roughly, the "power" at each frequency where there is a line is given by the square of the amplitude of the periodic term having that frequency.

A Fourier series model of raw biological time history data usually produces broad "peaks" rather than a line spectrum. One must then have a theory of peak broadening. One possibility is that the underlying process

---

[34] F. E. Yates, *in* "Dynamics and Thermodynamics of Complex Systems" (D. C. Mikulecky and M. Witten, eds.), in press. Pergamon, New York, 1992.

was *quasi-periodic* rather than purely periodic. Another notion would be that the underlying process was purely periodic but contaminated by *noise*. Still another hypothesis would be that the underlying dynamics were *chaotic*. (Chaotic dynamical systems can produce near-periodicity as well as other spectral pictures.) Spectral analysis alone cannot resolve the underlying nature of the generator of a time history. Many theoretical systems can start out quasi-periodic, but as excitation of the system is increased the motion may become "random" or, more exactly, *chaotic*. In the *quasi-periodic* regime motions can be decomposed into a Fourier series with a few fundamental frequencies and overtones. The spectrum will consist of a small number of possibly sharp lines. In the *chaotic* regime the trajectories sample much more (perhaps all) of the allowed phase space, and the spectrum becomes broad but may still manifest some frequency bands in which there are hints of "peakedness." But chaotic dynamics can also produce a $1/f$ spectrum of Mandelbrot *noise*.

Stability

Stability is as difficult to define as is *dimension*. It has multiple meanings. For linear systems we have a complete theory of stability, but for nonlinear systems there can be multiple interpretations. For example, if trajectories from nearby initial conditions stay close to each other and asymptotically approach a fixed point, or an orbit, or else wind themselves around the surface of an invariant 2-torus, we can easily imagine that the dynamical system is stable in the region of those initial conditions. The situation becomes more confusing if from nearby conditions trajectories diverge exponentially but nevertheless asymptotically find a low-dimension, *chaotic attractor* that provides the limit set for all points generated by the trajectories as time approaches infinity. We could then think of the attractor as defining a bounded behavior for the dynamical system and consider the system to be globally stable on the attractor.

The most useful concept of stability for nonlinear systems, in my opinion, is asymptotic orbital stability. It pictures a limit cycle in more than topological one dimension, and takes the view that nearby trajectories converge to an orbital process in which certain states (nearly) recur at (nearly) identical intervals. There may be wobble on amplitudes and wobble on frequencies, but same conditions are seen repetitively, even though the repetition is not precisely periodic. It is near-periodic. (I prefer this term to *quasi-periodic* because the latter has a precise meaning that may be too restricted to account for the observed recurrent behavior of most biological processes.) For the purposes of this chapter, a nonlinear dynamic system will be considered stable if it asymptotically approaches a limit point, a limit cycle, or a chaotic attractor as it evolves from its initial conditions. However, in a broader view, according to the homeodynamic heuristic,[31] complex dynamical systems express their stability in a global, limit cycle-like, near-periodic motion. (I have not achieved a formal proof of this conjecture.)

State

A state is a set of data that in the deterministic limit (of a state-defined system) gives us all that we need to know to predict future behavior. The present state is an input, some deterministic dynamic rule operates on it, and the future state is the output. The state is a vector whose values allows the estimation of future states. In the absence of noise, reconstruction of a state space can be accomplished even from a single time series as partial information. In other words, we can work with data whose dimension is lower than that of the true dynamics. In the presence of noise the reconstruction of the state space is not always possible. The concept of state and state-determined systems follows from the Newtonian–causal tradition, and it fails to deal adequately with *complexity*.[25,34]

Vectorfield

A vectorfield is a *mapping* that assigns a tangent vector or velocity to each point in some region of interest in a *manifold*. For each control bearing on a dynamical system, the dynamical system rule or mapping creates a particular vectorfield, giving the positions and motions for the state vector point of the system. The vectorfield is a model for the habitual tendencies of the situation to evolve from one state to another and is called the dynamic of the model. The vectorfield may have slow or fast regions in it, according to the velocity of the configurational point.

A vectorfield is a field of bound vectors, one defined at (bound to) each and every point of the state space. The state space, filled with trajectories, is called the phase portrait of the dynamical system. The velocity vectorfield can be derived from the phase portrait by differentiation in state-determined systems. The phrase *dynamical system* specifically denotes this vectorfield. For analytic tractability, we like to hypothesize that the vectorfield of a model of a dynamical system is smooth, meaning that the vectorfield consists of continuous derivatives with no jumps, no sharp corners.

Note: Much of dynamical systems theory addresses state-determined systems, whose velocities are specified by the state, for example, $\dot{X} =$

$f(X, \ldots)$. But complex biological systems may not be state-determined systems.[34] (This issue depends somewhat on how one defines "state.")

Acknowledgments

I thank Laurel Benton for valuable help with the manuscript, including clarifying discussions. This work was supported by a gift to the University of California, Los Angeles, for medical engineering from the ALZA Corporation, Palo Alto, California.

Author Index

Numbers in parentheses are footnote reference numbers and indicate that an author's work is referred to although the name is not cited in the text.

A

Abraham, R. H., 664
Abraham, R., 656
Abramowitz, M., 356
Ackers, G. K., 7, 123, 125, 372, 406, 407, 408, 409, 413, 416, 417, 418, 420, 421(15), 423(4), 425(20, 21, 22, 23), 472, 473, 475, 476, 477(9)
Ackley, D. H., 636
Acton, F. S., 111, 116(6)
Adamson, A. W., 280
Akaike, H., 64, 348
Alcala, J. R., 237, 303, 357
Alexander, S. L., 548, 555(10)
Allemand, C., 305
Alpert, B., 237
Altschul, S. F., 576, 585(8), 597, 598, 600, 601(49)
Ameloot, M. A., 483
Ameloot, M., 51, 117, 280, 281, 282(18, 20), 284(18), 285(18, 20), 286(20), 287(20), 288(29), 289(18), 291, 292(20, 49), 293(31, 49), 294(18), 295(18, 20), 298(20), 299, 300(20, 31), 303, 315, 316, 317, 319(18), 327(4, 11), 330, 336, 339(9), 360, 364, 370
Anderson, B. D. O., 312
Anderson, D. H., 316, 317(14), 321(14), 322(14)
Anderson, T. W., 136
André, J. C., 280, 357, 303
Andreasi Bassi, F., 75, 86(5)
Andrews, D. F., 267, 277(53)
Andriessen, R., 299, 303, 317, 319(18), 330
Ansari, A., 165
Antoulas, A. C., 312

Arcioni, A., 281
Armitage, P., 90, 93
Arratia, R., 600
Asadov, M. M., 311
Atkins, G. L., 82
Aubanel, E. E., 440
Aubard, J., 57, 58(7)
Auchet, J. C., 281, 291, 299(25)
Avery, L., 482

B

Badea, M. G., 257, 280
Bain, D. L., 420
Bajzer, Z., 203, 210(5), 212(5), 217, 225(5, 24), 227(5), 234(24), 280
Bak, P., 647
Baker, C. C., 447
Baker, S. H., 371
Bakhshiev, N. G., 363, 368(12)
Balter, A., 260
Bard, Y., 5, 31, 36(3), 62, 68, 71(1), 72(1), 92, 96(4), 119, 203, 348
Barkley, M. D., 370
Barnsley, M., 656
Baros, F., 357
Barr, J., 576
Barton, G. J., 577, 594(24), 595, 598
Bassingthwaighte, J. B., 648
Bau, H. B., 603, 605(2)
Baumann, G., 546
Bay, Z., 238
Beale, E. M. L., 53
Beckett, D., 418
Beechem, J. B., 381, 382(8)
Beechem, J. M., 51, 67, 118, 119(7), 129,

Rutihauser, H., 272
Ruysschaert, J.-M., 193

S

Saltz, J. H., 597
Sander, C., 627
Sandor, T., 222
Sankoff, D., 576
Santoro, M. M., 466, 467, 468(3, 4), 469(3), 471(4)
Sasaki, K., 371
Savitzky, A., 67, 378, 387(7), 441
Scarborough, J. B., 309
Schaffer, W. M., 658
Schlesinger, M. F., 644
Schmitz, A., 406
Schneider, S., 169
Schuster, T. M., 55
Schuyler, R., 256
Schwartz, R. M., 592
Schwartz, S., 597
Sedarous, S. S., 217, 225(24), 234(24), 280
Segel, I. H., 391
Sejnowski, T. J., 626, 636
Selinger, B. K., 202, 214(4), 215(3, 4), 280, 291, 292(47), 293(47), 370, 461
Selinger, B., 260, 328
Sellers, P. H., 576, 585(5), 589, 592, 598
Semendyayev, K. A., 207
Senear, D. F., 406, 409, 416, 421(15), 425(20), 472, 473, 476, 477(9)
Serra, M.-H., 305
Sharp, J. C., 203, 210(5), 212(5), 217, 225(5, 24), 227(5), 234(24), 280
Shaw, C. D., 664
Shaw, C., 656
Shaw, M. A., 546
Shchigolev, B. M., 306
Shea, M. A., 406, 408, 409, 413, 416(13, 14), 421(15), 472, 473, 475, 477(9)
Shepanski, J. F., 610
Shipley, G. G., 522
Shrager, R. I., 135, 136, 139(23), 160(23, 24, 30), 161(23, 24, 30), 168(23), 171(23)
Shrager, T. I., 442
Siemiarczuk, A., 206, 207(12), 218(12)
Small, D. M., 522

Small, E. W., 64, 117, 237, 239, 278(19, 20), 242, 243, 245, 248, 249(22), 250(21), 252(24), 253(24), 255, 256, 258(32), 260, 261, 262, 263(19), 264(19), 265(23), 274, 275(55), 276(55), 278, 280, 291, 292(46), 303, 370, 506, 507, 508, 509(5), 510(5), 512(5), 514(5), 515(5), 516(6, 7, 8), 517(7, 8), 518, 520
Small, J. R., 237, 239, 261, 262, 263(19), 264(19), 278, 280, 506, 507, 508(2, 5), 509(5), 510(5), 512(5), 514(5), 515(5), 516(6, 7), 517(7), 518, 520
Smith, B. T., 27, 37(28)
Smith, F. N., 413
Smith, F. R., 413
Smith, H. O., 406, 423(4)
Smith, R. E., 56
Smith, R. M., 597
Smith, R., 98, 100(7), 104(7)
Smith, T. F., 576, 577, 584, 585(6), 591(6), 598
Smith, W. R., 482, 483(3, 4, 5), 487(3, 4), 490(4), 491(4, 5), 494(3, 4)
Snyder, G. J., 509, 514(9)
Solie, T. N., 242, 278(20)
Sollenberger, M. J., 341, 349
Sommer, J. H., 130, 133, 165(10, 11, 12), 170(10, 11)
Somorjai, R. L., 576
Spall, J. C., 563
Stafford, W. F., 607
Staskus, P. W., 429, 431(9), 432, 433, 434
Steer, R. P., 291
Stegun, I. A., 356
Stein, D. L., 661
Steinberg, I. Z., 208, 280, 357, 358, 368, 461, 513
Stern, O., 448, 449(1)
Sternberg, M. J. E., 577, 594(24), 595(24), 598
Stewart, G. W., 159
Stewart, I., 656
Straume, M., 8, 21, 30(11), 34(11), 36(11), 118, 119, 120(8), 123, 125, 218, 382
Stroud, R. M., 133
Stuart, A., 217, 222(25)
Stubberud, A. R., 671
Sturtevant, J. M., 523, 525(6)
Subba Reddy, K. V., 135

Sudhof, T. C., 591
Suetler, C. H., 133
Sun, H. H., 647, 671(11)
Surewicz, W. K., 193
Susi, H., 193, 200
Sussman, J. L., 576
Sutherland, J. C., 453
Suzuki, H., 406, 420(1)
Sylvestre, E. A., 136, 147(26), 161(26), 161(26)
Szabo, A. G., 208, 209(16), 237, 260, 280, 291, 292(47), 293(47), 328
Szabo, A., 165

T

Tam, A. C., 506
Tang, C., 647
Tank, D. W., 634
Tenchov, B. G., 521, 524(2)
Terry, B. J., 400
Teukolsky, S. A., 66, 119, 159, 216, 219(22), 225(22), 313, 366, 367(18), 372(18)
Therneau, T. M., 203, 210(5), 212(5), 225(5), 227(5)
Thomas, G. B., 548, 555(9)
Thomas, M. P., 317
Thorner, M. O., 548, 565(16)
Tinoco, I., 429, 430(10)
Tinoco, I., Jr., 135, 427
Tirsch, W., 664
Todorova, I., 650
Torgeirsson, T. E., 133
Trotter, H. F., 142
Tukey, J. W., 267, 277
Turner, B. W., 125, 371, 483

U

Udupa, J. K., 306
Uhrig, R. E., 370
Ukkonen, E., 597, 598
Urban, R. J., 544, 547, 559(5), 561(5), 565(5), 570

V

Valdes, R., Jr., 418
Vale, W., 571
Valeur, B., 280
Vallee-Poussin, 308
Van Cauwelaert, H., 281
Van den Bergh, V., 299, 303(56), 317, 319(18)
Van der Donckt, E., 336
van der Meer, W., 281
Vance, M. L., 548, 565(16)
Vanderkooi, J. M., 280
VanLoan, C., 129, 132(1), 138(1), 139(1), 159(1)
Vecchi, M. P., 631
Veldhuis, J. D., 341, 349, 540, 542, 543, 544, 547, 548, 549, 550, 552(11), 554, 555(8), 556, 557, 559, 561(3, 4, 5, 8, 20), 565(3, 4, 5, 8, 11, 12, 13, 15, 16, 20), 570, 571, 572
Verral, R. E., 291
Vetterling, W. T., 66, 119, 159, 216, 219(22), 225(22), 313, 366, 367(18), 372(18)
Vincent, L. M., 280
Viovy, J. L., 303
Vogel, J. F., 169
Vogel, R. H., 260, 291, 292(45), 293(45), 328
Volmer, M., 448, 449(1)
von Hippel, P. H., 421, 443, 494

W

Wagner, B. D., 206, 207(12), 218(12)
Wahl, P., 238, 281, 291, 292(44, 48), 299(25)
Walbridge, A. G., 371
Walbridge, D. G., 299, 357, 364, 370
Waldrop, M. M., 238
Wang, M., 650
Ware, W. R., 206, 207(12), 218(12), 280, 335, 357, 363, 366(13)
Waring, A. J., 280
Warner, I. M., 317
Wastney, M. E., 392, 393(8), 394(8), 395(8), 397(8), 399(8), 401(8), 402(8), 403(8), 404(8), 405(9)

Subject Index

A

B

Back-propagation artificial neural network
applications of, 624–628
encoding schemes for, 624–626
NETtalk, 626
decision regions of, network and geometry of, 621–622
error surfaces with, 624
learning algorithm for, 622–623
neuron or processing unit in, 621
overtraining of, 623–624
generalization degradation due to, 623–624
procedure for, 623
protein secondary structure prediction with, 626–628
sigmoid output function in, 631
Backprop artificial neural network. *See*
Back-propagation artificial neural network
Bacteriophage λ *c*I repressor
gel chromatography of
characteristic large zone profile for, 418–419
representative dissociation curve for, 418–419
monomer–dimer stoichiometry for, 418–419
Bacteriophage λ *c*I repressor–operator
system or distribution of ligation species according to, 412
footprint titration of
compared with filter-binding titration, 422
microscopic Gibbs energies of, 414–416
free energy changes in, least-squares analysis of, 471–473
individual-site binding data for, simultaneous analysis of, 410–416
individual-site binding of, 416–417
individual-site loading free energies from, 413–414
individual-site titration data for, 474
microscopic configurations and free energy contributions for, 408–410
operator configurations and associated free energy states for, 472
thermodynamic linkage of, 424–425

Bad points. *See* Outliers
Bakhshiev approach, to solvent relaxation, 363
*Bam*HI hydrolysis, of superhelical plasmid DNA, 394–400
five-compartment model, 398–401
four-compartment model, 396, 399
model test, 400–405
six-compartment model, 401–402, 404
three-compartment model, 395, 397–399
Basin of attraction, definition of, 656–657, 664
Beer's law, 132
of normal absorption spectroscopy, 426
Bessel's interpolation function, 309
Bifurcation, definition of, 657
Binding constants
from ligand binding experiments, evaluation of, 1
resolved for *Hin*fI–DNA interactions, 424–425
Binding data, numerical analysis of, 481–505
advantages of, 483–487
computer programs for, 482–483
macromolecular model for, 492–497
confidence interval testing for free energy parameters of, 498–504
methodology for, illustration of, 491–497
observable mapping matrix for, 494–495
stoichiometric matrix for, 493
unique parameter recovery with, 497–504
Binding energies, for complex systems, determination of, by numerical analysis, 504
Binding expressions
for filter-binding technique, 420–423
for footprint titrations, individual-site, 408–410
for gel mobility shift assay, 418–420
Binding isotherms, for *Hin*fI restriction endonuclease interaction with pBR322 supercoiled DNA, 423, 425
Binding profiles, calculation of, constraints on, 484
Binding surfaces, for dimeric proteins that bind two ligand molecules cooperatively, simulated three-dimensional, 496–497

criteria for
    determination of, analysis of residuals
        for, 87–105
    for parameter estimation by least-
        squares, 7–8
    for hormone data, statistical tests of,
        348–349
    run test of, 96–98, 348–349
Gradient search method. *See* Steepest
    descent method
GraphiC (computer program), 266
Grid search method, of confidence interval
    estimation, 22–24
Grinvald and Steinberg recursion, 358
Ground-state complex model, of static
    quenching
    of multiple species, 455
    of single species, 451–452
        analysis of, 457–458
        cross analysis of, 458
Ground-state equilibrium constant, deter-
    mination of, 329–330
Growth hormone
    basal secretion of, increased, with GH-
        secreting pituitary tumor, 565, 567
    half-life, deconvolution-predicted, 553,
        556
    output of, regulation of, 471
    secretion waveform for, 548
    serum concentrations, assay characteris-
        tics, 546
    simulated serum concentrations, with
        random noise, multiple-parameter
        deconvolution fitting of, 562–563
Growth hormone-releasing hormone,
    function of, 571

H

Heart morphology, fractal branchings in,
    641–642
Hebbian model, for synaptic modification,
    611
Hemoglobin
    $\alpha\beta$ dimer, affinities between monomers
        in, 486
    human
        α chain, global and local similarity
            scores for, 590
        β chain, global and local similarity
            scores for, 590

oxygen binding to
    confidence probability distributions
        by Monte Carlo method for,
        122–127
    equation for, 6–7
    quaternary enhancement free energy
        difference in, 123
    statistical justification for quaternary
        enhancement in, 118
modified, time-resolved spectra after
    photodissociation of carbon monox-
    ide bound from, 129–134, 170–171
    global analysis of, 130–131
    model for, 170–171
    SVD of, 131–134
        rotated, 165–167
    structure matrix for, 445
Hermite formula, 312
Heterogeneity, fractal characterization of,
    648–650
Hexane, pyrene in, fluorescence decays of,
    compartmental analysis of, 330–336
Hill transformations, 1, 107
*Hin*fI–DNA interactions
    binding constants resolved for, 424–425
    binding isotherm for, 423, 425
His–Purkinje system, fractal branchings in,
    642
Homeodynamics, 661
    vs. homeostasis, 655
Hopfield artificial neural networks, 628
    energy of, description of, 629–630
    energy surface of, 630
    graded response, 631
        sigmoid output function in, 631
    probability of unit becoming active in,
        effect of temperature on, 632
    solution to six-city traveling salesman
        problem, 634–635
    three-unit, output states of, 629
    two-state model, 628–630
Hopfield–Boltzmann machine artificial
    neural network, setting weights of,
    633–636
    by content-addressable (associative)
        memories, 633–634
    by optimization, 634–635
Hormone(s)
    basal secretion of, estimation of,
        by deconvolution analysis, 565–
        568

I

Identifiability equations, 321–322
Identifiability theory, in compartmental analysis, 316–317
Impurity decays, 248. *See also* Contaminant decays
sources of, 255
IMSL (computer library), 328
Infrared spectral data
band narrowing of. *See* Infrared spectral data, Fourier resolution enhancement of
Fourier resolution enhancement of, 192–200
edge effects in, 199
estimation of input band shape, 197–198
practical considerations, 197–199
signal-to-noise ratio and K, 198
water vapor in, 198–199
Integrals, discretization of, by maximum likelihood analysis, 207–212
Integral transforms, for sums of exponentials, 57–58
Integrated rate equations, application to kinetic analysis, 378–380
Integration
Euler's method of, 379–380
iterative, for sums of exponentials, 56
available software, 56
Interpolation
applications of, 305
graphical techniques for, 306–307
methods for, 305–314
applications of, 314
polynomial, algorithm for, 307–311
with rational functions, 311–312
with spline functions, 312–313
advantages of, 313
applications of, 314
with trigonometric functions, 312
Intraspectral smoothing, methods for, 440–441
Ising model, 433
polymer melting with, stacking parameters and thermodynamic constants for, 434
Isooctane, pyrene in, rate constants for, 335

J

Jackknifing, of residuals, 370
examples, 370–371
Johnson noise. *See also* White noise
classification of, 644
Julia set, definition of, 638

K

KERNEL (computer program), for global analysis, advantages and disadvantages of, 51
KINDK data, evaluation of, 370
Kinetic analysis. *See also* Relaxation kinetics; Steady-state kinetics
application of rate equations to, 376–380
applications of, 375
of complex reaction profiles, 389–390
testing of, 390
direct method for, 376–377
of first-order reactions, 387–389
indirect method for, 377
initiating reactions for
methods for, 383–384
related to mixing reactants, 383–384
by mixing reactants, 383
by perturbation, 384
making complex reactions tractable for, 380–382
by global analysis, 382
by manipulating concentrations, 381–382
by manipulating rate constants, 382
methods for, 376–377
practical aspects of, 374–390
rate constants, observed or apparent, 385–386
of reaction profiles, 384–390
of second-order reactions, 389
temperature-jump experiments, 1, 3
data set for, 3
equation for, 3
evaluation of relaxation rate constants from, 1
of zero-order reactions, 386–387
Kinetic data, global analysis of, 382
Kolmogorov–Sinai (K-S) entropy, estimates of, applications of, with approximate entropy (ApEn) statistic, 653

Kolmogorov–Smirnov test, of residuals, 95–96

L

lac repressor protein, degree of inducer dissociation, calculation of, from fluorescence emission energy, 494

Lactate dehydrogenase, structure matrix for, 445

Lag_n serial correlation plots, of residuals, 98–99

Lagrange coefficient, generalized, 310

Lagrange formula, 310
  generalized, 313
  iterative, 311

Lagrange interpolation, computation of binodal stratification curves by, in mercury–dithallium chalcogenide systems, 311

Lagrange multipliers, use of, in solving chemical equilibrium problems, 490–491

Lagrange polynomial, 309–310
  generalized, 310

LAGUER (computer program), 66

Laguerre polynomials, 63

Laguerre's algorithm, for complex roots of polynomials with complex coefficients, available software, 66

λ-invariance plots, 245–246, 258–259
  definition of, 240
  interpretation of, with FLUOR program, 264
  for noiseless synthetic photoacoustic waveforms, 518
    recovered parameters from, 520
  rules for accepting, 264, 517–518

λ scans
  definition of, 240
  generation of, with FLUOR program, 273–274
  rules for accepting, 265

Lamp width errors, 248
  moment index displacement and, 253–254

Langmuir isotherm, 422

Laplace deconvolution
  against decay of reference solution, 291–293

of fluorescence decay surfaces, 279–304
  examples and applications of, 295–301
iterative method (LAP1)
  cutoff correction by, 285
  light scatter correction by, 284–285
  single-curve analysis with, 295–296
  theory of, 282–285
  time shift correction by, 284–285
noniterative method (LAP2)
  compartmental analysis with, 303–304
    implementation of, 304
  against decay of reference solution, 291–293
  global analysis with
    of anisotropy, 301
    of multiexponential decays, 296–299
    resolution of spectra by, 299–301
  implementation of, 293–295, 302
  light scatter correction by, 287
  linear, implementation of, 294–295, 302
  nonlinear, implementation of, 295, 302
  simultaneous analysis of multiple total intensity decay curves by, 287–288
  single-curve analysis with, 295–297
    of multiexponential decays, 296–298
  theory of, 285–287
  time shift correction by, 287
simultaneous analysis of multiple polarized intensity decay curves by, 288–291
simultaneous analysis of multiple total intensity decay curves by, 287–288
theory of, 282–293

Laplace transform
  forward, 57, 59. *See also* Padé–Laplace algorithm
  inverse, 56
  modified. *See* Laplace deconvolution
  Taylor series expansion of, 59–60

Least mean squares rule, 620

Least-squares analysis. *See also* Global analysis
  assumptions of, 4–8, 68, 407, 464
  better approximation in, definition of, 3
  χ^2 method, parabolic extrapolation of, 10–11
  description of, 3–4
  estimation of Fourier coefficients by, 343

for denaturant-induced unfolding of phenylmethanesulfonyl-α-chymotrypsin, simultaneous fit of, 469–471
for protein unfolding, 468
least-squares testing of, 465–471
Linear function, definition of, 4
Linear joint confidence interval method, of confidence interval estimation, 25–26
Lineweaver–Burke double-reciprocal transformations, 1, 38, 107
for steady-state kinetics, 73–74, 86
LINPACK (computer program), 159
Lipid–lipid interactions, in binary mixtures, 521, 523
Lipoprotein receptor, low-density, and mouse epidermal growth factor precursor, local sequence similarity between, 591
Loading free energies, individual-site, 475
calculation of, 475
comparisons of, 476
Logarithmic transforms, 1, 38
Luteinizing hormone
half-life, deconvolution-predicted, 553, 556
human, assays of, 340–341
plasma concentrations, increase in, mechanisms of, 568
radioimmunoassay characteristics, 546–547
secretion waveform for, 548
serum concentrations, assay, distribution and magnitude of measurement errors in, 545
Luteinizing hormone data
Akaike information content (AIC) index as function of order for, 349–350
Fourier coefficients as function of order for, 349–351
Fourier components for, amplitudes of, 353–355
run Z score as function of order for, 349–351
Lysozyme, structure matrix for, 445
Lysozyme c precursor, chicken, global and local similarity scores for, 590

M

Maclaurin series. *See* Taylor series expansion

MACSYMA (computer program), 603, 608
Mandelbrot noise, classification of, 644
Mandelbrot set, definition of, 638
Manifold, definition of, 669
Maple (computer program), 603
Maps (mapping)
with complex numbers, 670
definition of, 665, 669–671
Markov parameters
calculation of, 322
definition of, 321–322
Marquardt–Levenberg method, of nonlinear least-squares, conversion to global analysis, 45–49
Marquardt method, of parameter estimation, 9, 16–17
MathCAD (computer program), 604
Mathematica (computer program), 603
Mathematical model, appropriate. *See also* Model selection
definition of, 88
Matrices
autocorrelations of, as measure of signal-to-noise ratio, 161–162
of known data (noise-free)
estimation of, in presence of noise, 147–150
SVD of, 139–141
least-squares fitting with physical model. *See also* Global analysis
applications of, 168–169
with SVD-based analysis, 168–172
with noise, SVD of, 141–156
random
definition of, 137
singular values of, distributions of, 142–143
SVD of, 142–143
with rank-1 noise, SVD of, 150–156
signal-to-noise ratio of, 161
measurement of, 161–162
singular value decomposition
computation of, 159
effective rank of, 160–161
determination of, 147–150
preparation of, 157–159
smoothing of, 158
truncation of, 158
weighting of, 158–159